AF615057

Micropropulsion for Small Spacecraft

Micropropulsion for Small Spacecraft

Edited by
Michael M. Micci
Pennsylvania State University
University Park, Pennsylvania
Andrew D. Ketsdever
Air Force Research Laboratory
Edwards Air Force Base, California

Volume 187
PROGRESS IN ASTRONAUTICS AND AERONAUTICS

Paul Zarchan, Editor-in-Chief
Charles Stark Draper Laboratory, Inc.
Cambridge, Massachusetts

629.132 38 MIC
Ecole Centrale de Lyon
BIBLIOTHEQUE
36, avenue Guy de Collongue
F - 69134 ECULLY CEDEX
3940

Published by the
American Institute of Aeronautics and Astronautics, Inc.
1801 Alexander Bell Drive, Reston, Virginia 20191-4344

Copyright © 2000 by the American Institute of Aeronautics and Astronautics, Inc. Printed in the United States of America. All rights reserved. Reproduction or translation of any part of this work beyond that permitted by Sections 107 and 108 of the U.S. Copyright Law without the permission of the copyright owner is unlawful. The code following this statement indicates the copyright owner's consent that copies of articles in this volume may be made for personal or internal use, on condition that the copier pay the per-copy fee ($2.00) plus the per-page fee ($0.50) through the Copyright Clearance Center, Inc., 222 Rosewood Drive, Danvers, Massachusetts 01923. This consent does not extend to other kinds of copying, for which permission requests should be addressed to the publisher. Users should employ the following code when reporting copying from the volume to the Copyright Clearance Center:

1-56347-448-4/00 $2.50 + .50

Data and information appearing in this book are for informational purposes only. AIAA is not responsible for any injury or damage resulting from use or reliance, nor does AIAA warrant that use or reliance will be free from privately owned rights.

ISBN 1-56347-448-4

Progress in Astronautics and Aeronautics

Editor-in-Chief

Paul Zarchan
Charles Stark Draper Laboratory, Inc.

Editorial Board

John D. Binder
MathWorks, Inc.

Michael D. Griffin
Orbital Sciences Corporation

Lt. Col. Steven A. Brandt
U.S. Air Force Academy

Phillip D. Hattis
Charles Stark Draper Laboratory, Inc.

Luigi De Luca
Politecnico di Milano, Italy

Richard M. Lloyd
Raytheon Electronics Company

Leroy S. Fletcher
Texas A&M University

Ahmed K. Noor
NASA Langley Research Center

Allen E. Fuhs
Carmel, California

Albert C. Piccirillo
ANSER, Inc.

Vigor Yang
Pennsylvania State University

Table of Contents

III. Electrostatic Thrusters

V. Components

Preface

The launch of the first microspacecraft took the world by storm 43 years ago when the then-Soviet Union launched Sputnik in late 1957. The United States countered with an even smaller, slightly more capable microspacecraft, Explorer I, in early 1958. At launch, Sputnik weighed a little over 83 kg and Explorer weighed in at 14.5 kg. Explorer I was little more than a set of batteries, a radio transmitter, and a Geiger counter. Since the early days of space exploration, the mass, complexity, and capability of individual spacecraft have grown tremendously. There have been several critical advances that have allowed the dreams of highly functional spacecraft to come to fruition over the past four decades.

The advent of micromachining and microelectromechanical systems (MEMS) fabrication techniques has allowed the space community to dream again. This time the dreams turn to thoughts of armies of microspacecraft circling the globe, and other planets of the solar system, performing critical and highly complex tasks. Capable microspacecraft with distributed functionality are envisioned to take over the tasks of more massive and expensive platforms with increased survivability and flexibility. It is becoming increasingly evident that these microspacecraft will require efficient propulsion systems to enable many of the missions currently being investigated. The system constraints on mass, power, maximum voltage, and volume with which microspacecraft will undoubtedly have to contend pose several challenges to the propulsion system designer. Micropropulsion concepts that address these limitations in unique and beneficial ways will be of interest to the microspacecraft community.

For the purposes of this Progress Series volume, the definition of micropropulsion is any propulsion system that is applicable to a microspacecraft (mass less than 100 kg) mission. This definition allows the inclusion of a wide range of concepts from scaled-down versions of existing thrusters operating at reduced power levels to completely redesigned MEMS-fabricated thrusters with micron characteristic sizes.

Micropropulsion is an enabling technology for microspacecraft operations by making missions possible that otherwise could not be performed. For example, the formation and maintenance of platoons of microspacecraft will require a maneuvering capability to counter orbital perturbations. Microspacecraft missions involving large spacecraft resupply, repair, or surveillance will also require maneuverability. The mission requirements for microspacecraft will be varied and, in some cases, a large range of capability might be required on the same spacecraft. Micropropulsion systems must be extremely versatile to address these requirements. It is clear that there is a need for these systems—from high thrust chemical engines to high specific impulse electric thrusters—to fulfill specific missions, just as there is a need for larger spacecraft.

This volume was envisioned to show the state-of-the-art in micropropulsion concepts and activities at the early stages in the development of this new and exciting research area. It is the editors' hope that the task of updating the technological advances in micropropulsion be taken up some years from now and compared with this early work.

In closing, we would like to thank our reviewers for their time and efforts, without which this volume would not have been possible:

John Blandino
Iain Boyd
Ken Breuer
Rodney Burton
Frank Curran
Michael Dulligan
Alec Gallimore
William A. Hoskins
Siegfried Janson
Mary Kriebel
Lyle Long
Manuel Martinez-Sanchez
Keith McFall
Robert Melton
Juergen Mueller
E. P. Muntz
Bryan Palaszewski
James Polk
Robert Reinicke
John Schilling
Dino Sciulli
Ronald Spores
Peter Turchi
Dean Wadsworth
Ingrid Wysong

Andrew D. Ketsdever
Michael M. Micci
June 2000

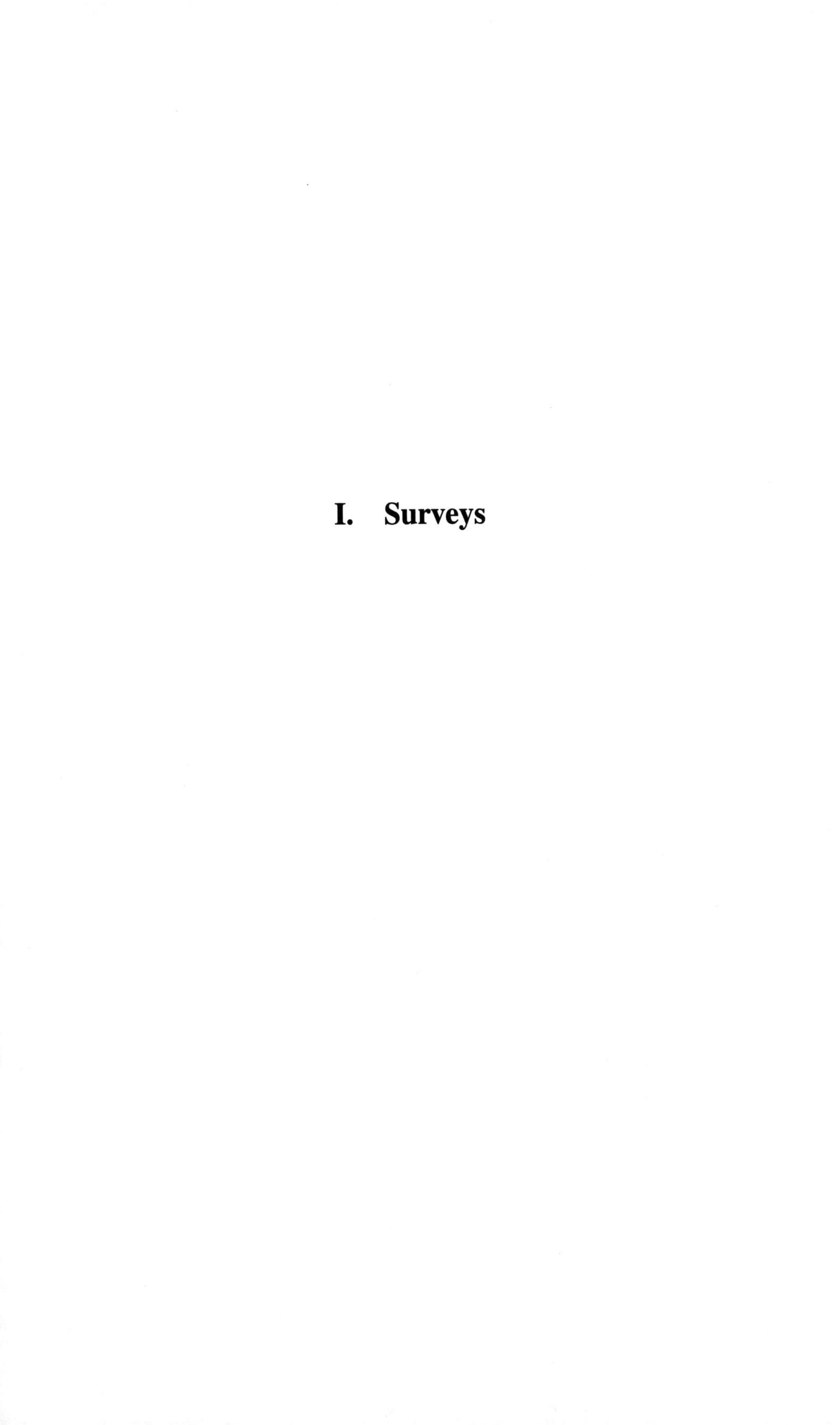

I. Surveys

Chapter 1

Micropropulsion Options for the TechSat21 Space-Based Radar Flight

John H. Schilling,* Ronald A. Spores,† and Gregory G. Spanjers ‡
Air Force Research Laboratory, Edwards Air Force Base, California

I. Introduction

A GREATER interest by government agencies in reducing the size of their satellites is evidenced by the recent increase in the number of government small-satellite programs. MightySat is a U.S. Air Force program utilizing small satellites for space experiments. The New Millennium Deep Space series of satellites and the Spartan bus for Shuttle-deployed satellites are examples of NASA small-satellite programs. The National Reconnaissance Organization (NRO) likewise has a strong program to reduce the size of its space assets,[1] while DARPA is funding a wide range of MEMS (micro electromechanical systems) programs that are applicable to microspacecraft. One of the newest government efforts employing small satellites is the Air Force Research Laboratory (AFRL) TechSat21 program,[2] which will demonstrate enabling technologies for a formation-flying constellation for space-based surveillance.

There are strong advantages for going to small satellites. One benefit is the substantial reduction in the overall life-cycle cost by making satellites less costly to construct, due to fewer components and the potential for mass production techniques. In addition, smaller satellites have greatly reduced launch costs. For the formation-flying concept of TechSat21, using small satellites enables the aperture of the system to essentially be the diameter of the constellation (~100 m), yielding much greater spatial resolution. Further, the utilization of many smaller satellites lends itself to a graceful degradation of the system capability as individual satellites are lost. The constellation can reconfigure itself for maximum resolution in range and Doppler shift of the target with the reduced number of satellites.

The Space Based Radar (SBR) mission objective is to detect moving ground targets and/or airborne targets from space. This concept has been proposed for over

This material is declared a work of the U.S. Government and is not subject to copyright protection in the United States.

*Research Engineer, SPARTA Inc. Member AIAA.

†Chief, AFRL Spacecraft Propulsion Branch. Member AIAA.

‡Leader, AFRL Electric Propulsion Group. Member AIAA.

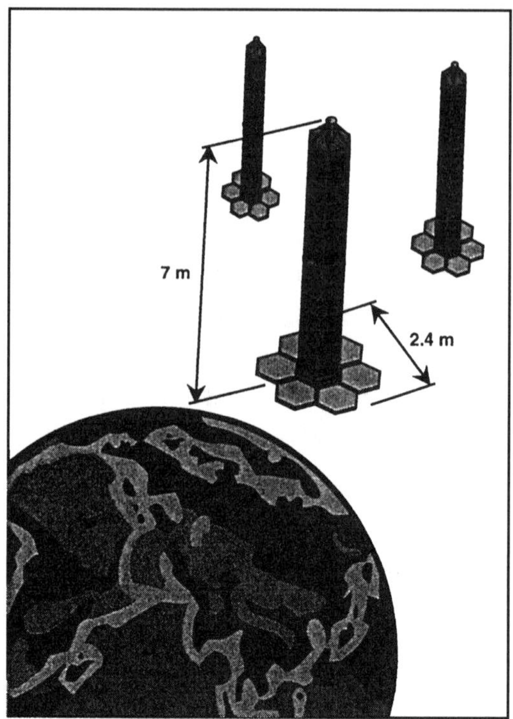

Fig. 1 Satellite formation for the 2003 TechSat21 flight using three spacecraft.

20 years, however, the cost of deployment has been considered prohibitive. The TechSat21 approach of employing small satellites could make this long-awaited Department of Defense goal a reality. The TechSat21 program plans to launch a three-spacecraft formation for critical technology demonstration in 2003, at an altitude of 650 km, shown in Fig. 1. This flight will be followed by a full demonstration mission of ~12 spacecraft in 2007 to demonstrate the space-based surveillance concept.[2] The critical technologies that will be validated are 1) ionospheric effects on radar, 2) interferometric radar signal processing from multiple transmitters/receivers, 3) orbital mechanics of a formation-flying constellation, and 4) spacecraft micropropulsion. Advanced micropropulsion concepts are critical for this mission due to the requirements for significant ΔV and minimal propulsion system mass.

Each TechSat21 spacecraft serves as one transmit/receive element of a distributed phased-array antenna, with an operational system consisting of a dozen or more spacecraft in a formation approximately a hundred meters across. In-space processing of the signals returned from each spacecraft allows detection of air and moving surface targets with search performance equal to a monolithic radar with a power-aperture product equal to the power aperture of a single component satellite times the square of the number of satellites in the formation and with a spatial resolution similar to that of a single antenna having a diameter equal to that of the entire formation.

This SBR mission represents a substantial deviation from the traditional constellation model, in which multiple satellites are used to provide global coverage,

Table 1 Proposed mass budget for TechSat21 spacecraft

Subsystem	Mass, kg
Radar antenna	46.8
Radar processor	5.5
Ionospheric sampler	6.8
Intersatellite comm	4.8
Attitude determination and control system	5.9
Propulsion	10.7
Navigation	2.0
Telemetry, tracking, and command	3.2
Command and data	6.3
Handling	
Structure	20.2
Power	20.4
Thermal	5.1
Total	137.7

but each spacecraft operates essentially independently within its coverage area and the distance between spacecraft is several thousand kilometers. The microsatellite formations proposed for the space-based radar application would involve multiple spacecraft operating in close proximity, as shown in Fig. 1. No single spacecraft has any independent mission capability. The requirement for cooperative action imposes constraints on many aspects of spacecraft design and operations. Of particular importance is the stringent stationkeeping requirement associated with maintaining the formation. The individual spacecraft within the formation each have slightly different orbital elements, and thus naturally respond differently to various perturbations. To maintain the relative positions of the spacecraft within the formation, these differential perturbations, which are principally from the orbital J2 perturbation, must be corrected by periodic stationkeeping maneuvers.

II. TechSat21 Design

The individual spacecraft for the TechSat21 mission are currently in the conceptual design phase. The proposed design, shown in Fig. 2, collapses into a 0.3-m^3 volume for launch, then deploys a 7-m boom and a 2.5-m antenna on orbit. As shown in Table 1, the total mass of the spacecraft is ~135 kg, of which ~10 kg is available for the propulsion system. Given the likelihood of weight growth, we will size the propulsion system for a 150-kg spacecraft. For attitude control purposes, we assume moments of inertia $I_{zz} = 50$ kg-m^2 and $I_{xx} = I_{yy} = 1000$ kg-m^2.

Approximately 350 W of electric power is produced by solar panels on the boom section, almost all of which is available for the propulsion system during the maneuver phases of the mission. No estimate is currently available for the power allotment for stationkeeping propulsion, as the propulsion system must compete with the radar transmitter for available power. However, as the stationkeeping thrust requirement is small compared to the maneuver requirement, any propulsion system capable of performing the maneuver mission with <350 W of power will almost certainly be able to perform the stationkeeping mission with minimal impact.

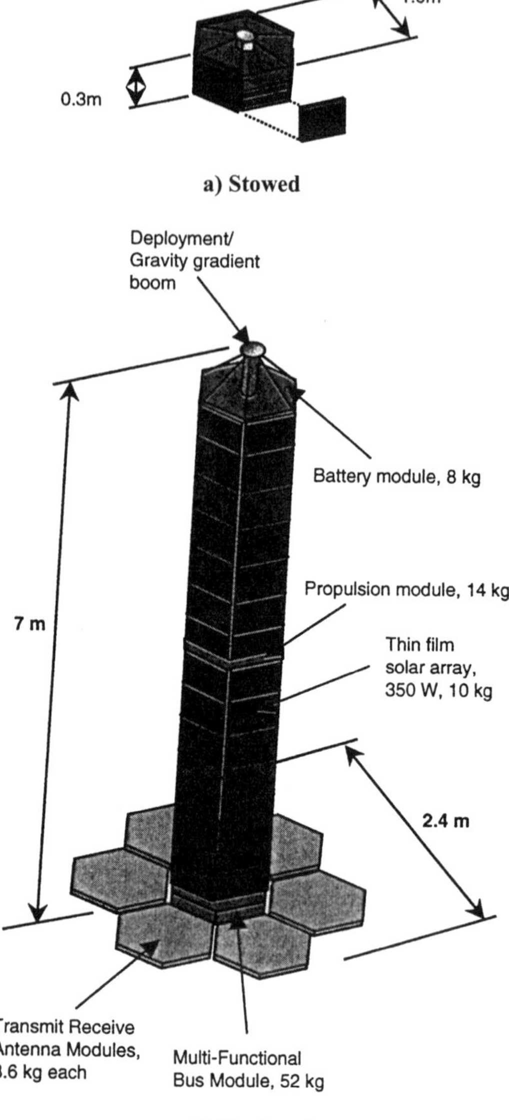

Fig. 2 Proposed design for the individual TechSat21 spacecraft: a) The spacecraft in the stowed configuration for launch and b) the spacecraft fully deployed.

The proposed TechSat21 spacecraft is passively stabilized by gravity gradient using the extended solar array boom, with magnetic torquers for attitude control. No momentum wheels or other precise three-axis attitude control system is currently planned, and the magnetic torquers provide only coarse attitude control with ±10-deg accuracy. This may ultimately be insufficient for propulsive maneuvers. The gravity gradient system, of course, provides no yaw control capability and limits the spacecraft to small deviations about a single pitch/roll orientation.

The propulsion requirement specified by the TechSat21 program office for the critical technology demonstration flight in 2003 is as follows: 1) total ΔV of 70 m/s, 2) formation maneuver ΔV of 30 m/s over a 30-day maneuver period, 3) stationkeeping ΔV of 40 m/s over a one-year life in a ~600-km orbit, 4) total of 5000–10,000 propulsive maneuvers, and 5) minimum impulse bit of 2 mN-s.

A series of formations with separation distances from 5 m to 5 km are to be tested during the first TechSat 21 flight experiment; the 30-m/s ΔV budgeted for maneuvering allows shifting between formations. A nominal 30 days is devoted to such maneuvers, which corresponds to a minimum thrust level of 2 mN. However, higher thrusts and shorter maneuvering periods are obviously preferable. Conversely, it may be possible to accept a lower thrust and longer maneuver time if substantial savings in propulsion system mass, cost, and complexity result.

The 2007 flight experiment is expected to have rather different propulsive requirements, with a longer mission duration and higher orbit altitude. This would also be true of any operational SBR system derived from TechSat 21 experience. Preliminary estimates for the TechSat 21 program, probably still appropriate to the 2007 experiment and to operational systems, call for a 390 m/s total ΔV broken down as follows: 1) orbit raising ΔV of 50 m/s, 2) drag makeup ΔV of 20 m/s over a 10-year life, 3) stationkeeping ΔV of 200 m/s over a 10-year life, and 4) deorbit ΔV of 120 m/s at the end of life. The first mission of the propulsion system is to increase carefully the orbital altitude to that of the planned constellation and insert the satellite in phase with the rest of the formation. This will dominate the thrust requirement, and again, we allow a nominal 30 days for the maneuver.

Because of the substantially higher ΔV requirement, the 2007 experiment may call for a different propulsion system than the 2003 mission. This paper focuses on propulsion system selection for the 2003 mission; however, in recognition of the fact that a common propulsion system for both missions will simplify spacecraft design, the suitability of various propulsion options to the later flight will be considered.

Stationkeeping and formation-forming maneuvers are expected to be performed once or twice per orbit, for a total of 5000–10,000 such maneuvers over the course of the mission. This corresponds to an average thrust impulse of 1–2 N-s per maneuver. In certain extreme cases involving spacecraft maneuvering in close proximity (~5-m separation), impulse bits as low as 2 mN-s may be required.

All of these maneuvers are expected to be confined to the spacecraft X–Y plane, so only yaw steering will be required of the propulsion system. There are three potential approaches to this requirement. First, the coarse yaw steering capability provided by the spacecraft's magnetic torque rods can be utilized, albeit with some sacrifice of precision maneuvering capability. As a second approach, multiple propulsive thrusters with canted nozzles can be used to provide yaw steering capability. Finally, a dedicated micropropulsion system can be used for yaw pointing and precision maneuvering.

The thrust requirement for such an ACS thruster system can be determined by considering the worst-case scenario. With maneuvers occurring twice per orbit, it could be necessary to rotate the spacecraft by a full 180-deg in one-half of an orbit period. Given an orbit period of ~6000 s and a spacecraft moment inertia $I_{ZZ} = 50$ kg-m^2, this requires a torque of 6.8×10^{-5} N-m. The configuration of the spacecraft favors ACS thrusters with a moment arm of 0.6 m, for 1.13×10^{-3} N of available thrust. If this is to be provided by a single pair of propulsive thrusters with nozzles canted at 30-deg, a thrust level of ~200 μN per thruster will be required; if by three opposed pairs of dedicated ACS microthrusters, 38 μN per microthruster will be required.

In most cases, of course, the spacecraft will not have to perform a full 180-deg rotation in half an orbit period. Maneuvers are more likely to be performed once per orbit and to require consistent and predictable pointing. In particular, approximately 90% of the stationkeeping mission consists of ascending-node correction for J_2 perturbation and drag makeup, both of which can be performed with burns in a single orientation once per orbit. Only the remaining 10% of the stationkeeping requirement would call for yaw pointing, presumably of random character. Over a one-year mission with 5000 stationkeeping maneuvers, this would require a total of 32 N-m-s of torque impulse.

The formation maneuvers are somewhat less predictable and will be assumed to be evenly divided between consistent and random orientation. This greater yaw pointing requirement will call for 113 N-m-s of torque impulse over a one-year mission. Also, it may be necessary to provide yaw stabilization against disturbance torques. The worst case in this regard would be an aerodynamic torque due to a c.g. misalignment of 3 cm (5% of vehicle characteristic width). This would produce a disturbance torque of ~0.75 μN-m, requiring 24 N-m-s of torque impulse per year for stabilization.

These combined attitude control requirements add to 169 N-m of torque impulse for the one-year mission. If provided by the main propulsive thrusters, again with canted nozzles at 30-deg, this corresponds to 560 N-s of total impulse, or 3.75 m/s of ΔV for a 150-kg spacecraft. Dedicated ACS microthrusters would have to provide only 281 N-s of total impulse, 47 N-s for each of six thruster units, corresponding to 1.9 m/s of ΔV. With three ACS thruster pairs to choose from, it will be possible to apply most (~81%) of this ΔV to the stationkeeping as well as yaw steering requirements, which is not the case if only a single pair of main propulsive thrusters must perform the yaw pointing requirement. Furthermore, canting the nozzles of the main propulsive thrusters by 30-deg reduces their efficiency for stationkeeping and formation maneuvers by 13%.

Thus, assigning the yaw steering requirement to the main propulsive thrusters requires the addition of a second such thruster and increases the net ΔV requirement to 85 m/s. If dedicated microthrusters are used, three pairs of thrusters, each of 38-μN thrust and 47-N-s impulse capability, will be required, and the ΔV requirement for the main thrusters is reduced to 68.5 m/s. If only the magnetic torque rods are used, there is no cost to the propulsion system, but precision pointing (and thus maneuvering) are sacrificed.

III. Micropropulsion Options

Traditionally, on-orbit propulsion for spacecraft has been provided by chemical rockets. Solid-propellant rockets are the simplest propulsion system available but cannot be used for stationkeeping due to the inability to repeatedly turn them on and off. Liquid propellant rockets are more versatile, and offer greater performance, albeit at the cost of increased complexity. Cold-gas thrusters offer sufficient versatility for stationkeeping in a simpler package than liquid rockets but have extremely limited performance. Small liquid and cold-gas systems will be considered for the TechSat21 mission in their conventional forms, with the liquid-propellant system serving as a baseline against which other candidate propulsion systems will be measured. In addition, a novel form of miniaturized solid rocket propulsion system, the digital MEMS thruster, is suitable for spacecraft in the 100 kg and under size class and is included in the analysis.

Unfortunately, no chemical propulsion system can offer a specific impulse greater than ~550 s, and systems suitable for use on small spacecraft are limited to approximately 220 s. This is adequate for the 2003 mission, but the low specific impulse is likely to result in unacceptably high propellant mass values for the more demanding 2007 mission. However, electric propulsion systems are available that offer much higher specific impulse at the expense of a substantial electric power requirement. As the proposed TechSat21 spacecraft has a substantial solar-electric power capability for the radar payload, while the mass budget for the propulsion system is rather tight, electric propulsion may be preferable for the TechSat21 full demonstration mission in 2007.

Two general categories of electrical propulsion system will be considered for the TechSat21 mission, electrostatic propulsion and electromagnetic propulsion. The former is characterized by the generation of charged particles, usually heavy ions, which are accelerated by an applied potential to velocities in excess of 10 km/s to produce thrust. One type of electrostatic system is the ion thruster, in which the applied potential is provided by a series of charged grids, with the ions provided by a separate discharge chamber. Conventional ion thruster designs do not readily scale down to the size and power levels required for TechSat21 due to discharge chamber physics, but novel ion sources currently under development may allow micro-scale ion thrusters [e.g., field-emission electrostatic propulsion (FEEP), micro-colloid thruster] to meet the TechSat21 mission requirements. Another type of electrostatic propulsion system is the Hall-effect thruster, also known as the stationary plasma thruster (SPT). In this design, the accelerating potential is provided by forcing a discharge current through a transverse magnetic field. There are several demonstrated Hall thruster designs suitable for the TechSat21 mission.

Electromagnetic propulsion involves the acceleration of a current-carrying plasma by an applied or self-generated magnetic field, rather than by an electrostatic potential. The requirement for a strong magnetic field limits steady-state electromagnetic thrusters to extremely high power levels, which is unacceptable for the TechSat21 mission. However, pulsed operation allows for arbitrarily low average powers, with the high peak power requirement being met by a capacitive discharge, and the pulsed plasma thruster (PPT) is a leading candidate for the TechSat21 mission. The PPT's use of an inert solid propellant with no moving parts is a particularly desirable feature for a system intended for use in a small, low-cost spacecraft.

Several interesting microthruster concepts are not addressed in this analysis, such as the vaporizing liquid microthruster (Jet Propulsion Laboratory), a free-molecular micro resistojet (AFRL), and other electrothermal devices. The high thrust of these devices makes them attractive for several microsatellite missions, however, their low specific impulse results in an excessive wet mass for the specific TechSat21 mission parameters. For an eventual full SBR deployment, this class of microthruster may be enabling for a fast deployment of the microsatellite formation.

A. Chemical Micropropulsion

Liquid-propellant rockets have been the standard for on-orbit propulsion throughout the history of space travel, and no introduction will be given. Due to the limited mass budget of TechSat21 and the implied requirement for simplicity, only monopropellant systems are considered, with the baseline monopropellant being hydrazine. The numerous handling difficulties of hydrazine notwithstanding, hydrazine monopropellant thrusters and propellant feed systems are mature, commercial products and can be integrated with the TechSat21 spacecraft with little difficulty. For the purposes of this analysis, the Primex MR-111 and MR-103M thrusters were specified, though numerous other manufacturers offer equally suitable systems.[3,4] These represent the smallest commercially available chemical thrusters and, while somewhat larger than optimal for TechSat21, are still reasonable for the mission.

While hydrazine monopropellant thrusters may offer insufficient specific impulse for this application, other monopropellant options are available. The Air Force Research Laboratory and NASA are currently developing a series of monopropellants based on hydroxyl ammonium nitrate (HAN), which promise to deliver up to 25% greater specific impulse than hydrazine. The high combustion temperature of these propellants requires the use of new materials for thruster construction, and there are also concerns regarding the long-term stability of the new propellants. Nonetheless, we will consider an AFRL advanced monopropellant formulation, RK-315A, for the TechSat21 mission. Thrusters designed for use with RK-315A will be assumed 25% heavier than comparable hydrazine thrusters due to design requirements imposed by the high chamber temperature.

Cold-gas thrusters are the simplest throttleable thruster available and, thus, the simplest propulsion system suitable for the stationkeeping and attitude control requirements.[5,6] Unfortunately, the combination of extremely low specific impulse and heavy, high-pressure propellant tanks results in unreasonably high total propulsion system mass values even though the thrusters themselves can be quite small. Cold-gas thrusters are wholly unsuited for the 2007 mission and are only marginally capable of performing the 2003 mission. However, they do have the advantage of using an inert, gaseous propellant and thus do not contaminate exposed spacecraft surfaces. Because of the threat of mutual contamination when spacecraft operate propulsion systems in close formation, the combination of cold-gas thrusters for attitude control and precision stationkeeping and a similarly noncontaminating main propulsion system will be considered in spite of the high propellant mass, with the Moog 58-102 thruster baselined for analysis.[7]

While all of the aforementioned chemical propulsion systems can be obtained in sizes suitable for the TechSat21 mission, they begin to suffer from scaling effects at that level. Chemical propulsion would be largely unsuitable in future microsatellite missions with mass budgets an order of magnitude smaller. To meet the microsatellite requirement, several institutions have proposed the Digital MEMS thruster.[8,9] This device uses semiconductor manufacturing techniques to etch thousands of extremely small ($\sim$500-μm) cavities and nozzles into a silicon wafer. Each cavity is filled with a propellant charge and serves as a one-shot microthruster at need.[10] The specific impulse and propellant mass fraction suffer in comparison with conventional chemical rockets, but the ability to scale down to extremely small sizes compared with conventional systems is desirable for the microsatellite application. The availability of small discrete thrust impulses is particularly advantageous for stationkeeping and attitude control. Both TRW and Honeywell presently have programs to fabricate digital MEMS thrusters and have tested the necessary igniter arrays. Test firings with full propellant loads are expected soon, and while there are still substantial technical challenges associated with the concept it will be considered as an option for TechSat21, using the performance estimated by TRW and Honeywell.

B. Electromagnetic Micropropulsion

The PPT[11] generates thrust through a surface discharge across the face of a solid Teflon propellant. The solid propellant is converted to vapor and partially ionized by ohmic and radiative energy from the arc. Acceleration is accomplished by a combination of thermal and electromagnetic forces to create usable thrust. The solid propellant is attractive for the small SBR satellites since it significantly reduces the thruster mass and volume by eliminating the propellant tankage and valves. In addition, eliminating the valve seals and flow regulation, which become increasingly problematic at small sizes, increases the engineering reliability of the PPT compared with gaseous or liquid propellant thrusters. The only moving part on the PPT is a spring, which passively feeds the propellant. The inherent engineering advantages of the PPT design have enabled the thruster to complete several space missions over the past 30 years with no failures.[12,13] Presently, the PPT is scheduled to fly in 2000 on the NASA EO-1 satellite[14] and is being considered for the New Millennium Deep Space 3 mission scheduled to launch in 2003.[15]

The last flight-qualified design,[16] for the LES 8/9 satellite in 1974, operating at 20 W of power achieved a thrust of 300 μN, a specific impulse of 1000 s, and a thrust efficiency of 6.4%. For the LES 8/9 PPT two electrode assemblies were used with their thrust vectors canted 30-deg to provide two thrust vectors for attitude control. A modern PPT, developed by Primex Aerospace for NASA, is presently being qualified for EO-1. The EO-1 PPT operates between 5 and 40 J per discharge to create a 100- to 700-uN-s impulse bit with a specific impulse near 1200 s. Two electrode assemblies are again used, however, the thrust vectors are pointed 180-deg apart.

For the main propulsion application, a somewhat different PPT electrode configuration is proposed with four electrode sets directed toward a common thrust vector. The four electrode sets are fed from one common capacitor as was done for both the LES 8/9 and the EO-1 PPTs. Each electrode would be fired at 20 J and 2 Hz for a total power level of 160 W. Use of four electrode sets eliminates the gimbal requirement, since the firing rate can be adjusted between the four units for thrust vector control. This minimizes the configuration mass and uses a design requiring minimal change from the commercially available design. The electrode redundancy also reduces the length of the breech-fed propellant to approximately 6 in., simplifying the spacecraft integration. Performance is conservatively assumed to be 1000-s specific impulse, 10% efficiency, and 700 μN at 40 W for each electrode set. This performance level is slightly degraded from that measured for EO-1 using a 40-J discharge and slightly better than that measured at 20 J for LES 8/9. Due to a recent resurgence in PPT funding, significant progress has been made in improving PPT performance and engineering. Laboratory model PPTs, with reasonably flight-like designs, have been demonstrated to achieve thrust to power levels three to four times above the LES 8/9 flight models.[17] The next generation of PPTs is also expected to use a coaxial geometry, which may reduce the radiated EMI and lessen the spacecraft interaction. These advanced PPTs could easily be available for both TechSat21 missions.

The micro-PPT is a simplified, miniaturized version of the PPT developed at the AFRL.[18] The micro-PPT uses a high-voltage discharge that is applied across the face of a coaxial propellant bar. Both pulsed and DC application of the high voltage has been tested. The discharge ignites through a self-breakdown, thus eliminating the PPT sparkplug and the associated mass, complexity, and energy requirements. Pulsed micro-PPT voltage application generally tends to require lower voltage than the DC micro-PPT and hence reduced shielding mass. DC voltage application eliminates the mass of the semiconductor switches and voltage amplification electronics. The propellant modules consist of annular Teflon propellant with inner and outer copper electrodes. Module diameters of 0.110, 0.140, and 0.250 in. have been tested. Typical breakdown voltages for the 0.110-in. propellant is under 3000 V for a pulsed discharge. Thus the micro-PPT can be energized from the trigger circuit of a standard PPT, effectively eliminating the PPU mass of the stationkeeping propulsion system. Micro-PPT discharges are typically in the 1-J regime and are estimated to create 25 μN of thrust at 1000-s I_{sp} for a 2-Hz firing rate. Thrust levels are estimated by extrapolating from full-scale PPT data and by operating the micro-PPT at higher pulse rates, since no reliable measurement capability presently exists at the low thrust levels required for microsatellite stationkeeping.

For applications on space-based radar microsatellites, the radiated EMI from the propulsion system poses a serious concern. For the PPT, gigahertz radiation from the spark ignitor discharge can interfere with the primary transceiver frequencies. EMI radiation from the main PPT discharge can interfere with the radar frequency shifts in the megahertz range. Although it is still a topic of current research, it is believed that the new generation of coaxial PPTs will better confine the EM radiation.

C. Electrostatic Micropropulsion

The high specific impulse, high efficiency, and modest mass have made Hall thrusters the electric propulsion system of choice for many future missions at power levels of 1 kW or above. Hall thrusters were developed in the former Soviet Union during the 1960s and 1970s and have been flown in over a hundred successful missions. Lightweight thrusters optimized for power levels of 50 to 200 W have been developed recently and would be suited for TechSat21 main propulsion.[19,20]

The relatively high mass of even the smallest Hall thrusters, though, renders them marginal at best for attitude control, despite the potential savings associated with the use of a common power processing unit. Also, the use of a Hall thruster requires a high-pressure propellant storage and feed system, further increasing the system mass and complexity. For purposes of this analysis the 200-W Hall thruster developed by Busek Corp. is considered for primary propulsion.

Another electrostatic thruster proposal under consideration is field-effect electrostatic propulsion (FEEP). This is an ion thruster using a field-emission ion source, in the form of a narrow slit anode through which cesium propellant is passed and ionized by the geometrically enhanced electric field.[21] This offers a more compact and efficient ion source than the traditional electron-bombardment ionization chambers used with ion thrusters, allowing the extension of electrostatic propulsion to smaller spacecraft than previously possible. FEEP systems have been demonstrated in the laboratory, and a FEEP thruster is scheduled for a Shuttle flight experiment in 2000. These systems offer extremely high specific impulse values at reasonable efficiency, but specific power and thrust are low. It may, therefore, be necessary to relax the 30-day maneuver time requirement for a FEEP-based propulsion system. For the purpose of this analysis, several combinations of the Centrospazio 120-W FEEP thruster with complimentary ACS thrusters will be considered.

The basic physical principles of the FEEP thruster can be scaled down to the true micropropulsion regime, resulting in the micro-colloid thruster. As with the digital chemical microthruster described earlier, this system consists of a large number of discrete thrusters micromachined into a silicon wafer. The thruster elements consist of microvolcano field-emission ion sources, in which a propellant is fed through a small, sharp needle, which serves as an anode similar to the Spindt-type microcathode. The field enhancement associated with the sharp tip of the needle results in the emission of charged micron-scale droplets of propellant. The droplets are accelerated by an applied electric field and neutralized by an external electron source or, perhaps, by a parallel microthruster element operating at the opposite polarity. Systems proposed by Phrasor Scientific and MIT use doped glycerol propellants and are predicted to achieve specific impulse values of order 1000 s at a reasonably high efficiency.[22] As yet, only limited progress has been made in fabricating thruster subassemblies and testing representative components, and no actual thruster has been constructed or tested, so there would be a high degree of technical risk associated with the use of micro-colloid thrusters in TechSat21.

Closely related to the micro-colloid thruster is the micro field ionization thruster (MFIT) from SRI, Inc. This also uses microvolcano field ion sources, but with a metallic propellant, typically gallium or indium. These materials melt at or near room temperature, which allows for a relatively simple feed system, while having a low ionization energy and surface tension, results in the emission of single ions rather than charged clusters or droplets.[23] The resulting high charge-to-mass ratio allows higher specific impulse values to be achieved with reasonable accelerating voltages, and the ionization mechanism is considerably more efficient than that of conventional ion thrusters. The current proposal calls for a specific impulse of 15,000 s or higher, with a correspondingly low thrust-to-power ratio, though it is likely that the specific impulse can be substantially reduced if necessary to meet thrust requirements. With a fixed ionization energy requirement, power efficiency will of course drop at lower specific impulse values, and the thruster is best used for moderate to high specific impulse missions. The development of MFIT thrusters is at a very early stage, and readiness for the 2003 TechSat21 mission is doubtful, but the potential benefits of such a compact, high specific impulse system for later stages of the program cannot be ignored.

D. Electrodynamic Tether

One further technology is considered for the primary propulsion requirements, the electrodynamic tether. However, due to the immaturity of this technology and lack of understanding of the fine control for maneuvering, the tether was considered only for the specified deorbit requirement for TechSat21. Several kilometers of thin wire can be deployed from a spacecraft and, by gravity gradient effects, oriented vertically with respect to the Earth. Orbital motion of this wire in the Earth's essentially stationary magnetic field will generate an electric potential along its length, and if a plasma contactor is placed at each end a current will flow. The $J \times B$ force produced by the interaction of this current with the magnetic field tends to decelerate the spacecraft and can perform the deorbit portion of the 2007 mission in a matter of months.[24]

Unfortunately, TechSat21 is likely to operate in a polar orbit, whereas electrodynamic tether systems require motion perpendicular to the Earth's magnetic field to produce thrust. Since the Earth's magnetic pole and the true North differ by 11.5-deg a small force is still generated, however, there is concern as to the length of time required to remove the satellite from the constellation. There is also a strong concern regarding deployment of the tether within the dynamic constellation and potential tangling with adjacent spacecraft. One approach that will be further investigated is the option of simply turning off the propulsion system of a failed satellite, allowing the J2 perturbation to drift the vehicle out of the constellation and then from a "safe" distance away from the constellation deploying the electrodynamic tether. These, plus general concerns regarding the technical maturity of tether systems at present, argue against recommending a tether deorbit system for TechSat21, but the option is considered for comparative purposes.

E. Electric Power Processing

It should be noted that for all of the electric propulsion systems described, dedicated power processing hardware is required. The TechSat21 spacecraft will have ample electric power available from the solar array boom, but the main bus can be expected to operate at less than 100 V, while the various electric thrusters require anywhere from 300 V to 10 kV. Also, few of the systems have been tested in a simple direct-drive configuration. Traditionally, voltage- or current-regulated switching power supplies are used, which tend to outweigh the thrusters they drive by a factor of two or three. We assume that such conventional power supplies will be used with the macro-scale Hall thruster, but for the various micropropulsion systems a solid-state DC–DC converter seems a more reasonable choice. A micro-PPT has been operated using such a system at AFRL, and it will be assumed that similar systems can be used with the colloidal and MFIT systems. With PPTs, a trigger unit and a discharge capacitor must also be provided and are included in the system weight estimate. Notwithstanding the requirement for a power processing unit, such systems are rightly considered separately from the thrusters themselves, as a single PPU can serve a substantial number of distinct thrusters in stationkeeping or attitude control operation. In the case of the PPT, it is fortuitously possible for the PPT trigger circuit to also serve as the entire power-processing unit for an associated micro-PPT system.

A comparison of the thruster proposals described above is given in Table 2. As can be seen, electric propulsion systems generally offer specific impulse values of 1000–1500 s and chemical systems approximately 200 s. Electric propulsion is thus quite likely to offer lower overall system mass, presuming that the thruster and PPU masses can be kept to acceptable levels. Thrusters can also be divided into micro and macro scales, with the macro thrusters being small versions of conventional

Table 2 Thruster performance comparison: Values for typical microsatellite installation

Type	I_{sp}, s	η, %[a]	Mass, kg	Thrust	Power, W
Cold-gas thruster	75	95+	0.01	5 mN	N/A
Solid rocket motor	185	90+	1.60	100+ N	N/A
Digital MEMS	200	~75	0.04	50 mN	N/A
Hydrazine monopropellant	220	95+	0.16	1 N	N/A
Advanced monopropellant	290	95+	0.20	1 N	N/A
Colloidal microthruster	500–1500	~50	0.08[b]	20 μN	0.2
Pulsed plasma thruster	800	~10	0.40[b]	500 μN	25
μPPT	1,000	~5	0.12[b]	40 μN	4
Hall thruster	1,500	~35	1.00[b]	10 mN	200
FEEP	8,000	~25	3.50[b]	800 μN	120
MFIT	15,000	~90	0.15[b]	1.5 mN	300

[a] η = total system efficiency (thrust power output to chemical or electrical power input).
[b] Values for EP systems do not include the power processing unit.

spacecraft propulsion systems and the microthrusters developed specifically for microsatellite applications, often using semiconductor-style, commonly referred to as MEMS, fabrication. The macrothrusters, with masses of order 1 kg, are generally suitable for the main propulsion application but are too heavy to be used in the numbers required for the stationkeeping/ACS application. The microthrusters, at approximately 100 g, can meet the stationkeeping/ACS requirement but lack the thrust needed for main propulsion of the 100-kg TechSat21 vehicle unless used in clusters. Some combination of the two is likely required.

IV. Analysis

Given the extremely tight mass budget set for the TechSat21 spacecraft, any comparison of propulsion options must center on predictions of propulsion system mass. To address this issue, detailed mass estimates for propulsion systems using the various proposed technologies were constructed. A total of 30 propulsion options was considered, with each of the potential main propulsion systems matched with one or more compatible stationkeeping systems. Mass estimates for each system were broken down into five categories—thruster, PPU, propellant, propellant feed, and miscellaneous—with one or more line items in each category as appropriate. Separate evaluations were made for the requirements of the 2003 and 2007 numbers, due to the different ΔV and propellant requirements.

The thruster category includes the main propulsive thruster or thrusters and the stationkeeping thrusters. The size and/or number of main thrusters was set by the ~2-mN thrust requirement for a total 30-day maneuver period, except in the case of the MFIT and FEEP systems, where a relaxed 60-day requirement was allowed due to their low thrust-to-power ratio. In some cases, such as the Hall thruster or any of the chemical systems, a single thruster of the smallest reasonable size provided a much shorter maneuver period. For attitude control and stationkeeping, a total of six ACS thruster elements with a minimum thrust of 40 μN was required. Any special mounting hardware required was also included in this category.

For electric propulsion systems, a high-voltage DC power-processing unit is invariably required, as previously described, and the use of semiconductor DC–DC conversion has been postulated in most cases. Also included in this category are any necessary high-voltage or high-current cables. In the case of PPTs, an energy storage capacitor and a pulse trigger generator are also required. If the main and ACS systems incorporate different electric propulsion technologies, separate power processing systems are generally required. The principal exception to this rule is the ability of a PPT trigger pulse generator to serve as the entire power processing system of a micro-PPT attitude control system, although additional switching hardware is required.

Sufficient propellant was provided to meet the specified ΔV requirements of the 2003 and 2007 missions, as described earlier. In some cases, it was deemed advantageous to provide separate propellant storage for each ACS thruster element rather than a feed system from a central tank. For these cases the requirement is set at 100-N-s total impulse per element for the 2003 mission and 1 kN-s for the 2007 mission to account for possible nonuniform propellant usage by the ACS. The propellant feed system includes tankage for main and ACS propellant, feed lines, valves, and flow control systems.

Ten percent of the propulsion system net dry mass is specified for structures and general mounting hardware. An additional 5% is specified for control systems and wiring harnesses using standard spacecraft design practice.[25] This is exclusive of any high-voltage distribution system incorporated in the PPU category. Finally, a 15% margin is set aside for unexpected system growth. This total of 30% net dry mass constitutes the miscellaneous category and includes the overall error margin in estimating subsystem-level dry masses. Propellant mass may be assumed to have an error margin of ±5% due to uncertainties in thruster specific impulse.

For each of the enumerated items, commercial off-the-shelf hardware was specified whenever possible, preferably space-qualified, but in the case of some PPU or propellant feed system components, ground or aviation hardware meeting relevant military specifications was used as a baseline. The intention is to estimate reliably the mass of a flight system rather than actually to design such a system. In some cases, commercial systems of different power levels were scaled linearly over a modest range to meet specific TechSat21 requirements. For experimental thruster concepts, flight-like laboratory test hardware was considered, and in the case of some technologies that have not yet reached even the test stage, the best estimates of the authors regarding developed system weights were used.

Space precludes giving the detailed mass breakdowns for all propulsion system options here, though a representative sample is given in Table 3. Figure 3 is a schematic layout of the same system, indicating the major components. While specific to the all-PPT propulsion option, other propulsion systems will have a similar configuration. Six ACS thrusters in a trilateral arrangement are specified for X–Y stationkeeping and yaw control, rather than the traditional eight-thruster orthogonal arrangement. While this does result in a small (<13%) reduction in efficiency due to cosine losses, the ease of integration with the hexagonal TechSat21 bus and the ~25% reduction in system dry mass due to the reduced number of thruster units more than compensates for nonorthogonal losses and leads to the recommendation of the trilateral system for this application.

Tables 4 and 5 provide comparative breakdowns of all the concepts included in this study, for the 2003 and 2007 missions, respectively. Most of the chosen propulsion systems can meet the specified 10-kg propulsion system mass

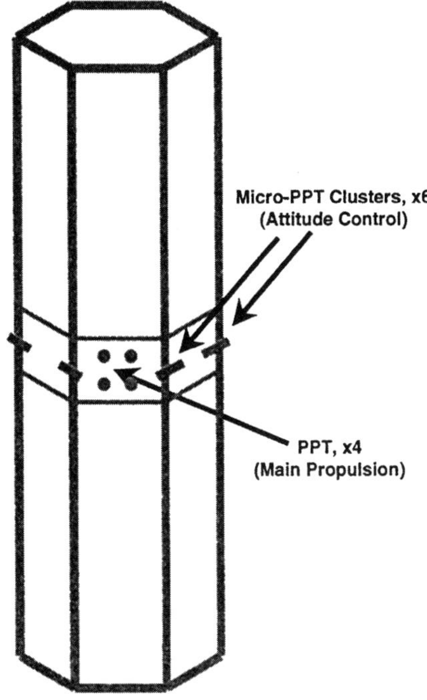

a) Spacecraft propulsion system layout

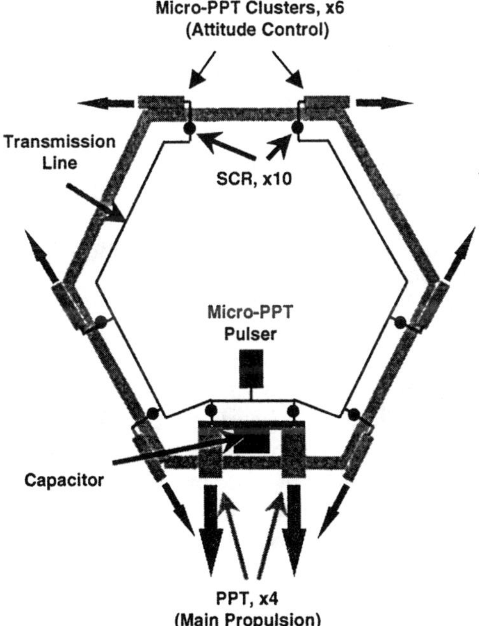

b) Propulsion module

Fig. 3 Representative propulsion system layout on a TechSat21 spacecraft. PPTs are shown for primary propulsion, with micro-PPTs for attitude control. a) Overall spacecraft propulsion system layout; b) detail of the propulsion module.

Table 3 Representative propulsion system mass estimate (PPT thruster with micro-PPT ACS, sized for 2003 mission)

Component	Type	No.	Unit mass	Total mass, g
Main thruster	CU Aerospace PPT-7	4	420 g	1680
Propellant	Teflon bars	8	165 g	1320
Capacitor	Unison	1	1460 g	1460
Power processing unit	Unison	1	2090 g	2090
Trigger pulse unit	Unison	1	330 g	330
ACS thrusters	AFRL μPPT modules	6	60 g	360
High-voltage cable	RG-58 or equivalent	2.5 m	40 g	100
SCR switch modules		6	20 g	120
Structures & mounts			10% of dry mass	600
Controls & wiring			5% of dry mass	300
Design margin			15% of dry mass	900
Total				9.25 kg

requirement for the 2003 mission. However, only systems incorporating the rather speculative MFIT thruster can meet mass budget for the 2007 mission. A number of electric propulsion systems offer total masses in the 12- to 15-kg range, which, in light of the technical immaturity of the MFIT system and the lack of other alternatives, must be considered acceptable. The conventional chemical monopropellant baseline, at more than twice the budgeted mass, is not a reasonable candidate for the 2007 mission.

Factors other than propulsion system mass must also be considered in comparing TechSat21 propulsion options. In particular, the technical maturity of the various systems is a major concern. Only the FEEP and chemical monopropellant systems could be constructed using existing flight-qualified hardware. The low-power Hall, PPT, and micro-PPT thruster systems have at least been demonstrated in the laboratory and are based on existing flight-qualified systems operating at higher power levels. Less-tested systems such as the MFIT, digital MEMS, and colloidal microthruster are higher risk and, therefore, may suffer increases in mass and cost while developing a flight unit.

Also relevant are the maneuver time and power requirement, though it is clear that any of the proposed systems can achieve acceptable performance in these regards. In particular, with a power budget of 350 W, the TechSat21 propulsion system is likely to be mass-limited rather than power-limited. Power processing units capable of handling the full available power would be excessively heavy (5+ kg for the PPU alone). All of the electric propulsion systems found to be competitive for TechSat21 operate at power levels of less than 200 W.

An additional factor that strongly influences the preferred propulsion option is the cost to integrate and space-qualify the system. Since the Space Shuttle is at present the most likely launch vehicle for the TechSat 21 critical technology demonstration mission, the required Shuttle safety review may dominate integration and qualification expense. For systems with reactive chemical propellants and/or pressurized tanks, Shuttle safety requirements (e.g., triple redundancy in valving) are particularly strenuous. Thus, systems with inert propellants stored in solid or low-pressure liquid form may be preferred. Especially advantageous are propulsion systems in which thruster, PPU, and propellant inherently constitute a single package.

Table 4 Propulsion system mass comparison, 2003 mission

Propulsion type		System mass breakdown, kg					
Main	ACS	Thruster	PPU	Propellant	Tankage	Misc.	Total
MFIT	MFIT	0.45	2.70	0.10	N/A	0.95	4.20
HAN	MEMS	0.60	N/A	4.30	0.70	0.40	6.00
HAN	Magnetic	0.40	N/A	5.05	0.80	0.45	6.70
HAN	μPPT	0.65	0.55	4.20	0.75	0.60	6.75
HAN	Colloid	1.00	0.35	4.20	0.75	0.65	6.95
N_2H_4	MEMS	0.50	N/A	5.65	0.85	0.40	7.40
Colloid	Colloid	3.30	0.90	1.75	0.25	1.30	7.50
HAN	HAN	1.05	N/A	4.80	1.15	0.65	7.65
Hall	Magnetic	2.00	2.00	0.75	1.35	1.60	7.70
N_2H_4	Magnetic	0.30	N/A	6.65	0.50	0.45	7.85
Hall	MEMS	2.15	2.00	0.90	1.35	1.55	7.95
HAN	Cold gas	0.45	N/A	5.45	1.80	0.75	8.45
N_2H_4	μPPT	0.55	0.55	5.50	0.85	0.60	8.05
N_2H_4	Colloid	0.90	0.35	5.50	0.85	0.65	8.25
Hall	μPPT	2.25	2.55	0.75	1.35	1.85	8.75
PPT	MEMS	1.85	3.75	1.45	N/A	1.75	8.80
N_2H_4	Cold gas	0.35	N/A	6.60	1.45	0.75	8.90
PPT	Magnetic	1.70	4.05	1.50	N/A	1.70	8.95
Hall	Colloid	2.60	2.35	0.75	1.35	1.90	8.95
N_2H_4	N_2H_4	0.95	N/A	6.30	1.25	0.65	9.15
PPT	μPPT	1.90	4.10	1.45	N/A	1.80	9.25
Hall	Cold gas	2.05	2.00	1.25	2.40	1.95	9.65
PPT	Colloid	2.25	4.10	1.55	N/A	1.90	9.80
MEMS	MEMS	3.55	N/A	5.30	N/A	2.05	10.90
PPT	PPT	3.35	4.30	1.55	N/A	2.30	11.50
FEEP	Magnetic	7.00	4.10	0.15	0.60	3.50	15.35
FEEP	MEMS	7.15	4.10	0.35	0.60	3.55	15.75
FEEP	μPPT	7.25	4.65	0.25	0.60	3.75	16.50
FEEP	Colloidal	7.60	4.45	0.25	0.60	3.70	16.60

Finally, the precision of the attitude control system may be an important consideration. Any of the systems listed can meet the preliminary requirements specified by the TechSat 21 program office, but increasing emphasis on close-proximity maneuvering is likely to tighten those requirements. Systems based on a single, large thruster with attitude control provided by magnetic torquers alone offer only coarse directional control, which may be problematic. Differential throttling of two or more main thrusters can offer greater precision but introduces coupling between yaw steering and X–Y plane translation. Only systems with a dedicated, microthruster-based attitude control system can provide a truly high precision.

Table 6 lists these parameters and issues for all studied propulsion system options. These systems fall into three general categories. The MFIT and, to a lesser extent, the colloidal microthruster are clearly preferable on virtually all technical grounds and are the only systems capable of meeting the mass budget for the 2007 mission. However, their low technical maturity is problematic for a 2007 launch

Table 5 Propulsion system mass comparison, 2007 mission

Propulsion type		System mass breakdown, kg					
Main	ACS	Thruster	PPU	Propellant	Tankage	Misc.	Total
MFIT	MFIT	0.45	2.70	0.40	0.10	0.95	4.60
Colloid	Colloid	3.30	0.90	5.00	0.60	1.30	11.10
Hall	Magnetic	2.00	2.00	4.50	2.10	1.85	12.45
Hall	μPPT	2.45	2.55	4.10	1.95	2.10	13.15
Hall	Colloid	2.60	2.35	4.05	2.05	2.10	13.15
PPT	MEMS	2.20	3.75	7.50	N/A	1.85	15.30
PPT	μPPT	2.10	4.10	7.50	N/A	1.85	15.50
PPT	Colloid	2.25	4.10	7.20	0.10	1.90	15.55
PPT	Magnetic	1.70	4.05	8.55	N/A	1.70	16.00
Hall	MEMS	3.70	2.00	5.95	2.10	2.35	16.10
FEEP	Magnetic	7.00	4.10	0.85	0.65	3.50	16.10
FEEP	MEMS	7.50	4.10	1.45	0.65	3.75	17.45
FEEP	Colloidal	7.60	4.45	1.00	0.75	3.75	17.55
FEEP	μPPT	7.25	4.65	1.15	0.65	3.80	17.70
HAN	μPPT	0.85	0.55	17.95	0.75	0.65	20.75
Hall	Cold gas	2.05	2.00	8.40	6.95	2.35	21.25
HAN	Colloid	1.00	0.35	18.00	2.25	1.10	22.70
N_2H_4	μPPT	1.20	0.55	18.30	2.10	1.15	23.30
HAN	HAN	1.05	N/A	20.10	2.35	1.00	24.50
HAN	MEMS	1.50	N/A	20.10	2.10	1.10	24.80
HAN	Magnetic	0.40	N/A	21.90	2.50	0.85	25.65
N_2H_4	Colloid	0.90	0.35	23.35	2.70	1.20	28.50
PPT	PPT	3.35	4.30	8.15	N/A	2.30	29.60
N_2H_4	N_2H_4	0.95	N/A	25.90	2.90	0.40	30.15
N_2H_4	MEMS	1.95	N/A	25.35	2.65	1.40	31.35
HAN	Cold gas	0.45	N/A	22.05	7.10	2.25	31.90
N_2H_4	Magnetic	0.30	N/A	28.15	2.65	1.10	32.20
N_2H_4	Cold gas	0.35	N/A	27.50	4.55	2.50	34.90
MEMS	MEMS	18.00	N/A	27.05	N/A	5.40	50.45

and absolutely rules out consideration for the 2003 mission. A variety of combinations based on demonstrated low-power Hall thrusters or PPTs teamed with a range of microthrusters for attitude control offer propulsion system masses in the 10-kg range for the 2003 mission and 15 kg for the 2007 launch, with acceptable performance in all other regards. Finally, chemical-based systems deliver adequate mass and performance for the 2003 mission but become excessively massive if called upon to meet the high ΔV requirements of the 2007 mission.

V. Conclusions

If only the 2003 critical technology demonstration mission is to be considered, the combination of a conventional hydrazine monopropellant thruster and micro-PPTs is the clear choice for TechSat21 propulsion. The hydrazine system can deliver adequate performance for main propulsion within the specified mass budget using off-the-shelf flight hardware, and the micro-PPT is the only demonstrated microthruster capable of providing precision attitude control for a 2003 launch.

Table 6 Propulsion system comparison

Propulsion type		Propulsion mass, kg		Maneuver, W	Maneuver time, days	Technology status	ACS precision
Ascent	ACS	2003	2007				
MFIT	MFIT	4.20	4.60	300	60	Research	High
HAN	MEMS	6.00	24.80	N/A	<1	Research	High
HAN	Magnetic	6.70	25.65	N/A	<1	Demonstrated	Low
HAN	μPPT	6.75	20.75	N/A	<1	Demonstrated	High
HAN	Colloid	6.95	22.70	N/A	<1	Research	High
N_2H_4	MEMS	7.40	31.35	N/A	<1	Research	High
Colloid	Colloid	7.50	11.10	15	30	Research	High
HAN	HAN	7.65	24.50	N/A	<1	Demonstrated	Medium
Hall	Magnetic	7.70	12.45	200	10	Demonstrated	Low
N_2H_4	Magnetic	7.85	32.30	N/A	<1	Flight-ready	Low
Hall	MEMS	7.95	16.10	200	10	Research	High
N_2H_4	μPPT	8.05	23.30	N/A	<1	Demonstrated	High
N_2H_4	Colloid	8.25	28.50	N/A	<1	Research	High
HAN	Cold gas	8.45	31.90	N/A	<1	Demonstrated	High
Hall	μPPT	8.75	13.15	200	10	Demonstrated	High
PPT	MEMS	8.80	15.30	100	30	Research	High
N_2H_4	Cold gas	8.90	34.90	N/A	<1	Flight-ready	High
PPT	Magnetic	8.95	16.00	100	30	Demonstrated	Low
Hall	Colloid	8.95	13.15	200	10	Research	High
N_2H_4	N_2H_4	9.15	30.15	N/A	<1	Flight-ready	Medium
PPT	μPPT	9.25	15.50	100	30	Demonstrated	High
Hall	Cold gas	9.65	21.25	200	10	Demonstrated	High
PPT	Colloid	9.80	15.55	100	30	Research	High
MEMS	MEMS	10.90	50.45	N/A	<1	Research	High
PPT	PPT	11.50	29.60	100	30	Demonstrated	Medium
FEEP	Magnetic	15.35	16.10	120	60	Flight-ready	Low
FEEP	MEMS	15.75	17.45	120	60	Research	High
FEEP	μPPT	16.50	17.70	120	60	Demonstrated	High
FEEP	Colloidal	16.60	17.55	120	60	Research	High

This combination minimizes both technical risk and system mass, which are the dominant concerns for the first TechSat21 flight.

However, chemical systems are completely unsuitable for the 2007 full demonstration mission, as the higher ΔV requirement results in excessive propellant mass. For the 2007 mission, the preferred option would be either a low-power Hall thruster or a pulsed plasma thruster for main propulsion with a micro-PPT system for attitude control. These systems meet all specified requirements save mass and exceed the mass budget by only a few kilograms. Furthermore, all of the technologies involved have been demonstrated in larger-scale flight systems, with hardware appropriate for TechSat 21 currently undergoing ground testing. The PPT-based system has the additional advantage of using only inert, solid propellants, substantially reducing integration and space qualification costs.

Since the Hall/micro-PPT and PPT/micro-PPT combinations can meet the requirements for the 2003 as well as 2007 mission, it may be preferable to design a single propulsion system for all TechSat 21 spacecraft. This would minimize development costs and allow early flight testing of the propulsive technologies needed for the later mission. The chemical system can be recommended only for the 2003 mission if aggressive scheduling mandates an absolute minimum of technical risk to meet integration deadlines.

If a chemical system is chosen for the earlier mission, research and development efforts must continue on either the low-power Hall thruster and/or the advanced pulsed plasma thruster, with the PPT presently being preferred due to ease of integration and space qualification, to meet the requirements of the 2007 mission. Furthermore, the field ionization microthruster and the colloidal microthruster both promise substantial improvements in mass and performance for the later mission and would then warrant serious attention. They are not, however, sufficiently mature to warrant a commitment to their use even for the 2007 flight.

References

[1]Ferster, W., "NRO Delays New Spy Fleet Amid Questions Over Cost," *Space News*, Vol. 9, No. 21, 1998, pp. 6–20.

[2]Das, A., "Overview of the AFOSR/AFRL TechSat21 Microsatellite Program," *Micro/Nanotechnology for Micro/Nanosatellites*," AFOSR, Albuquerque, 1998, Sec. 4.

[3]Mueller, J., "Thruster Options for Microspacecraft: A Review and Evaluation of Existing Hardware and Emerging Technology," AIAA Paper 97-3058, July 1997.

[4]Rocket Research Company, "Rocket Thrusters," *Hydrazine Handbook*, Olin Defense Systems Group, Redmond, WA 1995, Chap. 3.

[5]Janson, S. W., and Helvajian, H. "Batch-Fabricated Microthrusters: Initial Results," AIAA Paper 96-2988, July, 1996.

[6]Mueller, J., "Overview of Micropropulsion Workshop," Jet Propulsion Library, Pasadena, CA, 1997.

[7]Moog Space Products Division, "Cold Gas Thrusters," *Moog Space Products*, Moog, Aurora, NY, 1993, Sec. 4.

[8]Youngner, D., and Choueiri, E., "MEMS Mega-Pixel Microthruster Arrays for Microsatellites," *Proceedings, Formation Flying and Micro-Propulsion Workshop*, Air Force Research Lab., Lancaster, CA, Oct., 1998.

[9]Lewis, D., Antonsson, E., and Janson, S., "MEMS Microthruster Digital Propulsion System," *Proceedings, Formation Flying and Micro-Propulsion Workshop*, Air Force Research Lab., Lancaster, CA, Oct., 1998.

[10]Janson, S. W., and Helvajian, H. "Batch-Fabricated Microthrusters: Initial Results," AIAA Paper 96-2988, July, 1996.

[11]Burton, R. L., and Turchi, P. J., "Pulsed Plasma Thruster," *Journal of Propulsion and Power*, Vol. 14, No. 5, 1998, pp. 716–735.

[12]Brill, Y., Eisner, A., and Osborn, L., "The Flight Application of a Pulsed Plasma Microthruster; the NOVA Satellite," AIAA Paper 82-1956, Nov. 1982.

[13]Guman, W. J., and Williams, T. E., "Pulsed Plasma Microthruster for Synchronous Meteorological Satellite (SMS)," AIAA Paper 73-1066, Oct. 1973.

[14]Meckel, N. J., Cassady, R. J., Osborne, R. D., Hoskins, W. A., and Myers, R. M., "Investigation of Pulsed Plasma Thrusters for Spacecraft Attitude Control," 25th International Electric Propulsion Conf., Paper 97-128, Aug. 1997.

[15]Blandino, J. J., Cassady, R. J., and Peterson, T. T., "Pulsed Plasma Thrusters for the New Millenium Interferometer (DS-3) Mission," 25th International Electric Propulsion Conf., Paper 97-192, Aug. 1997.

[16]Vondra, R. J., "Flight-Qualified Pulsed Electric Thruster for Satellite Control," *Journal of Spacecraft*, Vol. 9, 1974, p. 613.

[17]Burton, R. L., Bushman, S. S., and Antonsen, E. L., "Arc Measurements and Performance Characteristics of a Coaxial Pulsed Plasma Thruster," AIAA 98-3660, July 1998.

[18]Spanjers, G. G., Schilling, J. H., et al., "The Micro Pulsed Plasma Thruster" (submitted for publication).

[19]Hruby, V., Monheiser, J., et al., "Development of Low Power Hall Thrusters," AIAA Paper 99-3534, July 1999.

[20]Khayms, V., and Martinez-Sanchez, M. "Design of a Miniaturized Hall Thruster for Microsatellites," AIAA Paper 96-3291, July, 1996.

[21]Gonzalez, J., Saccocia, G., and Von Rhoden, H. "Field Emission Electrostatic Propulsion (FEEP): Experimental Investigation on Microthrust FEEP Thrusters," AIAA Paper 97-3057, July 1997.

[22]Perel, J., Mahoney, J., and Sujo, C., "Micro-Electric Propulsion Using Charged Clusters," Proceedings, Formation Flying and Micro-Propulsion Workshop, Air Force Research Lab., Lancaster, CA, Oct. 20–21, 1998.

[23]Fehringer, M., Rudenauer, F., and Steiger, W., "Space-Proven Indium Liquid Metal Field Ion Emitters for Ion Microthruster Applications," IEPC Paper 93-157, 1993.

[24]Forward, R. L., and Hoyt, R. P., "Application of the Terminator Tether Electrodynamic Drag Technology for the Deorbit of Constellation Spacecraft," AIAA Paper 98-3491, June 1998.

[25]Larson, W. J., and Wertz, J. R., "Spacecraft Design and Sizing," *Space Mission Analysis and Design*, 2nd ed., Microcosm, Torrance, CA, 1993, Chap. 10.

Chapter 2

University Micro-/Nanosatellite as a Micropropulsion Testbed

Joyce Wong* and Helen Reed†
Arizona State University, Tempe, Arizona
and
Andrew Ketsdever ‡
Air Force Research Laboratory, Edwards Air Force Base, California

I. Introduction

STUDIES have shown that, by partitioning the functions of a single large satellite into a number of smaller satellites that orbit in close proximity and operate cooperatively, one could achieve cost and weight reductions.[1] Such ideas involve a cluster of several to many satellites that fly in formations from 10 to 1000 m in size. The satellites are in constant communication with each other. Each could perform a unique dedicated task, or the cluster could operate like a parallel computer, with each identical satellite contributing a small part to the whole. Hence, the cluster operates cooperatively to perform a function like a "virtual" satellite. These ideas have been applied to the TechSat21 radar mission, and preliminary estimates have indicated that there is merit to this approach.[1] In fact, the New World Vistas Space Technology Panel has advocated the use of networks and clusters of reconfigurable and adaptable micro-/nanosatellites to support cost-effective space missions.

A key element for microspacecraft operations is a practical micropropulsion system. Micropropulsion systems offer a wide variety of mission options, all relevant to formation flying, which include attitude control, station maintenance (especially in low Earth orbit; LEO), altitude raising, plane changes, and deorbit. Consider altitude raising, for example. Although the Hohmann transfer is the most efficient means for changing orbit, it requires substantial impulse instantaneously. This can translate into significant power requirements, propellant mass, and a more robust and massive structure. On the other hand, a near-circular spiral transfer, which

Copyright © 2000 by the authors. Published by the American Institute of Aeronautics and Astronautics, Inc., with permission.

*Propulsion Subsystem Leader—3CS, Ph.D. Candidate—Aerospace Engineering.

†Director—ASUSat Lab., Associate Director—ASU NASA Space Grant Program (National Space Grant College and Fellowship Program, Professor—Mechanical and Aerospace Engineering. Associate Fellow AIAA.

‡Propulsion Directorate, Advanced Concepts Division (PRSA). Senior Member AIAA.

requires a low-thrust constant burn, may pose less stringent requirements on the spacecraft. This orbit-transfer scheme is a more attractive alternative for micro-/nanosatellites, where the power, volume, and mass not only are limited, but might not be scaled proportionally from larger spacecraft. As another example, consider deorbiting. As individual satellites become useless, there is a strong interest in removing them from LEO to eliminate the growing problem of space debris. For this particular operation that takes place at the end-of-life (EOL) of a satellite, less stringent requirements for the micropropulsion system may be needed. For instance, power usage is generally not critical, pressure regulation may not be required, and lifetime testing can be unnecessary. Consequently, conventional propulsion systems might not be the optimal solution for micro-/nanosatellite missions.

The field of micropropulsion is still in its infancy, and further development of current concepts is very much needed. Nevertheless, there is a wide range of new concepts presently being investigated within government agencies, industry, and universities.[1,2] On the whole, the following issues that are generally associated with propulsion systems would require special attention when dealing with a "scaled-down" system on a micro-/nanosatellite.[3]

1) Materials compatibility between the propellant and the surface material. Even though the compatibility issue between the propellant and surface materials may not exist on larger spacecraft, it can be a problem on micro-/nanosatellites due to the materials used in microfabrication processes.

2) Contamination problems from propellant ablation and vaporization. This can be attributed to the close proximity of individual micro-/nanosatellites in constellation formation.

3) Valve leakage. As the propulsion system is scaled down to a microlevel, valve leakage must also decrease proportionally. The operation of microvalves is dominated by microscale transport phenomena that are fundamentally different from those in macroscale applications.

4) Passage clogging that results in single-point failures in micromachined devices. Micropropulsion systems can be more susceptible to passage clogging because of the microscales involved. Innovative approaches to filtering and nozzle design are necessary to ensure functionality and reliability.

5) System reliability and durability. As conventional propulsion systems are scaled to a microlevel, every microspacecraft part must perform under the same physical environment as larger spacecraft. The size of the part cannot compromise system reliability and durability. For example, the amount of propellant required by a micropropulsion system may be significantly less than that for a larger spacecraft; however, requirements for the containment of any hazardous propellant remain the same. Moreover, as the physical system is scaled, the mechanical integrity of the system must be preserved.

6) Manufacturing complexity. The scale of the product and the selection of materials would require different manufacturing methods. The feasibility and ease of manufacturing—the level of micromachining technology—can greatly affect the development of micropropulsion systems.

7) Integration complexity. There are several components; however, two of the main issues are considered here. First, the internal volume of a micro-/nanosatellite is much smaller than that of a larger spacecraft. Consequently, integration of the micropropulsion system with the entire spacecraft must be as simple as possible—involving very few steps, minimal removal of other components, and possibly a modular design to increase flexibility. Second, if part of the micropropulsion

system fails after integration, in situ replacement of the part may be so difficult that replacement of the entire micropropulsion module is preferable. This solution affects the number of backup systems required for a mission.

To be useful in micro-/nanosatellite operations, micropropulsion systems must be designed to overcome these challenges, while aiming to keep the unit lightweight, compact, low power, efficient, and inexpensive. As can be imagined, overall system considerations enter the selection of a micropropulsion system in addition to performance (specific impulse; I_{sp}) of the propellant.[4] The resources of mass, volume, and power available on micro-/nanosatellites will be much more limited than on larger satellites. Hence, additional research, analysis, and testing must be carried out to ensure functionality and mission success.

II. University Satellites as Technology Testbed

The development of micro-/nanosatellite technologies, including micropropulsion, can be achieved through a partnership between government and universities. A university satellite program with its industry and government partners can provide an inexpensive testbed and innovative solutions to satellite technologies. At the same time, such a program is essentially educating and preparing the next generation of scientists and engineers.[4] Another feature of such a program that strongly impacts the students' education is interaction with industry and government. This day-to-day contact brings the students closer to the industry environment and helps students establish a long-lasting network and identify future job opportunities. With the large amount of industry interaction associated with such a project, students also gain confidence in their abilities and develop effective public-speaking and human-interaction skills. Students acquiring these skills at the university level become even more valuable to their profession.

An example of such a program is the Arizona State University (ASU) Student Satellite Program. The design of ASU's first satellite, ASUSat1, began in October 1993, when a local launch vehicle company agreed to launch a small payload for the students if the satellite would perform meaningful science, weigh under 8 kg (including the release mechanism), and fit within an envelope 33 cm in diameter and 27 cm in height. Due to the size, power, mass, and funding constraints, things such as active control, radiation shielding, and many other complex systems were eliminated from the design. Students went through a series of invaluable lessons in search of feasible solutions. These lessons included problem definition, exploring design space, conducting trade studies, determining the feasibility of manufacturing, and quality control. It should be emphasized that these lessons were not taught in the classroom but were learned hands-on by participating in a real design project. In retrospect, the core objective of the project has been to explore the frontier of the "smaller, faster, cheaper" product space, which is the fundamental challenge to all micro-/nanosatellites and all the subsystems that support the mission.

ASUSat1 was one of the lightest satellites designed to do valuable science in space. The science payload included low-cost coarse-resolution spectral imaging, a global positioning system (GPS), innovative passive stabilization and damping, a 10-deg attitude determination system, autonomous operations, and an audio transponder for amateur radio (AMSAT) operators. The satellite's 14-sided cylindrical body was constructed of a lightweight carbon–fiber composite. There were 510 2×2 cm gallium–arsenide (GaAs) solar cells that provided 8 W of power on average for the satellite. The rest of the power system consisted of six nickel–cadmium batteries and high-efficiency DC–DC converters that supplied regulated 3 and 5 V to the spacecraft subsystems. Other components included dynamics and thermal sensors, a spherical fluid damper, a torque coil, and a gravity-gradient boom.[5–8]

The collective effort of over 400 students in the six years finally came to fruition on January 26, 2000, when ASUSat1 was launched on the maiden voyage of the Orbital/Sub-Orbital Program (OSP) Space Launch Vehicle. Approximately 50 min after launch, ASUSat1 was the first of the five payloads to be heard, when an amateur radio operator in South Africa received two beacons on its frequency. The contact confirmed that ASUSat1 had been successfully deployed from the rocket and that the satellite was functioning on orbit. Multiple contacts with ASUSat1 were made by amateur radio ground stations around the world. Initial telemetry received from these stations indicated that the satellite was healthy and functioning as expected, except for a possible charging problem. Unfortunately, this problem prevented the solar arrays from supplying power, and the operation of the satellite finally drained the batteries after 15 h.

Our mission objective was to show capability in a very low-mass, low-power, low-volume, and low-cost satellite. Even though the mission was brief, telemetry from ASUSat1 indicated that the majority of the student-designed satellite components operated as designed, including the receivers, transmitter, modem, computer, boot-loader software, data acquisition, carbon-composite structure, satellite deployment system, power storage and regulation, boom deployer, gravity-gradient stabilization, and thermal sensors. The signals to ground stations around the world were strong. Attitude sensor data were also obtained, but with only two frames available, it is not possible to draw any firm conclusions. Commissioning and analyses of the cameras, GPS, amateur radio repeater, and gravity-gradient fluid damper were scheduled for later in the mission, so no information on these components was available. The ASUSat1 experience was an incredibly positive one for the students. With the wealth of lessons learned from their first satellite, the students are only more eager and ambitious to complete their next mission, the Three Corner Sat constellation.

III. Three Corner Sat

Our next project is a joint effort among ASU, the University of Colorado at Boulder, and New Mexico State University and, thus, aptly named Three Corner Sat (3CS).[9] This constellation of three identical satellites, shown in Fig. 1, is expected to be inserted by the Shuttle at an initial altitude of 350 km. The spacecraft bus itself is a modular, easily configurable design that allows one to fly multiple science payloads with minimal modifications to the structure and configuration. Also, this program will emphasize student education, as students participate in design, manufacturing, integration, testing, mission operations, and program management.

A. Mission Description

The primary missions of 3CS include stereoscopic imaging, virtual formation flying, innovative communications, automated operations, and end-to-end command and data handling. The stereoscopic imaging mission will take pictures of dynamic atmospheric phenomena using the satellite formation created at deployment.

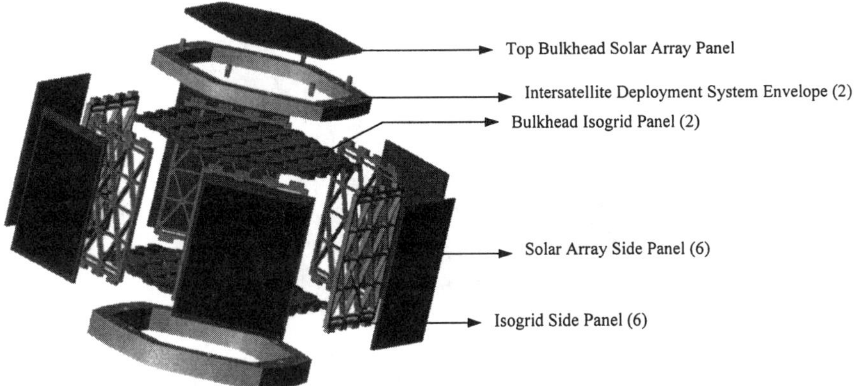

Fig. 1 Typical structure of an individual 3CS microspacecraft.

The constellation will make use of creative crosslink communications to aid in distributed and automated operation. This allows both the individual satellites and the formation to reconfigure for optimum data gathering, communications, and command and control.[10,11]

After deployment from the Shuttle, a micropropulsion system will be used to increase the altitude of the satellites, allowing an extended mission lifetime and greater data-gathering capacity. The primary mission requirements are given in the following section.

B. Spacecraft Description

The individual 3CS satellite will be hexagonal shaped, 45.7 cm across from point to point, and 25.4 cm in height. The primary load-bearing structure is an aluminum 6061-T6 isogrid frame as shown in Fig. 2. The panels and the bulkheads are interlocked (Fig. 3) to provide greater rigidity and flexibility in handling and testing. The top and all side panels will be covered with GaAs solar cells mounted

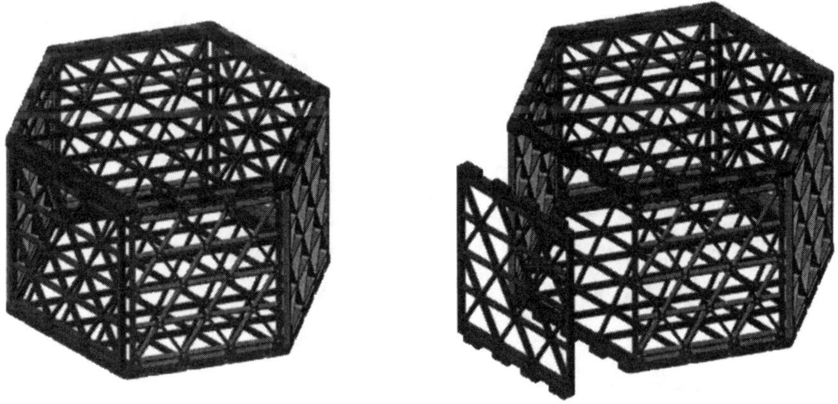

Fig. 2 3CS isogrid structure.

Fig. 3 Blow-up view of the satellite structure. The isogrid side panel (2) and the bulkhead (1) are interlocked to provide greater rigidity. The composite solar array substrate (3) will be mounted on the side panels in four places, with two mounts on the top and two on the bottom.

on kevlar and phenelic impregnated aramid honeycomb composite faceplates. The components that will be mounted on the exterior of the top bulkhead are the star mapper, the GPS patch antenna, and the top bulkhead solar array. The bottom panel will remain open for sensor access, the micropropulsion module, and the S-band patch antenna as shown in Fig. 4. The top and bottom panels will have a bolt pattern to accommodate either the separation system or a handling assembly. In addition to the imaging and micropropulsion experiments, other subsystems include the attitude and orbit determination and control subsystem, the communications subsystem, the end-to-end data subsystem, the electrical and power subsystem, and the structures, mechanisms, thermal, and radiation subsystem.

Current designs have the payload configured as in Fig. 5, so that the transceivers, flight computer, and electrical boxes will be attached to the top bulkhead and side panels. The imager, momentum-wheel assembly, propellant tank, and battery box will be placed on the inside of the bottom bulkhead and will have the capability of attaching to the side panels as well. The payloads in Fig. 5 are general

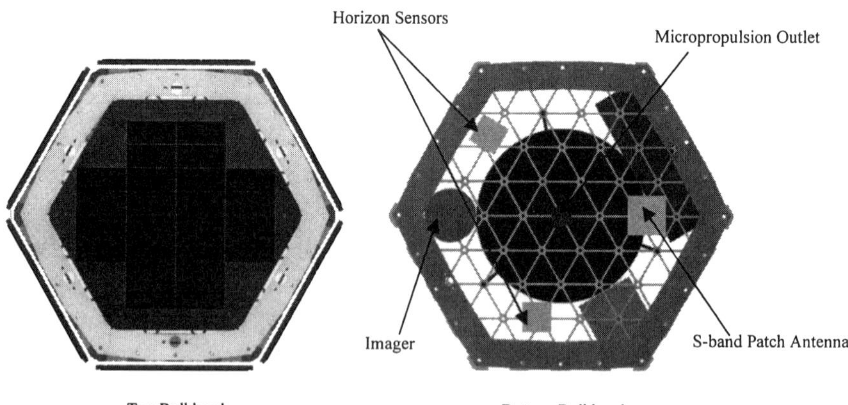

Fig. 4 Microspacecraft bulkheads.

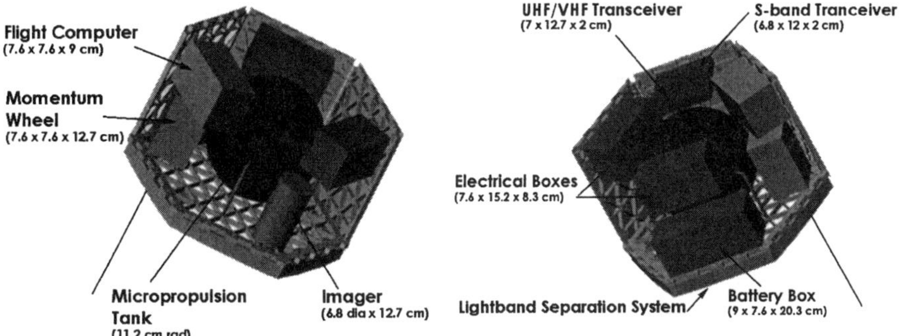

Fig. 5 Internal component placement.

envelopes, as the housings are yet to be designed. Although the propellant tank is modeled as spherical in the envelope drawing, the final shape is to be determined as well.

The 3CS stack will be composed of the three individual spacecraft with separation systems between them. The interface between the satellite stack and the Shuttle is the multiple satellite deployment system (MSDS). The basic configuration is shown in Fig. 6. The stack itself will weigh less than 50 kg and will stand 91 cm tall.

C. Operational Modes

The 3CS mission features eight operational modes. Following separation from the MSDS, the antenna on the bottom spacecraft will be deployed, communication with ground stations will be established, and the stack will go through a functional checkout mode. This mode will last approximately one week. The second mission mode is individual spacecraft separation. Once the correct orientation and pointing of the stack are verified, separation of the individual spacecraft will be initiated by

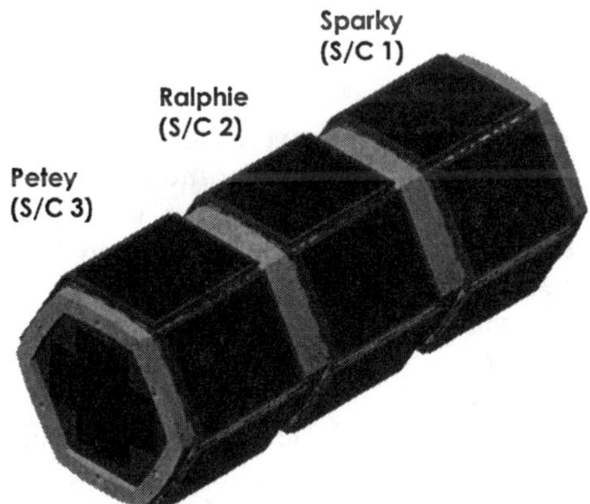

Fig. 6 3CS stack configuration. The satellites are named after the school mascots: Petey (NMSU), Ralphie (CU), and Sparky (ASU).

ground command. Upon separation, antennas will deploy from each spacecraft, and communication will be established with ground stations at each university. Telemetry data will be relayed to the ground, and health of the spacecraft will be determined accordingly. This phase will last less than one day. The third mode is individual spacecraft functional checkout. Completion of this mode will take approximately one week.

Following spacecraft functional checkout, there will be a performance evaluation mode for both the individual spacecraft and the entire 3CS formation. This mode will assess the formation flying and virtual communications capabilities. Intersatellite communication links will be used to transfer formation tasks that will best optimize performance. In addition, any shortcomings in individual spacecraft performance will be identified and compensated. This mode will take about two weeks.

The next mode will be the micropropulsion mode, in which the micropropulsion system becomes the primary experiment onboard the satellite. It will be used to raise the altitude of the satellites to prolong the on-orbit lifetime. The details of this mode are discussed in the following section.

Once the spacecraft is at an appropriate altitude, the main science operations mode will begin. This mode includes the primary 3CS mission of stereoscopic imaging. Furthermore, throughout this phase the virtual formation network will also be utilized. This primary mission will take approximately eight weeks, but will continue as long as the spacecraft are operational.

Finally, the micropropulsion experiment will become active again at the end of the proposed science mission. This additional micropropulsion mode will allow the developed thruster system to be fully tested and characterized during the thruster's maiden flight. The additional propellant required for propulsive maneuvers during this experimental mode will be kept within the mass requirements detailed in the following section.

In case of onboard emergency (e.g., the battery voltage falls below a critical value), the flight computer of the affected spacecraft will go into a safe mode. The spacecraft will orient itself into a maximum solar illumination position and transmit and listen at a preprogrammed rate until communication with the ground station is reestablished.

D. Mission Requirements

The objective of the micropropulsion experiment is to demonstrate the functionality of the system by prolonging the in-orbit life of the satellites. After Shuttle deployment, the satellite altitude should be at or above 350 km. If there is no attempt to offset the atmospheric drag, the satellite orbital life is expected to fall short of the requirements imposed by the science mission. As stated earlier, all of the operational modes, with the exception of the micropropulsion mode, require a minimum orbital lifetime of 85 days. This is the strictest requirement for the success of the primary mission. In the case of shorter orbital lifetimes, some of the scientific results will be compromised. Hence, a possible course of action is either to offset the drag continuously to maintain altitude or to raise the orbit to a higher altitude. The trade-off among these options is discussed in further detail later.

It should be stressed that the satellite mass and power budgets affect the micropropulsion experiment immensely. For instance, the average power produced by the solar array varies between 10 W, using high-efficiency silicon cells, and 15 W, with high-efficiency GaAs cells. Housekeeping electronics would require between 3 and 5 W of power, and an active attitude control system would need about 4 W of power. Obviously, not all components can be switched on at the same time. The micropropulsion experiment, consequently, would need to operate on a duty cycle like other electronic components onboard. Also, the internal volume and the mass of the satellite would limit the amount of propellant that can be accommodated and even affect the choice of propellant. Therefore, the method of demonstrating the micropropulsion system must be a compromise between the lifetime of the satellite and the resources available. Current estimates indicate that approximately 10 W of power and 4 kg of mass will be available for the micropropulsion system on the 3CS spacecraft.

E. Drag Estimates

To estimate realistically the drag force on the microsatellite, an accurate prediction of the neutral atmospheric density must be obtained. The most commonly used analytical model of the upper atmosphere is the mass spectrometer and incoherent scatter (MSIS) model.[12] The MSIS E90 model (1990 version) uses the $F_{10.7}$ flux and a_p geomagnetic index as input parameters. The $F_{10.7}$ flux is a measured quantity of the solar radio flux observed at a wavelength of 10.7 cm. Variations in the 10.7-cm wavelength are used by the model to estimate the long-term variations in solar activity that drives LEO density. Figure 7 shows the measured $F_{10.7}$ from 1991 to January 2000 and the predicted flux (with high and low estimates) through the anticipated 3CS mission time frame. The predicted values shown in Fig. 7 are

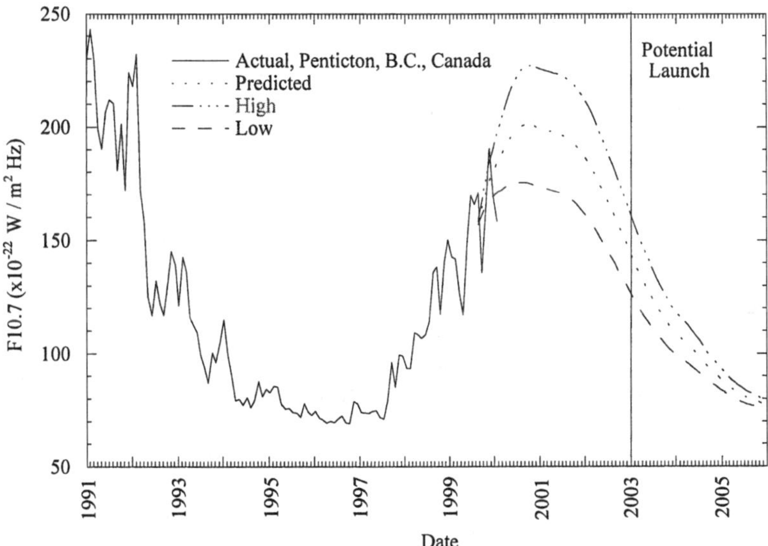

Fig. 7 $F_{10.7}$ flux during the potential launch time-frame solar cycle. Actual data from Penticton, British Columbia, Canada. Predicted data from the U.S. Department of Commerce, National Oceanographic and Atmospheric Administration.

used throughout the remaining calculations for the prediction of the drag force anticipated on the 3CS microsatellite.

Figure 8 shows the atmospheric total mass density as a function of orbital altitude for various values of the $F_{10.7}$ flux derived from the MSIS model, and Fig. 8b shows the MSIS results for the predicted, low, and high values of the $F_{10.7}$ flux during the launch time frame. The number density of the major atmospheric components as a function of altitude for a launch predicted value of $F_{10.7} = 139.5$ sfu (1 sfu = 10^{-22} W m^{-2} Hz^{-1}) is shown in Fig. 9. Figure 10 shows the total atmospheric number density throughout the anticipated initial orbit of the 3CS microsatellite assuming a Shuttle deployment. This density is used to estimate the initial drag force that the microsatellite will encounter.

A critical design requirement is that the thrust produced by the micropropulsion system must exceed the atmospheric drag imposed by the neutral atmosphere at the lowest deployable altitude (assumed to be 350 km from a Shuttle launch). To calculate the drag force on a spacecraft and subsequently the propellant requirement to overcome the drag, several spacecraft-dependent parameters are required. The ballistic coefficient is defined as

$$B = m/(C_D A) \tag{1}$$

where m is the total spacecraft mass, C_D is the coefficient of drag, and A is the total frontal area of the spacecraft (i.e., in the direction of the velocity vector). The drag coefficient ranges from 2 to 4 depending on the gas/surface interaction processes assumed.[13] A typical value of the drag coefficient for most spacecraft is approximately 2.2.[14] Table 1 shows the anticipated values of the ballistic coefficient for the 3CS microspacecraft.

The force due to drag on a spacecraft is given by

$$F_D = 0.5\, m\rho v^2 B^{-1} \tag{2}$$

where ρ is the atmospheric density derived by the MSIS model and v is the spacecraft orbital speed at a given altitude.

The maximum drag force extends from 0.04 to 0.14 mN for the range of ballistic coefficients and predicted atmospheric densities (Fig. 8b) at the lowest orbital altitude (350 km). Consequently, the minimum thrust from the micropropulsion system should be approximately 2–4 mN to overcome the expected drag force adequately.

The lifetime of an individual 3CS spacecraft as a function of initial deployment altitude is shown in Fig. 11. Figure 12 shows the orbit perigee altitude as a function of the number of orbits for various initial altitudes. The orbital decay and lifetime plots were generated using an industry-standard software package.* As mentioned earlier, the science portion of the 3CS mission requires a lifetime in excess of 85 days. The anticipated initial orbital altitude from a Shuttle launch (350 km) yields an expected lifetime of approximately 51 days based on the maximum ballistic coefficient (best case for lifetime calculations) in Table 1. Therefore, the micropropulsion system is required to ensure adequate on-orbit time for mission success.

The effects of drag on the individual 3CS satellites can be counteracted in two ways. First, a micropropulsion system can be used periodically to maintain orbit. Second, a micropropulsion system can be used to raise the spacecraft to a specific

*Web Site: http://www.stk.com, Satellite Tool Kit, Analytical Graphics, 2000.

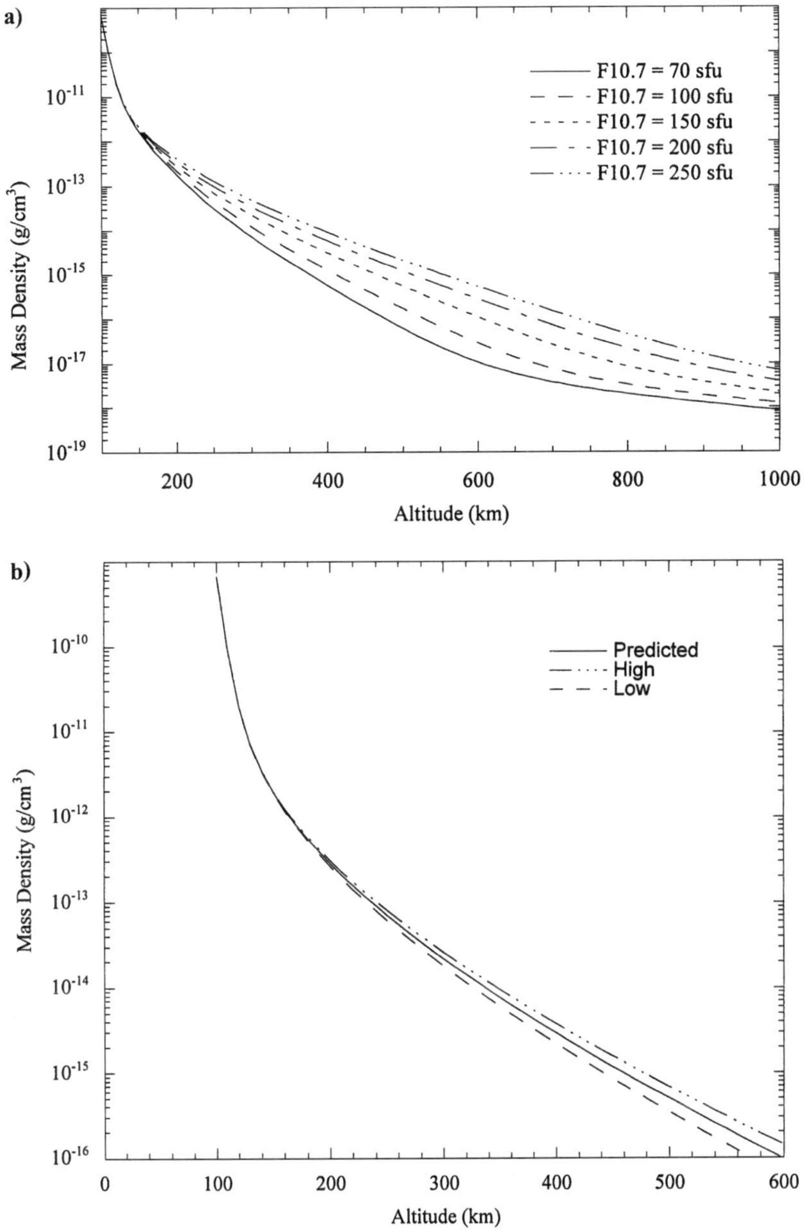

Fig. 8 Total atmospheric mass density as a function of LEO altitude. a) Various levels of solar activity. b) Predicted launch time-frame solar activity.

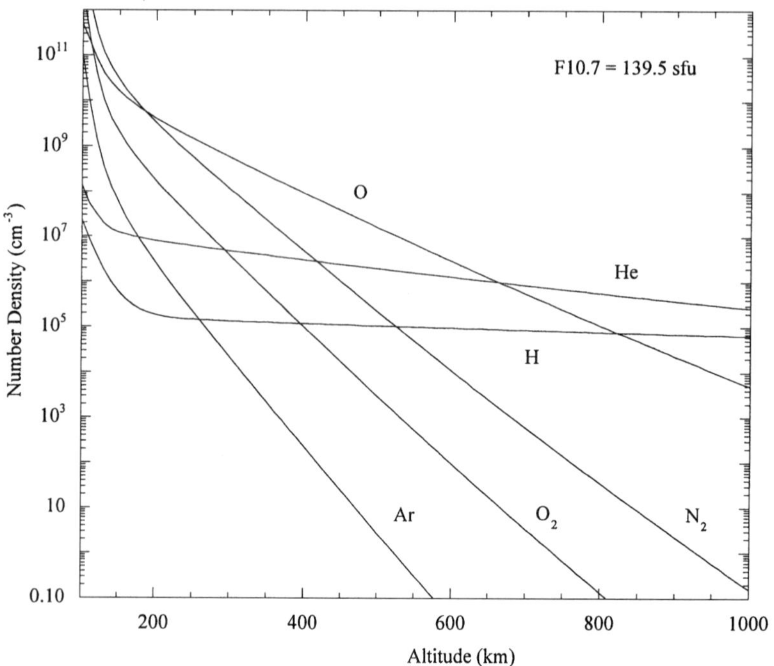

Fig. 9 Atmospheric number density of various constituents as a function of altitude.

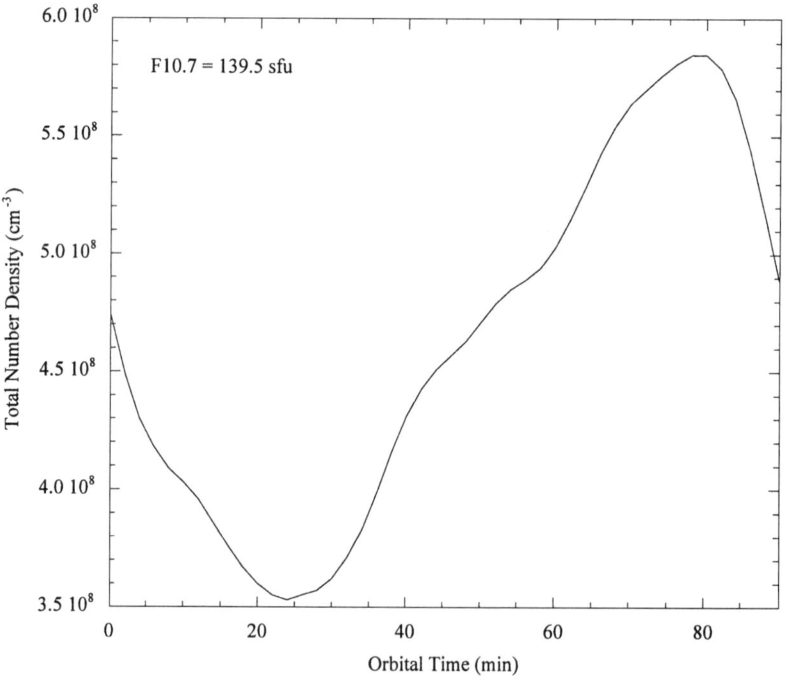

Fig. 10 Total atmospheric number density as a function of spacecraft orbit time for an initial orbital altitude of 350 km.

Table 1 Spacecraft parameters for the 3CS mission

Minimum total mass, kg	15
Minimum cross-sectional area, m^2	0.136
Minimum drag coefficient[14]	2.0
Maximum total mass, kg	18
Maximum cross-sectional area, m^2	0.150
Maximum drag coefficient[14]	2.7
Minimum ballistic coefficient, kg/m^2	37.0
Maximum ballistic coefficient, kg/m^2	66.2

altitude that can support the desired mission lifetime. A combination of these two methods can allow the spacecraft to be raised to an intermediate altitude that can then be maintained with reduced propulsive requirements.

Table 2 shows the estimated values of the spacecraft lifetime and propulsive requirements. Under the assumed mission operations, the spacecraft will be at an altitude of about 350 km at deployment. Maintaining the orbit at 350 km for approximately three months would require a prohibitive Δv, which translates to a large propellant budget. Therefore, the preferred technique would be to raise the orbit of the microsatellite to one that can support a nominal three- to four-month mission. For the $B = 66.2$ kg/m^2 case (best case), the optimum orbit raising

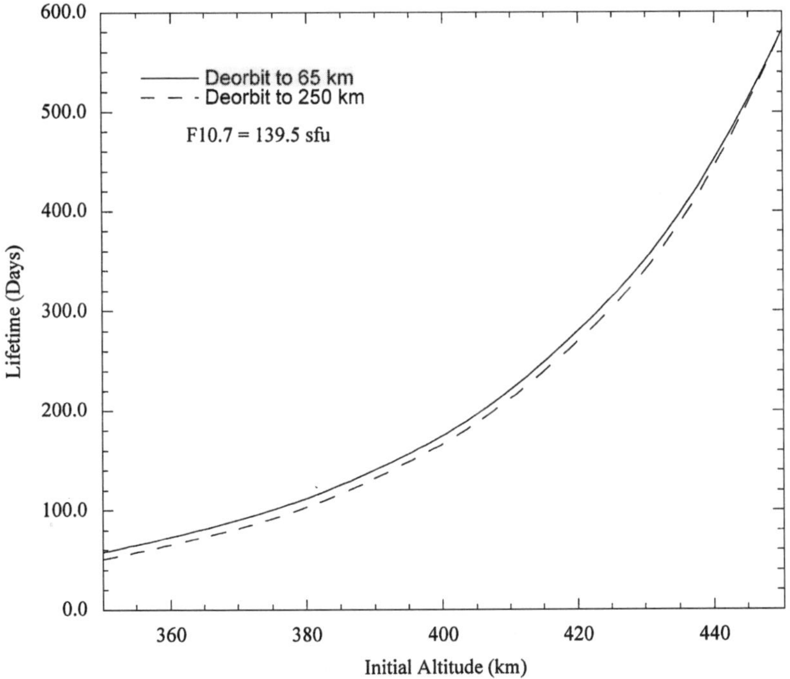

Fig. 11 Predicted lifetime vs initial orbital altitude for $B = 66.2$ kg/m^2 ($F_{10.7} =$ 139.5 sfu). Solid line: lifetime for final altitude of 65 km. Dashed line: lifetime for final altitude of 250 km.

Table 2 Lifetime and Δv for the 3CS spacecraft as a function of altitude

Final altitude, km	Lifetime for $B = 66.2$ kg/m^2, days	Δv to raise orbit from 350 km, m/s
350	58	0
375	100	14.3
400	174	28.6
425	312	42.9
450	584	57.2

maneuver would be to raise the orbit to between 375 and 400 km initially. Because there are several unknown quantities at this stage of 3CS development, such as actual solar activity and spacecraft ballistic coefficient, an initial orbit raise to 400 km will be opted to allow for error. Table 2 shows the Δv requirements to raise the initial orbit to various altitudes.

F. Estimated Δv Required

To achieve the initial science objectives of the 3CS mission, the micropropulsion system will be required shortly after Shuttle deployment to raise each of the

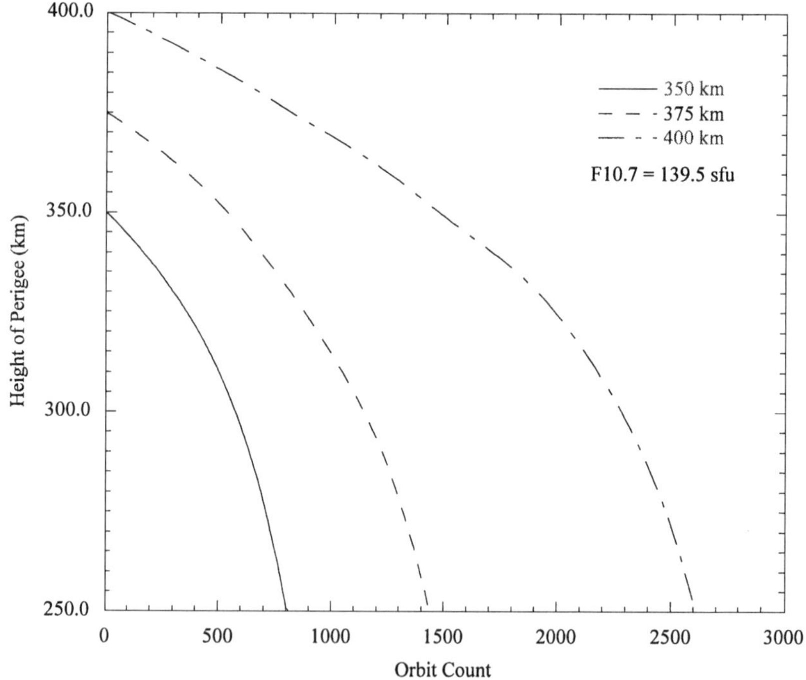

Fig. 12 Perigee height as a function of orbit number for various initial altitudes ($B = 66.2$ kg/m^2, $F_{10.7} = 139.5$ sfu).

Table 3 Total estimated mission-required Δv for 3CS

Δv, m/s	
Drag makeup—orbit raise	28.6
Deorbit	0
Other maneuvers	100
Pointing errors	3.9
Total Δv required for mission, m/s	132.5

3CS spacecraft to 400 km. Orbit-raising maneuvers performed with a low-thrust propulsion system will require a constant-thrust spiral transfer, which is subjected to pointing errors of the microspacecraft (see footnote, p. 34). The utilization of three-axis stabilization is expected to minimize the microsatellite pointing error to within 5-deg. This will increase the total propellant budget by 3% for the worst-case pointing configuration throughout the altitude-raising maneuver.

Since the initial and final altitudes of the 3CS microspacecraft are rather low, there are no propulsive requirements for deorbit. However, once the science mission is complete on the 3CS microsatellite, the micropropulsion experiment will begin again. This experiment will thoroughly test the selected micropropulsion system by performing a series of orbit-lowering and -raising maneuvers. The total propulsive budget is given in Table 3 for all potential maneuvers and compensation for losses. Additional maneuvers are desired to assess the micropropulsion system's ability to perform attitude control and demonstrate formation flying. However, they are expected to require minimal propellant.

IV. Potential Micropropulsion Systems for 3CS

Two micropropulsion systems are under consideration for flight on the ASU microspacecraft as a demonstration of unique technology that can be addressed within the prelaunch time frame. The systems currently being considered are the free molecule micro-resistojet (FMMR), which is described in detail elsewhere,[15] and a cold gas micronozzle thruster, which incorporates a laser-machined, three-dimensional conical nozzle with a throat diameter of 90 μm. Although these two systems do not produce as high a Δv for a given propellant mass as some electrical propulsion systems (e.g., Hall thrusters), their mass and power requirements are a better match for the 3CS constraints.

A. System Requirements for the Free Molecule Micro-Resistojet

The predicted performance characteristics of the FMMR are shown in Fig. 13 for a water propellant and a heated-wall temperature of 600 K. These results were derived from numerical simulations using the direct simulation Monte Carlo (DSMC) technique.[13,15] The FMMR will operate most effectively for the 3CS mission by utilizing a water propellant stored as ice on-orbit. For typical spacecraft temperatures in LEO (260 K), the vapor pressure of ice is approximately 195 Pa, which is an ideal stagnation pressure for the FMMR with a 100-μm slot width. This operating pressure gives a thrust per unit slot length of approximately 10 mN/m (Fig. 13), which implies that 40 slots with an individual length of 1 cm are required

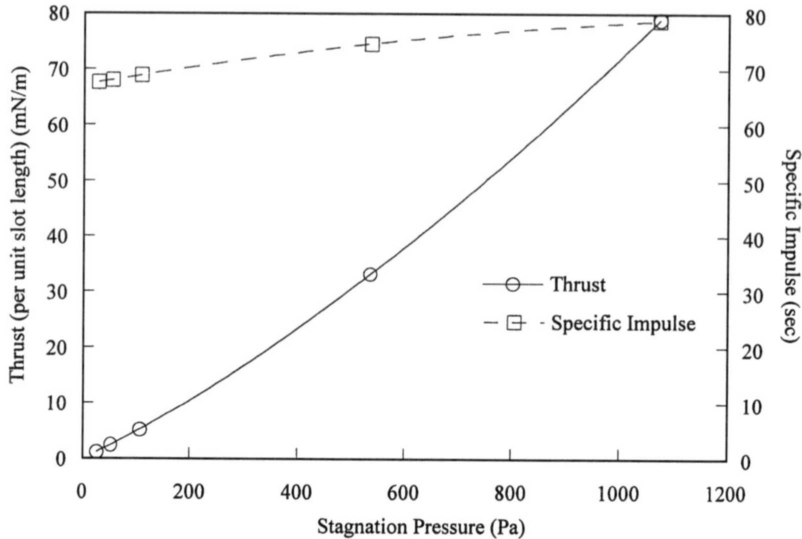

Fig. 13 Predicted FMMR performance characteristics as a function of operating pressure assuming a heated-wall temperature of 600 K and a water propellant.

to produce a 4-mN thrust. Although higher values of thrust can be obtained with higher stagnation pressures, there is a distinct advantage to operating the FMMR at low pressures.[15] The FMMR specific impulse at this stagnation pressure is approximately 70.25 s. As shown in Fig. 13, the smaller thrust required for attitude control can be obtained by reducing the FMMR stagnation pressure (or propellant storage temperature) without significantly compromising the overall efficiency.

1. *Propellant Mass Requirements*

The propellant mass required to perform Δv maneuvers is given by

$$m_p = m_0\{1 - \exp(-\Delta v / I_{sp} g_o)\} \tag{3}$$

where m_0 is the initial dry mass of the spacecraft. For the total required Δv given in Table 3, the propellant mass required for the FMMR is approximately 2.45 kg for a spacecraft dry mass of 14 kg. The volume required to store the water propellant would be approximately 0.0029 m^3, allowing a maximum 20% increase in volume as ice expands. Since the FMMR propellant is stored as a liquid at room temperature, the propellant tank need only be designed to survive the launch environment. The largest propellant volume could be contained in a spherical tank with a diameter of 17.8 cm. For a graphite propellant tank, the tank mass would be about 0.4 kg. The composite results are summarized in Table 4 for a Δv of 132.5 m/s and a dry spacecraft mass of 14 kg.

2. *Power Requirements*

The FMMR uses electrical power to heat the thin-film elements that transfer energy into the propellant gas through surface collisions. For the FMMR geometry and operating conditions described by Ketsdever et al.[15] approximately 6 to 8 W

Table 4 Micropropulsion system comparison

Thruster	FMMR	CG
Propellant	Water	N_2
Thrust, mN	4–6	5–10
I_{sp}, s	70.3	80.3
Propellant mass, kg	2.45	2.17
Empty propellant tank mass, kg	0.4	1.3
Full propellant tank mass, kg	2.85	3.47
Spherical tank radius, cm	8.9	15.8
Estimated power requirement, W	15	10–20

is required to heat the propellant gas to obtain the expected performance. Since the FMMR operates at very low pressures, the valve-sealing requirements are minimized, and the additional power required for valve operations should be minimized.[16] With the use of MEMS-fabricated isolation and actuating valves, the total power required to operate the FMMR can be maintained under 15 W. Pressure regulation inside the device can be achieved by controlling the propellant storage temperature (propellant vapor pressure) with waste heat from the microspacecraft.

3. *Overall System Structure*

The FMMR offers several additional benefits from a systems standpoint. First, the long expansion slots are not prone to catastrophic plugging by contaminants. Second, the propellant-feed-system mass and valving requirements are minimized. Third, the micromachined structure is lightweight and robust in construction. In addition, the entire slot assembly for the FMMR geometry of 40 slots with a width of 100 μm can be contained within a 2.5 $\times$ 2.5-cm area. Plus the added benefit of launching a benign propellant at atmospheric pressure makes the FMMR very attractive, especially in the case of the proposed Shuttle launch. Finally, the total FMMR system mass will be approximately 3.3 kg including propellant.

B. System Requirements for the Cold Gas Micronozzle

The cold gas (CG) micronozzle thruster has a throat diameter of 87.6 μm, an exit diameter of 257 μm, and a supersonic expansion angle of 15-deg. To provide a thrust of approximately 10 mN with a molecular nitrogen propellant, the CG thruster will be required to operate at a stagnation pressure of 10^6 Pa. At these conditions, the anticipated specific impulse for this thruster is 80.3 s.

1. *Propellant Mass Requirements*

Following the same analysis developed for the FMMR, the propellant mass required for the CG micronozzle thruster to perform the required mission is 2.17 kg. The minimum design operating pressure for the CG thruster is approximately 10^5 Pa, which indicates that some propellant will remain in the feed system at the spacecraft EOL. Based on the assumption that no propellant will be lost due to valve leakage, this implies that 0.7% more propellant mass will need to be stored to perform the mission based on the same Δv requirements. However, valve leakage can be a major concern with high-pressure systems.

The use of gaseous propellant on microspacecraft has two serious drawbacks. First, the relatively low density of the propellant requires large storage volumes on extremely space-limited microspacecraft. Second, gaseous propellants must be stored at high pressures, which requires relatively massive fortified propellant tanks compared to propellant mass. For example, a graphite propellant tank containing nitrogen stored at 20 MPa will require a mass approaching 1.3 kg. To reduce the storage volume, the storage pressure can be increased; however, the tank mass may increase to unacceptable levels.[15] The CG system requirements are summarized in Table 4.

2. *Power Requirements*

Unfortunately, the use of a CG micronozzle thruster does not come at a reduced power consumption. Since the propellant storage pressure is roughly 200 atm, a valve with an extremely low leak rate is required. Typically these valves require power to open of the order of 10 to 30 W.[17] However, lower power valves with increasingly lower leak rates are currently being developed even on the MEMS level.[18] In this general survey, it is assumed that the power supply mass for the CG thruster is equivalent to that required for the FMMR.

3. *Overall System Structure*

The CG micronozzle thruster has several disadvantages from an overall systems viewpoint; however, the technology has been previously demonstrated. The CG micronozzle system will require high-pressure feed lines, pressure regulation, and strict propellant filtering due to an additional concern of catastrophically plugging the nozzle throat. The total CG propulsion system mass will be approximately 4.5 kg including propellant.

V. Conclusions

The FMMR and CG systems are chosen among other micropropulsion technologies because their mass and power requirements fit well with the 3CS mission requirements. Moreover, the simplicity and maturity of the technology also promise a functional system to be completed within the two-year prelaunch time frame. Although both micropropulsion systems can satisfy similar operational requirements, the FMMR has several beneficial systems characteristics, which makes it the more attractive system for 3CS. For example, the propellant storage volume is greatly reduced over the high-pressure CG system, the total system mass is reduced, the geometry of the FMMR is easy to machine and quite robust, and the expansion slots are less susceptible to catastrophic clogging compared to the single point failure of the CG nozzle throat.

The mission presented is a worst-case scenario in which 3CS micropropulsion experiment begins at 350 km. With a higher Shuttle insertion of 400 km, the micropropulsion system requirements for mass, volume, and power will be reduced. Moreover, trading on-orbit lifetime for smaller resource usage provides another possibility. A number of system trades will be considered over the next several months of design to maximize the potential information derived from the 3CS mission. Based on the current study, the results for flight testing the FMMR are encouraging and suggest the success of the FMMR as a candidate for future microspacecraft propulsion.

Acknowledgments

The authors thank the original ASUSat Program Manager, Joel Rademacher, presently at Jet Propulsion Laboratory, having completed his M.S. with Helen Reed in July 1996; previous Program Manager, Shea Ferring, presently at Analex, having completed his M.S. with Dr. Reed in December 1998; previous Program Manager, Brian Underhill, who completed his M.S. with Dr. Reed in August 2000; current ASUSat1 Project Leader, Assi Friedman, who will complete his MBA in May 2001; current 3CS Project Leader, Hans Carlson, who will complete his M.S. with Dr. Reed in December 2000; Elaine Hansen, co-PI for 3CS at the University of Colorado at Boulder (CU); Tony Colaprete, Ph.D candidate at CU; Steve Horan, co-PI for 3CS at New Mexico State University (NMSU); Bobby Anderson, M.S. student at NMSU; Air Force Research Lab Space Vehicles Directorate at Kirtland for managing the University Nanosatellite Program; Goddard Space Flight Center and Johnson Space Center for their invaluable help and advisement on Shuttle safety; and *all* the students involved with the ASUSat1 and 3CS projects. Many thanks also go to Scott Webster (Orbital Sciences Corporation), Rich Van Riper (Honeywell Space Systems Group), and all the other faculty and industrial sponsors. Without their help this project would not be possible.

Support for ASUSat1 has been provided by Orbital Sciences Corporation, National Space Grant College and Fellowship Program (NASA Space Grant), AMSAT Organization, Honeywell Space Systems Group, Lockheed Martin Management and Data Systems—Reconnaissance Systems, National Science Foundation Faculty Awards for Women in Science and Engineering, Cogitec, Space Quest, Hughes Missile Systems, ORCAD, Solid Works Corporation, Zilog, Microchip, Dycam, Motorola (Satcom and University Support), SunCat Solar, PhotoComm, Inc., Eagle Picher Industries, Intel (University Support), Maxon, Universal Propulsion Company, Inc., ICI Fiberite Composites, DynAir Tech of Arizona (SabreTech), National Technical Systems, SpectrumAstro, Trimble Navigation, Bell Atlantic Cable, Lee Spring Company, Astro Aerospace, BekTek, Jet Propulsion Laboratory, Rockwell, Sinclabs, Inc., Applied Solar Energy Corporation, Gordon Minns and Associates, Communication Specialist, Advanced Foam and Packaging, XL Specialty Percussion, Inc., Simula, Inc., KinetX, Equipment Reliability Group, and ASU.

Support for 3CS has thus far been provided by AFOSR/DARPA/GSFC/JSC STP, NASA Space Grant Program, Lockheed Martin Management and Data Systems—Reconnaissance Systems, Honeywell Space Systems Group, Motorola, AFRL-Edwards, SpaceQuest, Intel, and Microchip.

References

[1]*Proceedings of the Workshop on AFRL Formation Flying and Micropropulsion*, Lancaster, CA, Oct. 1998.

[2]*Proceedings of the Workshop on AFOSR/DARPA AFRL/VSD Micro/Nanotechnology for Micro/Nanosatellites*, Albuquerque, NM, Apr. 1998.

[3]Ketsdever, A., Wadsworth, D., and Muntz, E., "Predicted Performance and Systems Analysis of the Free Molecule Micro-Resistojet," *Micropropulsion for Small Spacecraft*,

edited by M. Micci and A. Ketsdever, Progress in Astronautics and Aeronautics, Vol. 187, AIAA, Reston, VA, 2000, Chap. 5 (this volume).

[4]Ferring, S., Lenz, C., Friedman, A., and Reed, H., "ASUSat Lab—Past, Present, and Future," AIAA Paper 99-0802, Reno, NV, Jan. 1999.

[5]Friedman, A., Ferring, S., Lenz, C., Reed, H., and Underhill, B., "ASUSat1: The Development of a Low-Cost Satellite," *Proceedings of the 16th Space Symposium and AMSAT-NA Annual Meeting, UK*, American Radio Relay League, Newington, CT, July 1998, pp. 141–155.

[6]Rademacher, J., Reed, H., and Puig-Suari, J., "ASUSat 1: An Example of Low-Cost Nanosatellite Development," *Acta Astronautica*, Vol. 39, No. 1-4, 1996, pp. 189–196.

[7]Friedman, A., Underhill, B., and Reed, H., "ASUSat1—On-Orbit Operations and Satellite Profile," *Proceedings of the 17th Space Symposium and AMSAT-NA Annual Meeting, San Diego*, American Radio Relay League, Newington, CT, Oct. 1999, pp. 55–60.

[8]Underhill, B., Friedman, A., and Reed, H., "Dynamics and Control of Nanosatellite ASUSat1," American Astronautical Society, AAS 00-063, Breckenridge, CO, Feb. 2000.

[9]Underhill, B., Friedman, A., Wong, J., Reed, H., Hansen, E., Colaprete, T., Rodier, Horan, S., and Anderson, B., "Three Corner Sat Constellation—Arizona State Univ.: Management; Electrical Power System; Structures, Mechanisms, Thermal, and Radiation; Attitude/Orbit Determination and Control; ASU Micropropulsion Experiment; and Integration," SSC99-III-1, 13th Annual AIAA/USU Conference on Small Satellites, Logan, UT, Aug. 1999.

[10]Hansen, E., Colaprete, T., Rodier, D., Friedman, A., Underhill, B., Wong, J., Reed, H., Horan, S., and Anderson, B., "Three Corner Sat Constellation: C&DH, Stereoscopic Imaging, and End-to-End Data System," SSC99-IV-1, 13th Annual AIAA/USU Conf. on Small Satellites, Logan, UT, Aug. 1999.

[11]Horan, S., Anderson, B., Underhill, B., Friedman, A., Wong, J., Reed, H., Hansen, E., Colaprete, T., and Rodier, D., "Three Corner Sat Constellation—New Mexico State Univ.: Communications, LEO Telecommunications Services, Intersatellite Communications, and Ground Stations and Network," SSC99-VI-7, 13th Annual AIAA/USU Conf. on Small Satellites, Logan, UT, Aug. 1999.

[12]Hedin, A. E., "Extension of the MSIS Thermospheric Model into the Middle and Lower Atmosphere," *Journal of Geophysical Research*, Vol. 96, 1991, pp. 1159–1172.

[13]Bird, G., Molecular Gas Dynamics and the Direct Simulation of Gas Flows, Oxford Univ. Press, New York, 1994.

[14]Larson, W., and Wertz, J., Space Mission Analysis and Design, 2nd ed., Microcosm–Kluwer, Torrance, CA, 1992.

[15]Ketsdever, A., Wadsworth, D., and Muntz, E., "Performance, Systems, and Design Considerations for a Free Molecule Micro-Resistojet for Nano/Microspacecraft Thrust Generation," Presented at the Nanospace '98 Conf., Houston, TX, Nov. 1998.

[16]Janson, S., and Helvajian, H., "Batch-Fabricated Microthrusters: Initial Results," AIAA Paper 96-2988, July 1996.

[17]Bzibziak, R., "Miniature Cold Gas Thrusters," AIAA Paper 92-3256, July 1992.

[18]Sim, D., Kurabayashi, T., and Esashi, M., "A Bakable Microvalve with a Kovar-Glass-Silicon-Glass Structure," *Journal of Micromechanics and Microengineering*, Vol. 6, No. 1, 1996, pp. 266–271.

Chapter 3

Thruster Options for Microspacecraft: A Review and Evaluation of State-of-the-Art and Emerging Technologies

Juergen Mueller*
Jet Propulsion Laboratory, California Institute of Technology
Pasadena, California

I. Introduction

RECENTLY, a strong interest in micropropulsion devices has arisen within the space community. These devices will be required to deliver very low thrust values (millinewtons and below) and low impulse bits (tens of micronewton-seconds and below) and may feature engine masses and sizes orders of magnitude smaller than available with current thruster hardware. Interest in such devices is driven by the unique propulsion needs of some of the most advanced missions currently being studied by the scientific as well as military space communities. These needs range from primary propulsion and attitude control of microspacecraft having total wet masses of as little as a few kilograms[1,2] to precise positioning control of spacecraft constellations for interferometry missions, aiming to detect gravity waves or search for planets around distant solar systems,[3] and compensation of solar-pressure induced disturbance torques on large inflatable spacecraft.[3] Of these, microspacecraft applications may present some of the most stringent propulsion requirements, since, in addition to small impulse bits and low thrust levels, severe mass, volume, and power constraints, all discussed in the course of this study, may have to be adhered to.

It is the purpose of this study to review and evaluate existing micropropulsion hardware and emerging micropropulsion technologies in view of these constraints and point out future technology needs, primarily with respect to microspacecraft applications. This chapter is intended to serve as an introduction to more detailed descriptions of micropropulsion technologies contained in this book. This chapter addresses both propulsion engineers and spacecraft system designers and may serve as a reference and lead-in to more detailed accounts on micropropulsion

Copyright © 2000 by the American Institute of Aeronautics and Astronautics, Inc. The U.S. Government has a royalty-free license to exercise all rights under the copyright claimed herein for Governmental purposes. All other rights are reserved by the copyright owner.

*Advanced Propulsion Technology Group.

components and systems. For this purpose, a substantial References section appears at the end of this chapter.

The survey begins with a brief review of ongoing microspacecraft activities and, based on the microspacecraft material presented, a preliminary attempt to derive a set of micropropulsion requirements. This is followed by a series of self-contained sections intended to provide a thorough, yet concise, description of various propulsion technologies, the performances demonstrated with available hardware, an evaluation of the technology in view of microspacecraft applications, and an outline of future research needs and unresolved technical issues. Emphasis is placed on the smallest thruster hardware currently available. This survey concludes with an evaluation of existing propulsion technology and an attempt to identify future technology needs based on this evaluation. This work is based on earlier studies,[4,5] however, the material presented in this study has been significantly expanded and revised over earlier versions, made necessary by rapid current developments in the field of micropropulsion.

This chapter focuses on attitude control and delta-v maneuvers for microspacecraft only and does not address landing or takeoff operations. If landing and takeoff operations were considered for microspacecraft, a list of propulsion devices different from those reviewed in this study would likely result, emphasizing high thrust and, thus, high propellant flows. The need to sustain high propellant flow rates will not allow for a significant miniaturization of propulsion components beyond sizes already available today. (Note, however, that even in these cases a significant amount of research and development work will have to be devoted to areas such as component mass reduction and use of alternate propellants for low-temperature applications.)

Finally, micropropulsion systems will also require miniature feed system components, such as valves, pressure regulators, flow controllers, tanks, etc. In some of these areas considerable design challenges arise due to miniaturization, and advances in this area are essential for microthruster developments to be sensible. However, the thruster material reviewed in this study is so vast that surveying miniature feed system components could not be accommodated. However, a specific class of microelectromechanical systems (MEMS)-based microvalves has recently been surveyed and is presented in a companion chapter contained in this book.[6]

II. Recent Microspacecraft Developments

A. Background and Motivation

As alluded to in the Introduction, microspacecraft applications may present some of the most stringent design requirements for micropropulsion devices due to severe mass, volume, and power constraints. So as to obtain a better grasp of these design constraints, a closer examination of present microspacecraft systems, either in-flight or currently under design, appears useful.

There currently exists a wide variety of views regarding the appropriate definition of what a microspacecraft is. Within the U.S. Air Force, for example, any spacecraft with a mass of 100 kg or less is referred to as a microspacecraft.[7] Spacecraft with masses of 10 kg or less have also been termed "nanospacecraft" or "nanosats,"[7] and spacecraft with masses of 1 kg or less are sometimes referred to as "picosats." Similar definitions are being used in Europe.[8] At the Jet Propulsion Laboratory (JPL) a microspacecraft has been described as a spacecraft with a mass of 10–15 kg or less.[9] In one JPL study, different classes of microspacecraft have been distinguished, based on the approximate degree of required component miniaturization and level of integration required between components.[4,9]

Microspacecraft have been considered in the aerospace community since the early 1980s[10,11] and have gained increased attention ever since.[7,12–19] These modern microspacecraft differ from spacecraft of similar mass and size, constructed at the dawn of the space age, by higher degrees of capability targeted for their subsystems as well as their scientific payloads.[15] For example, higher degrees of maneuverability and improved attitude control, higher communication data rates, increased data handling and storage, and improved sensing abilities are desirable than were available on early small spacecraft designs. Improved capabilities may be achieved through novel microfabrication techniques to be used in the construction of spacecraft systems and components.[15,18,19] Such MEMS technologies have progressed significantly throughout the 1980s and 1990s[20] and are anticipated to be employed to a large extent in future microspacecraft designs.

One of the obvious key drivers in the pursuit of microspacecraft concepts is mass reduction. After decades of ever-increasing spacecraft masses and launch costs, novel microspacecraft concepts allow smaller and cheaper space missions to be envisioned. About 30% of the cost of a space mission is typically contained in the launch. These launch costs in turn are dependent to a large degree on the spacecraft mass. Reducing mass may thus significantly reduce mission cost.

This reduction in launch costs will have a particularly strong impact on a unique new class of missions, consisting of multiple spacecraft flying in constellations. Early on, Jones[15] pointed out that "there is a class of scientific and exploration missions that can be enabled by microspacecraft. This class of missions requires many simultaneous measurements displaced in position, as on a planet, a small body, or in a region of space." Constellation missions are currently being embraced by NASA in the sun–Earth connection theme, for example, for purposes of magnetic field mapping around the Earth, the sun, or other planets.[21] Microspacecraft constellations have also received significant attention in the military space community, where large, distributed antenna arrays are being envisioned, to be deployed on Earth-orbiting microspacecraft for high-resolution observations.[22,23] In the cases of such constellation missions, both the launch mass and the cost benefits of microspacecraft-based designs may prove crucial in the economic realization of such missions.

The recent loss of several high-profile space missions may also have contributed to the interest in microspacecraft mission scenarios. Mission risk may be reduced by off-loading scientific instruments from a large, single spacecraft onto a fleet of microspacecraft. Loss of a single, or even a few, microspacecraft may not jeopardize the entire mission. Mission scenarios may even be envisioned where some members of this fleet of microspacecraft are purposely sacrificed, or subjected to high risk, in destructive or particularly dangerous portions of a mission, such as "Ranger"-style planetary surface impacts[15] and in situ exploration of Saturn's ring system,[24] for example.

B. Recent Microspacecraft Design Trends

A list of several recent microspacecraft designs is given in Table 1.[25–39] This list is far from complete since a thorough review of microspacecraft systems was well beyond the scope of this study. However, Table 1 features some of the most recent systems, either flown in space, in planning, or under construction, and may serve

Table 1 Recent microspacecraft designs

Designation	Mission purpose/status	Lead	Mass, kg	Size, cm	Power, W	Voltage, V	Ref. No(s).
MightySat-1	Technology demonstrator; launched Dec. 1998	U.S. Air Force	64	48 × 69	≤32	—	25
Micro-Bus 70	Misc. missions; 14 launches prior to Apr. 1999	Surrey Space Centre, U. of Surrey, England	40–70	35 × 35 × 65	21–43	12	26
Orsted	Magnetic field and charged particle mapping; launched Feb. 1999	Danish Space Research Inst., Denmark	60.7	68 × 45 × 34	54 (EOL)	—	27
SNAP-1	Technology demonstrator, inspection of other spacecraft; awaiting launch June 2000	Surrey Space Centre, U. of Surrey, England	6.5	34 × 23	4 (avg.), 7 (peak)	7–9	28
PROBA	Autonomy demonstrator; in design phase	ESA	100	60 × 60 × 80	9	28	29
Falconsat	S/C charging; in design phase	U.S. Air Force Academy	50	46 × 46 × 43	24	12	30
New Millennium ST-5	Magnetic field mapping; in design phase	NASA Goddard Space Flight Center	20	42 × 20 (flat-to-flat)	7.5–8.5	5/0.25	21
ASU Sat 1	Awaiting launch	Arizona State U.	5	31 × 24	8.5–10	13	31

		University Nanosat Program					
3-Corner Sat	3 spacecraft, formation-flying demo, stereo imaging, cell-phone commun.; in design phase	Arizona State U., U. of Colorado, New Mexico State U.	10	45 × 25	33	3.3–5	31, 32, 33
ION-F	3 spacecraft, formation-flying demo, ionospheric studies, micro-PPT Exp.; in design phase	Utah State U., U. of Washington, Virginia Polytech. Inst.	10/13	45 × 12/45 × 25	18	28	32, 34, 35
Emerald	2 spacecraft, ionospheric studies, formation flying, micro-colloid thruster exp.; in design phase	Stanford U., Santa Clara U.	15	45 × 30	7	5/12	32, 36, 37
Constellation pathfinder	3 spacecraft, formation-flying demo, demo 1-kg S/C fab. and flight operations; in design phase	Boston U.	1	20 × 14	1	—	32, 38
Solar Blade Heliogyro	Solar sail demo; in design phase	Carnegie Mellon	5	—	28	—	39

as a list of examples allowing a first glance at microspacecraft design constraints. As a matter of fact, an examination of the microspacecraft systems listed in Table 1 allows several important observations to be made.

First, it can be noted that several of the microspacecraft missions listed, such as the NASA New Millennium ST-5 mission, as well as the majority of the university nanosatellite missions, focus on constellation missions. (The university nanosatellite program is jointly funded by the Air Force Research Laboratory (AFRL), Air Force Office of Scientific Research (AFOSR), Defense Advanced Research Project Agency (DARPA), and NASA.) In addition, not listed in Table 1, the Air Force-led TechSat 21 mission,[22,23] consisting of clusters of 100-kg spacecraft, as well as recent 1-kg-class "picosat" developments conducted by DARPA focus on microspacecraft constellation architectures as well. As mentioned above, in the case of spacecraft constellations, mass advantages and resulting launch cost reductions that may be achievable through the use of microspacecraft are particularly crucial due to the potentially large number of spacecraft involved.

Second, on inspecting Table 1 the observation can be made that microspacecraft masses keep decreasing dramatically. While the wet masses of earlier microspacecraft designs, many of them already in flight status, range around 50 kg or more, newer designs currently in the planning stage or under construction are pushing toward even lower masses, into the 20-kg range or well below. One example of such a newer design is the SNAP-1 spacecraft, a technology demonstrator developed at the Surrey Space Centre in England.[28] A photograph of this spacecraft is shown in Fig. 1. This spacecraft aims to demonstrate the ability to inspect other spacecraft using microspacecraft. It weighs about 6.5 kg and has a cold gas system with a 3-m/s delta-v capability to perform rendezvous maneuvers.[28]

Third, it can be noted on inspecting Table 1 that at present microspacecraft are extremely power-limited. Several of the earlier microspacecraft currently in flight status deliver less than 1 W/kg of spacecraft mass. Newer designs, mostly still in the planning stage, aim for values exceeding 1 W/kg by a factor of 2 to 3 in several cases. It should be noted that these values might not necessarily be a guide for all future microspacecraft missions. As in the case of conventional spacecraft, certain microspacecraft may be required to provide higher power levels if payload

Fig. 1 SNAP-1 spacecraft. Note human hand in upper-right corner for scale. (Courtesy of Surrey Space Centre, England.)

demands are such or may be equipped with dedicated power supplies for electric propulsion applications in the case of demanding high delta-v missions. However, it appears that microspacecraft will generate a definite demand for payloads or subsystems able to operate within extremely tight power budgets.

Finally, microspacecraft bus voltages will likely be less than the current spacecraft standard of 28 V. While some microspacecraft still feature the current 28-V bus, a trend toward lower bus voltages can clearly be noted on inspecting Table 1. Lower voltages will result in lighter-weight power conditioning equipment, benefiting mass-constrained microspacecraft designs. Currently, bus voltages for the newer microspacecraft designs range between 3.3 and 5 V. Future subsystem components, including propulsion, will likely have to operate within these voltage limits or provide their own dedicated power conditioning units.

Some of the spacecraft listed in Table 1 feature propulsion systems, either cold gas or experimental thruster systems for on-orbit demonstration, as in the case of some of the university nanosat missions. Propulsion will have a dramatic impact on the capability of future microspacecraft. Although in the past many microspacecraft have lacked propulsion systems altogether, future microspacecraft will likely require significant propulsive capability to provide a high degree of mission flexibility.[2] Attitude control in low Earth orbit, for example, may be possible with means other than propulsion, such as magnetic torquers, however, propulsive capability is required in higher Earth orbits, in interplanetary space, or around some of the other planets or moons in the solar system, either for direct attitude control or to off-load momentum wheels.[2] Constellation control also will require propulsive capability. Interplanetary microspacecraft will be in need for propulsive capability to perform midcourse trajectory course corrections. In addition, very small spacecraft are often launched in a "piggyback" configuration together with larger spacecraft to save launch costs. Propulsive capability may be required for the microspacecraft to adjust its trajectory according to the desired mission objective.[2] In Section III, an attempt is made to develop a preliminary set of propulsion requirements for microspacecraft based on the findings presented in this section.

III. Preliminary Set of Micropropulsion Requirements for Microspacecraft

Microspacecraft design envelopes pose unique requirements for all its subsystems with respect to mass, volume, power, and bus voltage as can be seen when inspecting today's early designs (Table 1). Propulsion will face additional requirements with respect to engine performance characteristics such as impulse bit and thrust, for example. These propulsion-specific requirements are difficult to predict accurately at this early stage of microspacecraft development and may vary significantly given the great variety of microspacecraft missions conceivable. Using the microspacecraft data reviewed above, a classification of micropropulsion requirements with respect to the level of system integration, impulse bit, and thrust level for various microspacecraft mass ranges was attempted. Note that this set of requirements has to be viewed as very preliminary, given the aforementioned early development status of microspacecraft systems and mission scenarios. As more concrete designs and missions emerge in the future, these requirements will likely have to be refined.

Table 2 Definition and classifications of microspacecraft for the purposes of the study

Designation	S/C mass, kg	S/C power, W	S/C dimension, m	Comments
Microspacecraft (AF/European definition)	10–100	10–100	0.3–1	Micropropulsion concepts beneficial due to weight/size savings, possibly enabling based on performance requirements (e.g., very small impulse bits for ultrafine spacecraft pointing). Low end of mass range: see below.
Class I microspacecraft (≤10-Kg, nanosat)	5–20	5–20	0.2–0.4	Use miniature "conventional" components, possibly MEMS/microfabricated. Conventional integration (e.g., feed lines) still possible; higher level of integration between components/subsystem desirable.
Class II microspacecraft	1–5	1–5	0.1–0.2	MEMS/microfabricated components; high level of integration between components and subsystems required (subsystems on a chip?).
Class III microspacecraft (picosat)	<1	<1	<0.1	All MEMS/microfabricated. Very high level of integration between subsystems and within subsystems required.

A. System Integration Requirements

In Table 2, an attempt was made to condense microspacecraft designs into several classes or mass categories, each believed to have similar requirements with respect to component miniaturization and system integration. Definitions listed in Table 2 are based on earlier studies and should be interpreted solely as a point of reference for further discussions in this chapter. No attempt has been made to set a particular standard in this area. Furthermore, mass, size, and power values used to define each category should be interpreted as approximate values.

According to Table 2, spacecraft with masses less than 100 kg, but larger than a few tens of kilograms, may still be characterized by subsystem architectures that follow traditional design approaches to a large extent, in both component design and integration. State-of-the-art propulsion hardware may be applicable in many cases and newly developed micropropulsion technologies may not necessarily be considered "enabling" for these spacecraft. However, miniaturization of propulsion components would clearly benefit the design, leading to weight and volume reductions. Thruster miniaturization may also be required if particularly small impulse bits are needed for very fine pointing of the spacecraft or out of reliability concerns—microfabrication may afford cost-effective batch fabrication of a multitude of redundant components.[40]

Somewhat smaller spacecraft, such as those labeled "Class I" in Table 2, ranging in mass between about 5 and 20 kg, may be characterized by the use of the smallest propulsion hardware either available today or currently under substantial development. These technologies may include conventional hydrazine attitude control thrusters to be used for primary propulsion, for example.[41] In some cases, however, this approach may impose severe limitations on spacecraft capabilities. As will be seen throughout this chapter, currently no hydrazine thruster hardware exists that could be used for attitude control of these types of spacecraft, since a multitude (typically a dozen units required for three-axis stabilization) of even the smallest hydrazine thrusters available today would vastly exceed mass and volume limitations of this class of microspacecraft. Cold gas thrusters, on the other hand, do exist in a miniaturized form already, but their use may lead to propellant leakage concerns and heavy, high-pressure propellant tankage. Thus, development of new propulsion hardware, taking miniaturization to new extremes and possibly incorporating advanced microfabrication techniques, such as MEMS technologies, may also be required.

For Class I spacecraft, this propulsion hardware may still be conventionally integrated through interconnecting conventional feed lines. However, higher levels of integration may be desirable. For example, in the so-called JPL Second-Generation Microspacecraft,[41] approximately 7–8 kg in mass, propellant feed lines also serve as the spacecraft bus structure. Note that this Second-Generation Microspacecraft was not designed for flight, but, rather, it is an evolutionary functional model of such a craft, with subsystem hardware constantly being upgraded to more "flight-like" versions.[41]

Microspacecraft with masses of between 1 and 5 kg have been categorized here as Class II microspacecraft. In the case of these types of microspacecraft, development of new, extremely miniaturized propulsion components will likely be required and may be considered "enabling," both in the case of primary propulsion and, in particular, also for attitude control, where multiple clusters of thrusters will be needed. These devices almost certainly will require some form of advanced micromachining technology in their fabrication. Because of severe volume constraints on such a spacecraft, a high level of integration between different propulsion components, and between the propulsion subsystem and other spacecraft systems, may be required. For example, in the case of MEMS-based technologies, several propulsion components, such as thrusters and valves, plus the required control electronics, may be integrated onto a single chip or a three-dimensional stack of chips. These requirements for an increased level of integration, in addition to a more pronounced degree of required component miniaturization, distinguishes this category of microspacecraft from the Class I microspacecraft discussed above.

Even smaller microspacecraft, with total masses of less than 1 kg, have been categorized as "picosats," or Class III microspacecraft. At the high end of this mass range, propulsion system requirements will be similar to those discussed for Class II microspacecraft, i.e., requiring an extreme level of component miniaturization and integration. At spacecraft masses substantially less than 1 kg, revolutionary new propulsion technologies may be required, and strong feasibility issues may emerge at this point. These propulsion systems would likely be based on significantly scaled-down versions of MEMS-based Class II systems. Given the large unknowns in the design of such a propulsion system, and indeed the overall architecture of a $\ll$1-kg spacecraft, these systems and their requirements have not been considered in this study.

B. Minimum Impulse Bit and Thrust Requirements

Minimum impulse bits for a thruster are determined by attitude control requirements. Some generic estimates can be made regarding attitude control requirements. Using spacecraft masses and sizes provided in Table 2 and assuming a cubical spacecraft bus, impulse bit requirements have been estimated assuming several different dead bands and time intervals between thruster firings, listed in Table 3. The estimates presented here are based on earlier calculations performed by Blandino.[42] A couple of thrusters have been assumed to fire in each case. Note that very small impulse bit requirements, well into the micronewton-second range and below, may result, in particular, for larger time intervals between thruster firings and tight dead bands. A larger time interval between thruster firings may be desirable since it would reduce propellant consumption by reducing the number of thruster firings and will allow for longer-duration quiescent spacecraft operation, enabling unperturbed scientific measurements, for example.

On the other hand, thrust requirements for slew maneuvers may be quite large by comparison. Assuming slew rate requirements of 180-deg/min with one couple of thrusters firing, which is not uncommon for state-of-the-art interplanetary (fly-by) spacecraft, required thrust levels may extend well into the millinewton range. Fulfilling both attitude control and spacecraft slew needs with the same propulsive control system may prove to be a very challenging task as will be seen throughout this chapter. However, performing multiple propulsive tasks with the same propulsion system will be an important design consideration to reduce propulsion system dry weight, complexity, and cost. It should be noted in this context, however, that these mission requirements may need to be revisited as concrete future microspacecraft missions develop. Slew rate requirements may be significantly different from the ones assumed here for missions other than fly-by, for example.

Table 3 Representative attitude control requirements for microspacecraft

S/C mass, kg	S/C typical dimension, m[a]	Moment of inertia, kg m^2	Required I_{bit}, N-s						Minimum thrust for slew, mN
			17 mrad (1-deg)		0.3 mrad (1 arcmin)		0.02 mrad (5 arcs)		
			20 s	100 s	20 s	100 s	20 s	100 s	
1	0.1	0.017	1.4×10^{-4}	2.9×10^{-5}	2.5×10^{-6}	5.1×10^{-7}	1.7×10^{-7}	3.4×10^{-8}	0.06
10	0.3	0.150	4.3×10^{-4}	8.5×10^{-5}	7.5×10^{-6}	3.0×10^{-6}	1.0×10^{-6}	1.0×10^{-7}	1.75
20	0.4	0.533	1.1×10^{3}	2.3×10^{-4}	2.0×10^{-5}	4.0×10^{-6}	1.3×10^{-6}	2.7×10^{-7}	4.65

[a]Assume cubical spacecraft shape.

As mentioned in the Introduction, micropropulsion devices may find applications in nonmicrospacecraft missions as well, such as space-based interferometry, for example. Although spacecraft used in this type of missions may not necessarily be microcraft, the tight attitude control requirements for these missions will require thrusters capable of delivering very small impulse bits and thrust levels. The Laser Interferometer Space Antenna (LISA) mission, for example, consists of three small spacecraft orbiting the sun on Earth-like orbits in a specific, precisely controlled formation.[3] The goal of this mission is to detect low-frequency gravity waves by detecting the relative motion exerted by these waves on the spacecraft. To maintain the spacecraft in precise formation, required thrust levels to offset solar disturbance torques have been estimated as approximately 2–20 μN, to be controllable within 0.1 μN. Other formation-flying missions have similar thrust requirements.[3] In addition, thrust levels as high as several millinewtons for spacecraft rearrangements in the constellation and impulse bits of the order of 10 μNs for attitude control have been estimated.[3]

Primary, i.e., delta-v, requirements for future microspacecraft missions do not depend on the size of the spacecraft and therefore requirements may cover wide ranges, as they do for conventional spacecraft missions today. At the low end, only a few to a few hundred meters per second may be required for a short-lived microspacecraft probe to be detached from a larger spacecraft to fulfill a certain portion of the mission. At the high end, challenging mission profiles, such as small-body (asteroid or comet) rendezvous or outer planet orbiters, may have substantially higher delta-v requirements, ranging up to several thousand meters per second. Within NASA, these more challenging mission profiles may be of considerable interest in the future, as they represent an important extension of missions already accomplished to date. Such considerations may impact microspacecraft designs as well, and these types of missions may necessitate the development of high specific impulse (I_{sp}) microelectric propulsion technologies if microspacecraft are to be used.

An additional requirement for chemical primary propulsion is the need to maintain appropriate thrust-to-spacecraft weight ratios. Values around 0.1–0.3 are typical for chemical thusters. Too high a thrust value may generate accelerations too large to be tolerated by the spacecraft structure, in particular, at times well into the mission when portions of the spacecraft structure may already be deployed. Too low a thrust-to-weight ratio, on the other hand, will lead to burn losses[43] and increase the required delta-v. This will always be the case for electric engines, however, in the case of electric engines the larger obtainable specific impulse will lead to substantial propellant mass savings easily offsetting any "burn" losses.

In the following three sections, chemical, electric, and newly developed micro-, or MEMS-based propulsion concepts are reviewed. The technologies are evaluated in terms of microspacecraft applications in Section VII, and further technology needs will be identified.

IV. Review of Chemical Propulsion Technologies

A. Bipropellant Engines

1. *Description of Technology*

Bipropellant engines are frequently considered for primary propulsion applications on conventional spacecraft, in particular for high delta-v maneuvers.

Advantages of bipropellant engines over other chemical systems, such as monopropellant thrusters, are their higher specific impulse, leading to propellant mass savings. Disadvantages are their relative complexity, with respect to both engine construction and feed system layout, resulting in a higher cost. Since separate feed systems for fuel and oxidizer, and possibly pressurant, are required, the component part count is high, leading to larger propulsion system dry masses than for monopropellant systems. Therefore, bipropellant systems are typically used on missions requiring high delta-v's (>1000 m/s) and large spacecraft.

Consequently, most bipropellant engines available today deliver fairly high thrust levels. Some smaller engines in the 5- to 22-N (1- to 5-lbf) thrust range have been built or are under significant development. Up to this point, applications for these engines were envisioned in the area of attitude control of larger spacecraft featuring bipropellant primary propulsion systems as well. Using a bipropellant attitude control system in these cases would allow the attitude control system to be tied into the primary propulsion system, thus eliminating the need for separate attitude control propellant tanks. Given these goals, considerable effort was devoted to fast thruster response times and short impulse bits.[44–48]

Challenges encountered in the design of bipropellant engines of such a small size include the potential of combustion efficiency losses due to reduced mixing and vaporization, and increased heat losses into the engine structure, resulting in thermal control issues of chambers, nozzle throats, and injector heads, as well as related material issues of these components. Additional challenges are accurate injector design at these small dimensions and related mixture ratio control issues as well as, possibly, spacecraft contamination issues due to the potential of incomplete mixing and vaporization inside the thrust chamber, potentially leading to the ejection of unburned propellant and its condensation on spacecraft surfaces.

Vaporization and mixing losses can occur in small engines due to the reduced chamber size. In general, better vaporization is achieved in longer chambers and for smaller injector orifice sizes,[49] while better mixing is achieved in combustion chambers having higher length-to-diameter ratios and a larger number of injector inlets.[49] Smaller-diameter engines also exhibit lower chamber flow Reynolds numbers, which leads to less turbulent chambers, thus reducing mixing.[49] Vaporization and mixing considerations and the associated combustion efficiency losses that are tolerable may thus limit the degree of miniaturization of a bipropellant engine.

Thermal control of small bipropellant engines is another key design issue. As chamber sizes decrease, surface-to-volume ratios increase, leading to higher heat losses per unit chamber volume, possibly making it more difficult to sustain required flame temperatures. Chamber wall cooling may pose new challenges as well. Film cooling, or boundary layer cooling (BLC), is often employed in bipropellant engines to keep the chamber wall within its thermal and structural design limits. Here, a fuel is injected close to the chamber wall. Since the propellant mixture is fuel rich, it does not burn completely and will shield the chamber wall from the heat output of the combustion reactions occurring closer to the center of the chamber. However, at the same time, combustion efficiencies are reduced due to incomplete combustion. While for more conventionally sized engines, about 15–30% of the fuel is commonly used for film cooling, these values may reach up to 30–40% for smaller engines in the 22-N class, causing performance losses[47] and possibly resulting in spacecraft contamination concerns due to possible ejection of unburned fuel that may condense on sensitive spacecraft surfaces (optical lenses, solar cells, etc.).

Elimination of film cooling was achieved in the small bipropellant attitude control engines developed by Rockwell for the Kinetic Energy Anti-Satellite (KE ASAT) program.[45,46] This resulted in increased combustion efficiency and decreased injector head complexity since no separate BLC holes were required. However, to survive the punishing thermal environment, high-temperature chamber materials had to be employed. In the case of the KE ASAT technology,[45,46] a carbon/silicon–carbide chamber was used. Despite the use of this high-temperature material, engine single-burn durations were limited to only a little over 20 s.

Other high-temperature chamber materials under significant investigation are rhenium–iridium composite materials. Rhenium is used as the substrate material because of its high melting point (3453 K)[47] and coated with an iridium layer for oxidation resistance. Iridium has a coefficient of thermal expansion (CTE) closely matched to that of rhenium and a high melting point, 2727 K.[47] Using this chamber material, specific impulses in excess of 300 s have been obtained in a 22-N thrust chamber over burn durations of 350 s.[47] Platinum/rhodium (Pt/Rh)[50] alloys have also been tested. Although the melting temperature of this alloy is lower than that of iridium-coated rhenium (Pt/Rh melts at 2171 K for a 80% platinum and 20% rhodium composition),[50] it avoids some of the fabrication difficulties, in particular, during joining procedures of different engine components, and requires no chamber coating procedures since the material itself provides oxidation resistance. Testing with a platinum/rhodium chamber is under way at Atlantic Research Corporation,[50] aimed at providing a thrust level of 22 N.

Injector design also requires careful attention in small bipropellant engines. Due to the small flow cross sections encountered in small engines, flow rate control, and thus mixture rate control,[51] may be affected since small deviations in flow cross sections may lead to large percentage variations in flow rate. Similarly, alignment of impinging propellant jets[47] require very close attention to eliminate poor engine performance repeatability or engine reliability problems. In addition, thermal management of the injector head is important to ensure that heat diffusion from the hot chamber material to the injector head is minimized to prevent vaporization of propellants in the injector. In addition, thermal limits of the injector material, which may be different from the high-temperature chamber material for machining reasons, have to be taken into account. Unlike-doublet injector types are favored[49,51] because of better mixing results and reduced heat load to the injector head by displacing the flame front away from the injector wall surfaces.[47] As mentioned above, more injector elements will lead to better mixing, however, limited engine size may limit the number of injector elements. In the case of the Rockwell KE ASAT engine discussed above, only a single unlike-doublet injector element is used.[45,46] (Combustion efficiencies are maintained at high levels due to the aforementioned elimination of the BLC layer, sacrificing engine lifetime.)

2. *Available Thruster Hardware*

Table 4 lists the smallest bipropellant engine technology available today. Note that not all engines listed in Table 4 are space-qualified at this point. Also, although commonly referred to as examples for the high degree of miniaturization achieved for bipropellant engines, the KE ASAT engines have been tested only up to 26 s in single-duration burns as mentioned above. These burn durations are too short for most interplanetary delta-v maneuvers. As mentioned above, the KE ASAT developments, as well as others, have focused on attitude control applications, rather than primary propulsion applications. As a result, pulsing performances

Table 4 State-of-the-art small bipropellant engines

Thrust, N	Manufacturer	Type	Fuel/oxidizer	I_{sp}, s	Weight, kg	Size (length max. diam.), cm	Comments	Ref. No(s).
4	DASA	—	MMH/MON-1	285	0.27	—	—	48
4.45	Marquardt	R-2/R-2B	MMH/NTO	280	0.43	$26.1 \times {<}9$	—	52
10	DASA	—	MMH/MON-1	290	0.3	—	Pt/Rh construction; regenerative cooled throat in previous version; 34 flight units built	48
10	Marquardt	R-53	MMH/NTO	290	0.4	16.5×5	Development completed	52
22	Marquardt	R-6C/R-6D	MMH or N_2H_4/NTO	289	0.67	$25 \times {<}13$	Flight applications for R-6C	51
22	Atlantic Research	A0809	MMH/NTO	290	0.55	21.7×5.4	Flight applications	54
22	Aerojet	SSD	MMH/NTO	280	0.59	18.5×6.9	—	55
22	Aerojet	—	MMH/NTO	313	—	—	Rh/Ir chamber; under development	47
22	Royal Ordnance	Leros 20H	N_2H_4/MON	285	0.85	20.9×6.6	Under development	53
30	Rockwell	—	MMH/NTO	287	0.1	—	Max. 26 s in single burn; max. accumulative burn, 77 s; 1.25 mixture ratio; developed for BMDO	45, 46
156	Marquardt	Divert	N_2H_4/NTO	—	0.1	—	20-s single burn demonstrated; developed for LEAP	44

rather than long-duration burns were emphasized and led to the currently exhibited design performances. Almost all engines use nitrogen tetroxide (NTO) and monomethylhydrazine (MMH) as oxidizer and fuels, respectively, due to storage reasons, acceptable performance values, and relatively benign mixture ratio sensitivities. Using an O/F mixture ratio of 1.6 results in equal propellant volumes for both fuel and oxidizer, so that identical tanks can be used (reducing development cost and time) and the spacecraft will experience no center-of-gravity (c.g.) shifts during burns.

3. *Evaluation, Issues, and Future Work*

Using a typical specific impulse value of 290 s as presented in Table 4, and delta-v mission requirements of 2500 and 3500 m/s, propellant mass fractions of 0.58 and 0.71 can be computed, respectively, for a bipropellant system. Thus, in the 2500- and 3500-m/s cases for a 20-kg spacecraft, merely 8 and 6 kg, respectively, of the spacecraft dry mass remain. Given the complexity and high part count of a bipropellant system, such systems may not be practical for Class I microspacecraft or smaller for delta-v requirements that high.

Smaller delta-v requirements around 1000–1500 m/s, on the other hand, would result in propellant mass fractions of 0.3–0.4 for a 290-s bipropellant system. However, a hydrazine monopropellant system with a specific impulse of 220 s (see below) would result in a propellant mass fraction of 0.37–0.5 in these cases. In the case of a 10- or 20-kg Class I spacecraft, this difference would be a mere ~1 or ~2 kg in propellant mass, respectively. Given the lower component part count of a monopropellant system, as well as the need for only a single tank, this higher propellant fraction may be offset by the simpler feed system, and lower cost, of the monopropellant option.

Bipropellant systems may thus not be too well suited for either high or low delta-v requirements onboard a microspacecraft with required wet masses of a few tens of kilograms or below. There may be applications for bipropellant systems toward the high end of the spacecraft mass range considered in Table 2, e.g., 100-kg class spacecraft. The possibility to use "dual-mode" engine technology, i.e., hydrazine/NTO bipropellant engines, where the hydrazine fuel is also used for attitude control in monopropellant thrusters, could reduce system disadvantages somewhat by eliminating the separate attitude control tank.

There also exists the possibility to provide separate chemical stages for very small microspacecraft requiring large delta-v maneuvers, increasing, of course, total injected spacecraft wet mass. An example of such a stage is given in Ref. 56, describing a hydrazine (N_2H_4)/chlorine pentafluoride (ClF_5) chemical upper stage, developed for the Lightweight Exo-Atmospheric Projectile (LEAP) program. Thrust levels provided by the LEAP stage are somewhat high (2056 N) for microspacecraft applications and there are concerns regarding the corrosivity and toxicity of ClF_5. Stages like these, or similar ones using more conventional propellants, however, may be required for orbit insertion maneuvers around distant planetary bodies, in particular, when these bodies are lacking an atmosphere (no aerobraking possible) or when they are located too far from the sun (solar power levels too low for use of solar electric propulsion). In addition, landing and takeoff operations will likely require bipropellant technology. However, those mission applications may require thrust levels well exceeding those obtainable with the engines listed in Table 4 due to high stage masses and large required vehicle accelerations to overcome the gravity of the respective planetary body.

Finally, there exists the potential for further, very aggressive miniaturization of bipropellant technology as currently pursued by the Massachusetts Institute of Technology (MIT), which is investigating microfabricated bipropellant engines.[57] However, at present ongoing work is still addressing basic feasibility issues of this engine concept.

As mentioned above, bipropellant engines have in the past been considered as attitude control applications on spacecraft where a primary bipropellant propulsion system was used to save the cost and additional mass and complexity associated with a separate monopropellant attitude control system. In the case of microspacecraft, however, existing engine technology (see Table 4) is far too large and heavy to be considered for microspacecraft attitude control. In addition, even though bipropellant thrusters do offer higher performances, reducing propellant requirements, the required propellant masses for attitude control are usually small, not providing an opportunity for large spacecraft mass reductions.

B. Monopropellant Thrusters: Hydrazine

1. *Description of Technology*

Hydrazine monopropellant thrusters combine engine technology substantially simpler than that of bipropellant engines with a high reliability, relatively simple feed systems, and intermediate performance characteristics (specific impulses are around 220 s for state-of-the-art hydrazine thruster technology). In a hydrazine thruster, the propellant is passed through a catalyst bed and decomposed. The decomposition products are nitrogen, hydrogen, and ammonia. The reaction takes place in two stages: hydrazine decomposes first through an exothermic reaction into ammonia and nitrogen. The ammonia then decomposes further through an endothermic reaction into hydrogen and nitrogen, however, leaving the overall reaction exothermic.[58] The degree of ammonia decomposition depends on many factors, among them feed pressure and catalyst type and geometry. Shell 405 is the standard catalyst used in the United States, consisting of 1.5- to 3-mm-diam alumina pellets coated with iridium. The catalyst pellets are contained within a mesh construction in a so-called catalyst bed. Upon contact with the iridium surfaces, the hydrazine decomposition reaction is initiated.

2. *Available Thruster Hardware*

Hydrazine thrusters have been used extensively on conventional spacecraft for attitude control as well as primary propulsion for intermediate to low delta-v maneuvers (up to about 1000 m/s or less). Development of this engine type began as early as 1949 at JPL, and the first space flight of this type was in 1966.[59] Hydrazine thrusters have been used ever since on all types of commercial, scientific, and military spacecraft and have become somewhat of the industry standard for attitude control thrusters on conventional spacecraft. Of interest here are the smallest available hydrazine thrusters in the 0.9- to 4.45-N range. These engine types could conceivably be used as main engines on microspacecraft, providing primary propulsion functions for low to intermediate delta-v maneuvers. These engine types are being manufactured in the United States by the Primex Aerospace, Kaiser–Marquardt, and TRW companies and abroad by Daimler Chrysler Aerospace in Germany.[60–62] Typical engine characteristics are listed in Table 5. An example of a 1-N hydrazine

Table 5 State-of-the-art U.S. hydrazine thrusters

Thrust, N	Manufacturer	Type	I_{sp}, s	Weight, kg	Size (length × max. diam.), cm	Comments	Ref. No.
0.9	Primex	MR-103	210–220	0.33	14.8 × 3.4	C/E, D, and G models; I_{sp} and thrust feed pressure dependent (340–<100 psia); considerable flight use	60
0.9	Marquardt	KMH S 10	226	0.33	14.6 × 3.2	I_{sp} and thrust feed pressure dependent; flight use	52
1.0	Daimler Chrysler	—	223	0.27–0.28	—	Derived from earlier 0.5- and 2.0-N designs	62
2.2	Primex	MR-111E	213–224	0.33	16.9 × 3.8	I_{sp} and thrust feed pressure dependent (370–60 psia); considerable flight use	60
4.45	Primex	MR-111C	226–229	0.33	16.9 × 3.8	I_{sp} and thrust feed pressure dependent (400–80 psia); considerable flight use	60
4.45	Marquardt	KMH S 17	230	0.38	20.3 × 3.2	I_{sp} and thrust feed pressure dependent; flight use	52
5	TRW	MRE-1	220	0.82	15.2 × N/A	Mass is for dual-thruster module; I_{sp} and thrust feed pressure dependent; considerable flight use	61
18	TRW	MRE-4	230	0.41	20.3 (length)	I_{sp} and thrust feed pressure dependent; considerable flight use	61

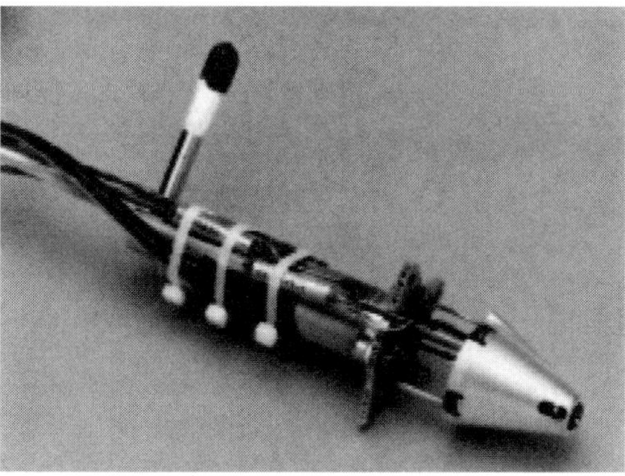

Fig. 2 Primex MR-103 C/D hydrazine thruster. (Courtesy of Primex Aerospace Company.)

monopropellant thruster is shown in Fig. 2, showing the Primex MR-103 C/D model.

A millinewton hydrazine thruster is currently under development at JPL. This thruster aims for impulse bit performances of 50–100 μNs, able to serve needs of fine attitude control on future small spacecraft and precise positioning control of spacecraft flying in constellation formations.[63] The goal is to provide impulse bits comparable to that of cold gas thrusters, however, using a liquid hydrazine propellant rather than a gaseous propellant, thus eliminating leakage concerns over potentially long mission lifetimes.[63] Such a thruster, if it could be successfully developed, may eliminate the need for reaction wheels otherwise used for fine attitude control, thus saving spacecraft weight, power, and cost, in particular, if this thruster could be integrated into an existing propulsion system for delta-v applications or coarse attitude control.[63] The millinewton hydrazine thruster is in early stages of its development, with hydrazine decomposition tests in miniature catalyst beds currently under way.[63]

3. Evaluation, Issues, and Future Work

The engine sizes, weights, and thrust levels of current hydrazine thruser hardware should allow for relatively easy integration into a Class I microspacecraft bus, mounted along the axis of the spacecraft for primary propulsion applications.[41] All thrusters listed in Table 5 have seen considerable flight use and potentially would require only minimal redevelopment for use as Class I main engines. Class II microspacecraft, in particular toward the lower end of their mass range (1 kg), and "nanosats," however, appear too small to take advantage of this existing technology.

One area of improvement in the use of state-of-the-art hydrazine thrusters as Class I main engines may be found in the valve area. Currently, a considerable weight fraction of a small hydrazine thruster is taken up by the thruster valve (greater than 50% for the small engines considered in Table 4). This fact may offer an opportunity for further weight reductions. Since the smallest hydrazine thrusters have been used mainly for attitude control purposes where fast valve action is essential (of the order of 15 ms on/off), these valves could possibly be replaced by slower valves, since primary propulsion applications as envisioned for microspacecraft may not require very short engine pulses. Slower valves, depending on design, may require less force to open the valve, which may reduce magnet masses and number of turns per solenoid coil, thus reducing the weight of the valve.

A disadvantage of hydrazine is its toxicity and flammability and the resulting complexity and cost of ground handling and propellant loading procedures. These procedures are obviously well established due to the extensive use hydrazine thrusters have seen on conventional spacecraft but may significantly contribute to the cost of a microspacecraft mission. In addition, as pointed out in the preceding section, a hydrazine propulsion system onboard a microspacecraft is practical only if small or intermediate delta-v maneuvers are required (i.e., <1000 m/s). In these cases, monopropellant systems will have an advantage over bipropellant systems due to the reduced system complexity and cost, smaller component part count, and, thus, smaller volume requirements. If higher delta-v's are required, monopropellant systems become increasingly heavy due to large propellant requirements. In these cases, bipropellant engines, likely to be mounted on a separate kick-stage, or electric thruster options (see below) will probably be required.

C. Monopropellant Thrusters: HAN-Based

1. *Description of Technology*

Recently, HAN-based monopropellant thrusters[64–68] have received increased attention. The monopropellant used in these thrusters is a mixture of an oxygen-rich component called HAN (hydroxylammonium nitrate; $NH_3OH^+ \ NO_{3-}^-$) and a fuel-rich component, both diluted in water. HAN is a solid at ambient conditions and the water is required to keep the propellant in solution. Different fuel-rich components have been considered. Originating with liquid gun propellant tests conducted by the U.S. Army, TEAN [triethanolammonium nitrate; $(HOCH_2 \ CH_2-)_3NH^+NO_3^-$][64,65] has been used. HAN/DEHAN mixtures have also been studied, consisting of HAN, water, and DEHAN [diethylhydroxylammonium nitrate; $(CH_3CH_2)HNOH{+}NO_{3^-}$].[64] Exposing the mixture to a catalyst causes a chemical reaction and exothermic decomposition of the components into CO_2, N_2, and H_2O.[64] HAN/TEAN mixtures in the past have experienced long-term storage problems when used as liquid gun propellants.[65] However, these storage problems apparently have been traced to impurities (copper, iron, and nickel) in the propellant mixture, remnants of the production process that caused decomposition of the propellant.[65] Improvements in the production process as well as proper storage vessel selection appear to have resolved these propellant stability issues.[65]

Many other HAN-based mixtures have also been studied, several of them with better propellant properties than HAN/TEAN. Recent tests at Primex[66,68] have included HAN/glycine, HAN/acetic acid, HAN/glycolic acid, HAN/methanol, HAN/ethanol, and others. In all cases, water acts as the diluent. Due to the water additive, both components can coexist in a mixture without detonation, as long as the water content is maintained at 10% or above.[67] Initial ignition and combustion tests at Primex indicated HAN/glycine as the preferred propellant blend.[68] Acid-based blends were unstable and led to overpressurizations upon ignition.[68] Alcohol-based blends appear more suitable for low-water-content, high-temperature, high-performance applications.[68] Indeed, water content has an important impact on thruster performance by influencing the decomposition temperature of the mixture. Meinhardt et al.[66] quote numerically computed and experimentally verified flame temperature values of 1977, 1671, and 1366 K for HAN/glycine compounds with water contents of 14.7, 21.2, and 26.0%, by weight, respectively. Specific impulses in these cases were 235, 215, and 193 s, respectively. Decreasing the water content will thus increase the flame temperature and specific impulse.

Using a "low-temperature" HAN/glycine (LTHG) mixture, catalyst performance, and lifetime experiments were performed.[68] Testing of various catalysts beds over various burn durations indicated that a specially thermally conditioned Shell 405 catalyst (see above) performed best, showing the least amount of degradation after repeated test firings.[68] Using this catalyst, a total accumulated test duration of 8000 s with 21 cold starts was achieved.[59,68] Preheating of the catalyst was shown to improve ignition response.[68] The final Shell 405 tests were performed at a catalyst preheating temperature of 426°C (800°F[68]).

2. *Evaluation, Issues, and Future Work*

HAN-based propulsion technology potentially offers some significant advantages over hydrazine systems. HAN-based propellants, as well as their reaction products, are nontoxic, neither mutagenic nor carcinogenic.[65] Vapor of LPXM46, a mixture consisting of 60.8% HAN, 19.2% TEAN, and 20% water, consists almost entirely of water up to temperatures of 65°C.[65] This will ease ground handling of the propellant and simplify its loading procedures, thus reducing associated cost. HAN-based propellants also have relatively high storage densities, approximately 40% higher than that of hydrazine, allowing for smaller, lighter-weight tanks, and HAN-based thrusters are able to operate at lower environmental temperatures. While hydrazine freezes at about 0°C, HAN/TEAN mixtures may be used at temperatures as low as about −33°C.[64] At this point, the viscosity of HAN/TEAN mixtures increases and propellant feeding will no longer be possible using conventional feed system technologies.[64] The freezing point of HAN/TEAN mixtures is about −42°C.[32] Similar values are found for other HAN-based mixtures. HAN/glycine, for example, can be used at temperatures as low as about −20 to −54°C, depending on water content.[66]

Higher storage densities and improved thermal operating capabilities of HAN-based thruster technology are beneficial for microspacecraft, since they allow for smaller and lighter storage tanks and the elimination of, or reduction in power for, tank and line heaters, reducing overall power requirements for the spacecraft. HAN catalysts, however, do require preheating. Reduced catalyst bed heating may lead to ignition delays and accumulation of liquid propellant in the catalyst bed, which in turn could lead to unpredictable engine starts.[66]

Tests conducted at Primex Aerospace also revealed that catalyst beds for HAN-based thrusters may need to be larger than for hydrazine systems and may result in reduced bed loading (mass flux through catalyst bed) than for hydrazine thrusters. The reason for these differences is the need to provide longer propellant residence times due to the more complex decomposition reactions believed to take place for HAN-based propellants. Reduced bed loading and increased propellant residence times could lead to heavier thrusters compared to hydrazine thrusters at comparable propellant mass flows.[66] However, it appears that this disadvantage may be offset by the higher HAN propellant density, which reduces tank masses, at least for missions demanding high propellant loads.

The flame temperatures of low-water-content HAN combinations are quite high and approach values common in small bipropellant engines. Thus, thermal design challenges similar to those found in the construction of small bipropellant chambers would have to be overcome by choosing appropriate chamber materials. However, as indicated above, flame temperatures may be lowered at the expense of specific impulse performance if the water content is raised.

Thus, HAN-based thrusters may be an attractive alternative to hydrazine thruster technology for Class I microspacecraft primary propulsion applications due to reduced toxicity, easier handling procedures, and increased propellant storage density. Heavier, 100-kg spacecraft may still use this technology for attitude control purposes. A substantial reduction in engine size over current test hardware would be required to meet Class II requirements. Considerable additional development work may still be required to bring current HAN-based thruster concepts to flight status, however, initial results appear promising.[66,68]

D. Monopropellant Thrusters: Hydrogen Peroxide

1. *Description of Technology*

Hydrogen peroxide (H_2O_2), when subjected to a suitable catalyst such as silver-wire mesh or liquid permanganate injection, decomposes into water and oxygen in an exothermic reaction that can be exploited for propulsive applications.[43,67,69–77] Indeed, hydrogen peroxide has been used extensively in this role in many well-known, and sometimes infamous, early aerospace projects. It has been used as an oxidizer in a bipropellant combination to power the German World War II Messerschmidt ME-163 Komet rocket plane,[72,73] as well as the British postwar "Black Knight" reentry test rocket, performing 22 successful launches.[72–74] It has powered the launch sled for the V-1 "buzz" bomb of World War II[72] and served as a gas generator propellant, using liquid potassium permanganate injection as a catalyst, in the fuel pumps of the V2 rocket.[72,73] It was used in a similar role in the postwar U.S. Redstone, Jupiter, and Viking missile/launcher projects.[72,73] As a monopropellant, hydrogen peroxide has also been used extensively for attitude control purposes. Projects in this application included the Mercury project to launch the first American into space,[73] the NASA Lunar Lander Simulator,[74] and the Bell X-1 rocket plane,[72,73] breaking the sound barrier for the first time in 1947, and the hypersonic X-15 research plane.[72,73]

Use of hydrogen peroxide waned as higher performing hydrazine monopropellant and storable MMH/NTO bipropellant combinations (see above) became available.[72] However, recently, the reduced toxicity and greater environmental friendliness (reaction products are water and oxygen) of hydrogen peroxide have gained increased attention. In particular, in microspacecraft applications, where cost reductions are being emphasized, eased propellant handling procedures that such a propellant might yield, feature prominently.

2. *Available Thruster Hardware*

Currently, Lawrence Livermore National Laboratories (LLNL) in the United States[76] and the Surrey Space Centre at the University of Surrey in England[77] are pursuing hydrogen peroxide thruster technology, in both monopropellant and bipropellant or hybrid thruster applications. LLNL used a 27-N (6-lbf) thruster provided by General Kinetics, Inc.[75,76] Under atmospheric conditions, this thruster provided 13–22 N (3–5 lbf) of thrust at an average I_{sp} of 95–100 s.[76] The vacuum specific impulse was estimated to be approximately 30% higher.[76] General Kinetics also provides a lower, 13-N (3-lbf) thruster.[75] Both thrusters have demonstrated total run times of 240 s.

LLNL also develops hydrogen peroxide warm gas jets for attitude control. A separate gas generator, which is also used to generate pressurized gas for propellant feed applications, decomposes hydrogen peroxide fuel. The decomposition products are then fed into a Moog cold gas thruster and are subsequently expanded to produce thrust.[76] Thrust levels of about 1.5–1.8 N at I_{sp} values of 65 s[76] were obtained under atmospheric conditions. The vacuum specific impulse was estimated at 85 s, i.e., higher than that of cold nitrogen.[76]

Hydrogen peroxide is envisioned to be used in multifunctional roles onboard microspacecraft. In the LLNL study peroxide is used for attitude control (hot and warm gas) as well as for tank pressurization to create feed pressure without having to resort to separate, heavy pressurization tanks. In addition, peroxide is envisioned to be used as oxidizer in a bipropellant engine. This dual-mode concept would allow for the elimination of a separate oxidizer tank, since attitude control monopropellant and oxidizer would be identical. A 22- to 45-N (5- to 10-lbf) bipropellant hydrogen peroxide engine is currently under development at LLNL.

A similar multifunctional role of hydrogen peroxide onboard small satellites is being envisioned at the Surrey Space Centre in England. There, peroxide is to be used as monopropellant in attitude control thrusters as well as an oxidizer in hybrid thrusters, thus saving tankage and cost.[77] (The hybrid work is described further below.) A 10-N monopropellant thruster is currently under development, anticipated to yield a 150-s I_{sp} and deliver minimum impulse bits of 0.1 Ns.[77] Virtually all present monopropellant hydrogen peroxide thrusters use silver-plated screens coated with samarium oxide as catalysts.[72–74]

3. *Evaluation, Issues, and Future Work*

The possibility of using hydrogen peroxide in multiple propulsive roles, such as for monopropellant attitude control or bipropellant or hybrid primary propulsion, is a very attractive feature of this propellant. Such concepts would lead to the elimination of separate oxidizer/attitude control propellant tanks, saving mass, volume, and cost. Cost reductions would likely also be achieved due to eased handling procedures for hydrogen peroxide, given the environmental friendliness of this propellant and its decomposition products.

On the other hand, long-term propellant storage remains a key feasibility issue to be addressed in thruster development for long-term space applications.[43,69–71,76,77] Hydrogen peroxide slowly decomposes when heated or exposed to a catalyst. Almost any organic substance can serve as such a catalyst.[71] If slow decomposition occurs in propellant tanks, as has been observed in the past,[43] tank pressure increases result over time and propellant is lost due to the slow conversion of propellant into its reaction products inside the propellant tank. Even more troubling is tank pressure rise, which could lead to tank rupture and catastrophic mission failure.[77] Under carefully controlled conditions (proper tank material selection, tank passivation, strict contamination control), peroxide decomposition of less than 1% of the total propellant mass over a one-year duration could be achieved. More recently, Whitehead et al.[76] performed hydrogen peroxide storage tests using various small test vessels of different materials. Both aluminum and anodized aluminum tests showed poor decomposition control. However, using a polyvinylfluoride (PVDF)-lined tank, significantly reduced decomposition rates were observed, leading to the conclusion that one-year unvented storage may be possible.[76] Other options to increase hydrogen peroxide storage durations may be the addition of stabilizers (e.g., Sn or phosphates[77]) to the solution.

Despite these encouraging results, very long-term hydrogen peroxide storage, such as multi-year or even decade-long storage onboard long-duration interplanetary missions still appears somewhat questionable at this stage. However, shorter term, near-earth microspacecraft missions may benefit from this thruster technology due to the reduced cost and dual mode capability associated with the use of hydrogen peroxide, provided that storage issues can be successfully resolved.

E. Cold Gas Thrusters

1. Description of Technology

Cold gas thrusters represent the smallest rocket engine technology available today.[78–81] Cold gas systems are valued for their low system complexity, their small impulse bit (I_{bit}), and the fact that, when using benign propellants (e.g., N_2), they present no spacecraft contamination problems. However, valve leakage may be a concern. Leakage is a potential result of the combination of small amounts of microscopic contaminants on the thruster valve seat, low propellant viscosity, and high-pressure propellant storage. Leaks that may pose no threat to liquid systems, due to the higher liquid viscosity of the propellant and lower-pressure liquid propellant storage may result in catastrophic loss of propellant in a cold gas application.

Cold gas systems are characterized by a low specific impulse, unless very light gases (H_2, He) are used. Table 6 lists typical cold gas performances, based on data found in Refs. 43 and 69. Neither hydrogen nor helium is commonly used, however, since storage problems due to large and heavy tankage would result as a consequence of the low gas densities, and additional leakage concerns would have to be considered due to the low molecular weight of these gases. However, while low specific impulses will eliminate cold gas systems from consideration for microspacecraft primary propulsion tasks, attitude control systems typically are much less sensitive to I_{sp} performance due to the fact that propellant budgets for this application are typically low, not offering much opportunity for mass savings even if a higher-performing propulsion system were used. Of the gases listed, nitrogen is by far the most frequently used cold gas propellant, due to a combination of reasonable propellant storage density, performance, and lack of contamination concerns.

2. Available Hardware

Table 7 lists some of the smallest cold gas thrusters available today.[79–83] The size, mass, and power requirements fit well within the Class I microspacecraft envelope. However, even a cold gas thruster of this size may perform only marginally with respect to impulse bit requirements (compare data in Table 7 with requirements in Table 3). The thruster built by Marotta Scientific Controls was developed for the ST-5 microspacecraft mission[81] (see Table 1). This thruster was designed with emphasis placed on power minimization. As mentioned in Sections I and II, future

Table 6 Cold gas propellant performances[69]

Propellant	Molecular weight, kg/kmol	Density (3500 psia, 0°C) g/cm^3	I_{sp}, s[a]	
			Theoretical	Measured
Hydrogen	2.0	0.02	296	272
Helium	4.0	0.04	179	165
Neon	20.4	0.19	82	75
Nitrogen	28.0	0.28	80	73
Argon	39.9	0.44	57	52
Krypton	83.8	1.08	39	37
Xenon	131.3	2.74[b]	31	28
Freon 12	121	—	46[c]	37
Freon 14	88	0.96	55	45
Methane	16	0.19	114	105
Ammonia	17	Liquid	105	96
Nitrous oxide	44	—	67[c]	61
Carbon dioxide	44	Liquid	67	61

[a]At 25°C. Assume expansion to zero pressure in the case of the theoretical value.
[b]Likely stored at lower pressure values (2000 psia) to maximize propellant-to-tank weight ratio.
[c]At 38°C (560 R) and area ratio of 100.

microspacecraft will be severely power constrained. The Marotta thruster is able to pull in with less than 0.35 W even at the highest operating pressures. The thruster requires a bus voltage of only 3.3 V. An example of a cold gas thruster developed by Moog for the Pluto Fast Fly-By mission is shown in Fig. 3, demonstrating its small size.

3. *Evaluation, Issues, and Future Work*

Using data from Tables 6 and 7, required leak rates for microspacecraft can be estimated and current cold gas thruster technology can be evaluated in this regard. Assuming an attitude control requirement of 50 m/s, which is typical for

Table 7 Small cold gas thrusters

Manufacturer	Moog[79]	Moog[83]	Moog[80]	Marotta[81,82]	Marquardt[52]
Type	58 × 125	58 × 141	58 × 115	—	—
Thrust, N	0.0045	0.0053	2.89	0.05–1.0	4.5
I_{bit}, N-s	10^{-4}	—	—	0.044	—
I_{sp}, s	65 (N_2)	N_2 propellant	—	N_2 propellant	—
Pressure, kPa	34.5	275	1460	350–6980	8840
Open response, ms	0.94	0.35	3.5 (spec.)	<5	<1.1
Power (pull-in), W	2.4	15	30	0.3 (3.3 V)	—
Weight, g	7.34	5.5	13	<50	5.4

Fig. 3 Moog Pluto fast-flyby thruster. (Courtesy of Moog Space Products Division.)

some smaller JPL interplanetary missions, and a specific impulse of 70 s (N_2), as well as an assumed microspacecraft mass of 10 kg (Class I), the required attitude control propellant mass would be 0.7 kg of nitrogen. At a storage density of 0.28 g/cm^3 for nitrogen at 3500 psia and 0°C, a tank volume of about 2500 cm^3 is required. Taking into account the possibility of propellant leakage, assume that an additional 10% of propellant is loaded onto the spacecraft, now requiring a tank volume of 2750 cm^3 at the same storage pressure. Assuming a spherical tank, this translates into an inner tank diameter of roughly 37 cm. This tank size is slightly larger than the envelope assumed for a 10-kg spacecraft but is within the right range. The tank, however, will dominate the spacecraft design layout. Assuming further that all of the additional 10% of the propellant may be lost over the course of the mission (corresponding to 250 cm^3 at 3500-psia storage pressure or almost 59,000 scc assuming zero compressibility of nitrogen in this rough estimate), the maximum allowable leak rates would be 9×10^{-4} scc/s for a two-year mission and 6×10^{-4} scc/s for a three-year mission. These low leak rate requirements are a consequence of the small spacecraft size. Since smaller spacecraft carry smaller onboard propellant supplies for the same attitude control requirements, less propellant can be lost due to leakage and valve leak rates for microspacecraft consequently have to be lower than for larger spacecraft.

Recent leak tests with cold gas thrusters have shown leak rates lower than those calculated above.[78,81] However, it is also important to consider "lifetime" effects on propellant leakage. The propellant tank is one of the major contaminant sources (microscopic metal flakes left over from fabrication, etc.). These contaminants may be carried along with the propellant flow, deposited onto the valve seat, and subsequently prevent the valve from sealing completely. As the mission wears on and more propellant flows through the system valves, the likelihood of valve seat contamination and propellant leakage may increase.

An interesting alternative to conventional cold gas propulsion using high-pressure gas tanks is the use of ammonia as a propellant. As pointed out by Nakazono,[84] ammonia has a vapor pressure of 33 psia (224 kPa) at -18°C. Thus, even without tank heaters, sufficient pressure could be provided to an ammonia cold gas thruster merely using the boil-off of the propellant. As can be seen from Table 6, the specific impulse obtainable with ammonia is higher than that achievable with nitrogen. The ammonia system would allow for liquid storage, reducing tank size and mass, and, due to the relatively low vapor pressures compared to the high propellant storage pressures of conventional cold gas systems, reduced leakage concerns. Depending on the available vaporization rates (dependent on the tank temperature), however, propellant flow rates may be limited.[84]

Because of the required large and heavy tankage and concerns of potential propellant leakage, the use of cold gas systems onboard microspacecraft has to be considered cautiously. Certainly, cold gas systems are not an option for primary propulsion applications. For attitude control applications, cold gas systems maybe considered in such cases where only limited spacecraft lifetimes are required, reducing leakage concerns. Such cases may be encountered in human-tended free-flyers used for spacecraft inspection, such as a space station, for example, reducing the time required for astronaut extra vehicular activity (EVA), or short-lived microspacecraft probes released from a larger "mother" craft. Applications of cold gas systems could be extended to longer-duration missions if current valve leak rates could be maintained over the course of the entire missions or reduced even further. Cold gas systems based on liquid storage of ammonia, on the other hand, appear to be a very attractive option for microspacecraft attitude control. In either case, obtainable impulse bits have to be reduced further, even for Class I microspacecraft applications. This requires the development of either faster valves or smaller nozzle throat areas. Fabricating nozzle throat diameters smaller than the ones obtainable today may require the exploration of new technologies, such as MEMS (see Section VI).

F. Tripropellant and Other Warm Gas Thrusters

In a tripropellant thruster (often referred to as a Tridyne™ thruster), a propellant mixture of hydrogen, oxygen, and an inert gas, such as helium or, more commonly for propellant storage reasons, nitrogen, is used.[85] The propellants are stored fully mixed; no separate tanks are required. The addition of the inert gas to the mixture renders the mixture noncombustible, until exposed to a suitable catalyst. Different catalysts are being studied, typically based on noble metal compounds.[70,86] Thruster performances range between 70 and 140 s of I_{sp}, depending on the gas composition.

Tripropellant systems, however, even though able to deliver higher performance than cold gas systems, suffer the same disadvantage of high-pressure propellant storage and the associated tank mass and weight, as well as leakage concerns, as cold gas systems. Thus, even though the required propellant masses may be reduced compared to those of conventional cold gas systems due to the higher I_{sp} performance, the required high-pressure propellant tanks will likely continue to dominate the spacecraft design as in the case of cold gas systems. Thus, the advantages gained with the use of tripropellant systems over cold gas systems onboard microspacecraft may be limited.

Cold gas technology could be adapted for use in another warm gas thruster option. In this case hydrazine propellant is decomposed in a separated gas generator, consisting in essence of a Shell 405 catalyst bed, and gaseous hydrazine decomposition products are fed to a plenum and, finally, exhausted through a "cold gas" thruster.[84] Such a system would not require separate propellant tanks if a conventional hydrazine thruster were used as the microspacecraft primary propulsion device. In addition, several cold gas thrusters in existence today already claim compatibility with hydrazine decomposition products. A separate heater would be required to heat the catalyst bed to allow for a sufficient number of starts, just as in conventional hydrazine thruster technology. By feeding the decomposition products into a plenum, from which they can be drawn to the various attitude control thruster clusters on demand, would lower the required number of catalyst starts.[84] Similarly, HAN-based or hydrogen peroxide-based[76] warm gas thruster options (the latter were discussed above) may be conceived. In the latter case, the aforementioned propellant storage issues would have to be addressed.[76]

A hydrazine warm gas system is a very attractive option for microspacecraft, in particular, when a hydrazine propellant supply is already onboard for primary propulsion purposes. Relatively high performance, comparable to the ammonia cold gas system described above, can be combined with compact propellant storage and the relatively near-term availability of the required propulsion components, drawing upon the cold gas heritage. As in the case of cold gas systems, either faster valves or smaller nozzle orifices are required to lower impulse bits to the requirements for both Class I and Class II microspacecraft applications.

G. Solid Rocket Motors

1. Description of Technology

Solid rocket motors have frequently been used in kick-stages for orbit raising or orbit insertion of spacecraft, beginning with the Explorer 1 spacecraft[43] and extending to the more recent Pioneer–Venus,[87] Magellan,[43,87] and Galileo missions, as well as numerous commercial missions (orbit raising). In solid motors, fuel (typically aluminum powder), oxidizer (typically ammonium perchlorate; NH_4ClO_4), and an organic binder (typically hydroxyl-terminated polybutadiene; HTPB) are combined into a composite to form the solid propellant.[43,87] The advantages of solid rocket motors are their compact size combined with a relatively high specific impulse performance—lower than that of bipropellant systems but higher than that of monopropellant systems.[88] For obvious reasons solid motors also do not suffer from propellant leakage concerns. Propellant sublimation by exposure to space vacuum through an open nozzle was considered a concern for the use of solid motors for deep-space applications in the past but has been found to have no impact on motor performance even after 10–15 months of in-space storage.[43,87] In the case of the Magellan mission, a Thiokol STAR 48B motor was fired for Venus orbit insertion after 462 days in space.[87] A Thiokol STAR 24 motor was fired after 6.5 months in space for the Venus orbit insertion of the Pioneer probe.[87]

Disadvantages of solid motors are that they are generally not restartable and, therefore, do not allow for orbit trimming. If several delta-v burns are required, it is necessary to stack multiple stages, leading to system complexities and higher propulsion system dry masses. The issue of orbit trimming is of particular importance for solid motors since exact prediction of delivered total impulse is difficult to estimate due to potential uncertainties in the grain temperature, the exact propellant composition, and the amount of inert material consumed.[43] Thus, a separate small liquid system may have to be provided.[87] A separate liquid system may also be required for despin of the satellite. Solid motors for space applications are usually not equipped with thrust vectoring capability. Although some larger motors with such a capability[89] have been tested, these nozzle gimbal systems may be too heavy and complex for very small motors, such as those required for microspacecraft applications. Thus, spacecraft generally are spin-stabilized before motor firings, and despin may be required after separation from the stage, depending on the mission.

2. *Available Motor Hardware*

Table 8 shows some of the smallest solid motors available today.[87–91] As can be seen by inspecting Table 8, envelopes and masses of the smallest available motors appear suitable for the Class I category of microspacecraft and specific impulse performances are quite good. However, thrust levels are much higher than desired and burn times are generally very short, which would lead to very large microspacecraft accelerations. For example, assuming a 20-kg overall spacecraft mass (including motor), using a Thiokol STAR 6B motor would result in accelerations of about 13g at the beginning of the burn and about 18g at the end of the burn. The spacecraft velocity increment that can be achieved with this motor on a 20-kg spacecraft would be 963 m/s. A similar calculation for a 10-kg spacecraft equipped with the STAR 5A motor would lead to an initial acceleration of 1.7g, an acceleration just prior to burnout of about 2.2g, and a delta-v of 641 m/s. Achievable delta-v's are limited by the reduced propellant mass fractions typically found for smaller motors. In addition to the motors listed in Table 8, PyroAlliance Company of France is also developing small rocket motors.[92] At present, a 40-N motor with a total impulse capability of 1 Ns is under development there.[92]

The high thrust forces and short burn times are a result of the intended use for most of these small motors, i.e., stage separation or use as missile divert engines (missile attitude control). In both cases it is essential to provide a relatively large thrust in a short amount of time. In the case of microspacecraft applications, this could lead to limitations of solid motor use due to the requirement of being able to fire only in a stowed vehicle configuration (no deployments) and possibly costly requalification of spacecraft components to account for these high accelerations. An exception to the fast-burning, high-thrust small solid motors shown in Table 8 is the STAR 5A motor. Even though accelerations in the example given above are still quite high, those values may be much more tolerable. The longer burn time and smaller thrust of this motor were achieved by using an end burner propellant grain. This grain type may be the grain of choice for microspacecraft applications. Even longer, lower-thrust burns could be accomplished if the length-to-diameter ratio of the motor case could be increased. It should be noted, however, that in end-burner configurations the motor casing downstream of the burning propellant face may see extensive heating and thus require special thermal protection.

3. *Evaluation and Issues*

Solid rocket motors may present an interesting alternative to more complex liquid systems if mission profiles are simple, require only single burns, and have intermediate delta-v requirements (<1000 m/s), a small liquid system is already onboard the microspacecraft for orbit trimming, or, alternatively, the required accuracy of the actually delivered delta-v is not too high. The absence of leakage concerns and the ability to compactly package solid motors will be attractive for

Table 8 State-of-the-art small solid rocket motors

Manufacturer	Type	Thrust, N[b]	Loaded weight, kg	Propellant weight, kg	Size (L×D), cm	Burn time, s[a]	I_{sp}, s
Thiokol[90]	STAR 5A[c]	169	4.7	2.3	22.5 × 13	32	250
Thiokol[90]	STAR 5C	1953	4.5	2.1	34 × 12	2.8	266
Thiokol[90]	STAR 5CB	2041	4.5	2.1	34 × 12	2.67	270
Thiokol[90]	STAR 6B	2513	10.3	6.1	40 × 18.6	5.9	273
Atlantic Research[91]	Marc 4E1	157	0.32	0.03	18.8 × 3.9	0.33	—
Atlantic Research[91]	Marc 4D	159	0.34	0.05	(11.1–15) × (3.9–4.2)[d]	0.64	199
Atlantic Research[91]	Marc 4B	168	0.36	0.07	(10.8–15)[d] × 3.9	1.07	196.5
Atlantic Research[91]	Marc 4C	197	0.33	0.04	(12.2–16.5)[d] × 3.9	0.6	225
Atlantic Research[91]	Marc 4A	202	0.37	0.08	(13.5–16.5)[d] × 3.9	1.07	232
Atlantic Research[91]	Marc 36A3	315	0.37	0.13	17.8 × 4.3	1.05	257
Atlantic Research[91]	Marc 36A1	325	0.37	0.14	17.4 × 3.9	1.04	252
Atlantic Research[91]	Marc 6A	916	1.53	0.46	25.9 × 7.8	1.08	218
Atlantic Research[91]	Marc 7E1	1316	2.63	0.97	37.3 × 7.4	1.0	189
Atlantic Research[91]	Marc 7C1	1793	2.32	0.97	37.3 × 7.4	1.05	219

[a]Ten percent thrust at ignition, 90% thrust at shutdown. [b]Burn time averaged. [c]End burner. [d]Depending on design version.

microspacecraft applications. Existing motor hardware appears to fit the envelope of Class I spacecraft, although longer burn times, lower thrust values, and, thus, lower vehicle accelerations should be aimed for. The benefits of using small solid rocket motors would be even more pronounced for smaller microspacecraft, such as types falling into the Class II category. Here, compactness plays an even greater role. Class II application of solid motors would require further miniaturization of solid motor technology and dedicated full development programs to achieve the desired reductions in size, weight, and thrust.

Solid motors have never been considered for spacecraft attitude control since each motor can be fired only once. However, as mentioned, many of these motors were designed for missile attitude control for the short flight durations typical for these vehicles, requiring only a limited number of motors. Interestingly, microfabricated thruster arrays featuring multiple solid motors have recently been proposed for microspacecraft attitude control. Work on this concept is in its earliest development phases, is currently being pursued under funding by the French space agency Centre National D'Etudes Spatiales (CNES),[93] and is discussed in greater detail in Section VI.

H. Hybrid Rocket Motors

1. *Description of Technology*

In a typical hybrid rocket motor a solid fuel is combined with a liquid or gaseous oxidizer, which is stored in a separate propellant tank and fed into the motor case.[70,94,95] As a result of this separation between solid fuel and liquid oxidizer, hybrid rockets exhibit some interesting characteristics. Hybrid rockets are restartable; are relatively safe compared with solid motors; offer appreciable specific impulse performances, up to about 300 s when using storable propellants; and still offer a higher degree of compactness than bipropellant systems by eliminating the feed system components required for the fuel side of the system. Hybrid rockets, at first glance, may thus appear as an attractive cross between high-performing bipropellant engines and compact solid motor technology.

Work on hybrid rocket motors started in the 1930s to 1940s in both Germany and the United States, with some of the very early work in Germany performed by rocket pioneer Hermann Oberth.[94,95] Development has continued on an on-and-off basis over the years. Focus was placed mainly on launch applications, leading to the development of the H-500 (312,000-N-thrust) and the H-250F (1,000,000-N-thrust) engines developed by the recently failed AMROC company.[94] Both of the latter two motors used HTPB and liquid oxygen (LO_2) as propellants. Research on hybrid rockets has also been performed at various university research laboratories around the world.[71] This work was performed on smaller test devices. The lowest quoted thrust value for a hybrid rocket engine is given by Sellers et al.,[71] at 10 N.

One of the disadvantages of hybrid motors, in particular, when viewed in terms of space applications with long mission durations (microspacecraft or otherwise), is the limited choice of storable propellant combinations available today. Typically, HTPB is being used as fuel, and LO_2 and hydrogen peroxide (H_2O_2) as oxidizers.[70,71,94–96] Although both oxidizers may be suitable for launch applications, they are not storable over long periods of time because they either are cryogenic (LO_2) or may slowly decompose over time in the propellant tank before use (H_2O_2; see above). Among storable oxidizer options, nitrogen tetroxide (NTO) has been used.[70] Chlorine fluorides, such as ClF_3 and ClF_5, have also been used as oxidants.[70] These substances, however, are highly toxic and corrosive. More recent tests involved nitrous oxide (N_2O).[77]

More recently, new hybrid development work was performed at the Surrey Space Centre at the University of Surrey in England,[77] specifically with microspacecraft and in-space applications in mind. Polyethylene (PE)-fueled hybrids using a hydrogen peroxide oxidizer are being investigated. The use of hydrogen peroxide ties into the previously mentioned application of this propellant for monopropellant attitude control, thus simplifying the system and reducing the system cost by eliminating the need for separate hybrid oxidizer/attitude control propellant tanks. Hydrogen peroxide, catalytically decomposed prior to injection into the hybrid motor, has also demonstrated autoignition (hypergolic) capabilities.[77] However, as mentioned above, storage issues need to be resolved for this propellant to be used on long-term space missions.[76,77]

A novelty of the Surrey hybrid motor is its unique geometry. Typically, hybrid motors consist of a long, slender cylindrical casing holding the fuel. Oxidizer is injected through an axial bore in the fuel and the fuel grain burns from the bore surface outward toward the casing wall.[70] Such long slender devices, however, may be difficult to integrate into microspacecraft and would require dedicated bus designs if such a motor were to be used, adding cost. The Surrey motor, on the other hand, features two flat "pancake" fuel disks separated by a given distance. Oxidizer is injected into the gap between the fuel disks tangentially and swirls inward toward a center bore in one of the fuel disks connected to an external nozzle. This motor type has been termed a vortex flow pancake (VFP) motor.[77] A picture of the VFP motor is shown in Fig. 4. Due to its short, flat cylindrical geometry, this motor can be attached externally to a microspacecraft, simplifying integration considerably. In addition, hydrogen peroxide catalyst beds may be mounted externally also while still remaining in close proximity to the motor, reducing heat loads into the spacecraft (hydrogen peroxide catalyst may operate as high as 600°C[77]). Initial testing using gaseous oxygen and PE have begun. Since film cooling was lower than expected, new higher temperature nozzles are currently being developed to avoid nozzle failure.[77]

2. *Evaluation, Issues, and Future Work*

Given that attention in hybrid rocket engine development was focused mostly on large launch motors, this technology has not yet been used for in-space applications. If recent test programs are successful, however, hybrid engine technology could fill a gap between high-performing, yet complex bipropellant engines and compact and simple, yet relatively inflexible solid motor technology. This technology could potentially be very beneficial for orbit injection or other high-thrust applications, possibly even takeoff and landing operations where throttling may be required.

Keeping in-space applications in mind, future work should focus on storable oxidizers. For deep-space missions, in particular, multiyear propellant storage may be required. One oxidizer option suitable for such missions may be NTO. Reverse hybrid options, where the oxidizer is solid and the fuel is liquid, may potentially allow for a greater choice of propellants to address storability concerns, as well as performance issues. Isp density considerations should be taken into account when selecting the liquid component to achieve a higher degrees of compactness and further improving the benefits of this technology for microspacecraft applications.

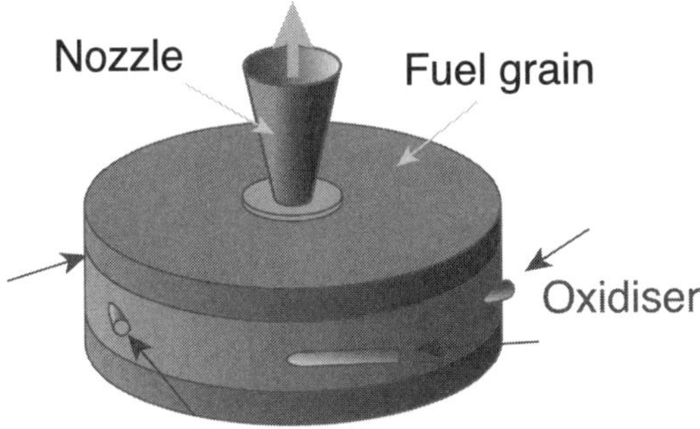

a) VFP hybrid motor concept

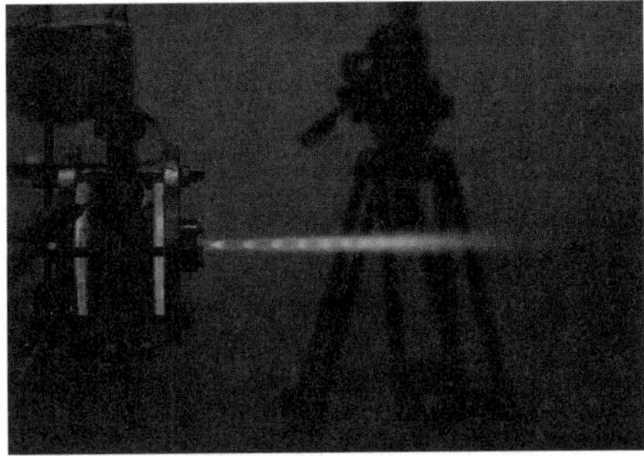

b) VFP hybrid motor firing

Fig. 4 Vortex flow pancake (VFP) miniature hybrid motor concept. (Courtesy of Surrey Space Centre, England.)

In this context, choosing a liquid component that could be used as an attitude control propellant as well could lead to potentially significant system mass and volume benefits. In the case of nonhypergolic propellants, ignition mechanisms need to be studied in greater detail to ensure restartability of the hybrid to maintain an advantage over solid motors in this regard.

V. Review of Electric Propulsion Technologies

A. Ion Engines

1. Description of Technology

In an ion engine, the propellant (typically xenon) is ionized in a gaseous plasma discharge. Ions are extracted from the plasma through an ion engine accelerator grid by means of electrostatic forces and accelerated across an electric potential difference of about 1.3 kV (see Fig. 5). In the process, xenon ions achieve a velocity of about 30,000 m/s, corresponding to a specific impulse of about 3000 s.

An ion propulsion subsystem consists of several components, all of which will have to be miniaturized for microspacecraft applications. These are the thruster itself, the power conditioning unit providing the required voltages to the engine, and the feed system. Within the thruster assembly, critical components include

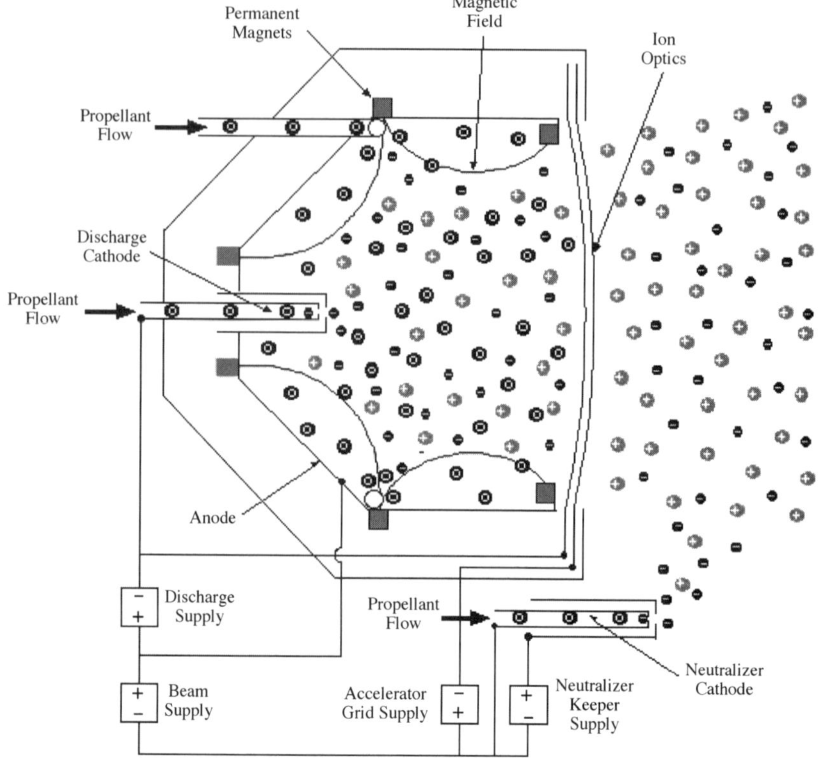

Fig. 5 DC electron bombardment ion thruster concept.

the engine cathode for certain engine types, the accelerator grid system, and the neutralizer (used to neutralize the ion beam to avoid charging the spacecraft). Different types of ion engines are being developed. DC electron bombardment types use an electron current emitted from a hollow cathode (discharge cathode) inside the engine body to ionize the propellant gas through collisions between the electrons and the propellant gas atoms. Radiofrequency (RF) electron bombardment engines use electrons accelerated in an inductive-coupled RF field to cause propellant ionization.

Advantages of ion engines are their high specific impulse, which may translate into significant propellant savings and spacecraft mass reductions. This fact is of particular importance for mass-constrained microspacecraft, especially for interplanetary missions, which may have high delta-v requirements. Using ion engine technology may lead to lighter overall spacecraft masses and shorter mission trip times compared with chemical bipropellant systems. In addition, xenon propellant, when stored at about 2000-psia pressure, takes on a supercritical state with a density about twice that of water. Reduced propellant requirements due to the high specific impulse of the engine and high propellant density will allow for compact propellant storage. Finally, xenon, an inert gas, is virtually noncontaminating to all spacecraft surfaces and components. Specific power values for state-of-the-art ion engine technology typically range around 30 W/mN.

However, propellant mass reductions due to higher specific impulse will have to be traded off against electric power requirements. Power requirements drive the power conditioning unit and power supply masses. To reduce the overall system wet masses, an optimum operating point must be selected for the engine, allowing for both significant propellant savings and low power system masses. For typical interplanetary mission requirements and current power system technology, this optimum is typically about a 3000-s I_{sp}.

2. *Available Hardware*

Table 9 lists some of the smallest ion engine technology available today.[97–102] As can be seen, all current ion engine systems are too large for use on Class I and sub-Class I microspacecraft, with respect to both mass (including the PPU) and power requirements, however, all may be applicable to 100-kg class spacecraft. These engines may also be used on separate electrical stages just like the bipropellant thruster technology discussed above and provide large delta-v increments for Class I-type microspacecraft in that configuration, however, at the cost of increasing the overall injected mass of the configuration. In the case of electric stages, a dedicated power supply will likely have to be provided. In addition to the thrusters listed in Table 9, small RF-type ion engines are being studied at Proel Technology Corporation in Italy.[103] Currently under laboratory development, a 9.5-cm-diameter engine delivered thrust levels of 5 mN with beam ion energy costs of 625 W/A. The specific impulse was 3000 s.

A very small xenon ion source was developed by Hughes for a U.S. Air Force/NASA Goddard spacecraft charging experiment in 1979.[104] This ion source was 3.6 cm in diameter and required 45 W of power to produce a thrust of 0.14 mN at a specific impulse of 350 s.[104] Calculating thrust power from these data to 0.24 W, the thruster efficiency can be determined as 0.5%, not accounting for any neutralizer losses. While these performances are low, note that this thruster had not been optimized to function as a propulsion device.[104]

NASA Glenn and Hughes in the late 1960s and early 1970s developed a 5-cm-diam mercury ion thruster.[104] This thruster had a power requirement of 72 W and delivered a maximum thrust level of 2 mN at a specific impulse of 3000 s. Using these data, one calculates an efficiency of 41%. Mercury ion thrusters are no longer being developed due to propellant handling, toxicity, and spacecraft contamination concerns.

Several cesium contact ionization ion thrusters have been built throughout the 1960s as well, typically ranging in size between 5 and 7 cm in diameter,[104] although even smaller devices (1.3 cm in diameter) have been built and tested.[105] In this type of thruster, cesium is brought in contact with a heated porous plug having a higher work function than the ionization potential of cesium, causing the ionization of cesium atoms upon contact. A 1.3-cm-diam engine (2.54 cm in length), using a permanent magnetic confinement field, was able to operate at 4.6 W and produce 12 μN (2.7 μlbf) of thrust.[105] The beam voltage was 1 kV and the beam current 0.23 mA,[105] from which a calculated beam power of 0.23 W and an estimated efficiency of about 5% can be determined. The same engine, when operated at a beam voltage of 3 kV, delivered a beam current of 0.38 mA at a total thruster power of 5.9 W.[105] Using these data, an efficiency of about 20% can be calculated. Thrust in this case was measured to 35 μN (7.8 μlbf).[105] Unfortunately, cesium is a very reactive propellant, easily forming solid deposits when exposed to moisture (such

Table 9 State-of-the-art small ion engines

Manufacturer	Hughes[99]	DASA[97,98]	JPL[100]	NASA Lewis[101]	Keldysh Research Center, Russia[102a]	Keldysh Research Center, Russia[102a]
Discharge type	DC	RF	DC	DC	DC	DC
Beam diameter, cm	13	10	15	8	5	10
Thrust, mN	17.8	5–15	21–31	3.6–10.9	1–5	6–19
I_{sp}, s	2585	3000	2500–3900	1760–2650	3100–3700	2500–3500
Power, W	439	240–600	500–900	100–300	50–140	150–500
Thruster mass, kg	5.0	1.6	2.5	—	—	—
PPU mass, kg	6.8	8.0 (PPU) 1.3 (RF gen.)	—	—	—	—
Comments	Flight hardware	Flight hardware	Laboratory development	Laboratory development	Laboratory development	Laboratory development

[a] Does not include neutralizer.

as in the prelaunch environment) and may also lead to spacecraft contamination (see below). Concerns such as these, low efficiencies due to the heater power required for the porous contact ionization plug, and the emphasis on ever-larger engines throughout the course of the 1970s may have led to the demise of this technology. Contamination and propellant handling concerns, as well as concerns related to low thruster efficiencies, still prevail today.

3. *Evaluation, Issues, and Future Work*

To integrate ion engine technology onboard a microspacecraft bus within the mass margins of a Class I microspacecraft, new technologies will have to be developed. Thruster sizes and power requirements will have to be reduced significantly. Challenges to be overcome here will be to maintain a plasma in a small, high-surface/volume ratio discharge chamber, where electron wall losses may be high. Cathodes and neutralizers will have to be miniaturized and cold cathode (field emitter array; FEA) technology may have to be explored. Micromachined grid systems, as well as miniaturized power processing units will be needed.

Currently, micro-ion engine projects are under way at the University of Southern California (USC), in collaboration with AFRL,[106] and at JPL,[107] partly in collaboration with the University of Michigan,[108,109] Previously, numerical studies on micro-ion engines were performed at MIT.[110] Work in the USC/AFRL collaboration focuses on a "hollow-anode" micro-ion engine body design not requiring any magnetic field confinement for the ionizing electrons,[106] recognizing the poor scalability of magnetic confinement fields to micro-engine sizes (requiring ever-higher field strengths to decrease the electron Larmor radius as engine sizes shrink). Hollow-anode concepts are also being studied at NASA Glenn Research Center.[57] JPL is exploring the feasibility of 1- to 3-cm-diam MEMS-hybrid ion engine technologies. Recent focus has been on feasibility studies of various crucial micro-ion engine components, such as micromachined accelerator grid systems and, in collaboration with the University of Michigan, FEA technology to be used as micro-ion engine cathodes and neutralizers.[107–109] Some of these approaches are discussed in greater detail in Section VI (Emerging Technologies). It should be noted, however, that at this point micro-ion engines have to be considered very advanced micropropulsion concepts that still have to overcome many feasibility concerns before they can be seriously considered for microspacecraft applications.

B. Hall Thrusters

1. *Description of Technology*

Hall thrusters[111] are electrostatic propulsion devices that often use xenon propellant. Plasma generation and ion beam acceleration are different from that in ion engines and lead to a more compact thruster technology. A schematic of a Hall thruster is shown in Fig. 6. In a Hall thruster,[111] electrons emitted from a hollow cathode external to the thruster are accelerated toward a positive anode located upstream and inside an annular discharge chamber. On their way to the anode, the electrons cross a radial magnetic field extending across the annular chamber. Due to Lorentz-force action, the electrons gyrate around the magnetic field lines and drift azimuthally through the annular channel, colliding with propellant gas atoms (xenon) and ionizing them. The ions are accelerated away from the engine

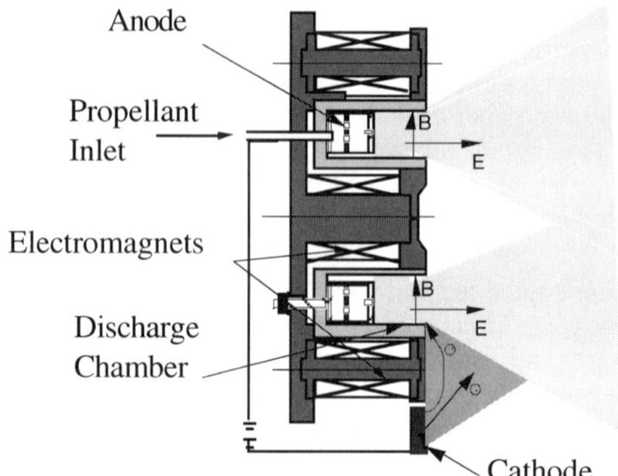

Fig. 6 Hall thruster concept.

by the same electric field that attracted the electrons. The ion beam is neutralized by additional electrons streaming off the cathode.

Due to the high electron density in the magnetic field region, a dense ion beam can be formed, overcoming space charge limitations found in ion engines. Hall thrusters are thus more compact for the same delivered thrust level than ion engines. On the other hand, Hall thrusters typically deliver specific impulses around 1500–2000 s, making them more suitable for primary propulsion on near-Earth missions (orbit transfer, repositioning, etc.), rather than interplanetary flights. A high specific impulse Hall thruster would be an attractive alternative to ion engines.

2. *Available Hardware*

Table 10 lists currently available, small Hall thruster hardware. As can be seen, current Hall engine technology appears relatively heavy and too power consuming to be used within the Class I microspacecraft design envelope and for smaller microspacecraft. Some of the smaller devices available today,[102,111–120] however, may find applications with 50- to 100-kg class spacecraft. With the exception of the SPT-50 and SPT-60 thrusters, all thrusters listed in Table 10 are still under laboratory development. SPT-50 and -60 thrusters have seen flight applications on Russian spacecraft in the 1970s.[113,117] As can be seen by inspecting Table 10, the smaller Hall thruster concepts operate between power levels of 50 and 500 W. At the more interesting low end of this power range, however, these thrusters provide substantially lower specific impulses than are typically associated with Hall thrusters, some as low as about 600 s. Thruster efficiencies also drop at these lower power levels into the 10–30% range.

The most aggressive miniature Hall thruster concept is currently under investigation at MIT,[116,118–120] and shown in Fig. 7. This device is about 4 mm in diameter. Recently obtained data, published in this book,[120] indicate performances of up to 865-s I_{sp} and 1.8-mN thrust and a thruster efficiency of 6%. The thruster was operated at a 300-V beam voltage, which is typical for larger thrusters as well. Using data provided in Ref. 120, thruster power can be calculated to 126 W in this case. It is being suspected that the low efficiencies may be due to low degrees of ionization, which in turn may have resulted in poor utilization of neutrals.[120] These low

Table 10 State-of-the-art small Hall thruster technology

	SPT-60[113]	SPT-50[113]	X-40[102a]	SPT-30[112]	BHT-200-X2B[114,115a]	K-15[102a]	MIT[120a]
Manufacturer	Fakel, Russia	Fakel, Russia	Keldysh Research Center, Russia	Moscow Aviation Institute, Russia	Busek Co, Inc.	Keldysh Research Center, Russia	MIT
Type	SPT	SPT	SPT	SPT	Tandem-style	TAL	SPT
Diameter, cm	6	5	4	3	2.1	1.5	0.4
Thrust, mN	30	20	5–35	5.6–13	4–17	5–16	1.8
I_{sp}, s	1300	1250	800–1750	576–1370	1200–1600	830–1718	865
Power, W	500	350	80–540	99–258	100–300	70–400	126
Efficiency	0.37	0.35	0.3–0.58	0.16–0.34	0.2–0.45	0.3–0.34	0.06
Comments	Flight hardware	Flight hardware	Laboratory prototype	Laboratory prototype	Laboratory prototype	Laboratory prototype	Laboratory prototype

[a] Does not include neutralizer.

Fig. 7 MIT 50-W Hall thruster. (Courtesy of MIT.)

degrees of ionization may have been the cause of magnetic field degradation due to excessive heating of the thruster permanent magnets.[120]

3. *Evaluation, Issues, and Future Work*

Hall thrusters offer attractive performances for near-Earth applications and are relatively compact in size compared to ion engines. Very small Hall thrusters, however, have to overcome obstacles due to poor scaling of magnetic confinement fields. Given the smaller channel dimensions, larger magnetic fields have to be provided to reduce electron gyration radii. Field strengths in the MIT device are 0.5 T.[120] For even smaller devices, magnetic field strengths would have to be increased further. Currently, samarium–cobalt permanent magnets are able to deliver about 1 T at their surface, thus posing limits to further miniaturization and making engine sizes substantially smaller than the ones listed in Table 10 and shown in Fig. 7 difficult to imagine. In addition, as for all electrostatic thrusters, miniature Hall thrusters will require small neutralizer technology. Current hollow-cathode type neutralizers are too large, heavy, complex, and power-consuming for such small thruster devices. Miniature Hall thrusters, like other electrostatic electric propulsion devices, may therefore benefit from neutralizer developments undertaken in the Field Emission Electric Propulsion (FEEP) programs or microfabricated FEA cathodes, both discussed further below.

C. FEEP

1. *Description of Technology*

In the FEEP (Field Emission Electric Propulsion) concept, thrust is generated through the acceleration of ions by means of electrostatic forces as in ion engines and Hall thrusters. However, and very importantly, unlike in ion engines and Hall thrusters, the ionization mechanism does not require a gaseous discharge. Instead, ions are produced by field emission from a liquid metal surface with the metal acting as the propellant, fed into the thruster by the action of capillary forces.[121–123] Typical FEEP concepts are shown in Fig. 8, showing pin or needle, capillary, and slit emitter geometries. The FEEP concept has thus several unique advantages

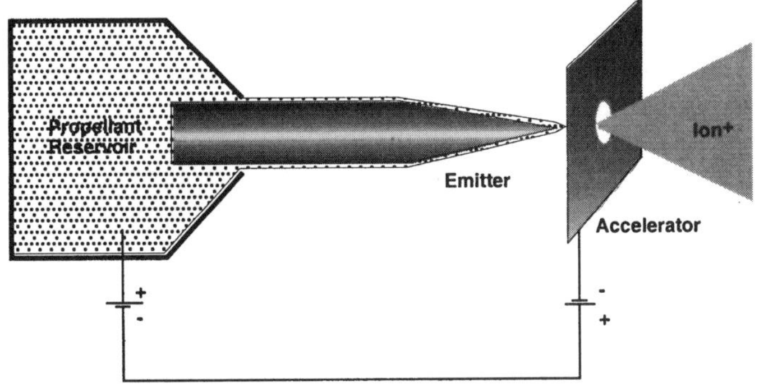

a) Pin- or needle-type emitter (external wetting)

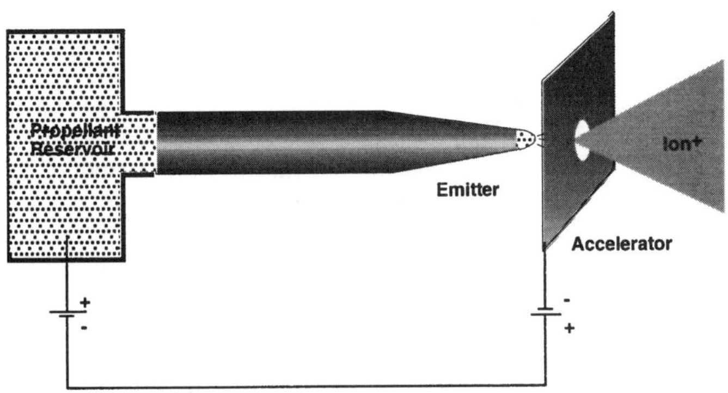

b) Capillary type emitter (internal wetting)

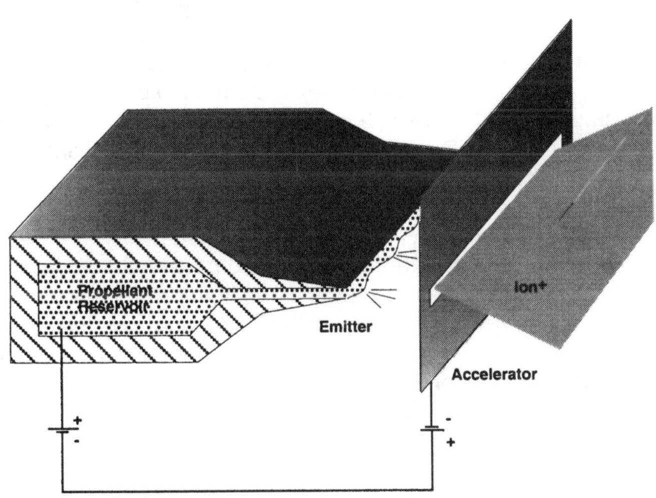

c) Slit emitter (internal wetting, multiple emission sites)

Fig. 8 Field emission electric propulsion (FEEP) emitter concepts.

over other electric propulsion concepts when considered for microspacecraft applications. By avoiding the use of gaseous discharges to generate ions and instead relying on a field emission process, the FEEP concept lends itself to a high degree of miniaturization, circumventing issues related to high-surface/volume ratio gaseous discharges, and associated high potential electron wall losses. The use of a capillary-fed propellant supply eliminates the needs for valves and a pressurant supply.[124] Also, FEEP thrusters, through the fact that they can be scaled to small sizes and are quick to actuate with thrust rise times as little as 3 ms,[125] can deliver extremely small impulse bits in the 10^{-8}-Ns range[126] and below.[127]

However, the field emission process requires high voltages, extending into the 10 kV range, leading to fairly high specific impulse values between 6000 and 10,000 s,[121] and thus relatively high specific power values result compared with other electric propulsion devices.[121–123] Specific power values vary with operating conditions, but a typical value for state-of-the-art FEEP technology is about 60 W/mN,[124] excluding the power required for the neutralizer. The use of liquid metal propellants has also repeatedly raised concerns regarding spacecraft contamination in the past,[121,124,128,129] in particular, for missions where instruments with sensitive optical surfaces may be involved, and this area will require additional investigation.

Field emission ion sources using liquid metals, namely, cesium (Cs), rubidium (Rb), or indium (In), have been studied since at least the late 1960s[130–133] and have almost immediately been proposed for space propulsion applications.[131,132] Early tests performed by Perel[131] and Perel et al.[132] employed capillary tubes (see Fig. 8) to characterize field strengths required for ion emission and explored capillary feed of Cs propellant. Beginning in 1972, FEEP thruster research has also been pursued vigorously in Europe.[121–123,126] Early tests were conducted with pin-, or needle-, type emitters,[123,133] (see Fig. 8), where capillary feed of the Cs propellant was achieved by the external wetting of a pin emitter,[121,133] as opposed to the internal wetting of Perel's capillary tubes. Work progressed from single pin emitters to linear arrays of stacked pins in 1975 to a linear slit emitter in 1979[123] (see Fig. 8). Slit emitters were being pursued to increase the emitting area of the thruster to yield higher thrust levels, a major focus in FEEP development at the time, and to avoid irregular emission characteristics observed for single emitters.[121,133,134] The bulk of the initial European work on FEEPs was performed at ESTEC in The Netherlands, with support from Culham Laboratory in England,[135] and subsequently in the 1980s under European Space Agency (ESA) funding at SEP in France and the University of Pisa in Italy.[126] Currently, work on linear slit FEEP systems is being performed predominantly by Centrospazio in Italy,[124–128,134,136–139] with additional support from ESTEC,[126] whereas work on pin, or needle, emitter types (also referred to as liquid metal ion sources; LMIS) is being conducted at the Austrian Research Center in Seibersdorf, Austria,[140] as well as the Technical University of Vienna, also in Austria. Here also a substantial amount of pioneering work on FEEP systems was performed.[121,129,133]

While there exist different field emitter configurations, such as the already mentioned needle, capillary, and slit emitter types,[121,133] the principle of operation is the same in all cases. In the slit emitter, for example, a liquid metal propellant such as Cs is fed by capillary forces through a narrow channel. Channel height is typically 1–1.5 μm[121,122,133,134,137,141] and the width is determined by the length of the slit, ranging from 1 mm[124] up to about 7 cm.[121,141] The narrow channel height is achieved by applying a typical micromachining technique, namely, thin-film deposition, in the fabrication of the FEEP thruster. The thruster consists of two

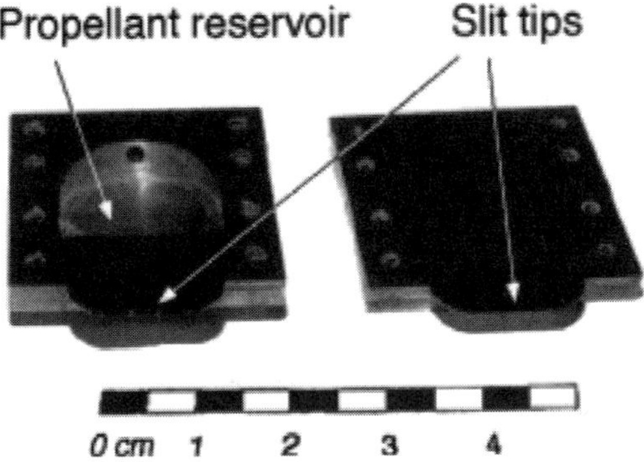

Fig. 9 FEEP slit emitter components. (Courtesy of Centrospazio, Italy.)

identical halves made from stainless steel or, more recently, Inconel[121,141] (see Fig. 9). Into one or both halves a recess is machined to form the propellant reservoir.[121] A nickel layer is then sputter deposited onto one of the thruster halves to a thickness equal to the desired channel or slit height.[121,141] Through proper masking the nickel layer outlines the desired channel contour. Clamping, or screwing, both thruster halves together now forms a channel between the two components equal in height to the thickness to the nickel layer.[121,124,141] In some cases, only a series of nickel dots is used to provide the proper spacing.[124]

The channel ends at the emitter tip formed by sharp edges that are located opposite a negative, (accelerator or extractor) electrode and separated by a small gap (about 0.6 mm) from the emitter tip.[121,123,133,134] An extraction voltage is applied between the two electrodes. The emitter carries a positive potential, while the accelerator is at a negative potential. The electric field being generated between the emitter and the accelerator now acts on the liquid metal propellant. The narrow slit width not only enables the capillary feed, but also, when combined with the sharp channel edges directly opposite the accelerator, ensures that a high electric field strength is obtained near the slit exit. The liquid metal column, when subjected to this electric field, begins to deform, forming cusps, which protrude from the surface of the liquid.[131] As the liquid cusps form ever-sharper cones due to the action of the electric field, the local electric field strength near these cusps intensifies. Once a local electric field strength of about 10^6 V/mm is reached, electrons are ripped off the Cs metal atoms. These electrons are collected through the liquid metal column by the channel walls, and the positive ions are accelerated away from the liquid through a gap in the negative accelerator electrode by the same electric field that created them. It has been observed that Cs ions are not emitted uniformly across the entire exposed liquid metal surface but rather from distinct emission sites formed by the liquid cusps.[142] Experiments and theoretical modeling have shown these emission sites to be spaced equidistantly within micrometers from each other, depending on slit width and radius of curvature of the emitter tips.[142]

2. *Available Hardware*

Table 11 lists representative FEEP thruster data obtained for slit emitter configurations. At present, all commercially available slit emitter FEEP devices are fabricated by Centrospazio in Italy. Pin-type emitters are currently under development

Table 11 State-of-the-art FEEP thrusters

Manufacturer	Centrospazio (Italy)	Centrospazio (Italy)	Centrospazio (Italy)	Centrospazio (Italy)
Propellant	Cs	Cs	Cs	Cs
Slit width, mm	2	70	5	70
Configuration	Single module	Single module	Cluster of 2 thrusters	Cluster of 4 thrusters
Nominal thrust, μN[a]	40	1400	2 × 100	4 × 1400
I_{sp}, s	9000	9000	9000	9000
Power, W[b]	2.7	93	13	370
Specific power, W/mN	66	66	66	66
Max. emitter voltage, kV	+5.5	+5.5	+5.5	+5.5
Accelerator voltage, kV	−5	−5	−5	−5
Thruster mass, kg	0.6	1.2	1	3.2
Thruster or cluster size, cm	8 × 6 × 8	13 × 7 × 9	10 diam. ×10	18 diam. ×15
PPU mass, kg	1	1.2	2	5.5
PPU size, cm	8 × 12 × 16	8 × 16 × 16	16 × 12 × 16	20 × 25 × 16
Comments	Qualification model	Qualification model	Under development	Under development
Ref. No.	143	143	143	143

[a]Maximum attainable thrust may be larger by a factor 2.
[b]Assuming all thrusters operating at nominal thrust.

at the Austrian Research Center in Seibersdorf.[140] Slit emitters with a wide variety of slit widths have been fabricated in the past, however, currently devices with slit widths between 2 mm and 7 cm are being offered.[143] These devices span a thrust range from 40 μN to 1.4 mN[143] and require power levels ranging from 2.7 W for the 2-mm unit to 93 W for the 7-cm unit, corresponding to a specific power value of 66 W/mN.[143] The mass of the 2-mm unit is 0.6 kg, and the 7-cm unit weighs 1.2 kg. A four 7-cm thruster module weighs 3.2 kg. Power processing unit masses are 1 kg for the 2-mm unit and only slightly higher for the 7-cm unit. In both cases, the same voltages are generated by this unit, being +5.5 V for the emitter and −5 V for the accelerator.

Typically an extraction voltage between the emitter tip and the accelerator electrode of 8–10 kV is required to achieve field emission using slit emitter geometries as discussed above.[122,134,137,141] There does exist some degree of freedom, however, in choosing the composition of this extraction voltage, i.e., the magnitude of the positive emitter voltage and the negative accelerator voltage, the sum of which equals the extraction voltage. The specific impulse of a FEEP device is determined by the emitter voltage. Higher emitter voltages will result in higher specific impulses. However, higher specific impulses result in lower thrust-to-power ratios, and thus, higher emitter voltages will reduce the thrust-to-power level of the device.[141] Some of the specific impulse capability may be sacrificed to achieve higher thrust-to-power ratios by decreasing the emitter voltage and increasing the magnitude of the accelerator voltage instead. However, accelerator voltages much higher (in magnitude) than -3 to -5 kV have been found to increase beam divergence and are thus not desirable.[123,127]

Table 11 does not include data on neutralizers. As with other electrostatic thruster concepts, FEEPs require neutralizers to emit a negative electron beam to neutralize the positive ion beam so as to avoid spacecraft charging. Neutralizers for FEEPs at this point remain an area of active development. Several approaches have been studied in the past and are in various stages of development. Concepts under investigation have included hollow cathode,[129,139] thermoelectric (hot filament),[127,139] thermionic (based on $BaCO_3$ technology),[128] and micromachined FEA[127] approaches. In the hollow-cathode type neutralizer, a Cs-containing compound (Cs_2CrO_4) is heated to about 660°C, releasing Cs and creating a Cs vapor that is ionized in a hollow cathode discharge, providing the electrons required for beam neutralization.[129] However, this concept has been abandoned after a few years of experimental investigation due to difficulties in obtaining control of Cs vapor flow.[143] In addition, neutralizer current-specific power consumptions were high, about 10 W/mA, due to the heater power required to release Cs vapor from the cesium compound.

The thermoelectric neutralizer concept consists of a hot, 1% thoriated tungsten filament. Electrons are released upon heating.[139] Neutralizer current-specific power consumption is extremely high for this device (100 W/mA) due to the high power levels required to heat the tungsten filament. In the thermionic device a barium paste is used to achieve a lower work function and electron emission at lower temperatures.[128] Neutralizer current-specific power consumption for such a device is about 0.16–0.33 W/mA. Micromachined field emitter array neutralizer concepts, if they can be realized, potentially may achieve neutralizer current-specific power levels of about 0.12 W/mA.[127] Field emitter array technology, however, still requires substantial development efforts. In these devices, electrons are being emitted from micromachined arrays of tips without the addition of heat (cold cathodes). These devices have been used in flat panel displays, however, at extremely low background pressure values. Conditions for these arrays when used as neutralizers, however, may be considerably more hostile, subjecting the emission tips to higher plume plasma pressures and ion sputter erosion. Given the small tip dimensions, sputter erosion may significantly alter emission characteristics over time if FEA designs are not properly adapted to the changed operating conditions inside the thruster plume. FEA cathodes may also be prone to shorting due to Cs propellant condensation. Investigations regarding the applicability of FEA designs as thruster neutralizers are currently being conducted at JPL in collaboration with the University of Michigan.[108,109]

3. *Evaluation, Issues, and Future Work*

FEEP thrusters, due to their ability to provide very low impulse bits as well as very low thrust values, in the micronewton range, have been considered for spacecraft fine attitude control, in particular, also for interferometry missions, such as LISA (see above), where precise relative spacecraft positioning is required.[144] Although these spacecraft typically exceed the mass ranges considered in Table 2, design considerations may still be similar enough for the 100-kg-class spacecraft considered in this study, and FEEP thrusters may thus be used for such spacecraft in a similar fashion. However, when considering even smaller spacecraft, such as the Class I microspacecraft in Table 1, the potential use of FEEP thrusters warrants closer inspection due to the potentially high power requirements resulting from the low thrust-to-power values, as well as mass penalties associated with power conditioning units, for example.

Following the example given in the cold gas thruster section above, for a typical Class I microspacecraft mission, 0.7 kg of nitrogen gas would be required to meet a 50-m/s attitude control delta-v budget on a 10-kg spacecraft. If a FEEP system with a specific impulse of 8000 s was used, the required propellant mass would be reduced to 6 g. While this is a substantial reduction, the required PPU mass, according to Table 11, of about 1 kg has to be taken into account as well. Assuming that only two PPUs would be required for the entire FEEP attitude control system (since typically only two thrusters need to be fired for attitude control maneuvers), and assuming further that this PPU can be switched to any thruster (further neglecting any mass associated with the switching units in this simple comparison), the total required PPU mass per spacecraft would be 2 kg. Assuming that 12 thrusters will be required in six pairs at 1 kg each (Table 11), the combined FEEP propulsion hardware mass would be 8 kg, neglecting any structural masses at this point. This is too heavy for a 10-kg spacecraft but may be applicable to larger craft if dry masses could be reduced further or if some of the thrusters could be arranged in quads saving mass as well.

Since Cs propellant can be stored in its liquid state, whereas nitrogen gas would have to be stored in high-pressure tanks, tank weight reduction will benefit the FEEP system. According to recent tank data,[145] a 3244-cm^3 tank (roughly the size required for nitrogen storage in the example above), capable of maintaining a maximum expected operating pressure (MEOP) of 10,000 psia (far more than required for a storage pressure of 3500 psia as assumed), weighs about 1.8 kg. These data happen to be based on a cylindrical tank. At about 10 g per cold gas thruster (including valve), 12 thrusters and the tank weigh a combined 1.92 kg. However, additional valves, regulator, and lines and fitting will be required for the cold gas system. Nonetheless, this simple comparison could leave the cold gas system competitive based on pure mass considerations. Other design considerations, however, such as cold gas leakage or the larger volume required for the cold gas propellant tank, will also have to be taken into account. There may also exist the potential for further mass reductions in FEEP system components, since devices listed in Table 11 were not mass-optimized for microspacecraft.

While FEEP thrusters appear to be able to meet the impulse bit requirements of a Class I microspacecraft, slew rate requirements as outlined in Table 2 may be impossible to meet within the power constraints considered for this type of spacecraft. As can be seen by inspecting Table 11, power requirements per thruster may be of the order of 90 W to achieve a thrust level of about 1.5 mN. Since typically two thrusters are fired during slew maneuvers, this power requirement would double. Furthermore, thrust levels this high can be obtained only with the larger-scale FEEP devices featuring several-centimeter-long emitter slits. Given that a multitude of these thrusters would be required for attitude control purposes, propulsion system dry weight could increase substantially and volume constraints may be encountered for Class I microspacecraft. Thus, FEEP systems will likely only be usable on microspacecraft if slew rate requirements can be relaxed significantly (or they do not exist at all) or a separate set of attitude control thrusters is provided onboard the spacecraft to perform such slew maneuvers. The latter option, however, would lead to increased propulsion system dry mass, system complexity, and cost.

Given the low microspacecraft masses, FEEP systems may also be considered for primary propulsion applications. Assuming a specific power requirement of 66 W/mN (Table 11), a current-specific neutralizer power requirement of 0.2 W/mA (based on a thermionic barium paste-type neutralizer and using a current-specific power value midrange of the data given above), and further assuming a thruster emitter current of 7 mA/mN (according to Petagna et al.[122]), one arrives at a specific power requirement per thruster of (66 W/mN + 0.2 W/mA∗7 mA/mN), i.e., 67.4 W/mN. Therefore, available thrust values for a Class I–type spacecraft with an onboard power supply of roughly 20 W will be about 0.3 mN. For this thrust value the thrust-to-spacecraft mass ratio will be about 0.015 mN/kg, assuming a 20-kg spacecraft mass. For the Europa orbiter mission currently being studied at JPL, solar electric propulsion (SEP) has been considered[146] with thrust-to-weight ratios of about 1.25 mN/kg at the beginning of the mission. Note that these values may change as mission plans are refined further. However, it appears that these values are clearly higher than those obtainable with a state-of-the art FEEP system onboard a microspacecraft. Using FEEP thrusters for microspacecraft primary applications will therefore require higher spacecraft power levels than those assumed in Table 2 or result in lower spacecraft accelerations than available with some of today's electric propulsion technologies on conventionally sized spacecraft. However, given unknown microspacecraft mission parameters, the option to resort to lower spacecraft accelerations remains open at this point.

In addition to microspacecraft power constraints and the need for improved neutralizer technology, propellant contamination issues will have to be considered for each mission. As mentioned, Cs has been used most frequently in FEEP thrusters in the past. Cesium, a solid at room temperature, was chosen as propellant due to its low melting point of 29°C, high atomic mass, low work function, and good wetting properties,[133] in particular, when used in conjunction with Inconel emitters.[141] However, the use of Cs results in potential contamination problems related due to the high reactivity of Cs and the fact that it may condense on spacecraft surfaces. Two types of contamination are being distinguished: self-contamination of the thruster due to reaction of Cs with environmental gases, mainly water vapor, leading to potential clogging of the emitter slit, and thruster-external contamination of other spacecraft surfaces due to deposition of propellant.

Self-contamination of FEEP thrusters has been studied fairly extensively in the past.[137,147] The concern during thruster self-contamination is the interaction of Cs propellant with environmental gases such as water vapor, carbon dioxide, and molecular and atomic oxygen, resulting in the production of crystalline substances with high melting points that may clog the narrow emitter slit. Reaction products (and their melting points) are cesium hydroxide (CsOH; 272°C), as a result of the interaction of Cs with water vapor; various cesium oxides (Cs_2O, 490°C; CsO_2, 433°C; Cs_2O_2, 594°C), due to Cs reaction with oxygen; and cesium carbonate (Cs_2CO_3; 610°C), as a result of Cs interaction with carbon dioxide.[147] Of the gases considered, water vapor poses the most serious contamination problem, building up a considerable higher degree of reaction products in the emitter slit than the other gases at comparable partial gas pressures.[147] Once incurred, these substances may lead to spotty ion emission from the thruster or lead to total device failure,

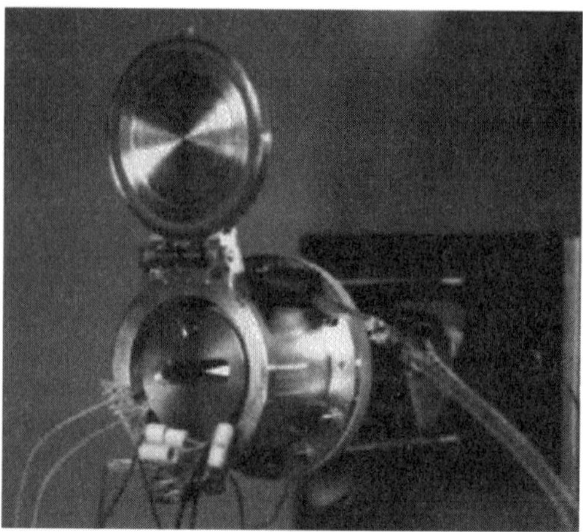

Fig. 10 Assembled FEEP thruster showing emitter containment housing. (Courtesy of Centrospazio, Italy.)

depending on the severity of the contamination. Here, slit emitters have shown a somewhat greater resiliency to contamination than capillary or pin emitters: as long as ion emission occurs even over short slit sections, any contaminating crusts that may have formed can eventually be blown away, provided that the thruster has undergone a proper priming procedure.[137]

Recognizing that most of the exposure of a FEEP thruster to the aforementioned gases will occur during ground handling, thruster priming procedures have been developed to avoid or minimize FEEP self-contamination.[123,128,134,137] These priming procedures require the thruster to be placed in an emitter container with a removable lid (see Fig. 10), which will be part of the thruster flight package.[128] The priming procedure consists of cleaning and bake-out of the thruster to remove adsorbed water and other contaminant layers and a subsequent cooling-down period and filling of the thruster with cesium still under vacuum conditions.[134] The emitter is then operated for approximately 2 h to remove any remaining contaminants in the slit area. After thruster shutoff, the emitter container is filled with an inert gas, the lid is sealed, and the setup is finally removed from the vacuum chamber. The emitter container will not be opened until in space (using a paraffin-based actuator, for example). Immediately after lid opening the thruster will be fired again to avoid any outgassing products from the launch vehicle or spacecraft to contaminate the thruster again.[128] It has been determined that following these thruster priming procedures, thruster initiation is possible at background pressures at least as high as 10^{-4} mbar and at partial water vapor pressures at least as high as 10^{-7} mbar.[137] These values are compatible with the local in-flight environment of even large spacecraft, such as the space shuttle, shortly after arrival in orbit.[128,147] However, to combat these thruster self-contamination issues successfully, the emitter container will have to be part of any FEEP thruster hardware and needs to be taken into account in the mass budget as well as the structural layout of the system.

Thruster-external contamination is still an area of concern. Unlike thruster self-contamination, it has not yet been as thoroughly studied, and additional work will be required in this area. Given that FEEP thrusters use liquid metal propellants that condense at room temperatures, cool spacecraft surfaces may obtain a mirror-coating as a result of thruster firings, which may be particularly detrimental for optical surfaces. In particular, for microspacecraft applications, with multiple spacecraft potentially flying in formation, spacecraft contamination could be a major concern, as thrusters on one spacecraft may fire into the direction of another spacecraft flying nearby. Mission scenarios may not always allow firings to be scheduled until other spacecraft have moved out of the line of sight. FEEP ground tests in large vacuum chambers or flight tests will be required, equipped with suitable diagnostics, such as quartz crystal microbalances, to perform measurements to quantify any potential spacecraft contamination.[128]

Concerns regarding the chemical aggressiveness of cesium propellant have also led to the study of other propellants for FEEPs. Indium (In) and rubidium (Rb) have been tested.[129–131,136,140] Indium is currently the standard propellant of the Austrian pin emitter-type LMIS sources.[140] Both Rb and In have higher melting temperatures, 39 and 156°C, respectively,[136] than Cs, with a corresponding value of 29°C. This will lead to additional power requirements to melt the propellant for thruster use, in particular, in the case of In. Indium LMIS sources, however, have shown impressive resiliency when exposed to humid environmental conditions. After the crash of an Ariane-5 rocket, LMIS sources where scattered at the crash site in a local Guyanese swamp.[140] These LMIS sources, which were part of an experiment flown on one of the payloads, were retrieved out of the swamps and then operated in vacuum again. Despite the adverse environmental conditions they had been placed under, they exhibited the same operating parameters as prior to the failed launch in most cases,[140] indicating less sensitivity to interaction with water than Cs.

For Rb, the heater power increase over Cs is limited, and the somewhat higher melting temperature may actually ease ground handling of the thruster, avoiding accidental liquefaction of propellant.[128] On the other hand, Rb has a lower atomic mass (85.4 AMU) than both Cs (132.9 AMU) and In (114.8 AMU),[141] resulting in a lower thrust-to-power ratio. Both In and Rb also have higher ionization potentials than Cs.[130,131] Although the chemical aggressiveness of In is somewhat reduced compared to that of Cs and Rb, similar handling and thruster packaging requirements typically apply, including the need for an emitter container.[140] Many other FEEP propellants have also been tested with both needle- and capillary-type emitters, however, not necessarily for propulsion applications. These include metals such as Ag, Au, Bi, Cd, Ga, Hg, K, Li, Na, Pb, Sn, and Tl, as well as nonmetals such as As, B, Be, Ge, and Si.[133] For reasons related to melting temperatures, ionization potentials, and atomic masses, however, the aforementioned three metals (Cs, In, Rb) are at present the only propellants considered for thruster applications.

Future work in field emission thruster technology is focusing on the use of microfabricated emitter arrays,[133,148] consisting of a series of micro-"volcano" structures on a wafer. Since extractable currents from each microemitter are much lower than those obtained with conventionally machined emitters, arrays of many of these emitters will be required to operate in parallel. The significance of such devices are their potentially significant mass and size reductions from conventional designs. However, unlike in the case of gaseous plasma devices, thruster efficiencies may scale well during miniaturization since even conventional FEEP designs already rely on micron-sized emitter dimensions. MEMS-based FEEP devices are discussed in greater detail in Section VI.

D. Colloid Thrusters

1. *Description of Technology*

Colloid thrusters were extensively studied in the late 1960s and early 1970s for spacecraft attitude control and drag makeup, preceding in part work conducted on FEEP devices.[130–132] However, like the latter, these devices fell out of favor with the space community due to their inability to produce high enough thrust values at reasonable power levels. With the advent of microspacecraft designs, however, a potential application for these devices may have arrived.

Colloid thrusters function similarly to FEEP devices, featuring emitter tips and accelerating electrodes resembling those used in FEEPs (compare with Fig. 8). However, unlike FEEP thrusters, they do not accelerate individual ions; rather, in a colloid thruster, thrust is produced by electrostatically accelerating fine charged liquid droplets ejected from a capillary[132,149] A strong electric field applied between the sharp-edged exit of the capillary and an external electrode causes charge separation inside the liquid propellant, which in most cases is doped with an additive to increase its electric conductivity. Through a combination of hydrodynamic instabilities, causing jet breakup into small liquid droplets, and the action of the applied field acting on the conductive liquid, charged droplets are extracted from the capillary at high velocities, producing thrust. Relatively high acceleration voltages, of the order of 4–20 kV,[132,150,151] applied between emitter and accelerator have been used. Although slit emitters have been tested, almost all past work on colloid thrusters has been performed with capillary[132,149] and pin[150] emitters. In the past, most colloid thrusters have used positive pressure feed to fuel the emitter tips,[152] although some of the later colloid concepts used capillary feed, similarly as in FEEP devices.[153]

Depending on the propellant used, either positive or negative liquid droplets can be produced. Most applications studied in the past used glycerol doped with sodium iodine (NaI) in a 20–30 g/100 ml solution[132,150] to produce positive droplets and glycerol doped with 2–10% sulfuric acid[132] to produce negative droplets. The concept of providing both positive and negative droplet emission in a single thruster array was termed a "bipolar thruster" by Perel et al.[132] Its significance is that it can potentially be self-neutralizing, provided that the same amount of current can be drawn from each set of capillaries, thus eliminating the need for a separate neutralizer. Given the composition of the propellant, colloid beams have been found to consist of several components, including charged droplets, the intended main component of the beam, and thermal glycerine molecules, molecular ions, and electrons.[151]

Several trade-offs have to be made in the design of a colloid thruster to optimize performance.[132] The specific charge, measured as coulombs per droplet mass, has to be high to obtain high specific impulses at reasonable voltages. The colloid thruster is an electric thruster, and as such, additional propulsion system masses associated with the power supply or conditioning will have to be offset by sufficient propellant mass savings to be obtained through high enough specific impulses. The specific impulse is also influenced by the so-called specific charge efficiency, which measures the distribution of specific charge in a droplet stream. A more "peaked" specific charge distribution will lead to higher specific impulses and higher propulsion system efficiencies.

Specific charge efficiency in turn depends on several parameters. Higher electric field strengths increase the specific charge but reduce the charge efficiency[132] since the ion component in the beam is increased.[152] Lower mass flow rates result in higher specific charge efficiencies, whereas a higher conductivity leads to a higher specific charge and a lower specific charge efficiency (again due to an increase in the ion component).[132] The capillary tip design affects the specific charge efficiency, mostly through its effect on the local electric field strength near the tip.[132] Some of these design considerations work against each other. A large potential drop caused by a strong electric field, for example, will create a higher specific charge and accelerate the droplets to a higher exhaust velocity (raising the I_{sp}), however, it will also decrease the specific charge efficiency (lowering the I_{sp} again to some extent). The decrease in the specific charge efficiency thus acts as a retardant in obtaining higher specific impulses.

In addition to these performance considerations, careful propellant selection is made to ensure proper thruster function and a long lifetime. High solvation (to take up dopants), low vapor pressure (to avoid precipitation of dopants on capillary walls near the tip, potentially clogging the system), a low freezing point (to avoid clogging), and low corrosivity (to ensure long thruster lifetime) are key parameters in the selection of the propellant.[132] Among the potential propellant solvents available, glycerol was found to have a low vaporization pressure and superior ability to dissolve dopants.[132] Platinum alloy (Pt/In) capillaries have been used because of their resistance to corrosion.[132]

2. *Available Hardware*

Table 12 lists data obtained with various colloid thruster configurations. Data are somewhat sketchy since, on the one hand, they have been assembled from information obtained from early colloid studies published in the 1960s, not always containing a complete set of colloid performance parameters. On the other hand, several newer colloid development efforts are still in very early stages, and data sets in some cases are not yet complete. However, Table 12 shows that colloid thrusters have been tested over a wide range of operating conditions. Specific impulses as high as 1450 s have been achieved, however, at emitter and accelerator voltages of 12.3 and −2.0 kV, respectively.[150] In this device, the thruster featured 432 emitter tips. Higher-thrust versions of colloid thrusters have been built featuring fewer, larger-scale capillary emitter tips with inner diameters as large as 2 mm. The large tip dimensions required even higher extraction voltages to achieve high enough electric fields and appreciable specific impulse values, ranging as high as +19 and −2.0 kV for a I_{sp} of 1382 s and a thrust of 159 μN per emitter. In a somewhat lower-thrust case of the same emitter (129 μN), likely achieved by lowering the mass flow rate, 1405 s could be obtained at voltages of +18.2/−2.0 kV[149] (note that lower flow rates increase the specific charge efficiency and thus the I_{sp}). A four-module thruster using these large-scale emitters was able to produce a thrust of over 1.3 mN, or about 0.33 mN per emitter.[149]

Noteworthy is also the aforementioned bipolar thruster concept developed by Perel et al. in the late 1960s.[132] This device featured 37 positive and 36 negative emitters, using sodium iodine- and sulfuric acid-doped glycerol propellants, respectively, and produced thrusts ranging between 0.2 and 0.56 mN at power levels of about 4.4 W/mN, requiring emitter voltages of +4.4 and −5.8 kV (accelerator electrode grounded), depending on the droplet polarity.[132] The thrust data listed in Table 12 for this device were obtained by dividing the total thrust by the total number of emitters and, thus, represent an average thrust value for the positive and

Table 12 State-of-the-art colloid thruster characteristics and performances

Source	Electro-Optical Systems[132]	TRW[150]	TRW/Edwards AFB[149]			Phrasor[153,154]	Stanford[155]
Propellant	20 g NaI/100 ml glycerol and 2 ml H_2SO_4/100 ml glycerol	30 g NaI/100 ml glycerol	—	—	—	1.5 *M* Ammonium Acetate/glycerol, 1.3 *M* NaI Glycerol[a]	NaI-Doped glycerol
Thrust/ emitter, μN	7.6	0.84	159	129	334.5	0.022	1[b]
Isp, s	450–700 (est.)	1450	1382	1405	1029	529	500[b]
Efficiency	—	0.69	0.77	0.77	0.7	—	—
Specific power, W/mN	4.4	—	—	—	—	—	10[b]
Emitter voltage, kV	+4.4/−5.8	12.3	19.0	18.2	—	4.0	6
Accelerator voltage, kV	0.0	−2.0	−2.0	−2.0	—	0.0	0
Charge-to-mass, C/kg	—	—	6370	—	—	4280	1000
Current/ emitter, μA	+150/−154	—	—	—	—	0.018	0.5–1.0
Emitter tip I.D., μm	200	127	2286	2286	2286	1	50
Comments	Bipolar concept, 37 pos. and 36 neg. emitters; total thrust, 0.56 mN	432 emitter tips per thruster; total thrust, 0.36 mN; 4350-h lifetime achieved with 36-needle module	100-h life test	—	4 emitters/thruster; total thrust, 1.338 mN	Laboratory development	Under development; target 0.5 kg/ 10 × 10 × 20 cm^3 mass/size of propulsion module

[a]Performance data recorded for this propellant. [b]Estimated performance goals.

negative emitters. In the case of the operating conditions presented in Table 12 for this device, specific impulses were estimated to range between 450 and 700 s.[132]

Most recently, after a several-decade-long hiatus in the development of colloid thrusters, interest in this thruster concept has increased again due to the advent of microspacecraft concepts and the fact that colloid thrusters lend themselves to a high degree of miniaturization by avoiding gaseous plasma discharges. At Phrasor Scientific Inc., tests were recently conducted with a single-emitter, 1-μm-I.D. capillary device.[153] At emitter voltages of 4 kV (accelerator grounded), a 529-s I_{sp} and 0.022 μN were achieved[153] using sodium iodine-doped glycerol propellant.[154] These tests, although conducted with conventionally fabricated glass emitter tips, are anticipated to be the prelude to a MEMS-based development effort for colloid thrusters (see below).

At Stanford University, in support of the Emerald university nanosat mission (see Table 1), developmental work on colloid thrusters is also currently under way.[155] The purpose of the colloid flight test onboard Emerald is to demonstrate colloid operation in orbit. The colloid system will attempt to change the rate of spin of the Emerald spacecraft, which can be detected via onboard attitude sensors.[155] Currently, 1- and 100-emitter capillary-type laboratory prototypes have been built.[155] The capillaries have an I.D. of 50 μm (0.002 in.).[155] The emitter–accelerator gap distance is 1 mm.[155] These prototypes use sodium iodine-seeded glycerol propellant.[156] The flight hardware is designed to operate in a bipolar mode. Targeted performance data for the flight unit are 0.1 mN at 1-W thruster power,[156] corresponding to a specific power of about 10 W/mN. The entire thruster package is targeted to weigh 0.5 kg[156] and is designed to fit within a $10 \times 10 \times 20\text{-cm}^3$ envelope.[155]

An interesting version of the original colloid thruster concept was tested in the early 1980s, termed the Capillaritron.[157] It operated on gases, such as argon, helium, hydrogen, and xenon. Plasma discharges formed inside the capillary near the tip.[152] Test results reported for argon indicate emission currents of 100 μA per tip for a 3-kV extraction voltage applied between the emitter tip and the accelerator electrode and an argon flow rate of approximately 2.4 sccm.[157] It is believed that ionization is not due to field emission, but due to electron bombardment ionization in the plasma sheath region near the tip.[157] However, efficiencies were low.[152]

3. *Evaluation, Issues, and Future Work*

Colloid thrusters appear to lend themselves well to miniaturization due to the absence of a gaseous plasma discharge, similar to FEEP devices. However, the power requirements for colloid thrusters are much lower than those for FEEP devices for the same thrust level. Thrust levels of 0.5 mN could be achieved using only about 2 W of power in a self-neutralizing, bipolar array based on existing data.[132] Using the same example of a 20-kg spacecraft as described in the FEEP section, a thrust-to-spacecraft mass ratio of 0.25 mN/kg could be obtained with only 10 W of power, half the power capability assumed to be available on a 20-kg spacecraft, yet providing more than 10 times the spacecraft acceleration as a FEEP thruster. Colloid thrusters therefore appear to fit well within Class I microspacecraft size and power constraints, possibly better than any other electric propulsion concept based on the currently available database.

However, the low specific power is the result of a fairly low specific impulse, limiting its use for primary propulsion applications. Unfortunately, despite these low specific impulse values, voltage requirements for this thruster are very high, ranging well into the multikilovolt range. Generating these high voltages on mass-constrained microspacecraft may be challenging since high-voltage power conditioning equipment is typically heavy due to heavy magnets. MEMS-based colloid thruster versions are currently being considered,[153] offering the potential of reducing the system weight and volume further and possibly reducing the system cost for multiple-emitter arrays due to the introduction of batch-fabrication processes.

Colloid thrusters may be considered for primary propulsion applications on Class I, microspacecraft where delta-v's are limited (due to the low I_{sp}), or for attitude control functions, where very small impulse bits are required and slew rates are below the values listed in Table 3. Future MEMS versions, if they can be successfully built, may be applicable to even smaller microspacecraft, i.e., spacecraft in the Class II category.

E. Pulsed Plasma Thrusters (PPTs)

1. Description of Technology

In a PPT, propellant is ionized and then electromagnetically accelerated between two parallel electrodes in a pulsed mode of operation.[158] In ablative PPT concepts, a solid Teflon bar is used as the fuel. The Teflon bar is pushed against a retaining lid between two electrodes by means of a negator spring (see Fig. 11). The electrodes are connected to a capacitor, which is unable to discharge because the vacuum and solid Teflon bar between the electrodes do not provide a conductive path. A spark plug located near the solid Teflon surface is fired, triggering the main discharge between the two electrodes feeding on ablated Teflon material from the fuel bar. As the capacitor discharges, high currents, in the kiloampere range, are momentarily conducted through the ablated propellant material and Lorentz forces, resulting from the interaction of this current with its self-generated magnetic field, act on the ionized components of the propellant material, accelerate the plasma, and expel it from the thruster. In the process, the Teflon bar is pushed forward toward the retaining lid and brought into position for the next pulse.[158]

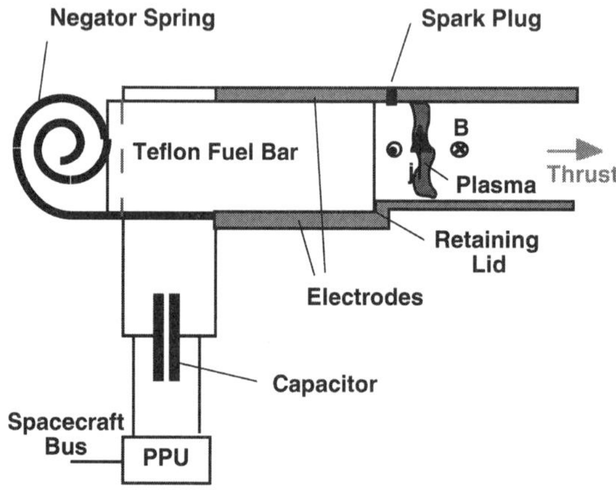

Fig. 11 Pulsed plasma thruster (PPT) concept. (Adapted from Ref. 179.)

Ablative PPTs have achieved a high degree of maturity over several decades of development, test, and flight applications. The first flight of a PPT took place on the Russian Zond 2 spacecraft approaching the planet Mars in 1964. Unfortunately the spacecraft was lost due to faulty radio communications.[158] Subsequent flights were conducted by the United States on the geosynchronous MIT Lincoln Laboratory LES-6 communications satellite in September 1968,[158,159] on a geosynchronous meteorological satellite (SMS) in 1974,[158,160] using PPTs for east-west stationkeeping and spin axis precision control, and on a series of U.S. Navy TRANSIT navigation satellites, called TIP/NOVA, beginning in 1981,[158,161,162] for drag compensation.

Ablative PPTs have been valued for their relative simplicity of operation, simple propellant feed system, and compact (solid) propellant storage, featuring no moving parts with the exception of the fuel bar.[158] PPTs are also able to provide minute impulse bits, of the order of micronewton-seconds, making them suitable for fine attitude control. Since microspacecraft attitude control requirements will push the required impulse bits to extremely low values, adequately miniaturized PPT concepts may find applications in this area. On the other hand, many PPT designs have been characterized by low thruster efficiencies and low thrust-to-power ratios. Gas-fed PPTs are also under investigation, using high-pressure xenon gas as a propellant. These devices are considerably less mature than ablative PPTs and are still in the laboratory development phase. Early research extends back well into the 1960s,[163] and the concept has recently found renewed attention.[164] Past research has indicated that these devices may deliver higher efficiencies and thrust-to-power ratios than ablative PPTs,[164] however, these findings still await verification. Gas-fed PPTs, unlike Teflon PPTs, do require a propellant tank and feed system. If run on xenon propellant, these devices could be tied into an existing ion propulsion system for primary propulsion applications, for example, using the same propellant tank.

Despite several decades of research and development on PPTs, there still exist today several focal areas of active research for this device, such as the study of propellant ablation and acceleration with the goal of improving PPT efficiency, using the same propellant tank, thruster lifetime testing, and thruster contamination studies. Plasma generation and propellant acceleration mechanisms in a PPT are complex. More detailed knowledge of these mechanisms, however, is required to understand current ablative PPT efficiency limitations. Obviously, considerations regarding the improvement of thruster efficiencies are of particular concern for applications of this technology on power-constrained microspacecraft, in particular if these devices were to be used for primary propulsion applications.

The propellant mass ejected from the PPT has been determined to consist of at least three components. First, there is the electromagnetically accelerated plasma component. The ablated propellant forms a plasma sheet in the ignition process, and large currents carried through the plasma interact with the generated self-magnetic field. Large current-carrying plasma columns are known to experience a "pinch effect" in a symmetrical electrode setup by which plasma particles moving with the current direction interact with the generated self-magnetic field and experience a Lorentz force pointing toward the center of the plasma column, "pinching," or narrowing, the column. In a PPT setup, the magnetic field surrounding the plasma sheet is larger on the upstream side (facing the solid Teflon surface) than on the downstream side (facing the thruster nozzle). This is due to the fact that the current passing through the plasma sheet and the electrodes forms a loop around the fuel bar (see Fig. 9). Magnetic field strengths inside this loop (on the Teflon side) are greater than on the outside of this loop (on the nozzle side). The resulting force imbalance, which can be expressed mathematically as the gradient of a "magnetic pressure,"[158] drives the plasma sheet toward the nozzle. Plasma velocities that can be achieved in this acceleration process are quite high and can range between 10 and 35 km/s.[158] This electromagnetic component is associated with a much slower thermal component, as the hot plasma expands thermally toward the nozzle exit as well.

An additional form of mass ejection from an ablative PPT thruster is particle emission.[165,166] Two distinct particle populations have been found in the exhaust of PPTs,[166] one consisting of metallic, <1-μm-diam particles believed to be the result of electrode erosion. The other component consists of particles ranging in diameter from 1 to 200 μm and containing fluoride. One hypothesis is that these particles are emitted from the solid Teflon surface as a result of high-pressure vapor generation beneath the Teflon surface. Vapor generation at this location may be a consequence of the absorption of UV radiation emanating from the plasma discharge adjacent to the Teflon surface. The significance of this observation is that the mass per pulse ejected in the form of microparticles from a PPT has been estimated in one recent experiment to be as high as approximately 40% of the total ejected propellant mass.[166] However, particle velocities have been estimated to be only of the order of a couple hundred meters per second.[166] Thus, particle emission contributes minimally to the thrust generated from a PPT device and may thus be a major contributor to the low observed thruster efficiencies. Spanjers et al.,[166] who performed these investigations, estimate that elimination of low-velocity particle emission could double PPT thruster efficiencies.

Finally, late cycle evaporation has been observed as another loss mechanism for PPTs.[165,167] Due to heating of the Teflon bar by the plasma discharge, propellant still evaporates from the Teflon surface even after the plasma sheet has departed. Since this portion of the propellant is no longer part of the plasma, it is not electrodynamically accelerated and, rather, expands at the relatively slow velocity of approximately 300 m/s[167] toward the thruster exit, thus not contributing significantly to PPT thrust either and therefore lowering the thruster efficiency. Spanjers et al.[167] were able to show that thruster efficiencies could be increased if thrusters were operated at lower power levels, decreasing thruster temperatures. Designs may therefore have to be developed resulting in cooler Teflon temperatures, thus reducing late cycle evaporation.[167]

Since PPTs are inherently pulsed devices, achieving high thruster cycle lifetimes is of critical importance. During ground testing, PPT thrusters have been operated at up to 34 million shots in the case of a thruster design intended for use on the LES 8/9 spacecraft.[168] The test was terminated voluntarily and thruster cycle lifetimes were extrapolated from existing data to as high as 1×10^{11} shots, although no actual test data exist to verify this extrapolation.[168] Potentially life-limiting components are the capacitor and the spark plug used to ignite the plasma. Obviously, concerns with respect to the failure of either component have largely diminished in view of the high number of cycles obtained so far. Spark plug erosion, however, was a concern in the past and has been studied in some detail.[169] The actual ignition mechanism is still not well understood.[158] The spark plug itself consist of an annular cathode surrounding a central anode. Both are insulated from each other by an insulating semiconductor material (not specified).[169] A DC surface discharge along the semiconductor surface is initiated that produces an initial plasma in the PPT electrode gap near the Teflon surface. This discharge leads to Teflon evaporation off the fuel bar, in turn triggering the main discharge. It has been observed that, to avoid unacceptable spark plug erosion by removing too much semiconductor material (which would threaten the lifetime), an optimal thickness of carbonaceous deposits on the spark plug, formed during previous PPT main pulses, is required.[169] In subsequent spark plug firings these deposits are then released, rather than ablating the semiconductor material itself, and form the trigger discharge.[169] However, too thick a deposit may electrically short the plug, lead to unpredictable firings, and cause plug embrittlement, possibly due to fluoride contained in the deposits.[169] Thus, proper spark plug ignition, although complex, is crucial to achieving proper thruster performances and lifetimes.

Contamination studies for PPT devices have been performed in the past at Fairchild Company,[170] JPL,[171,172] and recently NASA Glenn,[173] and in no case was substantial backflow contamination found. While these findings seem to indicate that spacecraft self-contamination is not an issue, it must be noted that microspacecraft may be deployed in constellations consisting of many spacecraft, potentially firing thrusters in the direction of another craft. Therefore, as would be required for other thruster concepts as well, further study regarding the potential and degree for mutual spacecraft contamination and the particular effects that PPT exhaust products may have on various spacecraft surfaces (in particular, also sensitive instrument optics) is likely needed.

2. *Available Hardware*

Table 13 lists performance characteristics of several representative PPT designs, past and present. As can be seen, PPT performances range between 2 μN and 4.5 mN in thrust and 200–5000 s in I_{sp}, with typical values of about 1000–1500 s, and impulse bits may be as low as a couple micronewton-seconds in the case of a newly developed micro-PPT design or as high as 22 mN-s in the case of the millipound thruster. Power levels range between 1 W in the case of micro-PPT versions[176] to 30 W, although higher power levels (>100 W) can be processed in PPT designs. Thruster masses (fueled) in the case of the older flight designs may be as high as 6–7 kg per thruster,[158,161] while micro-PPT designs may have masses as low as 0.5 kg.[176] For attitude control purposes, several PPT thrusters may be clustered, using the same capacitor, thus reducing the overall system mass.[177] Thruster lifetimes of up to 34 million shots were achieved in the case of the LES 8/9 thruster.[168] However, in the case of the millipound thruster design, requiring higher energies per shot of 750 J, capacitor failure at a substantially lower number of shots ($<1 \times 10^5$ shots) was observed.[158]

Most recent PPT developments include a dual-thruster module for the New Millennium Earth Observing-1 (EO-1) spacecraft,[178] a scaled-down version of this thruster for the "Dawgstar" spacecraft of the ION-F constellation pursued under the university nanosat program[179] (see above), and the aforementioned micro-PPT currently under development at AFRL,[176] having achieved unprecedented degrees of miniaturization. The EO-1 thruster provides a range of impulse bits from 60 to 860 μN-s over a range of power and specific impulses of 12–70 W and 650–1400 s, respectively.[178] So far, over 200,000 firings have been demonstrated. The "Dawgstar" thruster is currently under development and uses the EO-1 design as a baseline, however, it will feature reduced-cross-section fuel bars, a modified

Table 13 Performance parameters of state-of-the-art PPT designs

Type	Thrust, mN	I_{sp}, s	Power, W	I_{bit}, μN-s	Total impulse, N-s	Energy/ pulse, J	Thruster mass, kg	Fuel mass, kg	Comments
Zond-2[158]	2 (10 Hz)	410	—	—	—	50	5	0.4	10^6 shots qual. test. flight on Russian Zond 2 Mars Probe in 1964
LES 6[158,174]	0.027[180] (1 Hz)	200–590	2.5[175]	26	320[180]	1.85	—	—	Flight in Sept. 1968 on LES 6 satellite; 8900-h operation[162]; 12×10^6 shots[180]
NOVA[158,162]	0.3–0.37[161,179]	300[179]–850	30	375	2200,[161] 4400[162]	20	7.1[162]	0.45	13×10^6 shots; flight in early 1980s on NOVA satellites
SMS[160]	0.14–0.16 (0.8–1.7 Hz)	400	10–22 (0.8–1.7 Hz)	128–169 (1.7–0.8 Hz)	1720	8.4	—	—	13×10^6 shots[180]; flight on SMS satellite in 1974.
LES 8/9[158]	0.3[180]–0.6[179]	1000	25.5 (1 Hz)	297	7320	20	7.33[179]	0.75	34×10^6 shots,[168] also tested at 80 J per shot (0.25 Hz), $I_{sp} = 1450$ s and 1.5 mN-s I_{bit}[158]; not selected for flight

Millipound[158]	4.45	1210	—	22,300	—	750	—	—	Laboratory Model, capacitor failure at $<10^5$ discharges, spiral propellant bars
PPT-4[175]	—	850	—	150–450	—	5–15	—	—	Laboratory Model, U. of Illinois
PPT-5[175]	—	1000–5000	—	100–750	—	10–50	—	—	Laboratory Model, U. of Illinois
EOS-1[158,178,179]	—	650 (12 W)–1400 (70 W)	12–70 (1 Hz)	60–860	460	5–40	4.95	0.07 (×2)	Two PPT per module; I_{bit} varied by throttling capacitor charge duration; designed for New Millennium EO-1 S/C
μ-PPT[158,176]	0.002–0.03	—	1–20	2	—	<1	0.5	—	Laboratory prototype

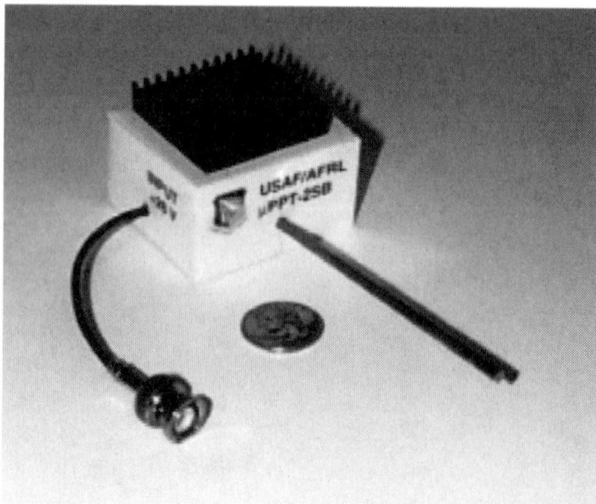

Fig. 12 AFRL micro-PPT. (Courtesy of Air Force Research Laboratory, Edwards AFB.)

electrode configuration, new capacitors (the EO-1 design provides much higher energies per shot than are required for the ION-F mission), and miniaturized electronics.[179] The AFRL micro-PPT provides impulse bits as low as 2 μN-s and thrust levels of between 2 and 30 μN at power levels of between 1 and 20 W.[176] A photograph of the micro-PPT is shown in Fig. 12. The thruster weighs as little as 0.5 kg[176] and has demonstrated 500,000 firings. These performances are achieved through a unique new design, placing the Teflon propellant into a coaxial electrode configuration comparable in size to standard TV coaxial cables. At present this thruster is still under laboratory development but has been targeted for use on the Air Force TechSat 21 constellation mission discussed in the introductory chapters.

3. *Evaluation, Issues, and Future Work*

As is the case for FEEP and colloid thruster designs, applications for both microspacecraft primary and attitude control may be contemplated for PPTs. Because of the small spacecraft masses, existing PPT designs may be able to provide sufficient thrust for primary propulsion applications. Assuming that for an input power of 20 W, roughly 0.3 mN of thrust may be produced by a PPT cycling at 1 Hz,[180] a thrust-to-spacecraft mass ratio of 0.015 mN/kg could be provided for a 20-kg Class I microspacecraft. This value is significantly lower than the beginning-of-mission value of a conceptual Europa mission using four DS-1 30-cm ion engines[146] based on estimations using preliminary data available at this stage of mission planning. Thruster lifetimes of about one year of continuous operation at 1 Hz could be expected if 30 million shots per thruster can be obtained as in the case of the LES 8/9 design. Using Table 13, a thruster with these design features may weigh about 5–7 kg fueled. Thus, PPT designs may be marginally suitable for Class I microspacecraft primary propulsion applications. For even smaller (Class II) spacecraft, miniaturized versions of PPT would be required, and at present it is not certain whether the required power levels would be compatible with Class II design constraints.

The ability of PPT thrusters to provide very small I_{bit} values, in the micronewton-second range, may make them candidates for microspacecraft attitude control as well. State-of-the-art PPTs appear to be applicable to 100-kg class spacecraft.

However, for Class I microspacecraft, miniaturized PPT versions will be required since thruster masses of conventional designs are too large. Even in the case of miniaturized PPT designs, sharing of capacitors between different thrusters in a cluster will likely be required to save mass. However, as in the case of FEEP thrusters, these PPT designs are not likely to meet slew rate requirements as outlined in Table 3. For conventional designs, a thrust level of 1 mN would require a PPT power input of 70 W at pulsing frequencies of 2–6 Hz, depending on the capacitor size.[180] These power levels can very likely not be afforded for attitude control purposes on any microspacecraft of Class I or lower. Thus, an additional thruster system would be required to perform higher-thrust slew maneuvers, adding system complexity, dry weight, and cost to the spacecraft design. A PPT attitude control system may, however, be sufficient if slew rate requirements could be relaxed or, in special cases, dropped, and sufficiently miniaturized hardware was available, such as thruster designs based on the AFRL micro-PPT technology, for example. In these cases, thruster contamination-related issues during constellation flying will likely require additional attention.

F. Resistojets

1. *Description of Technology*

In a resistojet a propellant, stored either in a gaseous or a liquid phase, is heated through conduction/convection from a heater element to vaporization, and the propellant is thermally exhausted through a nozzle. The simplicity of the device and the fact that liquid propellants may be used, allowing for compact propellant storage and reducing leakage concerns, make this concept attractive.

2. *Available Hardware*

Unfortunately, commercially available state-of-the-art resistojet technology[60] is far too heavy and requires far too much power to be useful for microspacecraft. Some work on small water resistojets was performed in the 1960s,[181–183] however, no heater power requirements were reported. More recently, however, new work on water and nitrous oxide (N_2O) resistojets has been conducted at the Surrey Space Centre at the University of Surrey in England, partly in collaboration with the U.S. Air Force.[77,184] Two systems, featuring the same heater configuration but using either water or nitrous oxide propellants, have recently been tested and delivered 150-s (H_2O) and 127-s (N_2O) impulse bits at a 100-W input power.[77] A nitrous oxide resistojet system developed for the 350-kg UoSat-12 spacecraft (built by the Surrey Space Centre as well) has delivered 125 mN of thrust at the 100-W power level.[77] The overall thruster weight is about 1.5 kg.[184] A photograph of the two resistojets is shown in Fig. 13.

A resistojet developed by the Russian Fakel Enterprise company was recently tested with nitrogen and xenon.[185] This device is targeted for attitude control on spacecraft where xenon propellant is used for other applications as well, e.g., primary ion or Hall thruster propulsion. It may not be too useful for very small microspacecraft (Class I and below) where cold gas propellant storage and leakage are concerns. The device was operated at power levels as low as 8 W. However, at these power levels I_{sp} values improved only marginally over those obtainable with cold gas propellants. For example, with nitrogen, 85 s was obtained,[185] vs about 65–70 s typically attainable with cold nitrogen propellant.

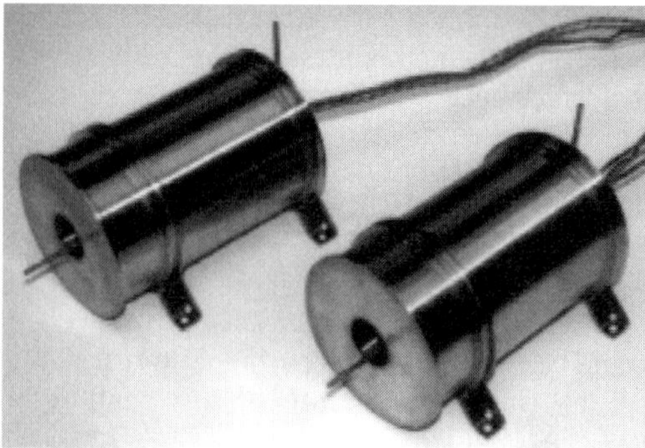

Fig. 13 Recent resistojet technology applicable to $\geq$100-kg-class spacecraft. (Courtesy of Surrey Space Centre, England.)

3. Evaluation, Issues, and Future Work

Currently available resistojet thruster hardware, such as the two Surrey concepts, may be applicable to 100-kg class spacecraft or larger where sufficient power levels (100 W) can be provided. To be used on smaller spacecraft (Class I and below), further miniaturization is required, requiring lower power devices. Power levels would have to be dropped by at least one order of magnitude for primary propulsion applications (drag makeup, small orbit changes) on Class I spacecraft. Use on Class II or smaller microspacecraft, or attitude control applications, would require still lower power levels, maybe as low as in the 1- to 5-W range.

However, if miniaturized resistojet thrusters can be successfully developed, they may represent an attractive attitude control option for microspacecraft. Liquid storage of propellants will reduce system weights from those of high-pressure cold gas storage systems and significantly reduce leakage concerns, which are two major advantages for microspacecraft applications. As an electrothermal thruster concept, these devices would likely produce higher thrust-to-power ratios than are obtainable from electrostatic or electrodynamic thruster options and thus may possibly be used for slew maneuvers as well. If combined with sufficiently fast valves, and outfitted with very small, microfabricated nozzle throats, very small impulse bits may be obtained. Work is currently under way on such microfabricated resistojet concepts at the Aerospace Corporation,[186] AFRL, in collaboration with USC,[187] and JPL,[188] and are discussed in Section VI.

Although water propellant is being used for test purposes in many cases due to ease of handling and safety concerns, power requirements for liquid propellant resistojets may be reduced by resorting to different propellants. Water is not the most suitable propellant for resistojet use due to its high heat of vaporization. Table 14, based on data from Ref. 69, lists several relevant properties of candidate propellants for a resistojet system. Of the propellants listed, ammonia and water immediately stand out due to their low molecular weight, which would result in relatively high I_{sp} performance. Of these two propellants, ammonia requires about half the heat of vaporization as water and would thus lead to lower power consumptions at comparable I_{sp} performances. An interesting observation was also made with regard to nitrous oxide propellant. When heating this propellant in a Surrey resistojet, nitrous oxide was shown to decompose thermally in a

Table 14 Properties of candidate resistojet propellants[69]

Propellant	Formula	Molecular weight, kg/kmol	Liquid density, g/cm^3	Heat of vaporization, kJ/kg
Ammonia	NH_3	17.0	0.6	1159.7
Propane	C_3H_8	44.1	0.49	339.3
Ethylchloride	C_2H_5Cl	64.5	0.92	388.1
Butane	C_4H_{10}	58.1	0.57	360.2
Freon 12	CCl_2F_2	120.9	0.98	141.8
Water	H_2O	18.0	1.0	2442.5
Hydrogenfluoride	HF	20.1	0.99	1505.9
Methanol	C_2H_3OH	44.0	0.79	1099.3
Methylchloride	CH_3OH	51.0	0.91	376.5
Ethane	C_2H_8	30.0	0.56	313.7
Ethylmethylether	$C_2H_5OCH_3$	60.0	0.8	350.9
Monomethylamine	CH_3NH_2	31.0	0.77	873.8

self-sustained decomposition reaction.[77] Gas temperatures as high as 1200–1600°C may be achieved in such a reaction without heat input required during steady-state operation[77] (heat would be required only to initiate the reaction). Nitrous oxide also has a high vapor pressure that may be exploited for self-pressurization in feed systems[77] similar to ammonia.[84]

VI. Emerging Technologies: MEMS and MEMS-Hybrid Propulsion Concepts

A. Case for MEMS Propulsion and Its Challenges

In the ongoing effort to reduce propulsion system weight and volume, several new MEMS or MEMS-hybrid propulsion concepts have recently emerged. Virtually all of these are still in very early stages of development, either still undergoing feasibility studies or only recently having passed proof-of-principle tests but requiring substantial additional development. These concepts rely on radically new design approaches, involving micromachining, or MEMS technologies. At present, most MEMS-based fabrication technologies are silicon based and thruster components may thus be placed on silicon "chips." However, non-silicon-based fabrication methods are being explored as well.

Although MEMS is not a prerequisite for such highly miniaturized micropropulsion concepts—indeed many concepts, such as the AFRL micro-PPT, the MIT mini-Hall thruster, and some of the new, miniaturized cold gas thruster technologies appear to meet microspacecraft design constraints without resorting to such technologies—MEMS does offer several unique advantages. First, MEMS allows extremely small and lightweight devices to be constructed. For microspacecraft in the very small mass ranges, such as Class II designs, these technologies may be a prerequisite to meet the stringent mass and volume constraints, in particular, for attitude control applications where multiple thruster units will be required.

However, it has been argued within the micropropulsion community, and justifiably so, that while MEMS-based components may achieve very high degrees of miniaturization, the finally assembled and packaged device, featuring the necessary protective covers, electrical interfaces, and propellant feeds, may not be significantly smaller and lighter than more conventionally machined, miniature thruster components already available today. For example, cold gas thrusters featuring miniature solenoid valves weighing as little as 7 g have been successfully fabricated using non-MEMS fabrication techniques.[78] However, using MEMS-based propulsion components will allow for an unprecedented degree of integration among different propulsion components as well as the required control electronics and, thus, offer the potential for significant additional mass and volume savings even over miniature non-MEMS components. For example, a MEMS-based thruster may be directly bonded to a MEMS-based valve or filter chip. Control electronics may be integrated on the flow component chips, or provided on separate chips banded to the former, and the whole unit may then be packaged into an extremely compact module with minimal external interfaces, easing, and therefore reducing, the cost of integration into the microspacecraft (see Fig. 14). The latter point will be of increasing importance for Class II and Class III microspacecraft. Here, it may no longer be feasible or practical, and certainly not cost-effective, to route miniature feed lines throughout the microspacecraft and individually weld and plumb components together.

Finally, for certain applications, MEMS-based components may offer performance advantages, such as in the case of attitude control thrusters. The need to provide extremely small impulse bits may be addressed by microfabricating very small nozzle throats through which a propellant may be thermally expanded. If paired with fast-acting, leak-tight MEMS-valve technology (in need of significant development[6]), achievable impulse bits may be reduced significantly over values obtainable with current chemical or cold gas technology. Such thermal expansion–based chemical or electrothermal thruster concepts may also exhibit higher thrust-to-power ratios than are available with electrostatic or electromagnetic thruster concepts and may, thus, be able to produce the thrust levels required for microspacecraft slew maneuvers within anticipated microspacecraft power constraints.

On the other hand, MEMS-based propulsion concepts will face many design challenges. Since silicon is the primary building material of choice in MEMS technology today, based on the considerable heritage obtained with this material, compatibility issues between silicon, or certain thin films deposited onto silicon (such as silicon oxide and silicon nitride), and various propellants will need to be explored. Note, however, that many metals may also be deposited onto silicon, likely alleviating many such concerns to a large extent. Non-silicon-based microfabrication methods may also need to be explored to avoid some of these design challenges. This approach, however, will require a substantial amount of additional basic fabrication process development.

Silicon is also a very good thermal conductor (with a thermal conductivity of about 150 W/mK) and its use may thus lead to thermal design challenges since thruster applications generate heat that will need to be contained to reduce thruster efficiency losses. Further, although silicon has very high yield strengths, approaching those of stainless steel, silicon is brittle, and internal pressurization, such as required in propulsion applications, will need to be examined carefully. Recent tests performed under static (nonvibrating) test conditions, however, have

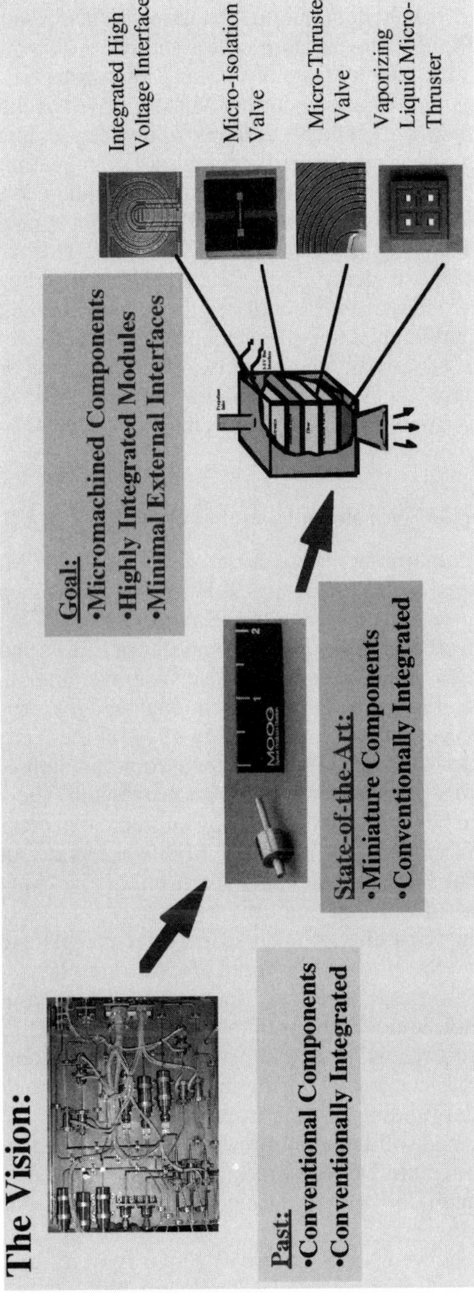

Fig. 14 Vision of novel integration/system miniaturization schemes enabled through the use of MEMS technologies.

yielded very high burst pressures of almost 3000 psig for an internally pressurized isolation valve concept.[189] Integration between silicon and nonsilicon (mostly metal) components will also need to be addressed since propellant tanks, due to the required size, will likely continue to be fabricated using more conventional metal-machining techniques for the foreseeable future.

Finally, MEMS-valve technologies are in need of substantial further technological improvements. For example, current MEMS valves exclusively are using hard seats (silicon-on-silicon), which, combined with lower sealing forces, leads to substantial leakage concerns, in particular, for gaseous applications.[6] More robust designs, featuring soft seats and fast actuation times, will be needed.

In the following, several currently ongoing MEMS-propulsion development activities will be introduced. As mentioned, many of these concepts are still in very early development stages. In many cases, these devices are undergoing feasibility studies at present and, if successful, will require substantial additional work to be considered for flight applications. Nonetheless, these devices currently represent some of the most advanced micropropulsion developments ongoing, pushing miniaturization to unprecedented levels, and may ultimately prove to be critical in the realization of very small microspacecraft concepts (Class II or III).

B. Brief History of MEMS Propulsion

According to the literature survey conducted in this study, MEMS-based propulsion was first introduced by Mitterauer in 1991[121] in the form of a proposed microfabricated FEEP thruster concept based on FEA technology. The purpose of that design study was to decrease further the mass and size of FEEPs and take advantage of the fact that critical thruster components, such as the emitter slit, were already sized in the micrometer range even for conventional designs. Apparently, no test units were built, however. Shortly thereafter, in 1994, at the Aerospace Corporation, Janson[18] extended the vision for MEMS-based propulsion concepts to other devices, such as MEMS-based resistojets and ion propulsion. These activities were part of a more comprehensive study to investigate microspacecraft designs based entirely on MEMS fabrication techniques.[19] Microspacecraft concepts had also been studied at JPL for some time at this point, primarily by Jones[12–17,41] as well as others.[10,11] As part of this ongoing activity, a study was conducted at JPL in 1995 to investigate the feasibility of microspacecraft concepts with masses of between 15 and <1 kg.[4,9] Several MEMS-based propulsion concepts were conceived and proposed in the course of that study, including MEMS-based phase-change thruster concepts using liquid[188] and solid propellants, as well as microvalves.[189] At about the same time, MEMS-based thruster concepts were conceived and pursued in Europe. ACR Electronic Company in Sweden began to develop cold gas thruster concepts under funding by the European Space Agency (ESA),[190–192] and in France micromachined solid motor arrays were being studied at the Laboratoire D'Analyse et d'Architecture Des Systèmes (LAAS) at the Centre National de la Recherche Scientifique (CNRS) under funding by the Centre National d'Etudes Spatiales (CNES).[93,193]

These early activities were soon followed by a flurry of different micropropulsion projects at various private companies, including TRW,[194] Marotta,[82] Phrasor Scientific,[153] Honeywell,[195] and SRI[196] in the United States and Centrospazio[136,148] in Italy; at university laboratories, such as MIT,[110,116,115,118–120,197,198] USC,[106] and

Princeton University[199]; and government institutions, such as AFRL[176,187] and NASA Glenn Research Center[57,200]; in addition to continued work performed by the aforementioned players, i.e., the Aerospace Corporation,[18,186,201] ESA,[190–192] LAAS/CNRS,[93,193] and JPL.[107–109,188,189] In the following sections microthruster concepts being investigated at these institutions are discussed. It should be noted that work on MEMS-based valve designs is currently also under investigation at various institutions,[6,189,190–192,201] however, it could not be accommodated in this survey. A view of these technologies is provided in Ref. 6.

C. MEMS-Based FEEP and Colloid Thruster Concepts

MEMS-based FEEP thruster versions have been considered for quite some time in Austria and Italy,[121,136] and, more recently, MEMS colloid thruster versions have also received attention.[153] Studies on micro-FEEP thrusters are currently being conducted by Centrospazio[136,148] in Italy and, at a very preliminary stage, by SRI[196] in the United States. Phrasor Scientific[153] is exploring micro-colloid designs, and work on these devices is also being performed at Busek Company in collaboration with MIT[116] and at Stanford.[155] In the MEMS versions of these concepts, FEA technology derivatives are being used. An FEA consists of an array of microfabricated, conical tips placed opposite a gate electrode (see Fig. 15). In a conventional FEA, a negative voltage is applied to the emitter tip and a positive voltage to the gate electrode. Through field emission from the very sharply pointed emitter tips (radius of curvature of a few tens of angstroms), electrons can be extracted from the tips, accelerated in the electric field between the tip and the gate electrode, and emitted from the array through an aperture in the gate electrode.

A MEMS-FEEP or colloid design will likely use modified FEA designs. In some cases, a so-called microvolcano configuration was proposed to replace the

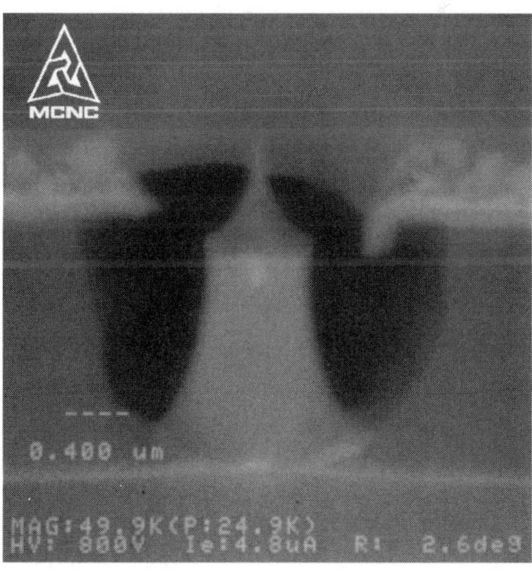

Fig. 15 Field emitter tip. Note 0.4-μm scale, center-left. (Courtesy of MSNC.)

emitter tip.[196] These structures, pioneered at SRI, are of dimensions comparable to conventional FEA emitter tips, however, they are hollow (hence the term "volcano"). By placing a gate electrode opposite these structures, very strong electric fields can be generated near the "rim" of these microvolcanoes. Capillary feed of the propellant through the volcano structure will expose the propellant to these strong fields at the emitter exit and lead to field emission as in conventional FEEP or cluster sprays as in colloid devices. To facilitate ion emission, voltages need to be reversed compared with a conventional FEA, i.e., the positive voltage now has to be applied to the emitter and the negative voltage to the gate electrode. "Needle"- or "pin"-type emitter arrays may also be envisioned using more conventional FEA emitter tips and external wetting of the tip with propellant, similar to the LMIS concept.[140]

MEMS-based FEEP or colloid designs offer the advantages of additional size and weight reductions, the ability to create modular architectures for these thruster types, and the potential for integration of these MEMS-based devices with other subsystems (e.g., control electronics) in future microspacecraft designs.[136,148] One of the major design challenges for a MEMS-based FEEP or colloid concept is the ability to stand off the required high voltages for particle acceleration. Note that while field emission (FEEP) or charged cluster generation (colloid thruster) may be achieved at significantly lower voltages than in conventional devices because the required electric fields are generated over shorter distances, these devices may still require high acceleration voltages to achieve desired specific impulses. Recent results seem to indicate that electric breakdown voltages for certain silicon oxides (which is a material typically used for electric insulation between the emitter and the gate electrode) is about 700 V/μm for electric breakdown through the bulk material and 200 V/μm along exposed oxide surfaces.[107] These breakdown field strengths vary with oxide thickness, and thinner oxides show increased breakdown field strengths but not necessarily increased breakdown voltages.[107] Careful accelerator electrode design will be required to deal with these design constraints.

Additional challenges to be overcome will be the investigation of material compatibility between typical FEEP or colloid propellants and the MEMS material of construction, as well as wetting properties of this material with respect to the propellant of choice. For these reasons, the microvolcano or emitter tip may have to be coated with a metal to result in wetting conditions similar to those of conventional devices.[136] Detailed studies of these structures with respect to sputter erosion due to ion bombardment will be required, and the possibility of shorting microfabricated field emitter arrays with conducting propellant needs to be investigated carefully.

D. Micro-Ion Engine Concepts

Although FEEP and PPT technologies exist and have reached high degrees of maturity, the development of other micro-electric propulsion technologies appears desirable. Micro-ion engine technology, if it can be successfully developed, will be characterized by a unique combination of high specific impulses, the use of inert, noncontaminating propellants, potentially higher thrust-to-power ratios than obtainable with FEEP or PPT systems, and, unlike in the case of the inherently pulsed PPT devices, a continuous mode of operation. Work on various micro-ion engine concepts and their components is under way at USC in collaboration with

AFRL[106] and at JPL.[107–109] Engine diameters currently considered are in the 1- to 3-cm range, after initial studies conducted at MIT indicated that much smaller engine dimensions may lead to excessive thruster efficiency losses.[110] Thrust levels in the submicronewton to few-micronewton range are being targeted. However, to arrive at a functional engine concept of this small a size, several feasibility issues will need to be addressed and overcome. Among these are 1) the sustainability and efficient operation of high surface-to-volume ratio plasma discharges, 2) the replacement of hollow-cathode technologies with lower-power-consuming and easier-to-miniaturize field emitter-based cathode systems to function as engine cathodes and neutralizers, 3) the fabrication and operation of miniature accelerator grid systems, 4) the feasibility of fabrication and operation of miniaturized power conditioning units, and 5) suitable miniaturized feed system components.

A "hollow-anode" ion thruster concept has been proposed by USC in collaboration with AFRL.[106] In its current configuration, the discharge chamber consists of a 1-cm-diameter hemispherical dome that is sealed by a planar orifice plate. A 1-mm-diameter orifice is located in the center of this plate. The dome and orifice plate are electrically insulated from each other. A negative voltage is applied to the dome, and a positive voltage to the orifice plate. An electric field is set up between the dome surface and the orifice plate, having its highest magnitude near the central orifice. Electrons emitted from the cathode are channeled into this maximum field region and ionize the propellant gas. The advantage of this concept is that no magnetic fields may be needed for electron confinement, since strong electric fields focus the electron paths. However, using argon as a test gas, discharges generated required a discharge voltage of several hundred volts. Gas pressures in the discharge chamber were 0.75 Torr or higher. Through Langmuir probe measurements in the plume, ionization fractions of about 0.1% were determined. Higher ionization fractions may be achieved inside the engine itself or by increasing the electron current emitted from the cathode surface through the use of FEA cathode arrays, for example, lining the inner dome surface. To complete the ion engine design, a grid system as well as a neutralizer will be required. Initial studies, however, were focused on the plasma generation process only. Hollow-anode concepts, but of a different design, are also being studied at NASA Glenn Research Center.[57]

The miniaturization of key ion engine components, such as cathodes and neutralizers, as well as grids, is crucial for any micro-ion engine concept to succeed. Studies on these micro-ion engine components are therefore being performed at JPL, including field emitters arrays to be used as ion engine cathodes and neutralizers,[108,109] and micro-ion engine grid technologies.[107] The challenge in using FEA technology in plasma discharges lies in the exposure of the fine, negatively biased emitter tips to the hostile plasma environment (see Fig. 15). Sputter erosion due to plasma ion bombardment may cause tip erosion and potentially change emission characteristics drastically. As the emitter tip is being blunted through erosion, required emission voltages may increase significantly. At JPL, partly in collaboration with the University of Michigan,[108] a feasibility study is under way to explore different emitter tip materials and their operating characteristics at various background gas pressures. In addition, MEMS-fabricated grid structures, to be integrated with the emitter array, are being studied. These so-called cathode lens and ion repeller (CLAIR) arrays are designed to prevent ions from reaching the emitter tip by means of repelling electrostatic forces and also provide control over the electron energy independent of the FEA gate voltage.[108,109]

E. MEMS-Based Microresistojet Concepts

MEMS-based microresistojet developments are being pursued by the Aerospace Corporation,[186,201] AFRL,[187] and JPL.[188] While the Aerospace and JPL concepts are laminar flow resistojet concepts, the AFRL concept is unique in that it relies on molecular flow in the heat exchanger region of the thruster and is thus termed the free molecular microresistojet (FMMR).[187] The JPL concept was termed the vaporizing liquid microthruster (VLM),[188] identifying it as a phase-change thruster concept that focuses on the use of liquid propellants, vaporizing them on demand to generate thrust, thus avoiding propellant storage and leakage concerns.

In the laminar flow concepts, the propellant is forced through a microfabricated channel etched into a silicon chip. In the Aerospace Corporation design, heaters are deposited on a free-standing membrane, with flow passing over and under this membrane. In the current layout of the VLM concept pursued at JPL, heaters are deposited onto two opposite channel walls (see Fig. 16). Simple, anisotropically etched, square-shaped nozzle profiles are being used at this point, serving as place-holders for more complex nozzle shapes explored elsewhere.[192,193] In the Aerospace concept, the heater element is made from polysilicon; in the VLM concept both polysilicon and gold heaters were explored. Gold heaters, due to their lower electrical resistances, will result in lower voltage requirements. Several thruster chips have been fabricated both at the Aerospace Corporation and at JPL, and at JPL vaporization of water propellant was recently demonstrated at input power levels as low as 2 W and voltages of 2 V, well within Class I microspacecraft capabilities.[188] However, no thrust and mass flow rate measurements have been performed yet due to previously unavailable diagnostics suitable for these devices. Efforts to address these issues are under way, in part in collaboration with Princeton University, providing thrust stand expertise.

In the AFRL FMMR[187] concept, the thruster is operated at very low gaseous plenum pressures, typically between 50 and 500 Pa. Provided that thruster dimensions can be kept small (1- to 100-μm channel width), a free molecular flow may be set up inside the thruster. The thruster dimensions are chosen in such a way that a gas molecule entering the thruster must impinge on a heated surface first before it can exit the nozzle (see Fig. 17). The nozzle expansion angle in turn is chosen large enough such that an exiting gas molecule cannot collide with the nozzle walls through direct line-of-sight movement. The exiting molecule will therefore maintain a kinetic energy equivalent to the heater temperature, being the maximum temperature inside the device, thus maximizing the achievable specific impulse. (Gas collisions inside the nozzle, which potentially could change the velocity vectors of gas molecules and lead to impact on nozzle walls, are eliminated by maintaining molecular flow conditions inside the nozzle.) The extremely low feed pressures will ease valve leakage requirements.

MEMS-based microresistojet concepts appear applicable to both Class I and Class II microspacecraft attitude control. Leakage and propellant storage issues will be avoided due to liquid propellant use, or low-pressure storage in the case of the FMMR. Lighter and smaller propulsion systems and, given the large impact that propellant tanks may have on spacecraft size and layout, in turn, smaller and lighter spacecraft may result. Reducing leakage concerns may increase mission reliability over cold gas-based attitude-controlled spacecraft. On the other hand, additional power will be required to effect phase transition and/or heating of the propellant, which may be of concern for extremely power-limited microspacecraft.

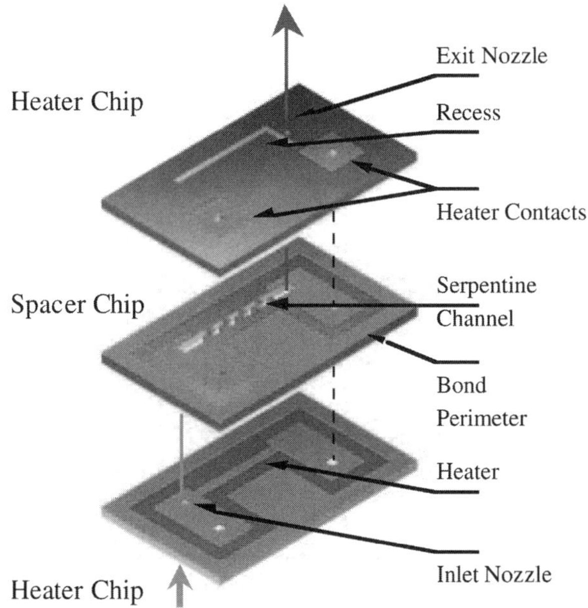

a) VLM concept

b) Cross section of VLM showing flow path (chip size is 1 × 1 cm^2)

Fig. 16 Vaporizing liquid microthruster (VLM) concept.

F. MEMS-Based Subliming Solid Microthruster Concept

Subliming solid thruster concepts have been studied in the past and substantial development work was performed in the 1960s.[197] Main contributors to the field were Rocket Research[202–205] (now Primex), Lockheed[206–211] (now Lockheed–Martin), and NASA Goddard.[212,213] Some work was also performed at Aerospace Industries,[214] NASA Glenn Research Center (formerly NASA Lewis Research

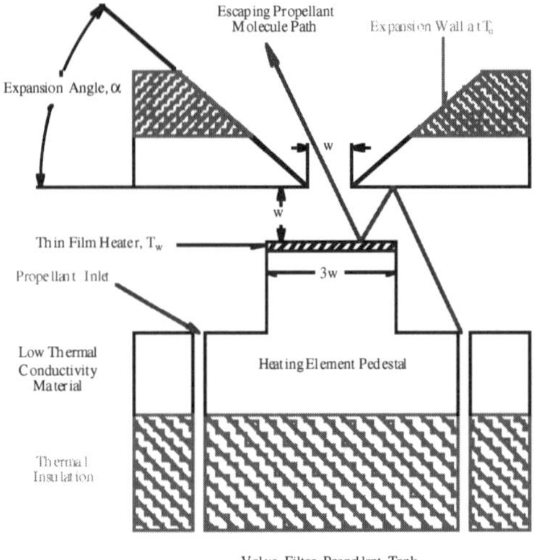

a) FMMR concept

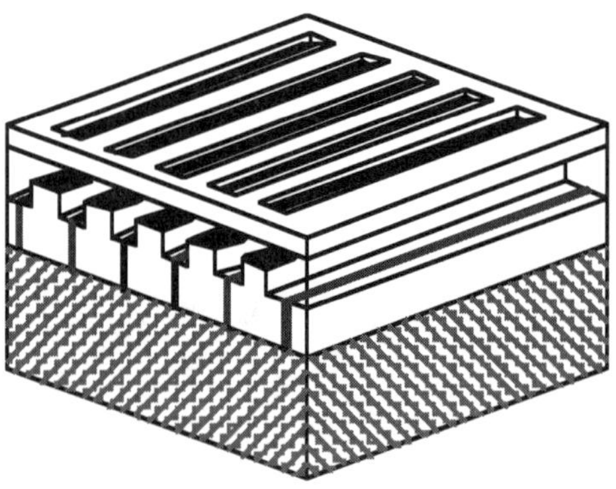

b) FMMR isometric view

Fig. 17 Free molecular resistojet concept. (Courtesy of Air Force Research Laboratory, Edwards AFB.)

Center),[215] and the Martin–Marietta Company[216] (now Lockheed–Martin also). In the subliming solid thruster concept, a solid propellant is chosen with a high sublimation pressure, such as ammonium hydrosulfide (NH_4HS) or ammonium carbamate ($NH_4CO_2NH_2$). Upon heating, gas pressure builds up inside the propellant tank and the vapor is vented through a valve and nozzle to produce thrust. An approximately 50- to 75-s I_{sp} may be obtained with the propellants above, suitable for attitude control purposes. Unfortunately, the aforementioned propellants are toxic and may be absorbed by touch through skin. Other subliming solid propellants exist, however, and may result in lower sublimation pressures.

The simplicity of this design and the solid storability of the propellant appear to lend themselves easily to miniaturization. Based on the 1960s work in this area, a subliming solid microthruster (SSM) concept was proposed at JPL using MEMS technology.[217] This concept features a very simple thruster chip design containing a micronozzle and a micromachined comb filter. This filter is designed to prevent solid propellant particles, which may drift into the thruster chip under zero-*g* conditions, from blocking the nozzle. Due to funding limitations, and focus on other micropropulsion concepts, work on the SSM concept is temporarily on hold. If it was successfully developed, this concept could serve Class I, II, and III microspacecraft attitude control.

G. MEMS-Based Cold Gas Thruster Concept

A MEMS-based cold gas thruster concept is being developed by ACR Electronic Company and Uppsala University in Sweden under ESA funding.[190–193] A module of four cold gas thruster with integrated piezoelectric valve assemblies is machined into a silicon wafer. By resorting to multilayered piezoelectric actuators, valve actuation voltages can be dropped to about 24 V,[191] significantly lower than the 100–200 V typically required for piezovalves.[6] The entire thruster quad assembly, including electronics and housing, is expected to weigh about 70 g and is about 40 mm in diameter. Thrust values of approximately 0.1 mN per thruster are being targeted.[190] Work on this cold gas thruster concept includes development of silicon-based micromachined nozzles featuring rotationally symmetric, cone-shaped contours (compare with Fig. 18).[192] Novel laser-based etching techniques are used in the machining process.[192] In this technique, a laser beam heats and partially melts the silicon surface to be processed. A chlorine gas then reacts with the molten part of the silicon and forms volatile silicon chlorides that can be removed from the etch site. Fabrication procedures allow only one half of the nozzle to be machined at a time. The two halves have to be joined through wafer bonding in the final fabrication step. The cold gas thruster concept is small enough to meet both Class I and Class II microspacecraft attitude control needs. Cold gas storage and leakage issues may continue to pose design challenges.

H. MEMS-Based Bipropellant Thruster Concept

A MEMS-based pump-fed bipropellant thruster concept is being studied at MIT.[57] This very aggressive project seeks to demonstrate the feasibility of a

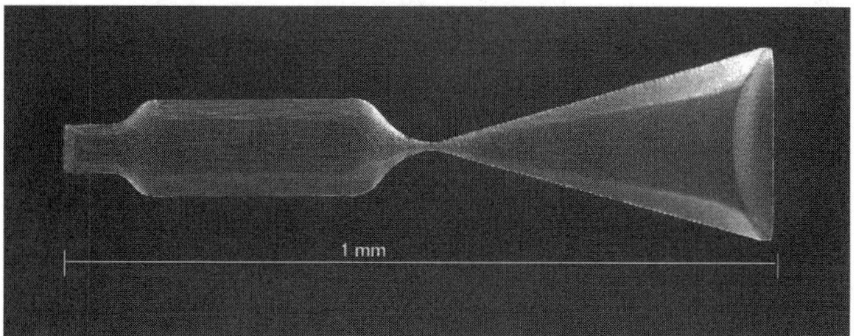

Fig. 18 Laser micromachined nozzle for MEMS cold gas thrusters (Courtesy Angstrom Space Technology Centre, Uppsala University, Sweden).

bipropellant thruster on a silicon chip. Thrust values of 15 N are targeted at mass flow rates of 5 g/s.[57] The fuel and oxidizer pumps would be required to provide about 75 W of pump power each.[57] Numerous design challenges need to be overcome including combustion on MEMS scales, heat losses into the chip structure, and resulting cooling needs, as well as issues related to bearing friction for the fuel and oxidizer pumps, given the required very high numbers of revolutions for these pumps. Benefits of the device would be extremely high thrust-to-weight ratios, estimated as 2000:1.

I. Digital Microthruster Array Concepts

So-called digital microthruster arrays, consisting of a multitude of microfabricated, single-shot thrusters placed onto a wafer, have recently been constructed. These devices may be used for attitude control, providing predetermined impulse bits by firing individual thrusters. Each thruster can be fired only once. No fewer than four digital thruster concepts are currently under active development. These are being pursued by the TRW Company in collaboration with the Aerospace Corporation and the California Institute of Technology,[194] the French LAAS at CNRS,[93,193] the Honeywell Company in collaboration with Princeton University,[195] and the NASA Glenn Research Center.[200] In addition, digital thruster array concepts have also been proposed and discussed independently with the author by Breuer,[218] Lang,[219] and Langmeier.[220] However, at present, due to the already substantial development effort in this area, no actual thruster development work has resulted in the case of the last three references.

In the French activity,[93,193] glycide azide polymer (GAP) propellant is loaded into individual cavities machined into a ceramic wafer. The cavities are sealed on one side with a Pyrex wafer, forming the back of the array. The front (nozzle) side of the cavities is sealed with a silicon nitride/oxide membrane featuring integrated polysilicon resistors. The resistors, one to each cavity, can be addressed individually and act as ignitors. Resistive heating of the ignitor causes the propellant to ignite. The burn progresses from the nozzle exit plane backward into the cavity. This approach prevents portions of unburned propellant to be ejected from the cavity, as might be the case had the burn progressed in the reverse direction. As the combustion pressure increases, the sealing membrane including the assigned heater element for that particular cavity is blown away, and the reaction products exit the nozzle. This nozzle is machined as part of the cavity and is initially filled with propellant as well. Upon removal of this portion of the propellant, the reaction products of the remainder of the propellant can expand through this nozzle. At present, up to 16 cavities have been placed onto a wafer and packaging densities of up to $3 \times 10^3/\text{cm}^2$ have been demonstrated.[193] Burn durations per cavity are fairly long, approximately 4 s. Peak thrust values between approximately 5 and 10 mN have been measured per cavity, depending on the throat diameter, typically between 1 and 2 mm in recent experiments. It is being estimated that the range of deliverable thrust values may be expanded to values between a few micronewtons and hundreds of millinewtons through proper selection of cavity and nozzle dimensions.

In the TRW/Aerospace work,[194] thruster arrays are targeted that may consist of up to 10^4–10^6 thrusters per 10-cm-diam wafer. At present, however, 19 cavities have been machined into a single chip (see Fig. 19). In current designs, silicon wafers are used in the array fabrication. Each individual thruster consists of a cavity,

Fig. 19 Digital thruster array by TRW/Aerospace Corp. (Courtesy of Aerospace Corporation.)

etched into the wafer, sealed on one side by a wafer containing a polysilicon heater element and on the other by a thin silicon nitride membrane. The cavities are filled with a liquid or solid propellant that will be vaporized or ignited upon heating. Once a sufficiently high chamber pressure is reached, the thin silicon nitride membrane bursts and the thruster produces an impulse bit, the magnitude of which will be determined by the amount of propellant loaded into the plenum and the propellant exit velocity. In present designs, heater elements are placed at the bottom of the cavity, opposite the nozzle and membrane.

In the Honeywell design,[195] which at the time of this writing consists only in a conceptual form, the thruster cavities are to be filled with a liquid propellant. However, both fuel and oxidizer cavities, separated by a thin silicon nitride membrane and stacked one on top of the other, are to be machined. Rupturing the thin separating membrane by pressurizing the fuel plenum due to heating causes the propellants to mix and react. As pressure builds due to the reaction, a second membrane, sealing the entire fuel/oxidizer cavity stack, ruptures and the reaction products are released to produce a discrete impulse bit. In this concept, up to 10^6 thrusters are to be placed on a 4-in. silicon wafer, each thruster unit providing an estimated impulse bit of about 3 μN-s. The propellant mass per cavity will be about 1.6 μg. It is estimated that about 10 mW will be required for individual thruster actuation. The total array mass has been estimated as 2.4 g. Due to the combustion of a fuel and oxidizer, it is hoped that relatively high specific impulses of about 200 s will be reached.

The thruster array being considered at NASA Glenn uses solid propellant pellets.[200] These pellets are to be placed into individual, microfabricated thrust chambers and ignited there. As the propellant, LAX 112, a gas generator compound consisting of 3,6-diamino-1,2,4,5-tetrazine-1,4-dioxide ($C_2H_4N_6O_2$), is currently being studied. Upon ignition, by means of heat addition through either a laser or a heated wire, this compound decomposes into nitrogen, hydrogen, and carbon monoxide.[200] The theoretical Isp performance for this compound is 244 s assuming frozen flow.[200] Several propellant decomposition tests have been conducted so far using wire and copper plate heating, as well as laser ignition.[200] At present, partial propellant decomposition has been achieved. Other compounds are currently under investigation to achieve higher degrees of propellant decomposition.[200] The fact that no sealing membrane is used would avoid the generation of debris around the spacecraft.

Advantages of digital thruster arrays are their relative simplicity, not requiring micromachined valves or complex feed systems. Thruster actuation is relatively simple. The arrays represent a complete propulsion system, including its propellant tanks. However, while individual thruster units are micromachined, overall array dimensions may be quite large. In some cases 4-in. silicon wafers are being envisioned. Obviously, smaller wafers could be fabricated, however, this would limit the number of available thrusters and impulse bits. To a large extent, the impulse bits delivered will depend on the cavity size. Thus a larger number of small impulse bits or a smaller number of large impulse bits can be provided per wafer. If packing densities of 10^6 thrusters per 4-in. wafer can be realized, the number of thrusters per unit surface area will be about 12,000/cm^2, neglecting edge effects. Thus, 10,000 units would require a surface area of about 0.8 cm^2, or a circular wafer about 1 cm in diameter. Ten thousand impulse bits may be sufficient for most missions and a 1-cm wafer coupon may easily be placed on Class I and Class II microspacecraft. However, if a packaging density of only 10^4 thrusters per 4-in. array could be realized, the required array area would correspondingly increase by a factor of 100. Such relatively large arrays, however, may cover substantial fractions of the microspacecraft surface area, competing with other subsystems, such as power (solar cells), communication (antennas), and scientific instruments. Thus, very high packaging densities will be essential for this concept to be applicable to microspacecraft, unless the required number of impulse bits could be reduced significantly, as may be the case for short-lived detachable microprobes.

Deliverable impulse bits may be very small for high-packaging density wafers and rapid slew maneuvers may not be possible. If additional thrusters can be provided on a larger wafer, a multitude of thrusters may be fired simultaneously in batches to generate larger, "quasi-steady" thrust values, which, however, may rapidly deplete the number of available thrusters. Since every microthruster in the array is located at a different moment arm with respect to the center of mass of the spacecraft, attitude control algorithms will need to take into account individual thruster locations. In the case of some digital arrays, debris will be generated around the microspacecraft due to fragmenting membranes and the potential impact of this debris on the mission will need to be studied on a case-by-case basis, in particular, for formation-flying applications.

VII. Evaluation of Existing Propulsion Technologies and Identification of Future Technology Needs

A. Evaluation of Existing Propulsion Technologies

In this section, the propulsion technologies surveyed in this study are evaluated in view of their application to Class I and II microspacecraft attitude control and primary propulsion. A coarse classification of the surveyed thruster technologies is attempted, placing each technology in one of three categories: those technologies that appear applicable to the task (the "yes" category), those that do not (the "no" category), and those that fall somewhat in between the latter two groups (the "maybe" category). The latter category serves to classify technologies that may fulfill some mission requirements, but not others, or those that have demonstrated

Table 15 Matrix of status of applicability of micropropulsion technologies to microspacecraft (status: spring 2000)

	Class I		Class II	
Technology	Primary	ACS	Primary	ACS
Bipropellant	No	No	No	No
Hydrazine	YES	No	Maybe	No
HAN-based	Maybe	No	Maybe	No
Peroxide	Maybe	Maybe	Maybe	Maybe
Monopropellant				
Solid	YES	No	Maybe	No
Hybrid	Maybe	No	No	No
Cold gas	No	Maybe	No	No
Warm gas	No	YES	No	Maybe
Tripropellant	No	Maybe	No	No
Ion	Maybe	No	Maybe	No
Hall	Maybe	No	No	No
FEEP	Maybe	Maybe	Maybe	Maybe
Colloid	Maybe	Maybe	Maybe	Maybe
PPT	Maybe	Maybe	Maybe	Maybe
Resistojet	No	Maybe	No	Maybe
Digital thruster arrays	No	Maybe	No	Maybe

concept feasibility but require additional development work. It was felt that this rather coarse classification is sufficient at this point, underscoring the relative uncertainty of this evaluation given the unknowns of future microspacecraft mission designs and requirements, as well as future research and development work still required for many micropropulsion technologies. Results are summarized in Table 15.

It is important to note that the classifications listed in Table 15 are expected to change as time progresses. The micropropulsion field is an area in considerable flux due to a large number of new developments, and these developments or new requirements may drastically change the potential of a technology over time. Some technologies that are placed in the "no" category today may move into the "maybe" or "yes" category in future years as new ideas and concepts are being developed. On the other hand, technologies that are in a "maybe" category today may dip back into the "no" category as anticipations for that technology may not have been fulfilled. To emphasize this point, Table 15 has been dated, i.e., spring 2000. Finally, given these uncertainties, a certain degree of subjectivity in performing this evaluation cannot be excluded, and it is entirely possible that different investigators may arrive at somewhat different conclusions. It is within these limitations that the results of this evaluation should be viewed.

Inspecting Table 15, the lack of suitable thruster technology for microspacecraft applications becomes strikingly obvious. This is not too surprising a result, given that most of the technologies reviewed were developed for spacecraft much larger than the microspacecraft considered here, and those technologies that are being developed specifically with microspacecraft applications in mind are still in their earliest stages of development. Notable, however, is the progress that has been made in this area in only the last three years as demonstrated when comparing the evaluation results shown in Table 15 with a corresponding table in Ref. 5, representing an earlier edition of this study. In particular, in the Class II microspacecraft categories many more propulsion technologies now appear in the "maybe" category, demonstrating the pioneering work that has been conducted in the development of microfabricated and miniature propulsion technologies. Nonetheless, many feasibility issues still remain to be addressed for these technologies (see Section VI above).

Some technologies stand out, however, and do appear applicable to microspacecraft even in, or close to, their current form. For Class I primary propulsion applications, the smallest available hydrazine attitude control thrusters may be a suitable option if only small to intermediate delta-v's are required. These thrusters could thus serve as microspacecraft main engines.[41] Solid motors also may provide a low-to-intermediate delta-v capability within their current design limits, although thrust values may need to be reduced further and burn durations should be lengthened.

Other thruster options considered for low to intermediate delta-v needs, such as hybrid motors or nontoxic HAN-based thrusters, currently are still under development, however, could prove very beneficial if available. Hybrid motors that use storable propellants, and would use a liquid component that could also be used as attitude control propellant, may have significant system advantages. High-I_{sp} primary thruster options currently remain limited and appear applicable only under certain conditions. These include FEEP, PPT, and colloid thrusters. FEEP and PPT thrusters, for example, deliver rather low thrust-to-power ratios, possibly leading to excessive power demands for primary propulsion applications. Colloid thrusters, although having higher thrust-to-power ratios than FEEPs and PPTs, deliver fairly low specific impulses compared with other electric propulsion concepts, yet still require very high voltages, which may result in high PPU masses. The development of higher thrust-to-power ratio, high-I_{sp} thruster options appears desirable.

For Class I attitude control applications, the need for new propulsion hardware development is even greater. Currently, cold gas and warm gas[84] systems offer the best near-term potential. However, cold gas options face concerns regarding propellant leakage and tank size and weight and may be used only for missions of limited duration. In addition, based on the requirements listed in Table 3, delivered impulse bits may have to be reduced even further through the use of new valve technology or even smaller (micromachined) nozzle throats. PPTs, FEEPs, and colloid thrusters may be applicable for Class I spacecraft attitude control and are indeed capable of delivering very small impulse bits. However, if slew rate requirements as listed in Table 3 are to be met as well, power requirements and thruster sizes may increase dramatically. However, given the aforementioned current uncertainty regarding future microspacecraft mission requirements, mission scenarios may arise where these thruster options may prove sufficient.

For Class II applications, virtually no state-of-the-art propulsion technologies appears suitable for either primary or attitude control propulsion. Considering primary propulsion applications first, solid motors may possibly be miniaturized further to provide high-thrust propulsion for low to intermediate delta-v applications or miniature hydrazine thrusters[63] may be developed for the same purpose.

FEEP, PPT, and colloid options could potentially be used for high delta-v primary propulsion. However, power constraints will limit thrust levels to very low values in the case of FEEPs and PPTs, limiting available spacecraft accelerations. Colloid thrusters, as discussed, will deliver only limited specific impulses. Once again, the development of higher thrust-to-power, high-I_{sp} electric thruster options appears desirable for Class II primary propulsion applications as well.

Class II attitude control may represent one of the greatest design challenges considered in this study. Thrusters will be characterized not only by mall impulse bit capability or low thrust levels, but also by very small size and weight to be able to fit a sufficient number of units (depending on attitude control requirements) onto the spacecraft. Highly integrated, modular propulsion architectures may have to be explored, involving the necessary valve technology as well (see Fig. 14). These thrusters will also have to be able to operate within very low power levels. Miniaturized FEEP, colloid, and PPT versions likely are able to deliver required impulse bits, and recent activities aimed at microfabricating these thrusters may result in thruster sizes suitable for this application. However, slew rate requirements will likely have to be relaxed considerably from the values considered in Table 3 (180°/min) for these electric thruster concepts to be able to operate within Class II power constraints. Various MEMS-based thruster concepts currently under development may eventually be able to provide both minimum impulse bits and adequate thrust levels for slew. However, these thrusters are in very early stages of their development, and any assessment of their applicability thus remains highly uncertain at this point. Advances in MEMS-based propulsion technologies, however, may prove critical for the vision of highly capable and maneuverable Class II (<5-kg) microspacecraft to be realized.

B. Identification of Technology Needs

Given the aforementioned observations, several micropropulsion technology needs may be identified.

1. Highly Miniaturized Thruster Modules for Microspacecraft Attitude Control

Attitude control will be of extreme importance for future microspacecraft. While miniature spacecraft have been built in the past requiring no attitude control at all, or have been relying on magnetic torquers for attitude control in near-Earth orbits,[2] these have to be considered special cases. Design flexibility, in particular, also for interplanetary mission applications, as well as a likely desire to build more sophisticated microspacecraft in the future, able to point cameras or other instruments with sufficient accuracy, will likely require a propulsion-based attitude control system on most spacecraft.

Given that microspacecraft will be extremely volume constraint and a multitude of thrusters will be required for attitude control, highly miniaturized thruster modules featuring a high degree of integration between thrusters and other components will be required (see, e.g., Fig. 14). Novel microfabrication techniques may need to be explored to realize such devices. Attitude control thrusters will be required that are able to meet both small impulse bit requirements into the micronewton-second range as well as minimum thrust requirements into the millinewton range for slew maneuvers. Although microspacecraft requirements will still need to be refined, it appears that extremely miniaturized thrusters able to deliver higher thrust-to-power ratios than available with current FEEP or PPT designs would be very desirable. Specific impulse will be of lesser importance for most attitude control applications.

2. *Low-Power Chemical Propulsion Options*

Low-power chemical propulsion options will be required for highly power constrained microspacecraft, fast orbit transfers (where power requirements for electric propulsion would be too high), or orbit insertions around distant planets (where solar electric propulsion is not feasible). These thruster options may also be used in cases where delta-v requirements are low to intermediate and the absence of electric propulsion power requirements may lead to system advantages. Deorbit maneuvers, for example, likely to be of major importance for future near-Earth microspacecraft constellations, may be more simply served by a small solid rocket motor. While hydrazine and solid motor options applicable to Class I microspacecraft may already exist, future design refinements, such as lower-weight, lower-power valves in the case of hydrazine thrusters or lower thrust levels and longer burn durations in the case of solid motors, would clearly be desirable. Other thruster options deserve close attention. For example, hybrid motor options would allow for thrust control unobtainable with solid motors yet be more compact than a bipropellant system, and HAN-based thruster options would simplify ground handling procedures, thus reducing mission cost.

3. *Efficient, Higher-Thrust/Power-Ratio, Noncontaminating Electric Propulsion Options*

In the past, only a few conventional spacecraft have made use of high-I_{sp} electric propulsion options. However, this trend appears to have been broken with the recent successes of electric propulsion technology on both scientific and commercial missions and increased recognition that many of the "easy" space missions have been flown, leaving more difficult-to-reach targets, requiring a higher delta-v capability, to explore. This new mindset will likely affect microspacecraft mission requirements as well. Here, the need to conserve propellant by resorting to high-I_{sp} electric propulsion options may be even stronger given the mass constraints of microspacecraft. Other applications, such as attitude control of constellations of spacecraft or large inflatable spacecraft, may result in a need for micro-electric propulsion systems as well, able to provide small thrust levels quasi-continuously over long periods of time to offset solar disturbance torques or residual atmospheric drag, for example.

Already, some electric thruster options exist that appear applicable, namely, FEEP and PPT systems. However, there appears to be an additional need for higher thrust-to-power electric engine technology to limit power needs for applications such as slew maneuvers, constellation repositioning, and simply higher spacecraft accelerations. Both FEEP and PPT systems exhibit relatively low thrust-to-power ratios. In the case of FEEP systems this is due to high-I_{sp} operation caused by the fact that field emission and ion acceleration are not separate processes. The required high voltages to cause field emission lead to high specific impulses, decreasing the thrust-to-power ratio. In the case of PPTs, thrust-to-power ratios are low due to the low thruster efficiency.

Colloid thrusters appear to offer higher thrust-to-power ratios than FEEPs, however, they have limited specific impulses. Colloid thrusters may therefore be applicable for repositioning or attitude control applications but may have limited potential for high delta-v applications. Ion engines also feature higher thrust-to-power ratios, offer high specific impulses suitable for delta-v maneuvers, and have the added benefit of using inert (noncontaminating) propellants. Thrusters operating on noncontaminating propellants would be very useful for obvious reasons. However, micro-ion engine developments will likely be very challenging, requiring gaseous discharges to be maintained in small discharge volumes, potentially reducing engine efficiency and therefore thrust-to-power ratios.

Microfabricated (MEMS) FEEP thrusters may also be an alternative. Imagine a micromachined field emitter generating the field strengths required for field emission with much lower voltages applied over much shorter electrode gaps. A separate electrode set would provide additional ion acceleration to the desired specific impulse. Since emitter and acceleration stage are now decoupled, greater flexibility exists to tailor the specific impulse to lower values, potentially resulting in higher thrust-to-power ratios than obtainable with conventional FEEP designs.

In addition to these technologies, additional needs exist in the development of extremely small and lightweight feed system components, such as valves, tanks, etc., as well as novel integration schemes for these components to reduce the cost of microspacecraft. Finally, if landing and takeoff operations are considered for microspacecraft, a separate class of lightweight, high-thrust propulsion components will be required. Depending on the spacecraft mass and gravity field of the targeted planet, the required thrust levels may be quite a bit higher than the levels considered in this study. Micropropulsion options may therefore not necessarily be applicable since high thrust levels necessitate high flow rates and, thus, larger flow cross sections. Emphasis in these cases will therefore likely focus on mass reduction, rather than size reduction, over state-of-the-art components.

VIII. Conclusions

Existing thruster technologies were reviewed in view of potential applications for microspacecraft. Only a few of the currently existing thruster technologies appear to be applicable for spacecraft of the sizes considered here. For primary propulsion applications, small hydrazine thrusters and solid motors may provide intermediate to low delta-v capability. Thrust values of solid motors may have to be reduced further to avoid excessive spacecraft accelerations. FEEP, PPT, and colloid thrusters may possibly be used as primary propulsion devices, although higher thrust-to-power values would be desirable than are obtainable with current FEEP and PPT devices, and the specific impulse may have to be raised for colloid thrusters.

For attitude control functions, currently available cold gas systems approach the performance requirements imposed by Class I (5- to 20-kg) microspacecraft designs with respect to minimum thrust and impulse bit values. Impulse bits, however, will have to be lowered for Class II (1- to 5-kg) microspacecraft. In addition, leakage concerns exist for cold gas systems and the required high-pressure storage tanks will dominate microspacecraft design with respect to both size and mass, even for relatively benign attitude control requirements. Ammonia cold gas thrusters or hydrazine warm gas systems may provide fairly near-term solutions to the propellant storage and leakage problem. FEEP, PPT, and colloid options may produce the small required impulse bits, however, they may not be able to meet the slew rate requirements within the anticipated microspacecraft power budgets. For spacecraft with masses in the 1-kg range, virtually no suitable propulsion hardware exists, either for primary propulsion or for attitude control. In many cases, existing thrusters are larger than the spacecraft in question. However, new, microfabricated thruster concepts are currently under investigation that may be applicable in these cases.

Future technology needs for both primary and attitude control propulsion were identified, requiring the development of extremely miniaturized, highly integrated, low-I_{bit} attitude control thruster modules, which will also have to be able to meet the slew rate requirements; low-power chemical thruster options for orbit insertion, fast orbit repositioning, or deorbit maneuvers; and high specific impulse, high thruster-to-pacer electric thruster options for high delta-v missions or special attitude control applications.

Acknowledgments

The author would like to thank Salvo Marcuccio of Centrospazio in Italy, John Mahoney of Phrasor Scientific, Inc., David Meinhardt of Primex Aerospace Corporation, Robert Shotwell of the Jet Propulsion Laboratory, and Greg Spanjers of the Air Force Research Laboratory for reviewing the FEEP, colloid, HAN-based propellant, hybrid motor, and PPT sections, respectively, leading to the incorporation of many useful comments made by the respective reviewers. Their efforts are sincerely appreciated. The author would also like to thank Indrani Chakraborty, Colleen Marrese, and James Polk of the Jet Propulsion Laboratory, David Gibbon and Gary Haag of the Surrey Space Centre in England, Siegfried Janson of the Aerospace Corporation, Andrew Ketsdever and Greg Spanjers of the Air Force Research Laboratory, Vadim Khayms of Massachusetts Institute of Technology, Olwen Morgan of Primex Aerospace Company, and Robert Reinicke of Moog Space Products Division for providing many of the photographs and sketches. Finally, the author would like to thank Colleen Marrese for her useful comments and the many new inputs she contributed to this study and John Blandino for performing the initial impulse bit and thrust calculations for microspacecraft on which the data presented in this chapter were based. The work described in this chapter was conducted at the Jet Propulsion Laboratory under contract with NASA.

References

[1]Collins, D., Kukkonen, C., and Venneri, S., "Miniature, Low-Cost, Highly Autonomous Spacecraft—A Focus for the New Millennium," IAF Paper 95-U.2.06, Oslo, Norway, Oct. 1995.

[2]Fleeter, R., "*Microspacecraft*," Edge City Press, Reston, VA, 1995.

[3]"Cosmis Journeys, Structure & Evolution of the Universe Roadmap 2003-2023, NASA, NP-1999-11-184-GSFC, 1999.

[4]Mueller, J., "Propulsion Options for Microspacecraft," Jet Propulsion Lab. Internal Document, JPL-D-13444, Pasadena, CA, 4 Dec. 1995.

[5]Mueller, J., "Thruster Options for Microspacecraft: A Review and Evaluation of Existing Hardware and Emerging Technologies," AIAA Paper 97-3058, 33rd Joint Propulsion Conf., Seattle, WA, July 1997.

[6]Mueller, J., "Review and Applicability Assessment of MEMS-Based Microvalve Technologies for Microspacecraft Propulsion," *Micropropulsion for Small Spacecraft*, Progress in Astronautics and Aeronautics, Vol. 187, edited by M. Micci and A. Ketsdever, AIAA, Reston, VA, 2000, Chap. 19 (this volume).

[7]Janson, S., "Micropropulsion Activities at the Aerospace Corporation," *Proceedings, Formation Flying and Micro-Propulsion Workshop*, Lancaster, CA, Oct. 1998.

[8]Contreras, P., "L'Approche Franàaise du Domaine Micro/Nano Satellites," *Proceedings, ESA WPP-132, 2nd Round Table on Micro/Nano Technologies for Space*, ESTEC, Noordwijk, The Netherlands, Oct. 1997.

[9]West, J., "Microelectromechanical Systems (MEMS)/Nanotechnology Studies," Jet Propulsion Lab. Internal Document, JPL-D-13302, Pasadena, CA, 12 Jan. 1996.

[10]Burke, J. D.,"Micro-Spacecraft," Jet Propulsion Lab. Publication 715-87, Pasadena, CA, 15 Oct. 1981.

[11]Staehle, R. L.,"Small Planetary Missions for the Space Shuttle," AAS Paper 79-288, Oct. 1979.

[12]Jones, R.,"Electromagnetically Launched Micro-Miniature Spacecraft for Space Science," Jet Propulsion Lab. New Technology Report NPO-17338/6846, Pasadena, CA, 18 June 1987.

[13]Jones, R. M., "Electromagnetically Launched Micro Spacecraft for Space Science Mission," *Journal of Spacecraft and Rockets*, Vol. 26, No. 5; Vol, 41, No. 10, Sept.–Oct. 1989, p. 338.

[14]Jones, R. M., "Coffee-Can Sized Spacecraft," *Aerospace America*, Oct.1988.

[15]Jones, R. M.,"Microspacecraft Missions and Systems," *Journal of the British Interplanetary Society*, Vol. 42, No. 10, Oct. 1989.

[16]Jones, R. M., "Think Small but in Large Numbers," *Aerospace America*, Vol. 42, No. 10, Oct. 1989.

[17]Jones, R. M., and Salvo, C. G., "Microspacecraft Technology for Planetary Science Missions," IAF-91-51, 42nd Congress of the International Astronautical Federation, Montreal Canada, Oct. 1991.

[18]Janson, S., "Chemical and Electric Micropropulsion Concepts for Nanosatellites," AIAA Paper 94-2998, 30th Joint Propulsion Conf., Indianapolis, IN, June 1994.

[19]Janson, S., "Mass-Producible Silicon Spacecraft for 21st Century Missions," AIAA Paper 99-4458, AIAA Space Technology Conference, Albuquerque, NM, Sept. 1999.

[20]Wolf, S., and Tuber, R. N., *Silicon Processing for the VLSI Era, Vol. 1. Process Technology*, Lattice Press, Sunset Beach, CA, 1986.

[21]Ticker, R. L., and McLennan, D., "NASA's New Millennium Space Technology 5 (ST5) Project," *Proceedings, IEEE Aerospace Conference*, Big Sky, MN, March, 2000.

[22]Martin, M., and Stallard, M. J., "Distributed Satellite Missions and Technologies—The TechSat 21 Program," AIAA Paper 99-4479, AIAA Space Technology Conf., Albuquerque, NM, Sept. 1999.

[23]Schilling, J., and Spores, R., "Comparison of Propulsion Options for TechSat 21 Missions," *Proceedings, Formation Flying and Micro-Propulsion Workshop*, Lancaster, CA, Oct. 1998.

[24]West, J., Personal communication, Jet Propulsion Lab., Fall 1995.

[25]Braun, B., Davis, R., Itchkawich, T., and Goforth, T., "MightySat 1: In Space," Paper SSC99-I-3, 13th Annual AIAA/USU Conf. on Small Satellites, Utah State Univ., Logan, UT, Aug. 1999.

[26]Zheng, Y., "The Gander Microsatellite Radar Altimeter Constellation for Global Sea State Monitoring," Paper SSC99-V-4, 13th Annual AIAA/USU Conf. on Small Satellites, Utah State Univ., Logan, UT, Aug. 1999.

[27]Thomson, P. L., and Hansen, F., "Danish Orsted Mission In-Orbit Experiences and Status of the Danish Small Satellite Programme," Paper SSC99-I-8, 13th Annual AIAA/USU Conf. on Small Satellites, Utah State Univ., Logan, Aug. 1999.

[28]Gibbon, D., Personal communication, Surrey Space Centre, May 2000.

[29]Teston, F., Creasey, R., Bermyn, J., Bernaerts, D., and Mellab, K., "PROBA: ESA's Autonomy and Technology Demonstration Mission," Paper SSC99-V-8, 13th Annual AIAA/USU Conf. on Small Satellites, Utah State Univ., Logan, UT, Aug. 1999.

[30]Chari, R., "Pre-Flight Characteristics of the US Air Force Academy's FalconSat-1," Paper SSC99-VII-8, 13th Annual AIAA/USU Conf. on Small Satellites, Utah State Univ., Logan, UT, Aug. 1999.

[31]Wong, J., "The University Micro/Nanosatellite as a Micropropulsion Testbed," Paper SSC99-VIII-5, 13th Annual AIAA/USU Conf. on Small Satellites, Utah State Univ., Logan, UT, Aug. 1999.

[32]Weidow, D., and Bristow, J., "NASA/DoD University Nano-Satellites for Distributed Spacecraft Control," Paper SSC99-V-5, 13th Annual AIAA/USU Conf. on Small Satellites, Utah State Univ., Logan, UT, Aug. 1999.

[33]Underhill, F., Friedman, A., Wong, J., Reed, H., Hansen, E., Colaprete, A., Rodier, D., Horan, S., and Anderson, B., "Three-Corner Sat Constellation—Arizona State University: Management; Electrical Power System; Structures, Mechanisms, Thermal, and Radiation; Attitude/Orbit Determination and Control; ASU Micropropulsion Experiment; and Integration," Paper SSC99-III-1, 13th Annual AIAA/USU Conf. on Small Satellites, Utah State Univ., Logan, UT, Aug. 1999.

[34]Hall, C. D., Davis, N. J., DeLaRee, J., Scales, W. A., and Stutzman, W. L., "Virginia Tech Ionospheric Scintillation Measurement Mission," Paper SSC99-III-2, 13th Annual AIAA/USU Conf. on Small Satellites, Utah State Univ., Logan, UT, Aug. 1999.

[35]Campbell, M., "UW Dawgstar: One Third of ION-F," Paper SSC99-III-4, 13th Annual AIAA/USU Conf. on Small Satellites, Utah State Univ., Logan, UT, Aug. 1999.

[36]Kitts, C., Pranajaya, F., Townsend, J., and Twiggs, R., "Emerald: An Experimental Mission in Robust Distributed Space Systems," Paper SSC99-VI-5, 13th Annual AIAA/USU Conf. on Small Satellites, Utah State Univ., Logan, UT, Aug. 1999.

[37]Bloomquist, R., Campbell, M., Das, A., Hall, C., Hansen, E., Horan, S., Kitts, C., Luu, K., Martin, M, Mistola, J., Pfeffer, A., Redd, F., Reed, H., Schlossberg, H., Twiggs, B., and Weidow, D., "Microsatellites and Formation Flying Technologies on University Nanosatellites," AIAA Paper 99-4535, AIAA Space Technology Conf., Albuquerque, NM, Sept. 1999.

[38]Rayburn, C. D., Spence, H. E., Petschek, H. E., Bellino, M., Vickers, J., and Murphy, M., "Constellation Pathfinder: A University Nanosatellite," Paper SSC99-V-1, 13th Annual AIAA/USU Conf. on Small Satellites, Utah State Univ., Logan, UT, Aug. 1999.

[39]Luu, K., Martin, M., Stallard, M., Schlossberg, H., Weidow, D., Blomquist, R., Campbell, M., Hall, C., Horan, S., Kitts, C., Redd, F., Reed, H., Spence, H., and Twiggs, B., "University Nanosatellite Distributed Satellite Capabilities to Support TechSat 21," Paper SSC99-III-3, 13th Annual AIAA/USU Conf. on Small Satellites, Utah State Univ., Logan, UT, Aug. 1999.

[40]Lang, M., Personal communication, ESA-ESTEC, Noordwijk, The Netherlands, Oct. 1997.

[41]Jones, R., "JPL Microspacecraft Technology Development (MTD) Program," Jet Propulsion Lab. Internal Document, Pasadena, CA, 31 May 1996.

[42]Blandino, J., Personal communication, Jet Propulsion Lab., Pasadena, CA, Spring 1997.

[43]Brown, C. D., "*Spacecraft Propulsion*," AIAA Education Series, AIAA, Washington DC, 1995.

[44]Ruttle, D., and Fitzsimmons, M., "Development of Miniature 35-lbf Fast Response Bipropellant Divert Thruster," AIAA Paper 93-2585, Monterey, CA, June 1993.

[45]Hodge, K. F., Allen, K. A., and Hemmings, B., "Development and Test of the ASAT

Bipropellant Attitude Control System (ACS) Engine," AIAA Paper 93-2587, Monterey, CA, June 1993.

[46]Craddock, J., and Janeski, B., "Design and Development of the Army KE ASAT ACS Thruster," AIAA Paper 93-1959, Monterey, CA, June 1993.

[47]Rosenberg, S. D., and Schoenmann, L., "New Generation of High-Performance Engines for Spacecraft Propulsion," *Journal of Propulsion and Power*, Vol. 10, No. 1, 1994. pp. 40–46.

[48]Schwende, M. A., Schulte, G., Dargies, E., Gotzig, U., and Scharli-Weinert, E., "New Generation of Low-Thrust Bi-Propellant Engines in Qualification Process," AIAA Paper 93-2120, Monterey, CA, June 1993.

[49]Chirivella, J., "Analysis of the Transient Behavior of the 4-Newton Attitude Control Thruster of the Comet Rendezvous Asteroid Flyby (CRAF) Spacecraft," Jet Propulsion Lab. Internal Document, JPL P.O. 958214, Pasadena, CA, 25 Nov. 1988.

[50]Driscoll, R., Yager, J., Roy, M., and Kammerer, H., "Development Tests on a 5-lbf Bipropellant Thruster Using a Platinum/Rhodium Thrust Chamber," AIAA Paper 98-3357, 34th Joint Propulsion Conf., Cleveland, OH, July 1998.

[51]Wichmann, H., and Fitzsimmons, M., "Miniature High Performance Delta V Engine," AIAA Paper 93-2582, Monterey, CA, June 1993.

[52]Kaiser–Marquardt, Company Information.

[53]Royal Ordnance, Company Information.

[54]Atlantic Research, Company Information.

[55]Aerojet, Company Information.

[56]Bryant, K., Knight, C., and Hurtz, R., "Planetary Lander Vehicles Utilizing LEAP Technology," AIAA Paper 94-2748, Indianapolis, IN, June 1994.

[57]Reed, B., "Micropropulsion Activities at NASA Lewis Research Center," Proceedings, Formation Flying and Micro-Propulsion Workshop, Air Force Research Laboratory, Lancaster, CA, 20–21 Oct. 1998.

[58]Rocket Research Company (now Primex), Hydrazine Handbook.

[59]Morgan, O. M. and Meinhardt, D. S., "Monopropellant Selection Criteria—Hydrazine and Other Options," AIAA Paper 99-2595, 35th Joint Propulsion Conf., Los Angeles, CA, June 1999.

[60]Primex, Company Information.

[61]TRW, Company Information.

[62]Fick, M., and Mütsch, T., "Low Thrust and Low Cost Monopropellant Thusters for Satellites and Satellite Constellations," AIAA Paper 99-2592, 35th Joint Propulsion Conf., Los Angeles, CA, June 1999.

[63]Parker, M., Thunnissen, D., Blandino, J., and Ganapathi, G., "The Preliminary Design and Status of a Hydrazine MilliNewton Thruster Development," AIAA Paper 99-2596, 35th Joint Propulsion Conf., Los Angeles, CA, June 1999.

[64]Jankowsky, R. S., "HAN-Based Monopropellant Assessment for Spacecraft," AIAA Paper 96-2863, Lake Buena Vista, FL, July 1996.

[65]Mittendorf, D., Facinelli, W., and Sarpolus, R., "Experimental Development of a Monopropellant for Space Propulsion Systems," AIAA Paper 97-2951, 33rd Joint Propulsion Conf., Seattle, WA, July 1997.

[66]Meinhardt, D., Brewster, G., Christofferson, S., and Wucherer, E., "Development and Testing of New, HAN-Based Monopropellants in Small Rocket Thrusters," AIAA Paper 98-4006, 34th Joint Propulsion Conf., Cleveland, OH, July 1998.

[67]deGroot, W. A., and Oleson, S. R., "Chemical Microthruster Options," AIAA Paper 96-2868, Lake Buena Vista, FL, July 1996.

[68]Meinhardt, D., Christofferson, S., Wucherer, E., and Reed, B., "Performance and Life Testing of Small HAN Thrusters," AIAA Paper 99-2881, 35th Joint Propulsion Conf., Los Angeles, CA, June 1999.

[69]Grossmann, I., Jones, I. R., and Lee, D. H., "Auxiliary Propulsion Survey, Part III: Survey of Secondary Propulsion and Passive Attitude Control Systems for Spacecraft," AFAPL-TR-68-67, Part III.

[70]Sutton, G. P., *Rocket Propulsion Elements*, 6th ed., Wiley, New York, 1992.

[71]Sellers, J. J., Meerman, M., Paul, M., and Sweeting, M., "A Low-Cost Propulsion Option for Small Satellites," *Journal of the British Interplanetary Society*, Vol. 48, 1995, pp. 129–138.

[72]Ventura, M., and Garboden, G., "A Brief History of Concentrated Hydrogen Peroxide Uses," AIAA Paper 99-2739, 35th Joint Propulsion Conf., Los Angeles, CA, June 1999.

[73]Ventura, M. and Mullens, P., "The Use of Hydrogen Peroxide for Propulsion and Power," AIAA Paper 99-2880, 35th Joint Propulsion Conf., Los Angeles, CA, June 1999.

[74]Morlan, P., Wu, P., Nejad, A., Ruttle, D., Fuller, R., and Anderson, W., "Catalyst Development for Hydrogen Peroxide Rocket Engines," AIAA Paper 99-2740, 35th Joint Propulsion Conf., Los Angeles, CA, June 1999.

[75]Wernimont, E. and Mullens, P., "Recent Developments in Hydrogen Peroxide Monopropellant Devices," AIAA Paper 99-2741, 35th Joint Propulsion Conf., Los Angeles, CA, June 1999.

[76]Whitehead, J. C., Dittman, M. D., and Ledebur, A. G., "Progress Toward Hydrogen Peroxide Micropropulsion," Paper SSC99-XII-5, 13th Annual AIAA/USU Conf. on Small Satellites, Utah State Univ., Logan, UT, Aug. 1999.

[77]Haag, G. S., Sweeting, M. N., and Richardson, G., "Low Cost Propulsion Development for Small Satellites at The Surrey Space Centre," Paper SSC99-XII-2, 13th Annual AIAA/USU Conf. on Small Satellites, Utah State Univ., Logan, UT. Aug. 1999.

[78]Strand, L., Toews, H., Schwartz, K., and Milewski, R., "Extended Duty Cycle Testing of Spacecraft Propulsion Miniaturized Components," AIAA Paper 95-2810, San Diego, CA, July 1995.

[79]Morash, D. H., and Strand, L., "Miniature Propulsion Components for the Pluto Fast Flyby Spacecraft," AIAA Paper 94-3374, Indianapolis, IN, June 1994.

[80]Bzibziak, R., "Miniature Cold Gas Thrusters," AIAA Paper 92-3256, Nashville, TN, July 1992.

[81]Gross, S. J., and Rhee, M. S., "Low Power Draw, 44 mN-sec Cold Gas Micro-Thruster and Driver System," AIAA Paper 99-2694, 35th Joint Propulsion Conf., Los Angeles, CA, June 1999.

[82]Marotta Scientific Controls, Inc., Company Information.

[83]Moog Space Products Division, Company Information.

[84]Nakazono, B., "Second-Generation Microspacecraft Propulsion Analysis and Alternatives," Jet Propulsion Lab. Internal Document, JPL-D-14038, Pasadena, CA, 8 Nov. 1996.

[85]Barber, H. E., Falkenstein, G. L., Buell, C. A., and Gurnitz, R. N., "Microthrusters Employing Catalytically Reacted N_2–O_2–H_2 Gas Mixtures, Tridyne," *Journal of Spacecraft and Rockets*, Vol. 8, No. 2, 1971, pp. 111–116.

[86]Mueller, J. M., and Mcfarlane, J. S., "Design of Tridyne Pressurization Systems for Liquid Oxygen Polybutadiene Hybrid Rocket Motors," AIAA Paper 91-2406, Sacramento, CA, June 1991.

[87]McGrath, D., "The History of Thiokol STAR™ Motor Missions," AIAA Paper 95-3129, San Diego, CA, July 1995.

[88]Heister, S., "Solid Rocket Motors," in *Space Propulsion Analysis and Design*, edited

by R. W. Humble, G. N. Henry, and W. J. Larson, Space Technology Series, McGraw–Hill, New York, 1995, p. 295.

[89]McGrath, D., "Motors with Movable Nozzles," AIAA 95-3020, San Diego, CA, July 1995.

[90]Thiokol Space Motors Catalog.

[91]Atlantic Research Co. Solid Motor Catalog.

[92]Amand, P. G., Maudet, N., and Faure, A., "Solid Propulsion Technologies for Spacecraft," Proceedings, Workshop on Low Cost Spacecraft Propulsion Technologies for Small Satellites, ESA-ESTEC, Noordwijk, The Netherlands, March 1998.

[93]Dilhan, D., "Small Rocket Motors for Micro Satellites," Proceedings, ESA Workshop on Low Cost Spacecraft Propulsion Technologies for Small Satellites, ESTEC, Noordwijk, The Netherlands, March 1998.

[94]Altman, D., and Humble, R., "Hybrid Rocket Propulsion Systems," in *Space Propulsion Analysis and Design*, edited by R. W. Humble, G. N. Henry, and W. J. Larson, Space Technology Series, McGraw–Hill, New York, 1995, p. 365.

[95]Helmy, A. M., "Chronicle Review of the Hybrid Rocket Combustion," AIAA Paper 94-2881, Indianapolis, IN, June 1994.

[96]Wernimont, E. J., and Meyer, S. E., "Hydrogen Peroxide Hybrid Rocket Engine Performance Investigation," AIAA 94-3147, Indianapolis, IN, June 1994.

[97]"RITA Ion Thruster Assembly," DASA Company Brochure.

[98]Bassner, H., Berg, H. P., and Kukis, R., "The Design of RITA Electric Propulsion System for Sat 2 (Artemis)," AIAA Paper 90-2539, Orlando, FL, July 1990.

[99]Beattie, J. R., Williams, J. D., and Robson, R. R., "Flight Qualification of an 18-mN Xenon Ion Thruster," IEPC Paper 93-106, Seattle, WA, Sept. 1993.

[100]Brophy, J. R., Pless, L. C., Mueller, J., and Anderson, J. R., "Operating Characteristics of a 15-cm-dia. Ion Engine for Small Planetary Spacecraft," IEPC Paper 93-110, International Electric, Propulsion Conf., Seattle, WA, Sept. 1993.

[101]Patterson, M., "Low-Power Ion Thruster Development Status," AIAA Paper 98-3347, 34th Joint Propulsion Conf., Cleveland, OH, July 1998.

[102]Gorshkov, O., "Low-Power Hall Type and Ion Electric Propulsion for the Small Sized Spacecraft," AIAA Paper 98-3929, 34th Joint Propulsion Conf., Cleveland, OH, July 1998.

[103]Noci, G., Capacci, M., Redaelli, R., Matucci, A., Matticari, G., Severi, A., and Sabbagh, J., "Development of Small Ion Thrusters for Lightsat Application," AIAA Paper 95-3071, 31st Joint Propulsion Conf., San Diego, CA, July 1995.

[104]Sovey, J. S., Rawlin, V. K., and Patterson, M. J., "A Synopsis of Ion Propulsion Development Projects in the United States: SERT I to Deep Space I," AIAA Paper 99-2270, 35th Joint Propulsion Conf., Los Angeles, CA, June 1999.

[105]Sohl, G., Fosnight, V. V., Goldner, S. J., and Speiser, R. C., "Cesium Electron Bombardment Ion Microthrustors," AIAA Paper 67-81, 5th Aerospace Sciences Meeting, New York, Jan. 1967.

[106]Young, M., Muntz, E., and Ketsdever, A., "Investigation of a Candidate Non-Magnetic Ion Micro-Thruster for Small Spacecraft Applications," AIAA Paper 98-3917, 34th Joint Propulsion Conf., Cleveland, OH, July 1998.

[107]Mueller, J., Pyle, D., Chakraborty, I., Ruiz, R., Tang, W., Marrese, C., and Lawton, R., "Electric Breakdown Characteristics of Silicon Dioxide Films for Use in Microfabricated Ion Engine Accelerator Grids," *Micropropulsion for Small Spacecraft*, Progress in Astronautics and Aeronautics, Vol. 187, edited by M. Micci and A. Ketsdever, AIAA, Reston, VA, 2000, Chap. 12 (this volume).

[108]Marrese, C. M., Polk, J. E., Jensen, K. L., Gallimore, A. D., Spindt, C., Fink, R. L., and Palmer, W. D., "Performance of Field Emission Cathodes in Xenon Electric Propulsion System Environments," *Micropropulsion for Small Spacecraft*, Progress in Astronautics and Aeronautics, Vol. 187, edited by M. Micci and A. Ketsdever, AIAA, Reston, VA, 2000, Chap. 11 (this volume).

[109]Marrese, C. M., Wang, J. J., Gallimore, A. D., and Goodfellow, K. D., "Space-Charge-Limited Emission from Field Emission Cathodes for Electric Propulsion and Tether Applications," *Micropropulsion for Small Spacecraft*, Progress in Astronautics and Aeronautics, Vol. 187, edited by M. Micci and A. Ketsdever, AIAA, Reston, VA, 2000, Chap. 18 (this volume).

[110]Yashko, G., Giffin, G., and Hastings, D., "Design Considerations for Ion Microthrusters," IEPC Paper 97-072, 25th International Electric Propulsion Conf., Cleveland, OH, Aug. 1997.

[111]Brophy, J.R, "Stationary Plasma Thruster Evaluation in Russia," Jet Propulsion Lab. Internal Document, JPL Publ. 92-4, Pasadena, CA, March 15, 1992.

[112]Jacobson, D., and Jankovsky, R., "Test Results of a 200 W Class Hall Effect Thruster," AIAA Paper 98-3792, 34th Joint Propulsion Conf., Cleveland, OH, July 1998.

[113]Atlantic Research Corp., Company Information.

[114]Hruby, V., Monheiser, J., Pote, B., Freeman, C., and Connolly, W., "Low Power, Hall Thruster Propulsion System," IEPC Paper 99-092, 26th International Electric Propulsion Conf., Kitakyushu, Japan, Oct. 1999.

[115]Hruby, V., Monheiser, J., Pote, B., Rostler, P., Kolencik, J., and Freeman, C., "Development of Low Power Hall Thrusters," AIAA Paper 99-3534, 30th Plasmadynamics and Lasers Conf., Norfolk, VA, June 1999.

[116]Martinez-Sanchez, M., "Advances in Micro-Propulsion: 50W Hall Thruster, Colloidal Thrusters," *Proceedings, Formation Flying and Micro-Propulsion Workshop*, Lancaster, CA, Oct. 1998.

[117]Martinez-Sanchez, M., and Pollard, J., "Spacecraft Electric Propulsion—An Overview," *Journal of Propulsion and Power*, Vol. 14, No. 5, 1998, pp. 688–699.

[118]Khayms, V., and Martinez-Sanchez, M., "Design of a Miniaturized Hall Thruster for Microsatellites," AIAA Paper 96-3291, Lake Buena Vista, FL, July 1996.

[119]Khayms, V., and Martinez-Sanchez, M., "Preliminary Experimental Evaluation of a Miniaturized Hall Thruster," IEPC Paper 97-077, 25th International Electric Propulsion Conf., Cleveland, OH, Aug. 1997.

[120]Khayms, V., and Martinez-Sanchez, M., "Fifty Watt Hall Thruster for Microsatellites," *Micropropulsion for Small Spacecraft*, Progress in Astronautics and Aeronautics, Vol. 187, edited by M. Micci and A. Ketsdever, AIAA, Reston, VA, 2000, Chap. 9 (this volume).

[121]Mitterauer, J., "Prospects of Liquid Metal Ion Thrusters for Electric Propulsion," IEPC Paper 91-105, International Electric Propulsion Conf., Viareggio, Italy, Oct. 1991.

[122]Petagna, C., von Rhoden, H., Bartoli, C., and Valentian, D., "Field Emission Electric Propulsion (FEEP): Experimental Investigation on Continuous and Pulsed Modes of Operation," IEPC Paper 88-127, International Electric Propulsion Conf., 1988.

[123]Gonzalez, J., Saccoccia, G., and von Rhoden, H., "Field Emission Electric Propulsion: Experimental Investigations on Microthrust FEEP Thrusters," IEPC Paper 93-157, International Electric Propulsion Conf., 1993.

[124]Marcuccio, S., Ciucci, A., Oest, H., Genovese, A., and Andrenucci, M., "Flight Demonstration Opportunities for FEEP," AIAA Paper 96-2724, Lake Buena Vista, FL, July 1996.

[125]Genovese, A., Marcuccio, S., and Andrenucci, M., "Experimental Characterization of FEEP Emitters," *Proceedings of the 2nd European Spacecraft Propulsion Conference*, ESA SP-398, ESTEC, Noordwijk, The Netherlands, 1997, pp. 243–250.

[126]Marcuccio, S., Genovese, A., Andrenucci, M., Bartoli, C., Gonzalez, J., and Saccoccia, G., "Field Emission Electric Propulsion (FEEP) System Study," IEPC Paper 93-156, International Electric Propulsion Conf., 1993.

[127]Marcuccio, S., Giannelli, S., and Andrenucci, M., "Attitude and Orbit Control of Small Satellites and Constellations with FEEP Thrusters," IEPC Paper 97-188, 25th International Electric Propulsion Conf., Cleveland, OH, Aug. 1997.

[128]Marcuccio, S., Paita, L., Saviozzi, M., and Andrenucci, M., "Flight Demonstration of FEEP on Get Away Special," AIAA Paper 98-3332, 34th Joint Propulsion Conf., Cleveland, OH, July 1998.

[129]Mitterauer, J., "Prospects of Microstructured Liquid Metal Ion Sources (MILMIS) for Field Emission Electric Propulsion (FEEP)," IEPC Paper 93-158, International Electric Propulsion Conf., 1993.

[130]Mahoney, J., Yahiku, A., Moore, R., and Perel, J., "Electrohydrodynamic Ion Source," *Journal of Applied Physics*, Vol. 40, No. 13, 1969, pp. 5101–5106.

[131]Perel, J., "Alkali Metal Ion Sources," *Journal of the Electrochemical Society*, Vol. 115, No. 12, 1968, pp. 343–350.

[132]Perel, J., Bates, T., Mahoney, J., Moore, R. D., and Yahiku, A. Y., "Research on Charged Particle Bipolar Thrustor," AIAA Paper 67-728, Colorado Springs, CO, Sept. 1967.

[133]Mitterauer, J., "Miniaturized Liquid Metal Ion Sources (MILMS)," *IEEE Transactions on Plasma Science*, Vol. 19, No. 5, 1991, pp. 790–798.

[134]Ciucci, A., Genuini, G., and Andrenucci, M., "Experimental Investigation of Field Emissions Electrostatic Thrusters," IEPC Paper 91-103, 22nd International Electric Propulsion Conf., Viareggio, Italy, Oct. 1991.

[135]Pfeffer, H., Bartoli, C., and von Rohden, H., "The Electric Propulsion Activities of the European Space Agency," AIAA Paper 78-713, 13th International Electric Propulsion Conf., San Diego, CA, April 1978.

[136]Marcuccio, M., Lorenzi, G., and Andrenucci, M., "Development of Miniaturized Field Emission Electric Propulsion System," AIAA Paper 98-3919, 34th Joint Propulsion Conf., Cleveland, OH, July 1998.

[137]Genovese, A., Marcuccio, S., Dal Pozzo, D., and Andrenucci, M., "FEEP Thruster Performance at High Background Pressure," IEPC Paper 97-186, 25th International Electric Propulsion Conf., Cleveland, OH, Aug. 1997.

[138]"Micro Thrust FEEP Systems," Centrospazio Company Brochure.

[139]Genovese, A., Marcuccio, S., Petracchi, P., and Andrenucci, M., "Neutralization Tests of a mN FEEP Thruster," AIAA Paper 96-2725, Lake Buena Vista, FL, July 1996.

[140]Fehringer, M., Rådenauer, F., and Steiger, W., "Space-Proven Indium Liquid Metal Field Ion Emitters for Ion Microthruster Applications," AIAA Paper 97-3057, 33rd Joint Propulsion Conf., Seattle, WA, July 1997.

[141]Bartoli, C., von Rohden, H., Thompson, S., and Blommers, J., "A Liquid Caesium Field Ion Source for Space Propulsion," *Journal of Physics D: Applied Physics*, Vol 17, 1984, pp. 2473–2483.

[142]Mitterauer, J., "Field Emission Electric Propulsion: Emission Site Distribution of Slit Emitters," *IEEE Transactions on Plasma Science*, Vol. PS-15, No. 5, 1987, pp. 593–598.

[143]Marcuccio, S., Centrospazio, Personal communication, Fall 1999, Spring 2000.

[144]Klotz, H., Strauch, H., Wolfsberger, W., Marcuccio, S., and Speake, C., "Drag-Free, Attitude and Orbit Control for LISA," *Proceedings of the ESA/ESTEC 3rd International Symposium on Spacecraft Guidance, Navigation and Control*, ESA SP-381, ESTEC, Noordwijk, The Netherlands, pp. 695–702.

[145]Lincoln Composites, Company Information.

[146]Brophy, J., and Noca, M., "Electric Propulsion for Solar System Exploration," *Journal of Propulsion and Power*, Vol. 14, No. 5, 1998, pp. 700–707.

[147]Mitterauer, J., "Contamination Test of a Cesium Field Ion Thruster," *Journal of Propulsion and Power*, Vol. 7, No. 3, 1991, pp. 364–366.

[148]Marcuccio, S., and Lorenzi, G., "Miniaturized Field Emission Electric Propulsion Systems," *Proceeding, Workshop on Low Cost Spacecraft Propulsion Technologies for Small Satellites*, ESA-ESTEC, Noordwijk, The Netherlands, March 1998.

[149]Hubermann, M. N., and Rosen, S. G., "Advanced High-Thrust Colloid Sources," *Journal of Spacecraft*, Vol. 11, No. 7, 1974, pp. 475–480.

[150]Kidd, P. W., and Shelton, K. H., "Life Test (4350 Hours) of an Advanced Colloid Thruster Module," AIAA Paper 73-1078, 10th Electric Propulsion Conf., Lake Tahoe, NV, Oct./Nov. 1973.

[151]Perel, J., and Wolfson, B., "Air Force Electric Propulsion Programs," AIAA Paper 76-1067, AIAA Electric Propulsion Conf., Key Biscane, FL, Nov. 1976.

[152]Mahoney, J., Personal communication, Phrasor Scientific, Inc., Duarte, CA, May 2000.

[153]Perel, J., Mahoney, J., and Sujo, C., "Micro-Electric Propulsion Using Charged Clusters," *Proceedings, Formation Flying and Micro-Propulsion Workshop*, Air Force Research Laboratory (AFRL), Lancaster, CA, Oct. 1998.

[154]NASA Internal Document, "Micro-Electric Propulsion Technology," Final Report for Contract NAS8-98176, NASA, Marshall Space Flight Center, Oct. 1998.

[155]Pranajaya, F. M., "Progress on Colloid Micro-Thruster Research and Flight Testing," Paper SSC99-VIII-6, 13th Annual AIAA/USU Conf. on Small Satellites, Utah State Univ., Logan, UT, Aug. 1999.

[156]Cappelli, M., Stanford Univ., Personal communication, Aug. 2000.

[157]Perel, J., Mahoney, J., Kalensher, B., and Forrester, A., "Investigation of the Capillaritron Ion Source for Electric Propulsion," AIAA Paper 81-0747, 15th International Electric Propulsion Conf., Las Vegas, NV, April 1981.

[158]Burton, R., and Turchi, P., "Pulsed Plasma Thruster," *Journal of Propulsion and Power*, Vol. 14, No. 5, 1998, pp. 716–735.

[159]Guman, W., and Nathanson, D., "Pulsed Plasma Microthruster Propulsion System for Synchronous Orbit Satellite," *Journal of Spacecraft and Rockets*, Vol. 7, No. 4, 1970, p. 409.

[160]Guman, W., and Williams, T., "Pulsed Plasma Microthruster for Synchronous Meteorological Satellite (SMS)," AIAA Paper 73-1066, 10th Electric Propulsion Conf., Lake Tahoe, NV, Oct./Nov. 1973.

[161]Brill, Y., Eisner, A., and Osborn, L., "The Flight Application of a Pulsed Plasma Microthruster: The NOVA Satellite," AIAA Paper 82-1956, 16th International Electric Propulsion Conf., New Orleans, LA, Nov. 1982.

[162]Ebert, W., Kowal, S., and Sloan, R., "Operational Nova Spacecraft Teflon Pulsed Plasma Thruster System," AIAA Paper 89-2497, 25th Joint Propulsion Conf., Monterey, CA, July 1989.

[163]Ashby, D., Liebing, L., Larson, A., and Gooding, T., "Quasi-Steady Pulsed Plasma Thrusters," *AIAA Journal*, Vol. 4, No. 5, 1966, pp. 831–835.

[164]Ziemer, J., and Choueiri, E., "Trends in Performance Improvements of a Gas-Fed Pulsed Plasma Thruster," IEPC Paper 97-040, 25th International Electric Propulsion Conf., Cleveland, OH, Aug. 1997.

[165]Spanjers, G., McFall, K., Gulczinski III, F., and Spores, R., "Investigation of Propellant Inefficiencies in a Pulsed Plasma Thruster," AIAA Paper 96-2723, 32nd Joint Propulsion Conf., Lake Buena Vista, FL, July 1996.

[166]Spanjers, G., Lotspeich, J., McFall, K., and Spores, R., "Propellant Losses Because

of Particle Emission in a Pulsed Plasma Thruster," *Journal of Propulsion and Power*, Vol. 14, No. 4, 1998, pp. 554–559.

[167]Spanjers, G., Malak, J., Leiweke, R., and Spores, R., "Effect of Propellant Temperature on Efficiency in the Pulsed Plasma Thruster," *Journal of Propulsion and Power*, Vol. 14, No. 4, 1998, pp. 545–553.

[168]Vondra, R., and Thomassen, K., "A Flight Qualified Pulsed Electric Thruster for Satellite Control," AIAA Paper 73-1067, 10th Electric Propulsion Conf., Lake Tahoe, NV, Oct./Nov. 1973.

[169]Aston, G., and Pless, L., "Ignitor Plug Operation in a Pulsed Plasma Thruster," AIAA Paper 81-0711R, *Journal of Spacecraft*, Vol. 19, No. 3, 1982, pp. 250–256.

[170]Guman, W., and Begun, M., "Exhaust Plume Studies of a Pulsed Plasma Thruster," AIAA Paper 78-704, 13th International Electric Propulsion Conf., San Diego, CA, April 1978.

[171]Rudolph, L., and Jones, R., "Pulsed Plasma Thruster Contamination Studies," AIAA Paper 79-2106, Oct. 1979.

[172]Rudolph, L., Pless, L., and Harstad, K., "Pulsed Plasma Thruster Backflow Characteristics," AIAA Paper 79-1293, 15th Joint Propulsion Conf., Las Vegas, NV, June 1979.

[173]Myers, R., Arrington, L., Pencil, E., Carter, J., Heminger, J., and Gatsonis, N., "Pulsed Plasma Thruster Contamination," AIAA Paper 96-2729, July 1996.

[174]MacLellan, D., MacDonald, H., and Waldron, P., "Lincoln Experimental Satellites 5 and 6," AIAA Paper 70-494, 3rd Communications Satellite Systems Conf., Los Angeles, CA, April 1970.

[175]Burton, R., "Pulsed Plasma Thrusters for Satellite Micropropulsion," *Proceedings, Air Force Research Laboratory Formation Flying and Micro-Propulsion Workshop*, Lancaster, CA, Oct. 1998.

[176]Spanjers, G., "Micro-Propulsion Research at the Air Force Research Laboratory," *Proceedings, Air Force Research Laboratory Formation Flying and Micro-Propulsion Workshop*, Lancaster, CA, Oct. 1998.

[177]Meckel, N., Cassady, R., Osborne, R., Hoskins, W., and Myers, R., "Investigation of Pulsed Plasma Thrusters for Spacecraft Attitude Control," IEPC Paper 97-128, 25th International Electric Propulsion Conference, Cleveland, OH, Aug. 1997.

[178]Benson, S. W., Arrington, L. A., Hoskins, W. A., and Meckel, N. J., "Development of a PPT for the EO-1 Spacecraft," AIAA Paper 99-2276, 35th Joint Propulsion Conf., Los Angeles, CA, June 1999.

[179]Cassady, R. J., Hoskins, W. A., Campbell, M., and Rayburn, C., "A Micro Pulsed Plasma Thruster (PPT) for the 'Dawgstar' Spacecraft," Proceedings, IEEE Aerospace Conf., Big Sky, MN, March 2000.

[180]"Pulsed Plasma Thruster," Primex Company Information.

[181]Miksch, R. S., and Heller, K. G., "Design and Development of a Vaporjet Attitude-Control System for Space Vehicles," ATL-D-648, Contract AF 33(616)-7044, Wright-Patterson AFB, OH, Dec. 1961.

[182]Tinling, B. E., "Measured Steady-State Performance of Water Vapor Jets for Use in Space Vehicle Attitude Control Systems," NASA TN-D-1302, NASA, Washington, DC, May 1962.

[183]Kanning, G., "Measured Performance of Water Vapor Jets for Space Vehicle Attitude Control Systems," NASA TN D-3561, NASA, Washington, DC, 1966.

[184]Lawrence, T., Sweeting, M., Paul, M., Sellers, J., LeDuc, J., Malak, J., Spanjers, G., Spores, R., and Schilling, J., "Performance Testing of a Resistojet Thruster for Small Satellite Applications," AIAA Paper 98-3933, 34th Joint Propulsion Conf., Cleveland, OH, July 1998.

[185]Jankowsky, R., Sankovic, J., and Oleson, S., "Performance of a Fakel K10K Resistojet," AIAA Paper 97-3059, 33rd Joint Propulsion Conf., Seattle, WA, July 1997.

[186]Janson, S., "Batch-Fabricated Resistojets: Initial Results," IEPC Paper 97-070, 25th International Electric Propulsion Conf., Cleveland, OH, Aug. 1997.

[187]Ketsdever, A., Wadsworth, D., Vargo, S., and Muntz, E., "The Free Molecule Micro-Resistojet: An Interesting Alternative to Nozzle Expansion," AIAA Paper 98-3918, 34th Joint Propulsion Conf., Cleveland, OH, July 1998.

[188]Mueller, J., Chakraborty, I., Bame, D., and Tang, W., "The Vaporizing Liquid Micro-Thruster: Proof of Principle and Preliminary Thermal Characterization," *Micropropulsion for Small Spacecraft*, Progress in Astronautics and Aeronautics, Vol. 187, edited by M. Micci and A. Ketsdever, AIAA, Reston, VA, 2000, Chap. 8 (this volume).

[189]Mueller, J., Vargo, S., Forgrave, J., Bame, D., Chakraborty, I., and Tang, W., "The Micro-Isolation Valve Concept: Initial Results of a Feasibility Study," *Micropropulsion for Small Spacecraft*, Progress in Astronautics and Aeronautics, Vol. 187, edited by M. Micci and A. Ketsdever, AIAA, Reston, VA, 2000, Chap. 17 (this volume).

[190]Stenmark, L., and Lang, M., "Micro Propulsion Thrusters and Technologies," Proceedings, Second European Spacecraft Propulsion Conference, ESA SP-398, ESTEC, Noordwijk, The Netherlands, May 1997, pp. 399–405.

[191]Stenmark, L., Lang, M., Köhler, J., and Simu, U., "Micro Machined Propulsion Components," Proceedings, 2nd Round Table on Micro/Nano Technologies for Space, ESA WPP-132, ESTEC, Noordwijk, The Netherlands, Oct. 1997, pp. 69–76.

[192]Stenmark, L., "Micro Machined Cold Gas Thrusters," Proceedings, Workshop on Low Cost Spacecraft Propulsion Technologies for Small Satellites, ESA-ESTEC, Noordwijk, The Netherlands, March 1998.

[193]Rossi, C., Estève, D., Fabre, N., Do Conto, T., Conedera, V., Dilhan, D., and Gnélou, Y., "A New Generation of MEMS Based Microthrusters for Microspacecraft Applications," Second International Conference for Micro/Nanotechnologies for Space Applications, MNT 99, Pasadena, CA, April 1999.

[194]Lewis, D., Antonsson, E., and Janson, S., "MEMS Microthruster Digital Propulsion System," *Proceedings, Formation Flying and Micro-Propulsion Workshop*, Air Force Research Lab. (AFRL), Lancaster, CA, Oct. 1998.

[195]Youngner, D., and Choueiri, E., "MEMS Mega-Pixel Microthruster Arrays for Micro-Satellites," *Proceedings, Formation Flying and Micro-Propulsion Workshop*, Air Force Research Laboratory (AFRL), Lancaster, CA, Oct. 1998.

[196]Pearson, E., and Schwoebel, P., "Microfabricated Field Ionization Thrusters," *Proceedings, Formation Flying and Micro-Propulsion Workshop*, Air Force Research Lab. (AFRL), Lancaster, CA, Oct. 1998.

[197]Bayt, R., Ayon, A., and Breuer, K., "A Performance Evaluation of MEMS-Based Micronozzles," AIAA Paper 97-3169, 33rd Joint Propulsion Conf., Seattle, WA, July 1997.

[198]Breuer, K., and Bayt, R., "Viscous Effects in Supersonic MEMS-Fabricated Micronozzles," Proceedings, Formation Flying and Micro-Propulsion Workshop, Air Force Research Lab. (AFRL), Lancaster, CA, Oct. 1998.

[199]Fisch, N., Raitsos, Y., Fruchtman, A., and Choueiri, E., "Micro Hall Thruster," Proceedings, Formation Flying and Micro-Propulsion Workshop, Air Force Research Lab. (AFRL), Lancaster, CA, Oct. 1998.

[200]de Groot, W., Reed, B., and Brenizer, D., "Preliminary Results of Solid Gas Generator Micro-Propulsion," AIAA Paper 98-3225, 34th Joint Propulsion Conf., Cleveland, OH, July 1998.

[201]Janson, S., and Helvajian, H., "Batch-Fabricated Microthrusters: Initial Results," AIAA Paper 96-2988, 32nd Joint Propulsion Conf., Lake Buena Vista, FL, July 1996.

[202]Sutherland, G. S., and Maes, M. E., "A Review of Microrocket Technology: 10^{-6} to 1 lbf Thrust," *Journal of Spacecraft and Rockets*, Vol. 3, No. 8, 1966, pp. 1153–1165.

[203]"Subliming Solid Control Rocket," Final Report, NASA CR-711, Contract NAS 5-3599, Rocket Research Corp., June 1965.

[204]"Development of the Subliming Solid Control Rocket Phase II," NASA CR-712, Contract NAS 5-9070, Rocket Research Corp., Seattle WA, March 1967.

[205]"Application of the Valveless Subliming Control Rocket to the NRL Gravity Gradient Satellite," RRC-66-R-59, Contract Nr-5123(00)(X), Naval Research Lab., May 1966.

[206]Hardt, A. P., Foley, W. M., and Brandon, R. L., "The Chemistry of Subliming Solids for Micro Thrust Engines," *Astronautica Acta*, Vol. 11, No. 5, 1965, pp. 340–347.

[207]Owens, W. L., Jr., "Design Aspects of Subliming Solid Reaction Control Systems," AIAA Paper 68-516, Atlantic City, NJ, June 1968.

[208]Kindsvater, H. M., "Simplified Space Mechanisms Using Subliming Solids," Proceedings of the First Aerospace Mechanisms Symposium, Univ. of Santa Clara, Santa Clara, CA, May 1966.

[209]Owens, W. L., "An Experimental Study of Superheated Subliming Solid Thruster Performance," AIAA Paper 70-210, 8th Aerospace Sciences Meeting, Jan. 1970.

[210]Hardt, A. P., Foley, W. M., and Brandon, R. L., "The Chemistry of Subliming Solids for Micro-Thrust Engines," AIAA Paper 65-595, Colorado Springs, CO, June 1965.

[211]Kindsvater, H. M., "Design Criteria for Subliming Solid Applications," ICPRG/AIAA Solid Propulsion Conf., Anaheim, CA, June 1967.

[212]Forsythe, R. W., "Impulse and Thrust Stand Test of a Subliming Solid Micropropulsion System," NASA TN D-3245, NASA Goddard Space Flight Center, March 1966.

[213]Federline, F., "Development of the Subliming Solid Control Rocket—Contributions from the Explorer XXXV (AIMP-E) Satellite Program," Document X-723-68-227, NASA Goddard Space Flight Center, June 1968.

[214]Greer, H., and Griep, D. J., "Dynamic Performance of a Subliming Solid Reaction Jet," Report No. TR-1001 (2230-33)-I, Aerospace Industries, Dec. 1966.

[215]Berkopic, F. D., "Performance of Two Subliming Solid Propellant Thruster Systems for Attitude Control of Spacecraft," NASA TN D-3841, Lewis Research Center, 1967.

[216]Maycock, N., and Pai Vemeker, V. R., "A Photochemical Microrocket for Attitude Control," *Journal of Spacecraft and Rockets*, Vol. 6, No. 3, Mar. 1969.

[217]Mueller, J., Muller, L., and George, T., "Subliming Solid Micro-Thruster for Microspacecraft," New Technology Report NPO-19926/9525, Jet Propulsion Lab. Technology Utilization Office, Pasadena, CA, April 1996.

[218]Breuer, K., Personal communication, Massachusetts Inst. of Technology, Spring 1996.

[219]Lang, M., Personal communication, ESA/ESTEC, Jet Propulsion Lab., Spring 1997.

[220]Langmaier, J., Personal communication, Jet Propulsion Lab., Spring 1998.

Chapter 4

System Considerations and Design Options for Microspacecraft Propulsion Systems

Andrew D. Ketsdever*
Air Force Research Laboratory, Edwards Air Force Base, California

Nomenclature

A = area, m^2
a = speed of sound, m/s
B = magnitude of magnetic field, T
c^* = characteristic velocity, m/s
c_p = gas specific heat at constant pressure, J/(kg K)
D, d = diameter, m
h_g = gas heat transfer coefficient, W/(m^2 K)
I_{sp} = specific impulse, s
Kn = Knudsen number
k = material thermal conductivity, W/(m K)
L^* = combustion chamber characteristic length, m
M = mass flow rate, kg/s
m = mass, kg
n = number density, m^{-3}
p = pressure, Pa
q = electron charge, C
q_c = heat transfer due to conduction, W
q_r = heat transfer due to radiation, W
Re = Reynolds number
R_g = radius of gyration (Larmor radius), m
T = temperature, K
V_c = combustion chamber volume, m^3
v = flow velocity, m/s
v_p = velocity component perpendicular to magnetic field, m/s

This material is declared a work of the U.S. Government and is not subject to copyright protection in the United States.

*Senior Research Engineer, Propulsion Directorate, Propulsion Sciences and Advanced Concepts Division (PRSA). Senior Member AIAA.

γ = ratio of specific heats
ε = emissivity
λ = mean free path, m
κ = gas thermal conductivity, W/(m K)
μ = gas viscosity, kg/(m s)
ρ = gas density, kg/m^3
σ = Stefan–Boltzmann constant, 5.67051×10^{-8} W/(m^2 K^4)
σ_1 = electron ionization cross section, m^2
$\Im$ = thrust, N

Subscripts

e = electron
FM = free molecule flow
L = limit isentropic flow
n = neutral
o = stagnation
t = nozzle throat

I. Introduction

THE growing interest in the use of microspacecraft within government and industry is driving a critical need for new propulsion systems capable of fulfilling a wide range of mission requirements. A thorough review of the current state of micropropulsion concepts has been compiled by Mueller.[1] The concepts described are diverse and exhibit a wide range of operating characteristics and performance values. The applicability of each propulsion system is highly mission dependent and subject to system constraints. The purpose of this chapter is to describe some of the design, fabrication, and microspacecraft system limitations associated with the scaling of micropropulsion devices.

A. Microspacecraft

The feasibility of constellations or platoons of microspacecraft performing the functions of relatively large spacecraft in a distributed way is currently being investigated within the U.S. Air Force and the National Aeronautics and Space Administration (NASA).[2–5] Although it is not clear that these concepts inherently come at reduced cost, they do lend themselves to increased survivability, flexibility and functionality. In an attempt to standardize the definition of microspacecraft, the Air Force Research Laboratory (AFRL) has proposed the standard detailed in Table 1.[6] Although the AFRL definition has not been completely adopted throughout the community, a broad definition of microspacecraft having a mass less than 100 kg will form the basis of further discussion in this work.

The systems constraints on mass, power, maximum voltage, and volume with which microspacecraft will undoubtedly have to contend pose several challenges to the propulsion system designer. In general, microspacecraft propulsion systems will need to be as efficient as, or perhaps even more efficient than, their large spacecraft counterparts to maximize the limited resources provided.

Table 1 AFRL proposed satellite classification standard

Total spacecraft mass	Description
100–1000 kg	Small spacecraft
10–100 kg	Microspacecraft
1–10 kg	Nanospacecraft
<1 g	Picospacecraft

B. Micropropulsion

Just as there are varying standards for the definition of microspacecraft, there are many definitions of what the term "micropropulsion" really describes. Some definitions use characteristic size, while others use producible thrust level. Perhaps the most general definition of micropropulsion is any propulsion system that is applicable to a microspacecraft (mass less than 100 kg) mission. This definition allows the inclusion of a wide range of concepts, from scaled-down versions of existing thrusters operating at reduced power levels to completely redesigned microelectromechanical (MEMS)-fabricated thrusters with micron characteristic sizes. Although this definition is rather broad, it appears to be the most widely accepted. However, the objective of this work is to address the systems-related issues and design considerations in the limit of MEMS devices since the ultimate goal of micropropulsion systems (even for relatively large microspacecraft) lies in this direction.

Micropropulsion is an enabling technology for microspacecraft operations by making possible missions that otherwise could not be performed. For example, the formation and maintenance of platoons of microspacecraft will require a maneuvering capability to counter orbital perturbations. Microspacecraft missions involving large spacecraft resupply, repair, or surveillance will also require maneuverability. The mission requirements for microspacecraft will be varied, and in some cases a large range of capability might be required on the same spacecraft. Micropropulsion systems must be extremely versatile to address these requirements. It is clear that there is a need for micropropulsion systems from high-thrust chemical engines to high-specific impulse electric ion thrusters to fulfill specific missions. Applications of micropropulsion systems for every mission need to be studied carefully to ensure that the propulsion system has sufficient benefit.

The design of micropropulsion systems is also complicated by the scaling of systems-level aspects with spacecraft size including power, mass, and volume. One systems-level aspect that may not scale with spacecraft size is the expected propulsion system lifetime. Although microspacecraft will be designed to be replaceable, propulsion system failures will not be tolerated any more than on larger spacecraft. To maintain a platoon of microspacecraft, members of the platoon will need to be removed from the general platoon "neighborhood" at the end of the spacecraft's useful life or in the event of premature failure of a spacecraft subsystem. In this regard, the micropropulsion system may be required to function even after a major spacecraft failure or after several years on orbit.

To ensure a long lifetime, micropropulsion systems will need to be robust and durable. Issues will arise from corrosive propellant usage, plugging of small orifices with contaminants, thin-film degradation, MEMS component mechanical and thermal cycling, and environmental interactions. The lifetime of a microspacecraft may also be limited by contamination of spacecraft sensors and other surfaces as a direct result of micropropulsion systems. Although there is no evidence that the contamination issue with microspacecraft is increased over large space structures, contamination is a concern for nearly all spacecraft operations. Obviously, the potential for contamination from a given micropropulsion system should be addressed when considering the appropriate system for a particular mission.

Early indications suggest that micropropulsion systems in general may experience decreased performance efficiency to due losses associated with small characteristic sizes, limitations on system mass and power, and the lagging development of adequate micromachined support hardware. Propulsion engineers must address these issues through the use of novel approaches that utilize small-scale properties to the overall system's benefit. In the limit of MEMS fabrication scales, simply scaled-down versions of existing thrusters may not perform as expected.

Careful attention should be paid to the characteristics of propulsion systems that scale favorably with reduced size. As discussed in the following sections, there are several large-scale thruster characteristics that do not scale favorably with reduced size. Those characteristics that do scale favorably may hold the key to the design of efficient micropropulsion systems.

II. Micropropulsion Scaling Issues

The recent trends in spacecraft technology have indicated an increasing emphasis on the miniaturization of subsystems including propulsion systems. Although simple macroscopic approaches describing physical processes can solve many engineering problems, they are often not sufficient to understand and solve complex problems involving microengineered systems. With this in mind, the following sections describe macroscopic and microscopic approaches to understanding some of the fundamental processes associated with micropropulsion devices.

A. Micronozzle Expansions

A large class of micropropulsion concepts will require the expansion of propellant gases through micronozzle geometries. These concepts include electrothermal and chemical systems that are capable of relatively high thrust levels. The flow-through micromachined structures is inherently rarefied, leaving some continuum descriptions of the flow invalid. For this reason, it is important to understand the loss mechanisms and performance issues related to gas flows at very low Reynolds numbers.

1. Flow-Through Micronozzles

To obtain the reduced thrust levels and size requirements needed for microspacecraft missions, nozzle operating characteristics and physical dimensions must be scaled. In the limit of continuum isentropic flow through a large pressure drop, reducing the operating pressure, the throat area (diameter), or both can scale a nozzle's thrust. For this case, the thrust is proportional to

$$\Im \propto p_o A_t \propto p_o d_t^2 \tag{1}$$

The Reynolds number gives a measure of nozzle efficiency in terms of viscous flow losses. The Reynolds number at the nozzle throat is given by

$$Re = \rho a d_t / \mu \propto p_o d_t / T_o^x \tag{2}$$

where x is a positive value between 1.2 and 1.5 depending on the gas.[7] For a nozzle's viscous losses to scale favorably, the Reynolds number, which is proportional to $p_o d_t$, must remain constant or increase. For a constant stagnation pressure and temperature, the thrust in a micronozzle can be reduced by a factor of 100 by reducing the throat diameter by a factor of 10, which reduces the throat Reynolds number by only a factor of 10. The recent availability of micromachined nozzles with throat diameters approaching 20 μm has created a new possibility for relatively efficient low-thrust systems by reducing the physical dimension of the nozzle.[8] Reducing the thrust level by a factor of 100 by reducing the stagnation pressure alone reduces the throat Reynolds number by a factor of 100, resulting in higher viscous losses.

To maintain a constant level of viscous losses (constant Re), a reduction in thrust by a factor of 100 requires a reduction in throat diameter by a factor of 100 and an increase in stagnation pressure by the same factor. Since pressure increases on this order are not desired for microspacecraft operations (see Section III), the operational Reynolds numbers for micronozzles may in fact range from 10^2 to 10^4, indicating a potential for high viscous and rarefaction losses. In addition, most propulsion concepts that would utilize micronozzle expansions rely on an increased theoretical specific impulse through stagnation temperature increases (done either chemically or electrically). Equation (2) indicates that a significant fraction of high-temperature micronozzle flows will be at an increasingly low Reynolds number.

The flow-through low-Reynolds number nozzle has been studied by several authors both experimentally and computationally.[8–13] In these studies, it was found that thick viscous boundary layers develop at Reynolds numbers below 1000, indicating poor nozzle efficiencies. The inefficiencies arise from the adverse interaction of the subsonic boundary layer with the core of supersonic flow causing the flow not to expand fully in the diverging nozzle section. In some cases, the viscous layer can occupy most, if not all, of the diverging nozzle section.

More recently, flow-through micromachined nozzles with characteristic throat diameters of less than 200 μm have been studied.[8,14,15] The Reynolds numbers expected for these micronozzle flows range from 10^2 to 10^4 for achievable stagnation pressures and temperatures on-orbit (i.e., consistent with spacecraft limitations, current materials, and fabrication processes). Therefore, viscous losses are expected to be significant in a reasonable percentage of micronozzle applications. Figure 1 shows Mach contours for a 20-μm-throat diameter nozzle with a Reynolds number of approximately 802 obtained using a commercially available Navier–Stokes (NS) code.[15] The viscous nature of the flowfield is apparent from the large subsonic boundary layer, and the inviscid core is restricted to a small portion of the flow near the centerline. For comparison, the ideal, one-dimensional isentropic expansion gives an exit Mach number of 6.5. Figure 2 shows NS results for the specific impulse through the same micronozzle geometry for two Reynolds numbers and stagnation temperatures. As expected, the efficiency of the micronozzle expansion decreases with Reynolds number.

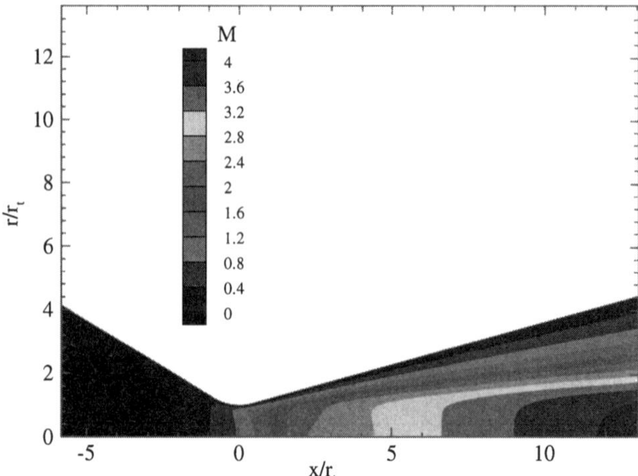

Fig. 1 Micronozzle Mach contours for stagnation temperature $T_o = 600$ K, stagnation pressure $p_o = 0.5$ MPa, and an argon propellant (Re = 802).[15]

Because a propellant's viscosity increases with temperature, viscous losses will increase for high-temperature applications for micronozzles such as chemical thrusters, arcjets, and resistojets. Although the performance from high-temperature systems is expected to be better than from cold gas systems, the overall propulsive efficiency may be significantly reduced. Other inefficiencies within the nozzle are not addressed by simple Reynolds number scaling including propellant heat transfer to the walls, finite-rate excitation of internal energy modes (frozen flow losses), propellant condensation, losses associated with nozzle expansion, and losses associated with micronozzle fabrication (e.g., surface roughness).

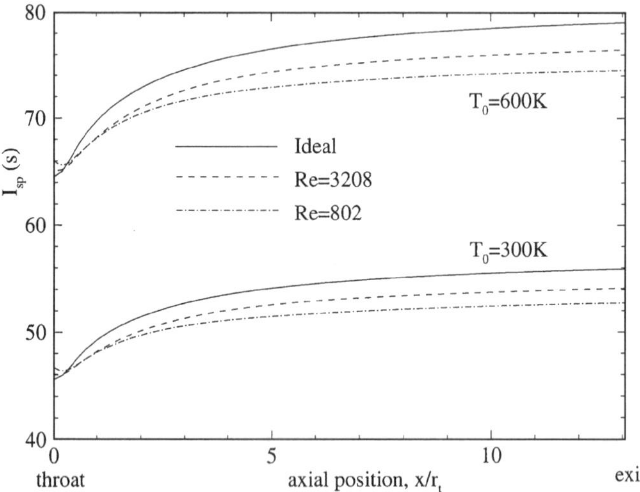

Fig. 2 Micronozzle specific impulse as a function of normalized position for various Reynolds numbers and stagnation temperatures with an argon propellant.[15]

As the throat Reynolds number decreases (or the Knudsen number increases), the nozzle inefficiencies continue to increase until the limiting case of free molecule flow is reached. In this case, molecule–molecule collisions are negligible; instead, molecule–surface interactions will dominate the flow. The ratio of the specific impulse for limit isentropic flow to that for free molecule flow is[16]

$$(I_{sp})_L/(I_{sp})_{FM} = \{4\gamma/\pi(\gamma - 1)\}^{\frac{1}{2}} \tag{3}$$

The ratio from Eq. (3) is equal to 2.11 and 1.78 for γ equal to 1.4 and 1.67, respectively. This gives a lower bound to flows through small characteristic dimensions where the local Knudsen number can be of the order of one. In the case of $Kn \sim 1$, it has been demonstrated by Ketsdever et al.[16] that expansion through a long, narrow slot may be advantageous for some applications.

On a microscopic level, detailed information is required on the processes of gas–surface interactions to understand loss mechanisms in micronozzle geometries. At a low Reynolds number, the molecular mean free path increases, making the physics of gas–surface collisions increasingly important. Although a great deal of information is known about gas–surface interactions between simple gases and common materials, little is known in terms of microscopic properties for most complex propellant molecules interacting with surfaces of interest in micropropulsion devices. For example, the level of molecular dissociation or internal energy state distributions of propellant molecules after colliding with a surface is important for frozen flow losses within the nozzle.

2. *Micronozzle Modeling Issues*

Traditional (NS) continuum-based computational techniques for the simulation of micronozzle flows can often provide erroneous or misleading results. These inaccuracies generally result during the computation of molecular transport effects. The macroscopic properties of any fluid flow may be identified with average values of the appropriate molecular quantities at any location within the flow. These properties may be correctly identified as long as there is a sufficiently large number of molecules within the smallest significant volume of that flow. With this continuum condition satisfied, transport terms can be calculated using macroscopic variables, such as temperature, rather than microscopic variables, such as the molecular velocity distribution function. Such continuum approaches comprise the vast majority of computational and analytical tools for studying fluid behavior and are based on the Euler or NS equations. When this condition is not met, there is a limit imposed on the range of validity of these continuum equations. This limit occurs when gradients of the macroscopic variables become so steep that the scale length is of the same order as the mean free path of the gas.

Because the flows through very small throat diameters (of the order of tens of microns) even at large stagnation pressures (several atmospheres) result in relatively small Reynolds numbers ($Re < 200$), the predicted performance results obtained from NS solutions may be inaccurate.[11,15] The limitations of NS solvers in predicting the flow structure and performance of micronozzle flows has been shown by recent comparisons[15] between full NS solutions and those obtained by the direct simulation Monte Carlo (DSMC)[17] technique. The differences in specific impulse between the two numerical techniques for the $Re = 1300$ ($p_o = 10$ atm) case is approximately 2.5 to 3% as shown in Fig. 3. At lower Reynolds numbers, the difference in specific impulse can be as large as 15% as demonstrated for

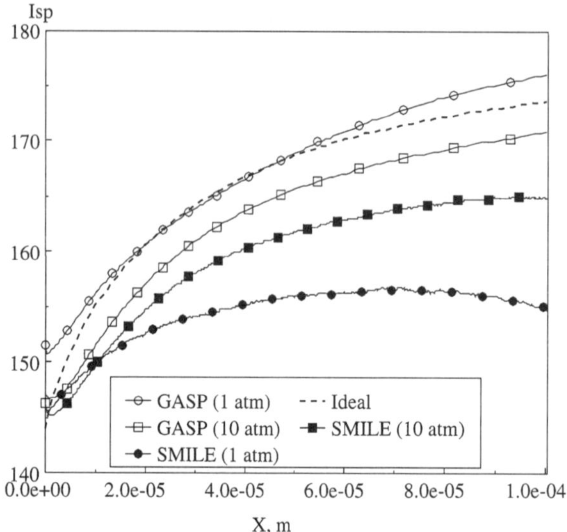

Fig. 3 Specific impulse along the axis of a MEMS-fabricated micronozzle.[15]

a micronozzle flow with $Re = 130$ ($d_t = 27.7\,\mu$m, $p_o = 1$ atm, $T_o = 300$ K, helium propellant). In fact, Fig. 3 shows that the NS solution at $Re = 130$ gives a predicted performance larger than the ideal isentropic value. General trends in the flow such as density and velocity distributions also do not compare well between the two methods for low Reynolds numbers. For micronozzle flows with $Re < 250$, it appears that the DSMC method gives more accurate results for macroscopic performance characteristics. This implies that numerical design optimization of micronozzles operating at very low Reynolds numbers may be a rather complex and expensive process unless NS codes can be modified to give acceptable results.

In all aspects of numerical modeling, the boundary conditions used in the model can dominate the development of the flowfield. This is especially the case for flow-through micronozzles since the effects of the boundaries transmit throughout the entire simulation due to the very small characteristic lengths involved. The NS solutions described above may be improved by the addition of velocity slip along the nozzle walls and improved outflow boundary conditions. The areas of applicability of the NS solutions must be seriously considered due to the reduced computational complexity and cost. Since the Knudsen number can be large for a majority of micronozzle flows, the effects of gas–surface interactions becomes increasingly important, especially at elevated temperatures. For $Kn > 1$, gas–surface interactions can dominate gas–gas collisions, making the physical surface models in simulation tools critical.

3. *Micronozzle Design Considerations*

The advantages to reducing the thrust obtained by a micronozzle system by decreasing the throat diameter and increasing the stagnation pressure are obvious compared to thrust reduction by lowering the stagnation pressure alone. Micronozzles will inherently operate at a low Reynolds number, indicating a need for increased understanding of viscous losses and viscous interactions with surfaces.

Figure 3 depicts a very interesting result for micronozzles operating at a rather low Reynolds number ($Re \sim 130$ for $p_o = 1$ atm). As can be seen for the micronozzle case for $p_o = 10$ atm, the local I_{sp} increases throughout the nozzle expansion. However, the I_{sp} reaches a maximum short of the nozzle exit for the $p_o = 1$ atm case. Under sufficiently low-Reynolds number operating conditions, the long diverging section length leads to large viscous losses, creating a peak in the nozzle performance. Shorter expansion lengths are therefore advantageous at very low Reynolds numbers for improving performance and reducing thruster size and weight. Since the flow through a shorter nozzle exhibits a thinner boundary layer near the nozzle exit, thruster-induced backflow contamination concerns may also be decreased.

B. Ion Formation at Small-Scale Lengths

1. Containment of Electrons

Electron losses to discharge chamber walls will limit the efficiency of microscale ion devices. In large-scale ion engines and Hall thrusters, magnetic fields are used to contain the electrons in the discharge volume and minimize losses to interior surfaces. Through the use of magnetic fields, the path length of the electrons is increased to a few ionization mean free paths (the average distance an electron travels before ionizing a neutral molecule) within the discharge chamber, allowing for a high probability of ionization. The ionization mean free path for electron ionization of a neutral propellant molecule (in most cases xenon) is

$$\lambda = 1/(n_n \sigma_i) \tag{4}$$

The electrons spiral along the magnetic field lines with a radius of gyration (also known as the Larmor radius) given by

$$R_g = m_e v_p/(qB) \tag{5}$$

To minimize wall losses, the radius of gyration should be some reasonable fraction of the discharge chamber diameter. For example, a 10-cm-diam ion engine would require an approximately 0.1-T magnetic field. As the thruster size is scaled downward to the millimeter size, a 10-T magnetic field would be required to contain the electrons in the discharge region effectively. The weight and power requirements of solenoids or permanent magnets would be prohibitive for this size device. Yashko et al.[18] give a rigorous review of the issues associated with this type of ion engine scaling.

2. Grid Acceleration and Breakdown

For micro-ion engines, ion extraction and accelerating grids will be required if performance similar to that of large-scale ion thrusters is expected. To achieve a high specific impulse, micro-ion grids will have to hold off voltage differences of the order of 1 to 1.5 kV. For MEMS-fabricated grids, limited insulating materials are available that can be easily packaged with the rest of the propulsion system. The breakdown of these materials due to high potential differences between the accelerating grids is an issue due to the very close (tens of microns) spacing of the grids. Figure 4 shows the anticipated breakdown modes for micro-ion engine grid assemblies.[19]

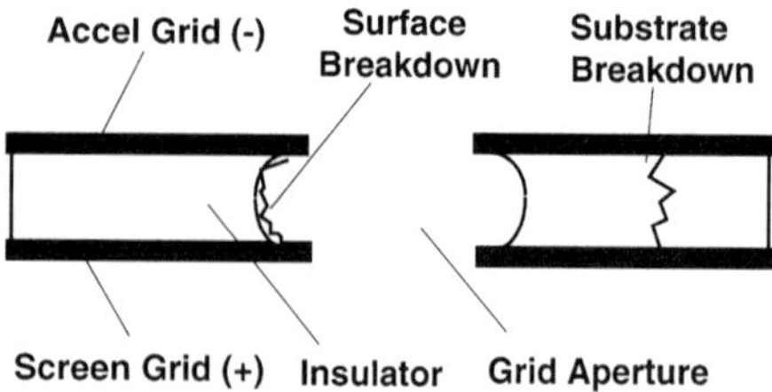

Fig. 4 Anticipated grid breakdown modes.[19]

Initial studies have looked at the potential of surface and substrate breakdown in MEMS-fabricated accelerating grid systems.[19] These tests used silicon oxide as the grid insulating material. The breakdown field strengths for various oxides deposited in different manners are shown in Fig. 5. As shown in this figure, there is a general trend toward a reduction of the breakdown field strength as the oxide thickness decreases. It should be noted that the breakdown field strength is highly dependent on the properties of the oxide layer and its operational environment. Defects in the oxide, surface temperature, and operating pressure are all important factors in determining the appropriate hold-off potentials for a given insulating surface.

3. *Micro-Ion Thruster Modeling Issues*

As with neutral flow models utilizing DSMC, plasma models have been developed to incorporate the "molecular" properties of plasmas. Particle in cell (PIC) codes have been successfully demonstrated for a wide range of problems and seem ideal for the calculation of ion propulsion system properties.[20,21] However, there are some improvements and subsequent validations that need to be made for

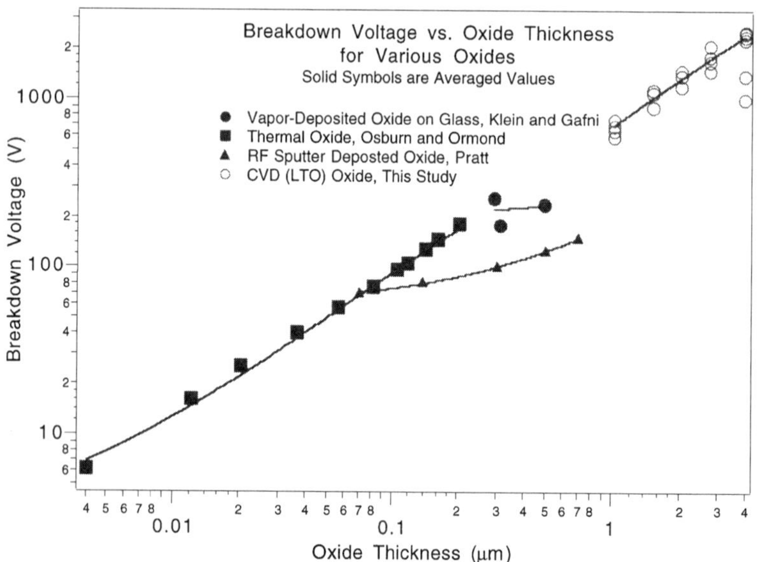

Fig. 5 Breakdown voltages vs insulator thickness for various oxides.[19]

micro-ion thrusters in general. With the expected decrease in degree of ionization in small-scale ion thrusters comes the increased importance of neutral flows in these devices. Most PIC plasma codes currently in use for large-scale ion thrusters assume some uniformly varying neutral flowfield that seems to provide adequate results for a large class of problems in the thruster plume. To characterize micro-ion thrusters fully from the discharge chamber through the far plume, current modeling techniques are attempting to incorporate neutral particle flow models such as Monte Carlo collisional schemes with traditional plasma codes.[22] The validity of these codes remains an open question for many micropropulsion systems. One key feature of micro-ion thrusters may be extremely low degrees of ionization (from 1 to 10%), which may allow for decoupled DSMC–PIC analysis.

C. Micron-Scale Combustion and Mixing

Advanced liquid and solid propellants are targeted at filling high-thrust and high-total impulse requirements for microspacecraft. As mentioned in previous sections, the use of propellants stored on-orbit in either liquid or solid form have several systems advantages over the use of gaseous propellants stored at high pressure. Liquid propellants for use in combustion processes, however, have additional concerns in micropropulsion systems. Typical liquid combustion propellants such as hydrazines (N_2H_4, MMH), nitrogen tetroxide (N_2O_4), methyl amine (CNH_5), and chlorine trifluoride (ClF_3) are volatile and rather toxic, leading to survivability issues for some common micromachined materials in the presence of these propellants. Also, the combustion process releases a great deal of chemical energy in the form of heat or require catalysts at elevated temperatures. The high temperatures associated with these combustion processes (usually greater than 1300 K) will need to be addressed by micromachinable materials, thereby adding complexity. The increased cost of handling liquid combustion propellants could be undesirable for low-cost microspacecraft missions.

1. *Mixing Length Requirements*

One critical requirement for bipropellant liquid combustion systems is for the fuel and oxidizer to be fully mixed inside the combustion chamber. The level of mixing has a direct correlation with the performance of the thruster system. Mixing can be particularly troublesome even in rather large-scale devices, and the length required to mix a fuel and oxidizer properly will not easily scale to very small sizes. Therefore, as the critical length scale of the combustion chamber decreases, the efficiency of chemical thrusters requiring mixing is expected to decrease rather dramatically. There may be additional concerns with injector technologies operating at low flow rates.

As discussed earlier for micronozzle flows, the reduction in characteristic dimension usually comes at the expense of a lower Reynolds number. In the case of the combustion chamber, a reduced Reynolds number implies a less turbulent flow. Turbulent flow is generally desired for optimum mixing of fuels and oxidizers. Again, the reduced Reynolds number characteristic with microscale devices is not desirable from a performance standpoint.

Premixing combustion partners can alleviate some of the problems with mixing propellants in microscale combustion chambers. However, these premixtures can be inherently unstable and require careful handling. Another way to reduce the inefficient mixing of components in chemical microthrusters is to use monopropellants or solid propellants that do not require mixing.

2. Residence Time for Combustion Chemistry

Every chemical process requires a finite time in which to react that in a microscopic sense is of the order of a few times the mean collision time of species in the gas phase. Since intermolecular collisions are required for the chemical process to occur, the stagnation region must have a characteristic length of the order of the mean free path at stagnation pressures. This limit in scaling will affect the minimum allowable size of chemical thrusters. The residence time requirement will hold for all chemical thrusters including monopropellant systems. In the case of monopropellants, the time required for the propellant to be in contact with the catalyst for reaction to occur will drive the minimum length scale of the device.

3. Combustion Modeling Issues

The same modeling issues hold for micronozzle flows from combustion chambers as mentioned in the previous section. The temperatures in the combustion region being high will undoubtedly cause low-Reynolds number flows through converging–diverging micronozzles. The modeling of multiphase combustion is a topic of current research for macroscopic systems. The effects of unburned fuel droplets or soot may begin to dominate at small scales due to the inherent inefficiencies in the overall combustion process. Kinetic approaches will undoubtedly be required, as the characteristic dimensions of the combustion chamber are reduced and chemical equilibrium is not applicable. Gas–surface interaction models will become increasingly important for monopropellant systems to ensure the efficient coupling of catalytic energy into the propellant.

4. Combustion Design Considerations

A combustion chamber characteristic length can be defined as[23]

$$L^* = V_c / A_t = V_c p_o / M c^* \tag{6}$$

A combustion chamber's characteristic length depends on several factors including the propellant type, chamber geometry, and injector or catalyst configuration. For microchemical propulsion systems, the design criterion is for L^* to be as small as possible while maintaining acceptable performance.

Presently, monopropellants appear attractive for micropropulsion applications since mixing of fuel and oxidizer components is not required. However, monopropellant concepts require the use of catalysts at elevated temperatures (700–1800 K). Silicon-based micromachined components exposed to this extreme temperature environment will not survive. Catalyst degradation is a problem plaguing current monopropellant systems. This degradation may be somewhat alleviated in micropropulsion systems since the majority of the degradation appears to be mass flow dependent, and the mass flow rates from microchemical systems are expected to be quite small.

For microchemical thrusters, combustion instabilities may increase as the scale size decreases due to the expected lower-pressure operation, poor liquid pumping, poor mixing, and possibility of propellant flow interruption due to contaminants. The flow-through MEMS-fabricated components upstream of the combustion chamber will need to be characterized fully to minimize these instabilities.

Fig. 6 Proof of a concept nickel micronozzle with a throat diameter of 750 μm. The minimum feature size for this technique can be as small as 50 μm.

The expansion of high-temperature gases through micronozzles has been addressed in previous sections. Propellants that are designed to burn much hotter than traditional propulsion propellants may begin to see diminishing returns after the micronozzle expansion. The low Reynolds number expected at these high temperatures will create a highly viscous flow that will act to lower the overall efficiency of the system. MEMS-fabricated micronozzles will also not survive the extreme temperature environments of the expanding flow. New techniques to fabricate metal nozzle and components capable of surviving higher temperatures will be of great interest. Figure 6 shows a metal micronozzle fabricated by a multistep laser machining and metal coating process.[24]

D. Micro-Heat Transfer

Heat transfer in microscale devices often depends on the microstructure of the material being considered. For example, data obtained on a material's thermal conductivity used to calculate the heat transfer through a material by conduction can be quite different depending on the process used to form the material (e.g., diamond thin films). Material defects and impurities are also critical factors in determining a material's heat transfer properties. The effects of defects and impurities are heightened in microstructures whose feature sizes may be of the same order as the affected site. Point defects can strongly influence the energy transport in a solid since they act as scattering centers for energy carriers such as electrons and phonons.[25] Similar effects are encountered with polycrystalline structures at grain boundaries. Typically, grain boundaries or interfaces between single crystals contain many imperfections and defects. The grain boundary plays a critical role in the mechanical, electrical, chemical, and thermal properties of a material.

1. Thermal Radiation

Infrared radiation with wavelengths between 1 and 100 μm is emitted by every surface. On the macroscopic level, the calculation of heat transfer due to thermal radiation is given by the Stefan–Boltzmann law as

$$q_r = \varepsilon \sigma A T^4 \tag{7}$$

where ε is the emissivity of the surface, σ is the Stefan–Boltzmann constant, A is the material surface area, and T is the material temperature. For a two-body problem, Eq. (7) becomes

$$q_r = \varepsilon \sigma A \left(T_1^4 - T_2^4\right) \tag{8}$$

where the subscripts 1 and 2 denote the surface temperatures of surfaces 1 and 2, respectively.

As mentioned earlier, the concept of a uniform surface property for a microsystem may not be valid since defect sites that contain impurities can also affect the emissivity. For example, point defects containing chemical impurities can produce color centers in crystals that are otherwise transparent. On the microscopic level, the interaction between photons and phonons, electrons, and atoms in the material lattice needs to be investigated further. In microstructures, there is the potential of adjacent microstructures affecting the local electromagnetic field near the element of interest, which could modify the radiation interactions. However, the physics involved with these potential perturbations is well understood.[25]

The amount of radiative heat lost in a micropropulsion device scales favorably as the characteristic length is decreased since it depends on the surface area or l_c^2 at a constant surface temperature. However, radiative heat loss can still represent a major performance loss mechanism for micropropulsion systems which operate at elevated temperatures. For example in a micro-electrothermal propulsion system, the input power to the device is also assumed to decrease over larger systems. Therefore, the ratio of radiative power lost to input power may not scale as favorably as imagined.

2. *Conduction*

A simple one-dimensional form of Fourier's law for heat conduction can be written

$$q_c = kA(\mathrm{d}T/\mathrm{d}x) \tag{9}$$

where $\mathrm{d}T/\mathrm{d}x$ is the temperature gradient across the material. If the thermal conductivity is assumed to be constant throughout the material, Eq. (9) can be written

$$q_c = kA(T_2 - T_1)/\Delta x \tag{10}$$

where $T_2 - T_1$ is the temperature difference across Δx, the material thickness in the x-direction.

A relatively minor issue associated with a simple macroscopic approach to conductive heat transfer was mentioned earlier in terms of the assumption of constant thermal conductivity of the material. This is a minor issue since a spatial and temperature dependency of the thermal conductivity can be included in the integral solution of Eq. (9). However, one critical problem with the entire formulation lies in the assumptions used in deriving Fourier's law from basic principles. The above approach is valid only for systems in local thermodynamic equilibrium (LTE) in some small volume compared to the overall system. When the characteristic size

Table 2 Thermal conductivity at $T = 300$ K and melting temperature of common materials[27,28]

Material	Thermal conductivity (k), W/m K	Melting temperature, °C
Silicon	157	1350
Silicon dioxide (SiO_2)	1.38	1200
Silicon nitride (Si_3N_4)	30.1	1200–1400
Silicon aerogel	0.017	1200
Silicon carbide (SiC)	75–155	1400
Aluminum (6061-T6)	104.7	660
Nickel	90	1440
Titanium	180	1670
Stainless steel (304)	15	1425
Copper	368.5	1084
Silver	407.1	962
Teflon (PTFE)	0.35	340
Pyrex	1.1	230–490

of the system becomes of the order of the volume over which local thermodynamic equilibrium is assumed, the LTE formulation breaks down and additional formulations based on nonequilibrium thermodynamics must be utilized.[26]

Despite the inherent problems with the preceding formulation that make it inadequate to describe complex microsystems, it can lend some insight into the conductive heat transfer behavior at small characteristic dimensions. For example, it can be seen from Eq. (10) that large temperature differences are quite difficult to maintain through microscale thermal insulation unless the thermal conductivity of the insulator is extremely low. The thermal conductivity and melting temperature of some common metals and MEMS materials are given in Table 2. As the linear dimensions of a device scale down, the heat conduction at a constant ΔT decreases as the characteristic dimension (l_c) assuming that all dimensions scale equally. This is an interesting result considering that the radiative heat transfer scales with l_c^2 as described in the previous section.

3. *Micronozzle Heat Transfer*

The heat transfer between a high-speed, high-temperature gas and its flow channel is given by[29]

$$\frac{h_g D}{\kappa} = 0.026\left(\frac{Dv\rho}{m}\right)^{0.8}\left(\frac{mc_p}{\kappa}\right)^{0.4} \tag{11}$$

The term on the left-hand side of Eq. 11 is known as the Nusselt number, $Dv\rho/\mu$ is the Reynolds number, and $\mu c_p/\kappa$ is the Prandtl number. From Eq. (11), it is evident that the smallest flow dimension (the throat in the case of nozzle flow) corresponds to the maximum heat transfer coefficient.

For constant-temperature operation, Eq. (11) varies as $Re^{0.8}/D$ (or $\rho^{0.8}/D^{0.2}$) as the characteristic dimensions of the flow channel are scaled. Therefore, in micronozzle flows, the amount of heat transfer between the gas and the nozzle walls is a competition between the characteristic scaling factor (decrease D) and the operating pressure. Although reducing the operating pressure can reduce the heat

Table 3 Thermal expansion for several materials[28]

Material	Thermal expansion, $10^{-6}/°C$
Silicon (single crystal)	2.35
Silicon dioxide (SiO_2)—bulk	7.1
Silicon nitride (Si_3N_4)	0.8
Silicon carbide (SiC)	3.3
Aluminum (pure)	25
Aluminum oxide (Al_2O_3)	5.4–8.7
Diamond (singe crystal)	1.0
Polysilicon (annealed)	2.8
Tungsten	4.5
Stainless steel (316)	17.3

transfer to the nozzle walls, the effect of decreasing pressure (decreasing *Re*) can adversely affect the micronozzle efficiency in terms of increased viscous losses. For constant-*Re* scaling of micronozzle flows, the heat transfer coefficient varies as $1/D$ at a constant operating temperature. Therefore, attempting to maintain performance for micronozzle expansions could have a significant effect on the heat transfer. Surface roughness can also have a large effect on the amount of heat transfer in a micronozzle. Experiments have indicated that the heat transfer can be increased by a factor of two by surface roughness.[29]

4. *Thermal Expansion*

The thermal expansion or contraction of material can become a design challenge for micron-scale devices. Micromachined structures that have components made of different materials or large temperature gradients are particularly vulnerable to adverse thermal expansion effects. Different thermal expansion can cause problems such as gasket leakage and stress fractures. The thermal expansion for several materials is given in Table 3.

5. *Thermal Modeling Issues*

There are a number of thermal models that can accurately predict heat transfer characteristics of engineering systems. The complications of microsystems can have a major impact on the thermal modeling of conduction as described above. Kinetic approaches have become necessary for the modeling of micro-fluid flows and similar approaches may be required for specific micro-heat transfer problems where scattering of energy carriers in a material matrix can dominate the energy transfer mechanisms.[25] As with any modeling technique, whether applied to fluid, gas, ion, or energy (heat) transport, traditional models may not be valid in predicting even macroscopic properties for microsystems. The areas of applicability of more traditional analytical approaches need to be investigated, and new models need to be developed when appropriate.

Vast improvement has been experienced by radiation models through optically thick and thin gases in recent years.[30–32] Finite volume, discrete ordinates, and line-by-line methods are the most common approaches to the prediction of radiative transfer in absorbing, emitting and scattering in media and at surfaces. These methods are only as good as the individual models that they incorporate and the boundary conditions that are set by the user.

6. *Thermal Design Considerations*

The design of thermal systems for MEMS-fabricated devices can be quite challenging. Silicon has a rather high thermal conductivity (in some cases higher than that of some aluminum alloys), and heat conduction through micropropulsion devices can be rather large. This can be either convenient or a design challenge, depending on the particular system. For example, in compact devices, the thickness of insulating material can be quite restricted. As seen in the above formulation for conduction, extremely thin insulating layers will not support a large thermal gradient unless the thermal conductivity of the material is very low. On the other hand, very small contact points can be used between surfaces to minimize the amount of heat transfer that can take place.

A concept that has received an enormous amount of attention recently is the use of aerogels for their attractive thermal properties.[33] As shown in Table 2, typical silica aerogels have a total thermal conductivity of approximately 0.017 W/mK.[34] The thermal conductivity of silica aerogels is limited by the energy transport of gas molecules within the porous material and the minimum density requirements for structural integrity. Lowering the gas pressure in the environment in which the aerogel is located can reduce the thermal conductivity.

Thermal expansion and contraction of materials at temperature extremes will be a major issue in micropropulsion systems, where small-scale changes in a material's dimensions can be critical. This is especially true for systems with several components micromachined from different materials that may experience extreme temperature differences. It is clear that thermal issues will be very important in micropropulsion devices, as they are in a wide range of microdevices. The particular properties of micro-heat transfer can be of enormous potential if exploited in an imaginative manner.

E. MEMS Device Considerations

An important problem that has occupied the attention of researchers for many years is the flow of rarefied gases through tubes, nozzles, and ducts. This flow problem offers the researcher a relatively simple experimental geometry with a wide variety of analytical solutions available for the continuum and free molecular flow limits. Additionally, the application of a sufficiently high pressure ratio in a device can result in the simultaneous existence of continuum, slip, transition, and free molecular flow regimes along the flow path within the device due to the pressure gradient. An appropriately detailed experimental study of this environment, for example, could provide the background to assess the accuracy and limits of validity of the NS equations of continuum gas dynamics when approaching the rarefied gas limit. MEMS devices such as micro sensors, actuators, valves, motors, accelerometers, and other devices with dimensions measured on the micron scale employ the transport of gases and liquids where the small physical dimensions provide a unique flow environment that has received very little study to date. Surface, viscous, and rarefaction influences, for example, become more important as the size scale becomes smaller, even when the working gases are at atmospheric pressures.

1. Knudsen Number Regimes in Microscale Devices

The Knudsen number is the ratio of the mean free path in a gas to a characteristic dimension of the flow. Using the Knudsen number as a parameter, fluid mechanics may be subdivided into the following four flow regimes.

$Kn < 0.01$:	Continuum flow
$0.01 < Kn < 0.1$:	Slip flow
$0.1 < Kn < 3$:	Transition flow
$Kn > 3$:	Free molecular flow

Flows with small values of Kn are the best known from a traditional fluid mechanics standpoint. The Knudsen number can approach unity due to a low molecular density or very small characteristic dimension. For many MEMS devices, the Knudsen number is outside the continuum regime even at atmospheric pressure conditions due to the extremely small feature size of these devices (of the order of 1 μm). For example, typical microtube designs can easily yield diameters in the range of 1–2 μm. The mean free path of air at standard conditions is approximately 60 nm. For a MEMS device utilizing a 1-μm-diam microtube at atmospheric pressure, a Knudsen number clearly in the slip flow regime ($Kn \sim 0.06$) would result. Noncontinuum effects could have a significant effect on the device's operation. Careful fabrication, experimental investigation, and modeling will be necessary to ensure desirable behavior within these devices.

2. Microvalve Operation

High propellant leak rates through MEMS valves is known to be an important problem facing their utilization in micropropulsion systems.[1,14,16] MEMS valve technology is progressing to the point where acceptable leak rates for long-term missions may be realized in the near-future.[35] However, an often overlooked problem with MEMS valves is the characteristics of the gas flow through these devices.

Many MEMS valves under consideration have very small actuation distances (of the order of tens of microns). The flow through these devices must be characterized to determine the effects of the valve flow on the performance of the overall propulsion system. For example, the flow from the valve may be sent to a micronozzle with a throat diameter larger than the valve flow channel.

Some microspacecraft missions will require low-impulse bit (integrated thrust with time) maneuvers. For example, a slew maneuver on a microspacecraft may require on the order of 1 μN-s. There are two ways of obtaining this with a propellant valve-controlled micropropulsion system. First, operate the thruster at very low thrust levels and actuate the valve relatively slowly. Second, operate the thruster at a high thrust level and actuate the valve rapidly. For example, a thruster that produces a minimum thrust level of 1 mN, requires a valve actuation time of the order of 1 ms. Current MEMS valve technology that utilizes piezoelectric actuation can perform open/close cycles of the order of 1 ms to tens of milliseconds. At any rate, the flow from this actuation will most likely be highly nonuniform. If it is to be expanded through a micronozzle, the Reynolds number range of operation during the I-bit maneuver can vary greatly, causing the nozzle to operate very inefficiently for a significant fraction the maneuver time. The predictability and reproducibility of the transient gas flow from the valve and other components are also questionable. Highly detailed experimental and numerical analysis of these MEMS propulsion system components will be necessary.

III. Micropropulsion System Considerations

It is evident that an overall system level approach will be required for microspacecraft and micropropulsion designs. For example, analyses that separate the power supply mass of an electric propulsion system when determining the appropriate thruster may no longer be valid in the microspacecraft arena. A given micropropulsion system must be designed with the general limitations inherent with microspacecraft in mind. The performance of micropropulsion systems must be viewed in light of the systems concerns generated or solved by their particular design. For example, a thruster with a relatively low specific impulse might be the most attractive system for a particular mission given that it solves many system level concerns. A brief discussion of these limitations as they relate to the micropropulsion system is given.

The first generations of microspacecraft will be severely mass, power, and volume limited. Micropropulsion systems not only will have to address the mission requirements, but also will have to fit within the system constraints of particular spacecraft. The acceptable mass for a micropropulsion system is very mission dependent, but it is imaginable that restrictions on the total wet mass of the propulsion system may be approximately 10–20% (somewhat lower than acceptable limits on large spacecraft). The need for MEMS-fabricated, fully integrated propulsion packages is relatively clear. General rules of thumb indicate that, during spacecraft operations, approximately 1 W per kg of spacecraft mass will be available for the propulsion system.[1] Of course, higher power levels can be supplied to the propulsion system during periods of critical orbital maneuvering or when other systems are not required. Perhaps one of the most severe restrictions on micropropulsion systems is the volume restriction of microspacecraft. Most gaseous propellants (even relatively high-density xenon for use in ion thrusters) may require too much storage volume, thus eliminating them from the smallest microspacecraft.

Microspacecraft will also be voltage-limited in that the maximum voltage acquired from a small solar array may be only a few volts. Voltage levels of this order are typically too low for most spacecraft systems, thus requiring heavy power processing units. For microspacecraft, the limits in power and voltage arise from the surface area available on the spacecraft structure for mounting of the solar cells. Deployable solar arrays will need to be compact and lightweight (relative to the spacecraft) to be useful for microspacecraft.

A. Micronozzle System Considerations

As discussed previously, micronozzle expansions must operate through relatively high pressure ratios to be efficient. This indicates that micronozzle systems will need to operate at pressures ranging from 0.5 to 10 atm. For cold gas systems, a gaseous propellant is typically stored at very high pressures (>20 MPa). This implies a relatively thick-walled and heavy propellant tank, which can require an excessive volume.[16] High-pressure gas storage also places restrictions on the propellant feed lines, valves, and pressure regulators. Although MEMS valves are currently undergoing a great deal of development, MEMS shut-off valves are currently not available with acceptable leak rates for very high pressures.[14]

Due to the systems-related problems with high-pressure gas storage, micropropulsion systems that can operate on propellants stored as liquids or solids on orbit can be beneficial. Liquid or solid storage not only alleviates valve leakage concerns but also significantly reduces the required tank volume over high-pressure gas storage. An additional concern, for nonchemical thrusters, with using liquids and solids as propellants is the power required to vaporize the propellant in some fashion to operate efficiently. This is typically done by resistive heating (vaporization or sublimation) or by electrical discharge.

One final system consideration for micronozzle configurations is the potential of catastrophically plugging the nozzle throat with contaminants. Strict filtering will be required to protect against this possibility, which adds system complexity and mass. Although filtering can reduce the probability of catastrophic plugging, this single point failure must be addressed in the design phase of any micronozzle configuration.

B. Micro-Ion Thruster System Considerations

It seems clear that current gaseous propellant ion propulsion systems are not easily scaled much below the millimeter size region. New approaches will need to be identified to improve performance while satisfying microspacecraft system-level considerations. For example, increasing the potential on the accelerating grids can compensate for a decreased ionization flux from a micro-ion thruster due to electron wall losses. Obviously there are severe limits to this practice on microspacecraft in terms of power supply mass. However, a complete system analysis for each individual mission taking into account propellant mass savings vs power supply mass should be investigated.

There are obvious system benefits to micro-ion thruster concepts that do not use magnetic fields or accelerating grids or that utilize solid or liquid propellants. One nonmagnetic concept that has been investigated incorporates a geometrical configuration in an attempt to increase electron current densities in specific regions of the discharge.[36,37] In this case, the hemispherical electrode shown in Fig. 7 is used to focus electrons at the exit orifice of the device (where ion formation is the most beneficial). Figure 7 also shows the potential contours inside the discharge region. Although these systems-derived thrusters may have limited performance relative to large-scale ion thrusters, their implementation onto microspacecraft can be significantly improved.

Micro-ion thrusters that address systems concerns will have several inherent benefits such as valveless operation, liquid or solid propellant storage, and low integrated power modes (perhaps achieved with pulsed operation). However, there will undoubtedly be some drawbacks to their use due to the potential for low efficiency or poor performance compared to other more complex systems. For example, the extremely low thrust levels expected from micro-pulsed plasma thrusters makes them unattractive for missions requiring a large Δv due to the prohibitive number of firings required; however, they are very attractive from an overall systems viewpoint. At any rate, one micro-ion thruster system will not address all of the mission requirements for microspacecraft operations. A variety of thrusters of all types will be needed to fill specific applications.

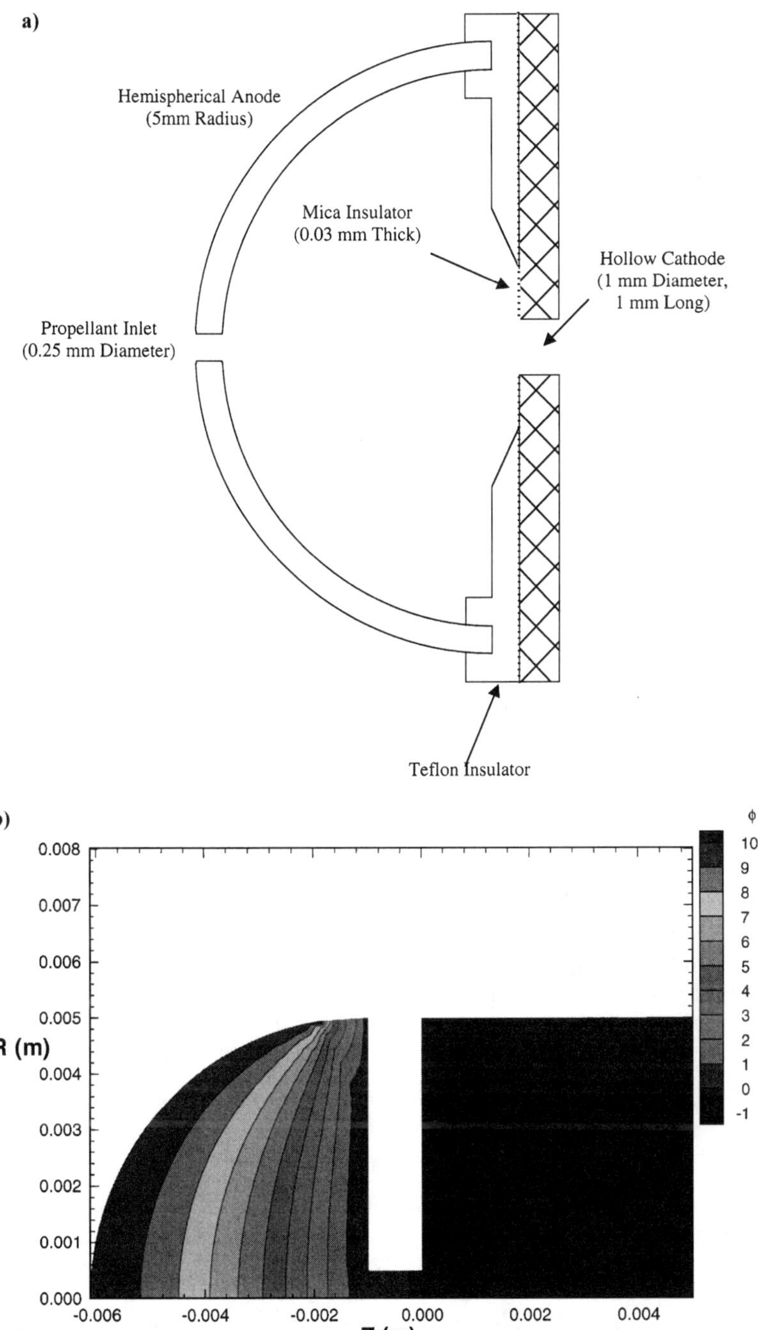

Fig. 7 Hemispherical electrode: a) nonmagnetic, hollow cathode configuration; b) axisymmetric calculation of the potential contours for the electrode geometry. The electric field lines are perpendicular to the potential lines.

C. Microchemical Thruster System Considerations

Chemical micropropulsion systems have similar drivers for design as many of the other concepts discussed in the previous sections. Obviously, propulsion systems that incorporate microspacecraft limitation considerations directly into their design are of much interest. The use of propellants that are easily handled, not corrosive to most surfaces, and store easily on-orbit (i.e., no high-pressure gases or cryogenic fluids) may not have the highest specific impulse available but will make up for this shortcoming through better integration. Concepts that use subliming solids are of particular interest since valveless concepts can be envisioned. Microvalves and fluid pumps will play a significant role in other chemical micropropulsion systems where precise mixtures of fuels and oxidizers will be required for optimum operation.

Many chemical systems may also require pressurant gases to be stored to assist with propellant management. This high-pressure gas storage will need to be addressed in microspacecraft systems studies to assess the added volume, mass, and complexity. Microchemical thrusters that can operate without the need for high-pressure blow-down systems or that might be able to use the stored pressurant gas for a dual purpose (e.g., propellant for a cold gas thruster system) will be of interest.

Solid propellants have several system-related benefits due to their stable nature, ease of storage on-orbit, and minimum valve requirements. However, most solid combustion fuels have a controllability issue with throttling and restart. For this reason, there is a growing interest in the use of subliming solid propellants in micropropulsion systems. These solid propellants decompose at reasonable temperatures into low molecular mass components, typically CO_2, NH_3, or H_2O.[38] Because of the small characteristic dimensions available from micromachined nozzles, adequate vapor pressures may be obtained to provide acceptable performance by absorbing heat from the surrounding spacecraft structure.

Propellants that sublimate at temperatures higher than nominal spacecraft temperatures will require active energy sources to provide the heat of sublimation for adequate pressures to be produced. The energy required for sublimation can come from sources such as resistive heaters and diode lasers. A subliming solid thruster can be throttled by adjusting the energy input to the propellant enhancing the controllability. A potential problem with subliming solid systems is the tendency of the propellant to recondense on cold surfaces within the thruster. Additional heat sources may be required in the feedlines and near valves to prevent condensation in undesired locations.

IV. Conclusions

Novel approaches and concepts will be critical in determining the future applications of micropropulsion systems. As mentioned earlier, simply scaling down existing large-scale thrusters has limited application as the characteristic size grows ever smaller. Spacecraft with masses in the range of 50–100 kg can benefit from existing thrusters and scaled-down versions of existing thrusters, but the resource-limited microspacecraft with a mass lower than 10 kg is going to require micropropulsion systems developed from overall systems approaches.

It is becoming evident that micropropulsion systems that can be fully integrated with other MEMS subsystems or that serve a dual purpose may have significant benefits in microspacecraft applications. For example, a propulsion system that can operate at the same voltage level as other subsystems (payloads, for example) or can act as a structural member of the spacecraft has obvious mass savings.

A great deal of work is required to produce acceptable micropropulsion systems that will have many of the beneficial characteristics described throughout this work. There are many areas of current research where great strides can be made by those in several communities. For example, one topic receiving a large amount of attention in the MEMS community is the possibility of micromachining in other than silicon-based materials. This work has tremendous implications to MEMS propulsion concepts in that it could make available materials with significantly improved thermal, electrical, and mechanical properties. MEMS propulsion designers sorely need materials that are better electrical or thermal insulators.

Other areas where investigators might have a significant impact are high-accuracy micro-Newton thrust measurements, very low-mass flow (fluid and gas) measurements, and the development of diagnostic tools that have sufficient spatial resolution. Currently, many micropropulsion concepts cannot be performance tested due to the lack of facilities capable of making thrust measurements of the order of tens of micro-Newtons. Diagnosing micropropulsion systems also becomes problematic due to the spatial resolution of most diagnostic techniques currently in use. Typical laser or electron beam spatial resolution is limited to something of the order of 0.1 mm. Systems capable of achieving spatial resolutions better than this are typically expensive and complex.

Acknowledgments

The author would like to thank E. P. Muntz, Dean Wadsworth, Juergen Mueller, Ingrid Wysong, and Jay Levine for their helpful suggestions and contributions to this work.

References

[1]Mueller, J., "Thruster Options for Microspacecraft: A Review and Evaluation of Existing Hardware and Emerging Technologies," *Micropropulsion for Small Spacecraft*, Progress in Astronautics and Aeronautics, Vol. 187, edited by M. Micci and A. Ketsdever, AIAA Reston, VA, 2000, Chap. 3 (this volume).

[2]Janson, S., Helvajian, H., and Robinson, E., "The Concept of 'Nanosatellite' for Revolutionary Low-Cost Space Systems," IAF 93-U.5.573, 44th International Astronautical Congress, Graz, Austria, Oct. 1993.

[3]Collins, D., Kukkonen, C., and Venneri, S., "Miniature, Low-Cost, Highly Autonomous Spacecraft—A Focus for the New Millennium," IAF 95-U.2.06, 46th Inernational Astronautical Congress, Oslo, Norway, Oct. 1995.

[4]Fleeter, R., *Microspacecraft*, Edge City Press, Reston, VA, 1995.

[5]Cobb, R., "TechSat 21: Advanced Research and Technology Enabling Distributed Satellite Systems," *Proceedings of the AFRL Formation Flying and Micropropulsion Workshop*, Lancaster, CA, Oct. 1998.

[6]Das, A. (ed.), *Proceedings of the AFOSR/DARPA/AFRL/VSD Micro/Nanotechnology for Micro/Nanosatellites Workshop*, Albuquerque, NM, Apr. 1998.

[7]Kauzmann, W., *Kinetic Theory of Gases*, W. A. Benjamin, New York, 1966.

[8]Bayt, R., Ayon, A., and Breuer, K., "A Performance Evaluation of MEMS-Based Micronozzles," AIAA Paper 97-3169, July 1997.

[9]Rothe, W., "Electron-Beam Studies of Viscous Flow in Supersonic Nozzles," *AIAA Journal*, Vol. 9, No. 5, 1971, pp. 804–811.

[10]Rae, W., "Some Numerical Results on Viscous Low-Density Nozzle Flows in the Slender-Channel Approximation," *AIAA Journal*, Vol. 9, No. 5, 1971, pp. 811–820.

[11]Boyd, I., Penko, P., Meissner, D., and DeWitt, K., "Experimental and Numerical Investigations of Low-Density Nozzle and Plume Flows of Nitrogen," *AIAA Journal*, Vol. 30, No. 10, 1992, pp. 2453–2461.

[12]Zelesnik, D., Micci, M., and Long, L., "Direct Simulation Monte Carlo Model of Low Reynolds Number Nozzle Flows," *Journal of Propulsion and Power*, Vol. 10, No. 4, 1994, pp. 546–553.

[13]Chung, C., Kim, S., Stubbs, R., and DeWitt, K., "Low-Density Nozzle Flow by the Direct Simulation Monte Carlo and Continuum Methods," *Journal of Propulsion and Power*, Vol. 11, No. 1, 1995, pp. 64–70.

[14]Janson, S., and Helvajian, H., "Batch-Fabricated Microthrusters: Initial Results," AIAA Paper 96-2988, July 1996.

[15]Ivanov, M., Markelov, G., Ketsdever, A., and Wadsworth, D., "Numerical Study of Cold Gas Micronozzle Flows," AIAA Paper 99-0166, Jan. 1999.

[16]Ketsdever, A., Wadsworth, D., and Muntz, E. P., "The Free Molecule Micro-Resistojet: An Interesting Alternative to Nozzle Expansion," AIAA Paper 98-3918, 34th Joint Propulsion Conf., Cleveland, OH, July 1998.

[17]Bird, G., *Molecular Gas Dynamics and the Direct Simulation of Gas Flows*, Clarendon Press, Oxford, UK, 1994.

[18]Yashko, G., Giffin, G., and Hastings, D., "Design Considerations for Ion Microthrusters," IEPC 97-072, 25th International Electric Propulsion Conf., Aug. 1997.

[19]Mueller, J., Tang, W., Li, W., and Wallace, A., "Micro-Fabricated Accelerator Grid System Feasibility Assessment for Micro-Ion Engines," IEPC 97-071, 25th International Electric Propulsion Conf., Aug. 1997.

[20]Wang, J., Biasca, R., and Liewer, P., "Three-Dimensional Electromagnetic Monte Carlo Particle-In-Cell Simulations of Critical Ionization Velocity Experiments in Space," *Journal of Geophysical Research*, Vol. 10, No. A1, 1996, pp. 371–382.

[21]Longo, S., and Boyd, I., "Coupled PIC/MCC and State-to-State Continuum Model of a Parallel-Plate RF Hydrogen Discharge," AIAA Paper 98-2984, June 1998.

[22]Oh, D., and Hastings, D., "Three Dimensional PIC-DSMC Simulations of Hall Thruster Plumes and Analysis for Realistic Spacecraft Configurations," AIAA Paper 96-3299, 32nd Joint Propulsion Conf., Lake Buena Vista, FL, July 1996.

[23]Humble, R., Henry, G., and Larson, W., *Space Propulsion Analysis and Design*, McGraw–Hill, New York, 1995.

[24]Ketsdever, A., Wadsworth, D., Wapner, P., Ivanov, M., and Markelov, G., "Fabrication and Predicted Performance of DeLaval Micronozzles," AIAA 99-2724, 35th Joint Propulsion Conf., Los Angeles, CA, June 1999.

[25]Tien, C., Majumdar, A., and Gerner, F. (eds.), *Microscale Energy Transport*, Taylor and Francis, Washington, DC, 1998.

[26]Rohsenow, W., Hartnett, J., and Cho, Y., *Handbook of Heat Transfer*, 3rd ed., McGraw–Hill, New York, 1998.

[27]Madou, M., *Fundamentals of Microfabrication*, CRC Press, New York, 1997.

[28]*Handbook of Chemistry and Physics*, edited by D. Lide, CRC Press, New York, 1998.

[29]Sutton, G., *Rocket Propulsion Elements*, 6th ed., Wiley, New York, 1992.

[30]Murthy, J., and Mathur, S., "A Finite Volume Method for Radiative Heat Transfer Using Unstructured Meshes," AIAA Paper 98-0860, Jan. 1998.

[31]Liu, J., Shang, H., Chen, Y., and Wang, T., "Investigation of Rocket Plume Radiation by Discrete Ordinates Method," AIAA Paper 96-0348, Jan. 1996.

[32]Menart, J., Heberlein, J., and Pfender, E., "Theoretical Radiative Transport Results for a Free-Burning Arc Using a Line-by-Line Technique," AIAA Paper 98-0992, Jan. 1998.

[33]Schaefer, D., Olivier, C., Ashley, C., Richter, D., Farago, B., Frick, B., Hrubesh, L., Vanbommel, M., Long, G., and Krueger, S., "Structure and Topology of Silica Aerogels," *Journal of Non-Crystalline Solids*, Vol. 145, 1992, pp. 105–112.

[34]Bernasconi, A., Sleator, T., Posselt, D., Kjems, J., and Ott, H, "Dynamic Properties of Silica Aerogels as Deduced from Specific-Heat and Thermal-Conductivity Measurements," *Physical Review B—Condensed Matter*, Vol. 45, 1992, pp. 10363–10376.

[35]Mueller, J., "A Survey of the State-of-the-Art MEMS Valve Technology," AIAA Paper 99-2725, 35th Joint Propulsion Conf., Los Angeles, CA, June 1999.

[36]Young, M., Muntz, E. P., and Ketsdever, A., "Investigation of a Candidate Non-Magnetic Ion Micro-Thruster for Small Spacecraft Applications," AIAA Paper 98-3917, 34th Joint Propulsion Conf., Cleveland, OH, July 1998.

[37]Young, M., Muntz, E. P., and Ketsdever, A., "Unique Hollow Cathode as a Candidate Non-Magnetic Ion Thruster," AIAA paper 99-2854, 35th Joint Propulsion Conf., Los Angeles, CA, June 1999.

[38]Hardt, A., Foley, W., and Brandon, R., "The Chemistry of Subliming Solids for Micro Thrust Engines," *Astronautica Acta*, Vol. 11, No. 5, 1965.

II. Electrothermal Thrusters

Chapter 5

Predicted Performance and Systems Analysis of the Free Molecule Micro-Resistojet

Andrew D. Ketsdever*
Air Force Research Laboratory, Edwards Air Force Base, California
Dean C. Wadsworth†
ERC, Inc., Edwards Air Force Base, California
and
E. P. Muntz‡
University of Southern California, Los Angeles, California

Nomenclature

A	= expansion slot or orifice area, m^2
$\bar{c}'$	= propellant molecule average thermal speed, m/s
F_{tu}	= material ultimate tensile strength, GPa
g	= acceleration of gravity, 9.81 m/s^2
I_{sp}	= thruster intrinsic specific impulse, s
I_{sp},eff	= thruster effective specific impulse, s
k	= Boltzmann constant, 1.38×10^{-23} J/K
M	= mass, kg
$\dot{M}$, $\dot{m}$	= propellant mass flow rate, kg/s
m	= propellant molecular mass, kg
n	= number density, m^{-3}
p	= pressure, Pa
R	= universal gas constant, k/m
T	= temperature, K
$\Im$	= thrust, N
t_L	= time required to leak propellant through valve, s
u_{limit}	= limit velocity through high pressure ratio, m/s
V	= volume, m^{-3}

This material is declared a work of the U.S. Government and is not subject to copyright protection in the United States.

*Senior Research Engineer, Propulsion Directorate, Advanced Concepts Division. Senior Member AIAA.

†Principle Scientist.

‡A. B. Freeman Professor, Department of Aerospace Engineering. Fellow AIAA.

w = expansion slot width, m
x = axial position, m
y = tangential position, m
α = slot expansion angle, deg
γ = propellant ratio of specific heats
Δv = orbital maneuver velocity increment, m/s
ε = factor of safety
κ = valve leak rate, m^3/s
ρ = density, kg/m^3
τ = shear, N
ζ = fraction of propellant lost due to valve leakage

Subscripts

des = designed condition
FM = free molecule
i = initial
MN = micronozzle
p = propellant
r = required to perform mission
s, stor = stored (propellant tank)
t = propellant tank
w = heated wall
x = along axial direction
0 = stagnation condition

I. Introduction

IT is clear that in many cases micropropulsion systems will not be simply scaled-down versions of existing macroscale thrusters.[1,2] As recently discussed by Muntz and Ketsdever,[3] the domain of small (in the limit micromechanical) propulsion systems offers many opportunities if the distinctive characteristics that dominate at small scales are exploited in an imaginative manner. One example of a unique microthruster, designed to show the utility of micrometer-scale technology application, is the free molecule micro-resistojet (FMMR) illustrated in Fig. 1.

The FMMR operates with unusually low stagnation pressures (50 to 500 Pa) and correspondingly large slot apertures. The design requirement is to arrange a surface held at the required stagnation temperature to be the last surface contacted by a propellant molecule before it exits the expansion slot. This requirement suggests that the spacing between the heating element and the expansion slot be of the order of the mean free path of the stagnation gas to reduce undesired intermolecular collisions that act to limit the overall efficiency. Since several configurations can be imagined, the particular stagnation chamber arrangement shown in Fig. 1 is only one possibility. The FMMR can be either a microelectromechanical systems (MEMS) device or a somewhat larger mesoscale device.

The FMMR offers several distinct advantages over conventional microthruster concepts for attitude control and station keeping maneuvers. The FMMR combines MEMS fabrication techniques with simple, lightweight construction that consists

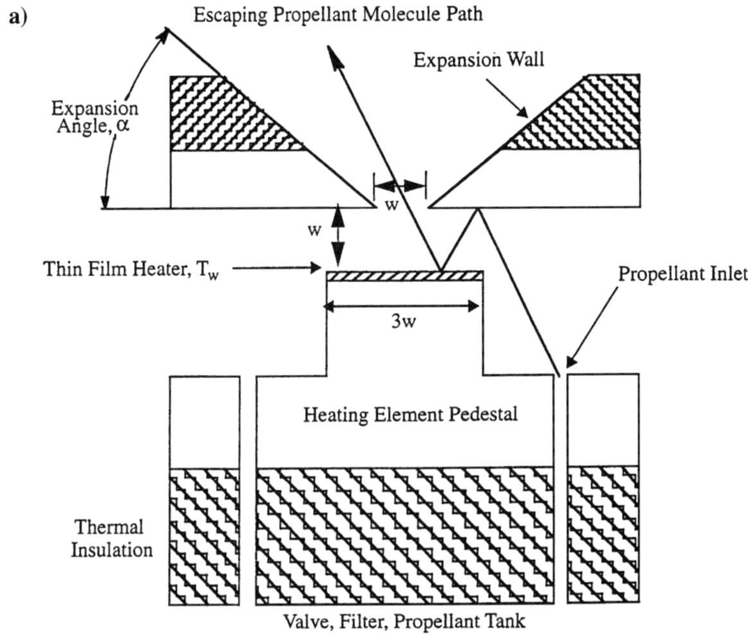

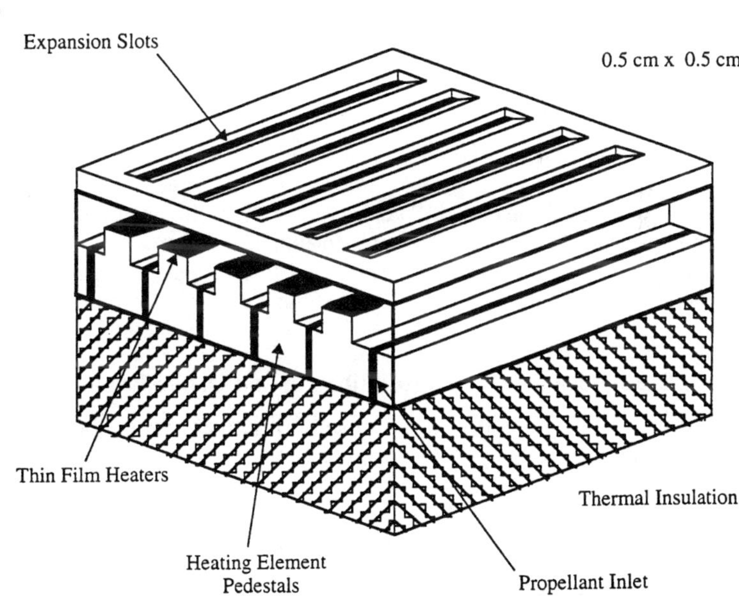

Fig. 1 FMMR: a) Schematic cross section showing the heating element arrangement with the expansion slot; b) multislot ($w = 100\ \mu$m) configuration with a 0.5×0.5-cm cross-sectional area.

of a thin-film heating element, at a temperature T_w, and an exhaust slot. A long (~1-cm), narrow (1- to 100-μm) slot is chosen over a small nozzle expansion because of the possibility of catastrophically plugging a nozzle throat (typically of the order of 20 μm in diameter) with contaminants and ease of manufacture. The free-molecule condition is chosen for the additional benefit of reduced propellant storage pressure, therefore easing the propellant tank mass and valve leakage requirements.

The aim of this research was to identify, through numerical computations, an optimized design (based on performance) for the FMMR geometry. Once an optimized design was identified, the FMMR performance was compared to the performance obtained with a typical micronozzle expansion with an overall systems level approach.

II. Theory

A. Specific Impulse from Free Molecule Flow

In free molecule flow, the thrust or flux of normal momentum from an orifice with stagnation pressure p_0 is

$$(\Im)_{\mathrm{FM}} = \frac{p_0}{2}A = \frac{n_0 k T_0}{2}A \tag{1}$$

and the free molecule mass flow from the orifice is

$$(\dot{m})_{\mathrm{FM}} = m\frac{n_0 \bar{c}'}{4}A = m\frac{n_0 A}{4}\sqrt{\frac{8kT_0}{\pi m}} \tag{2}$$

Therefore, the specific impulse of the FMMR thruster operating at a stagnation temperature T_0 is

$$(I_{\mathrm{sp}})_{\mathrm{FM}} = \sqrt{\frac{\pi}{2}\frac{k}{m}T_0}\Big/ g \tag{3}$$

In general, the FMMR heated wall temperature T_w is higher than the thruster stagnation temperature T_0 near the expansion slot. The difference is due to incomplete accommodation at the heated surface and line-of-sight escape through the expansion slot from relatively cold regions of the stagnation chamber.

B. Specific Impulse from Limit Equilibrium and Orifice Expansion

The intrinsic specific impulse for a limit equilibrium expansion through a high-pressure ratio nozzle (infinite exit Mach number) is given by[4]

$$(I_{\mathrm{sp}})_{\mathrm{L}} = \frac{u_{\mathrm{limit}}}{g} = \left\{\frac{2\gamma}{\gamma - 1}\left(\frac{k}{m}\right)T_0\right\}^{\frac{1}{2}}\Big/ g \tag{4}$$

The intrinsic specific impulse for a continuum gas expansion through an orifice (exit Mach number equal to unity) into low pressure is given by

$$(I_{\mathrm{sp}})_{\mathrm{OR}} = \left\{\frac{2(\gamma + 1)}{\gamma}\frac{k}{m}T_0\right\}^{\frac{1}{2}}\Big/ g \tag{5}$$

Therefore, the ratio of the specific impulse for limit equilibrium expansion to continuum orifice expansion into low pressure is

$$\frac{(I_{sp})_L}{(I_{sp})_{OR}} = \left\{\frac{\gamma^2}{(\gamma+1)(\gamma-1)}\right\}^{\frac{1}{2}} \tag{6}$$

The ratio from Eq. (6) is equal to 1.43 and 1.24 for γ equal to 1.4 and 1.67, respectively.

The ratio of the specific impulse for limit expansion through an ideal nozzle to that for free molecule flow through an orifice or slot is

$$\frac{(I_{sp})_L}{(I_{sp})_{FM}} = \left\{\frac{4\gamma}{\pi(\gamma-1)}\right\}^{\frac{1}{2}} \tag{7}$$

The ratio from Eq. (7) is equal to 2.11 and 1.78 for γ equal to 1.4 and 1.67, respectively. If nozzle losses (viscous and radial flow) are taken into account, the differences between the free molecule expansion and the nozzle expansion become even less pronounced.

Since I_{sp} variations are frequently traded for other system advantages, the system implications of the FMMR compared with micronozzle expansions for very small spacecraft are of interest. The performance losses associated with the free molecule slot expansion can be countered from a systems point of view through trade-offs such as design simplicity, reduced propellant storage pressure, and minimization of catastrophic nozzle clogging due to particulate contaminants. The effective or systems-derived specific impulse for a micronozzle expansion is addressed in the following sections.

III. Calculations

The analytical free molecular results from the previous section are expected to be useful in the basic design of the FMMR. However, the actual micro-resistojet involves multidimensional transitional rarefied gas flow, and more sophisticated analysis methods are required for accurate performance predictions. The direct simulation Monte Carlo (DSMC) method[5] provides a means to simulate the flow of a general rarefied gas at the molecular level. Many of the input models required in DSMC, such as gas–surface interactions, are topics of current research. Even with these uncertainties, the detailed information available from direct parametric simulations is indispensable to the design process. Parametric variation of input model features can also be used to assess the sensitivity of performance predictions to these uncertainties.

The present DSMC implementation has been applied to several problems related to micro-resistojet design, where the key feature to be resolved is the development of a bulk motion and the generation of thrust in a confined rarefied gas due to differential heating of a bounding surface.[6–9]

The results presented in the following sections use fully accommodating, diffusely reflecting surfaces characterized by a temperature T_w (the heated wall temperature). More accurate simulations, which iteratively couple the gas and structural heat transfer properties and thus eliminate the need for an input surface temperature distribution, will be reported at a later time.

The DSMC code is instrumented to sample and spatially resolve the flux of molecular mass, momentum, and energy through an arbitrary flowfield plane, chosen here typically to be the slot throat or the slot exit. These quantities are also sampled at all surface elements. These flux data allow division/assignment of thrust contributions and losses to various components of the structure, such as the slot walls. With the exception of the heating element wall temperature, all surfaces in the simulations are held at a stagnation temperature of 300 K.

The fidelity of the simulations depends on two features, the degree to which constraints of the DSMC method are met and the realism of the models and assumptions used. The former requirement was easily met for this rarefied flowfield. The nominal grid contained 2000 cells, with cells sized to be typically much less than 25% of the local mean free path. The nominal time step was approximately one-half of the plenum collision frequency.

The nominal simulation was run for 2000 unsteady time steps and for 10^4 additional steady-state time steps during which sampling of the flowfield occurred. The unsteady portion of the solution corresponded to approximately 200 acoustic times based on the plenum gas mean thermal speed and the slot width. The nominal simulation contained 2×10^5 particles at steady state. Nominal local cell sample sizes were of the order of 2×10^5, and flux plane sample sizes were 10^6. Statistical scatter in the specific impulse is estimated to be much less than 1% due to the large sample sizes obtained.

Since experimental performance measurements of this device are unavailable, a large variety of sensitivity studies has been carried out to estimate simulation accuracy. The largest potential source of error in the present simulations is expected to be in the modeling of the gas–surface interaction process. For nonideal surfaces, parametric studies using several phenomenological models show variations in predicted specific impulse of the order of 5%.[10]

IV. Results

Design parametrics were carried out by varying the heating element surface area, heating element wall temperature, T_w, slot divergence angle, propellant, and stagnation pressure. Each case was for a gas expansion slot width, w, of 100 μm and a slot depth of 250 μm. The slot width was not varied in the parametric study since an optimized design based on a given slot width can be easily scaled in the FMMR. The gas stagnation temperature far removed from the heating element and the temperature of all nonheated surfaces is 300 K.

The nominal operating conditions for the FMMR were $p_0 = 53.6$ Pa (corresponding to $Kn \sim 1$), $T_w = 600$ K, $\alpha = 40$ deg, and an argon propellant. Throughout the parametric study only one of these parameters was varied per calculation.

The performance of the FMMR is driven by molecular interactions with the thin-film heaters. Therefore, a design input was to determine the optimum sizing of the heated surface. Since the heated surface runs the entire length of the expansion slot, the FMMR performance was calculated varying only the thin-film heater width as shown in Fig. 2. The optimum heater element width is approximately 300 μm or $3w$. The optimum width is based on two factors. First, the FMMR specific impulse increases only marginally above 300 μm. Second, the marginal increases in I_{sp} above a width of 300 μm comes at a cost of increased power usage, which is proportional to the heater element area.

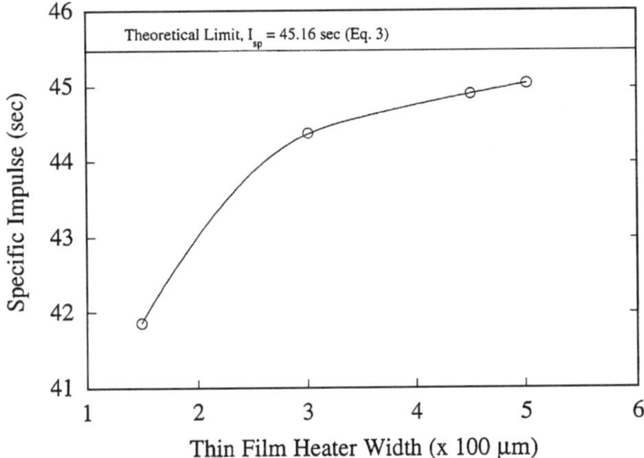

Fig. 2 Intrinsic specific impulse as a function of thin-film heater width for T_w = 600 K.

The effect of pedestal surface temperature on performance for an argon propellant gas is shown in Fig. 3. Although argon does not make a particularly attractive propellant for the FMMR, it is used in the numerical simulations for simplicity. The trend toward an improved I_{sp} as the temperature is increased is less than, but still comparable to, that predicted by the free molecular results [Eq. (3), varying as $\sqrt{T}$]. The differences from the theoretically predicted values are due to propellant molecules colliding with the expansion walls of the slot.

The major loss mechanism in terms of performance for the FMMR is expected to arise from hot propellant molecules (at a temperature close to T_w) colliding with the cooler slot expansion walls (at a temperature T_0). The effect of slot divergence angle, α, on performance is shown in Fig. 4. The case of $\alpha = 54.74$ deg corresponds to that etchable for the $\langle 100 \rangle$ plane of crystalline silicon. Figure 5 shows the net axial force along the slot expansion walls for several expansion angles. For the smaller expansion angle cases ($\alpha < 60$ deg), the expansion walls act to convert tangential momentum from the expanding flow into axial momentum,

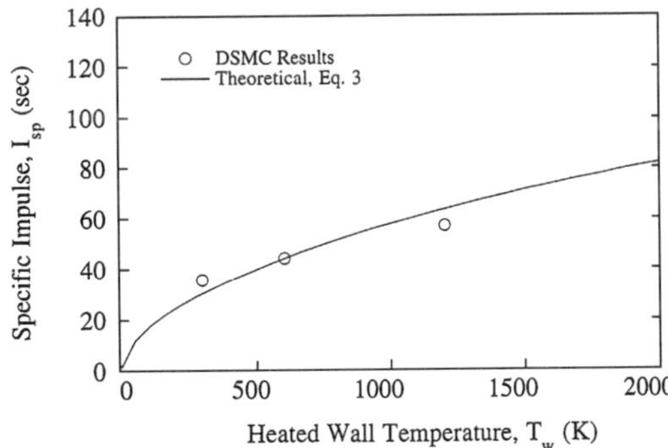

Fig. 3 Intrinsic specific impulse vs T_w for argon propellant. Theoretical line from Eq. (3).

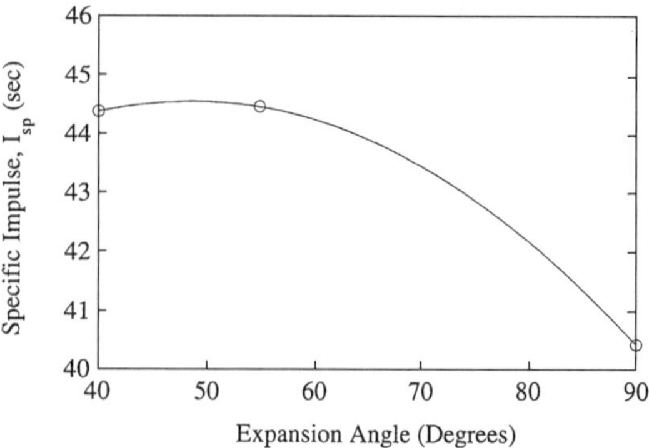

Fig. 4 Effect of slot expansion angle on intrinsic specific impulse.

which improves the overall performance of the FMMR. The magnitude of the increase in thrust due to the expansion walls is given by the integral under the curves in Fig. 5. The benefit of this momentum conversion is diminished for expansions through thick slots. For larger expansion angles (i.e., approaching 90 deg), the shear forces will dominate the flow at the walls, which act to degrade significantly the thruster's performance. The limiting case of 90 deg results in a constant width slot, where only shear losses arise. From this analysis, thin expansion walls with expansion angles less than 60 deg appear to be preferred from a performance standpoint. The relatively large shear losses indicate that, for these highly rarefied flows, low expansion ratios (or short slot heights) are preferable. This same design criterion arises in continuum analyses of less rarefied nozzle flows.[11]

The effect of propellant gas (molecular mass, m) on performance is shown in Fig. 6. The trend is also comparable to that predicted by the free molecular results [Eq. (3), varying as $\sqrt{1/m}$]. The polyatomic ammonia and water vapor have been modeled assuming zero internal energy. The gas–surface interaction process is expected increasingly to influence performance for more complex molecules.

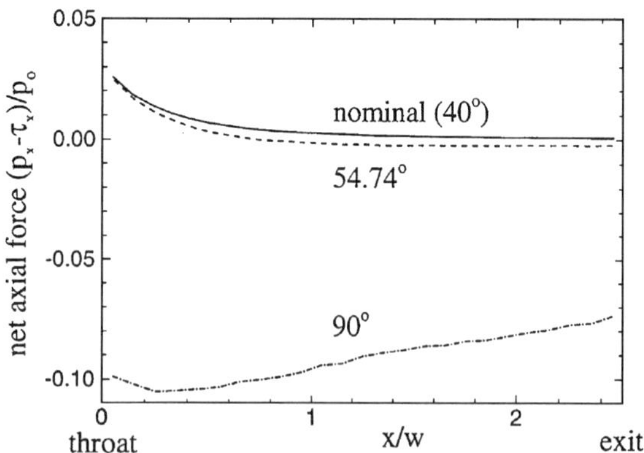

Fig. 5 Effect of slot expansion angle on axial force contribution.

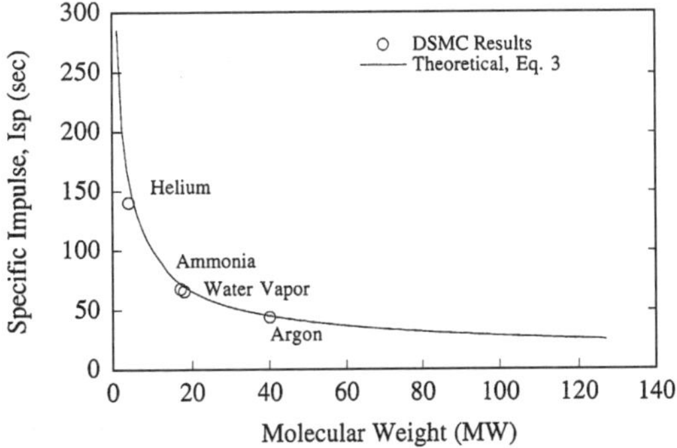

Fig. 6 Intrinsic specific impulse for various propellants for T_w = 600 K. Theoretical line from Eq. (3).

Typical flowfield results for the nominal operating conditions and a thin-film heater element width of 300 μm are given in Fig. 7. The contours consist of raw data, with each "pixel" corresponding to a flowfield cell, giving an indication of grid resolution. The left side of Fig. 7 shows translational temperature contours. At this level of rarefaction, slip phenomena are expected to be large. This feature is confirmed by the peak temperature in the gas near the pedestal remaining much lower than the pedestal wall surface temperature of 600 K. The right side of Fig. 7 shows axial velocity contours. The acceleration of the gas due to the slot expansion is evident, while near the slot wall large velocity slip occurs.

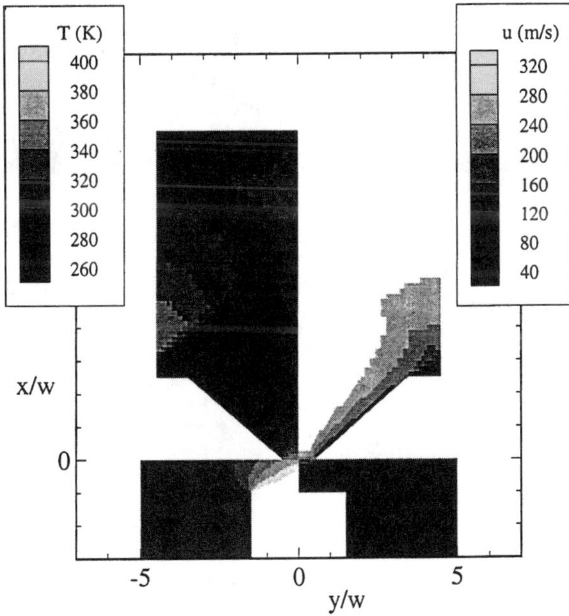

Fig. 7 Flowfield contours for nominal operating conditions. Left: Translational temperature. Right: Axial velocity.

V. Discussion

A. FMMR Estimated Thruster Performance

For a nominal-case thruster ($p_0 = 53.6$ Pa, $T_w = 600$ K, $\alpha = 40$ deg, $w = 100$ μm, argon) with a slot length of 1 cm, the FMMR design produces a thrust of approximately 0.025 mN per slot using an argon propellant as derived from the DSMC results. Therefore, a thruster arrangement of 10 slots produces a total thrust of 0.25 mN. For applications that require large thrust levels, the stagnation pressure, heated wall temperature, slot length, and total number of slots can all be increased to achieve the desired thrust level.

Although argon was used as the propellant for the FMMR for simplicity in the DSMC calculations presented in the previous section, water actually makes a very attractive propellant for the FMMR. An optimum configuration would be to have the propellant stored on-orbit as a solid with a temperature around 245 K. This results in a vapor pressure for ice near 50 Pa, which is the nominal FMMR operating pressure for a 100-μm slot width ($Kn \sim 1$). If higher operating pressures are desired for higher thrust applications, waste heat from the spacecraft could be used to increase the vapor pressure in the storage tank. In this way, the pressure regulation scheme for the FMMR is greatly simplified. If a desired temperature cannot be achieved on-orbit, the FMMR slot width can be designed to accommodate the expected propellant storage temperature (vapor pressure).

B. FMMR Scaling

The FMMR can be scaled to higher thrust levels by increasing the pressure (smaller Knudsen number for a constant slot width of 100 μm) in the stagnation region as shown in Fig. 8 for the nominal operating conditions. The thrust (mN) is calculated per slot and normalized per unit length (m) of the slot expansion. For constant Knudsen number, the thrust per unit slot length is constant, implying that as the slot width decreases, the stagnation pressure increases. Therefore, the small slot size tends to increase the thrust generated from the FMMR per unit area since more slots (operating at the same thrust level) can be fabricated on a given surface.

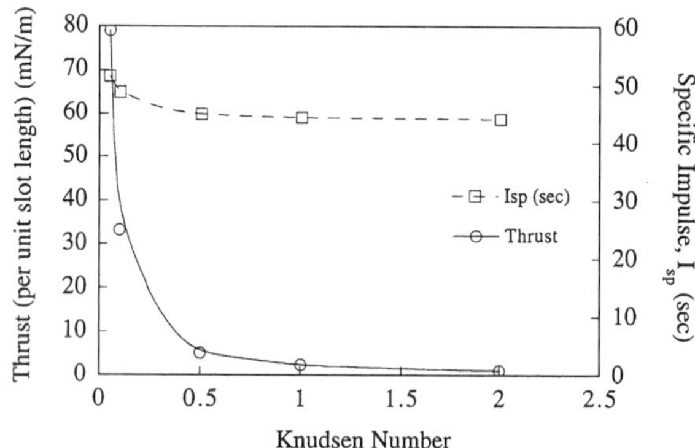

Fig. 8 FMMR performance as a function of operating Knudsen number (argon pressure).

Extremely low values of thrust, which are important for incremental maneuvers and critical pointing, can be achieved by simply reducing the pressure in the stagnation region without significantly sacrificing specific impulse as shown in Fig. 8. Arbitrarily small impulse bits (I-bits) are possible by this simple strategy with easily achievable valve cycle times.

The impulse bit is based on the thrust and the valve actuation time, implying that the higher thrust level requires a shorter valve response time for the same impulse bit. Since microspacecraft will require impulse bits of the order of micronewton-seconds, reducing a microthruster's thrust level, without sacrificing much in specific impulse, is essential to maintain easily achievable valve actuation times. Figure 8 indicates the flexibility of the FMMR in meeting several mission requirements on the same spacecraft.

Since a micronozzle must operate at relatively high thrust levels (high *Re*) to minimize losses in I_{sp},[12] very small I-bit maneuvers will place strict requirements on valve actuation times ($\leq$1 ms). At the MEMS level, valve actuation times below several milliseconds have yet to be consistently demonstrated with acceptable leak rates.

C. Power Usage and Heat Transfer Considerations

Besides the power utilized to heat the propellant molecules, power is used in the form of heat loss mechanisms in the FMMR, namely, conduction and radiation. An estimate of the radiative heat loss from the heating element to its surroundings for blackbody radiation (worst case) at $T_w = 600$ K gives a heat flux of nearly 6900 W/m^2. For a typical heating element with a width of 300 μm and a length of 1 cm, this translates into a heat loss of 0.02 W per heating element.

Heat conduction through the pedestal from the thin-film heaters can be minimized by suspending the heaters above the pedestal surface, which is a well-practiced MEMS technique.[2] If it is assumed that the heaters are in full contact with the top pedestal surface as a worst case, then the heat loss due to conduction through the pedestal ($T = 300$ K) is approximately 0.15 W per heating element for the same heating element area used above. This assumes that an appropriate thermal insulation material with properties similar to those of Teflon can be developed for use in MEMS applications.

The level of heat loss in the FMMR is not expected to differ substantially from that of a MEMS-scale micronozzle with similar operating temperatures and fabricated from similar materials. In the case of the micronozzle, which requires relatively high operating pressures, the heat of vaporization for the propellant needs to be supplied for propellants stored as a liquid or solid. Additional power (in the form of heat) is also required to ensure that the propellant does not condense within the micronozzle (e.g., near relatively cold walls).

VI. Systems Analysis

In an attempt to quantify the fundamental viability of the FMMR, an initial systems comparable or effective specific impulse has been estimated. The effective I_{sp} modifies the thruster's intrinsic I_{sp} by taking into account such system losses as high-pressure propellant storage, MEMS valve leakage, and propellant left over in the storage system at the spacecraft end of life (EOL).

A. Mass of Stored Propellant

It is well-known that the mass of propellant needed for a typical microsatellite mission depends on the total Δv required to perform the mission. However, additional propellant mass is required as EOL propellant once the thruster's design minimum pressure, $p_{0,\text{des}}$, has been reached. Additional mass is also required to account for MEMS valve leakage in microthruster systems when the storage pressure is high.[13] Therefore, the total mass of the stored propellant, $M_{p,s}$, is given by

$$M_{p,s} = \frac{M_{p,r}(1+\zeta)}{[1-(p_{0,\text{des}}/p_{i,\text{stor}})]} \tag{8}$$

B. MEMS Valve Leakage

The propellant leak rate (mass loss per unit time) through a given MEMS valve is given by

$$\dot{M} = \kappa p_s \tag{9}$$

$$\frac{\mathrm{d}p_s}{\mathrm{d}t} = -\frac{\dot{M}}{V_t} R T_s \tag{10}$$

At time $t = 0$, $p_s = p_{i,\text{stor}}$. For this boundary condition, Eq. (10) is solved by

$$\frac{p_s}{p_{i,\text{stor}}} = \exp - \left(\frac{\kappa R T_s}{V_t} t \right) \tag{11}$$

Therefore, the lifetime of the system to lose all of its usable propellant is given by

$$t_L = \ell n \left(- \frac{p_{i,\text{stor}} + p_{0,\text{des}}}{2\kappa (p_{i,\text{stor}})^2} \right) M_{p,s} \tag{12}$$

For a system to lose a fraction ζ of its usable propellant, the time required is

$$t_{L,\zeta} = \ell n \left(- \frac{(1-\zeta) + \zeta\{p_{0,\text{des}}/p_{i,\text{stor}}\}}{2\kappa p_{i,\text{stor}}} \right) M_{p,s} \tag{13}$$

C. Propellant Storage Tank Mass

For an overall system analysis of microthruster systems, it is important to consider other sources of mass in addition to the propellant. One of these major sources is the propellant storage tank. For the assumption of a thin-walled spherical storage tank, the mass ratio is

$$\frac{M_t}{M_{p,s}} = \frac{3}{2} \frac{\rho_t}{\rho_{p,s}} \left(\frac{\varepsilon p_{i,\text{stor}}}{F_{t,u}} \right) \tag{14}$$

As can be seen in Eq. (14), the thin-wall approximation ($F_{t,u}/\varepsilon$ large) gives a mass ratio independent of the initial storage pressure since the propellant storage density is proportional to $p_{i,\text{stor}}$. A similar expression can be derived for cylindrical thin-walled tanks.[14]

For a thick-walled spherical tank, the expression for the tank-to-propellant mass ratio is

$$\frac{M_t}{M_{p,s}} = \frac{1}{2}\frac{\rho_t}{\rho_{p,s}}\left[\left(1 + \frac{\varepsilon p_{i,\mathrm{stor}}}{F_{t,u}}\right)^3 - 1\right] \tag{15}$$

Since the typical value of $F_{t,u}/\varepsilon$ is large compared with the initial propellant storage pressure over the tank surface area,[15] the values obtained from Eq. (15) are relatively constant for typical storage pressures up to 10 MPa.

For a nitrogen propellant stored at 20 MPa and 300 K in a spherical titanium ($F_{t,u} = 1.23$ GPa) tank, the ratio of the tank mass to the propellant mass is approximately 1.0 using a safety factor $\varepsilon = 2$. Similar analysis for an graphite tank yields a mass ratio of 0.483. For microthruster systems that use propellant stored as high-pressure gases, the storage tank mass can be of the same order as the propellant mass required to perform the mission.

D. Effective Specific Impulse

The effective specific impulse of a thruster system in terms of the extra mass associated with minimum operating pressure, propellant loss due to valve leakage, and storage tanks can be derived from Eqs. (8) and (15). In this formulation, the effective I_{sp} is given by

$$I_{\mathrm{sp,eff}} = \frac{I_{\mathrm{sp}}[1 - (p_{0,\mathrm{des}}/p_{i,\mathrm{stor}})]}{(1 + \zeta)(1 + (M_t/M_{p,s}))} \tag{16}$$

where I_{sp} is the intrinsic specific impulse of the thruster [Eq. (3) or (4)].

Figure 9 shows the sensitivity of Eq. (16) to various parameters. In this formulation, $p_{0,\mathrm{des}}$ is 0.1 MPa, and the tank-to-propellant mass ratio is calculated from Eq. (15). The propellant tank is a titanium tank storing molecular nitrogen propellant with $\varepsilon = 2$. Figure 9 indicates that, even for no leakage through the MEMS valve (i.e., $\zeta = 0$), the ratio of the effective I_{sp} to the intrinsic I_{sp} is only 0.5. This represents a major overall system performance loss. The optimum propellant storage pressure leading to the maximum effective I_{sp} is a balance between meeting

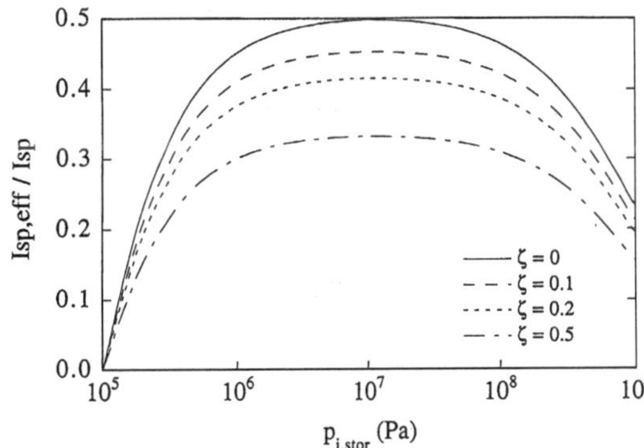

Fig. 9 Ratio of effective specific impulse to intrinsic specific impulse for a stored high-pressure nitrogen gas propellant.

the minimum required operating pressure and the effect of the storage tank mass at very high pressures.

E. Effective Specific Impulse Comparisons of the FMMR with a Cold Gas Thruster

As mentioned earlier, the operating characteristics of the FMMR appear very attractive using a water propellant stored on-orbit in solid form. Figure 6 shows the intrinsic I_{sp} of the FMMR operating on water propellant with $Kn = 1$ and $T_w = 600$ K to be approximately 68 s.

The propellant storage pressure (i.e., the vapor pressure for a solid propellant) being extremely low (of the order of 10^2-Pa maximum) has several advantages from a systems viewpoint. First, the storage tank need be designed only to handle launch stresses since $p_{i,\text{stor}} \ll F_{t,u}(M_t \ll M_{p,s})$. In this case, standard materials such as titanium, aluminum, and graphite can be replaced by much lighter materials. Second, the reduced storage pressure lowers the leak rate through MEMS valves to trivial levels ($z \approx 0$). Third, since $p_{0,\text{des}}$ is unusually small, nearly all of the propellant is used by the FMMR at EOL ($p_{0,\text{des}} \approx p_{i,\text{stor}}$). From these arguments, the effective I_{sp} for the FMMR is very close to its intrinsic value of 68 s.

Table 1 gives a general comparison between the FMMR operating with a water propellant and micronozzle systems that store high-pressure helium, nitrogen, and argon propellants as a function of the operating temperature. The effective I_{sp} results for the micronozzle cases are also shown for various values of ζ. In the case of the FMMR, the operating temperature is T_w, and in the case of the micronozzle, the operating temperature is the stagnation temperature. For a cold gas micronozzle thruster (20-μm-diam throat, outlet-to-throat area ratio of 20, and expansion angle of 15 deg) operating on a gaseous nitrogen propellant, the ideal intrinsic specific impulse is approximately 80 s from Eq. (4). Assuming that there are no losses due to viscous forces in the micronozzle, the minimum design pressure will be 0.1 MPa ($Re \approx 400$). A graphite propellant tank is assumed with

Table 1 Comparison of effective specific impulse for the FMMR and cold/warm gas micronozzle (MN) cases

Thruster (propellant)	T_0, K	$M_t/M_{p,s}$ (graphite)	ζ	Intrinsic I_{sp}, s	Effective I_{sp}, s	$V_t/V_{t,\text{FMMR}}$
FMMR (H_2O)	300	0.0015	0	48.1	48.1	1
FMMR (H_2O)	600	0.0015	0	68.0	68.0	1
MN (He)	300	6.8	0	253.7	32.7	129.8
MN (He)	300	6.8	0.1	253.7	29.7	142.7
MN (He)	300	6.8	0.2	253.7	27.2	155.7
MN (He)	600	6.8	0	358.7	46.2	91.8
MN (He)	600	6.8	0.1	358.7	42.0	100.9
MN (N_2)	300	0.48	0	80.3	54.1	5.6
MN (N_2)	300	0.48	0.1	80.3	49.2	6.2
MN (N_2)	600	0.48	0	113.6	76.6	4.0
MN (Ar)	300	0.34	0	56.7	42.4	5.0
MN (Ar)	300	0.34	0.1	56.7	38.5	5.5

a storage pressure of 20 MPa for all of the gaseous propellants listed in Table 1. If it is assumed that $\zeta = 0$, the effective I_{sp} for the nitrogen cold gas micronozzle is approximately 54 s.

As indicated in Table 1, the FMMR can operate with a water or ammonia propellant in a cold gas mode (i.e., with the heating elements off) without the problem of the propellant recondensing in the device. This is due to the low operating pressure and short residence time of water on surfaces with a temperature of 300 K. Cold gas micronozzle flows, on the other hand, will experience some degree of condensation in the stagnation region, which acts to degrade performance seriously.

It is evident from Eq. (16) that the mass of the propulsion system's power supply has not been included in the calculation of the effective specific impulse. It is assumed that the power supplies and related plumbing required for the FMMR and the cold and warm gas micronozzles will be roughly the same. Cold gas thrusters that operate at high-pressures typically require heavy, high-power valves, thick-walled tubing, and pressure regulators, which tend to balance out the weight of the FMMR power supply.

Finally, as the last column in Table 1 shows, there is an additional benefit in storing solid or liquid propellants over gaseous propellants in terms of reduced storage volume. This becomes a critical factor in volume-limited microspacecraft.

F. Propellant Storage Volume Considerations

Although microspacecraft may be mass and power limited, perhaps the most critical obstacle for the propulsion subsystem is the severe volumetric limitation. The volume required to store the propellant is a function only of the mission requirements (i.e., Δv required) and the density of the propellant. However, the mass of the propellant stored at the beginning of the mission $M_{p,s}$ depends on the minimum operating pressure of the thruster and the valve leak percentage from Eq. (8). The propellant mass required to perform a single or series of maneuvers is given by[4]

$$M_{p,r} = M_{0,\mathrm{SC}}\left(1 - \exp\left(-\frac{\Delta v}{I_{sp}g}\right)\right) \tag{17}$$

where $M_{0,\mathrm{SC}}$ is the initial dry mass of the spacecraft and I_{sp} is the thruster intrinsic specific impulse.

Therefore using Eq. (8) yields the volume of the stored propellant as

$$V_{t,s} = \frac{M_{0,\mathrm{SC}}\{1 - \exp[-(\Delta v/I_{sp}g)]\}(1+\zeta)}{\rho_{p,s}[1 - (p_{0,\mathrm{des}}/p_{i,\mathrm{stor}})]} \tag{18}$$

If it is assumed that $M_{0,\mathrm{SC}}$ differs between the high-pressure micronozzle operation and the FMMR only in the mass of the propellant storage tank, then in the limit of small Δv the ratio of propellant storage volume becomes

$$\frac{V_{t,\mathrm{MN}}}{V_{t,\mathrm{FMMR}}} = \frac{\rho_{p,\mathrm{FMMR}}}{\rho_{p,\mathrm{MN}}}\left\{\frac{I_{\mathrm{sp,eff,FMMR}}}{I_{\mathrm{sp,eff,MN}}}\right\} \tag{19}$$

where the terms with subscript MN refers to the micronozzle values.

In the case used previously for a cold gas nitrogen micronozzle expansion with the propellant stored at 20 MPa and $\zeta = 0$, the propellant density ($T_0 = 300$ K) and effective specific impulse are 224.5 kg/m^3 and 54.1 s, respectively. For the FMMR operating on water vapor (from propellant stored as ice) with $T_w = 600$ K, the effective specific impulse is 68 s. Therefore, the ratio from Eq. (19) is approximately 5.6 for a graphite storage tank. For a titanium propellant tank, the storage volume ratio is approximately 22. Although factors of 2–3 reduction in the volume ratio can be envisioned by storing the nitrogen propellant at higher pressure, the results plotted in Fig. 9 show the limitations of this approach.

VII. Conclusions

The FMMR exemplifies how a novel concept is applicable to small-scale thrusters. System accommodations often outweigh more narrowly focused performance issues when propulsion systems are considered for spacecraft operations. This will be extremely relevant for highly integrated microspacecraft. Although the FMMR's intrinsic performance is not as high as that of an ideal continuum expansion nozzle, the FMMR outperforms continuum nozzle expansions from an effective or systems performance perspective. The FMMR offers several advantages over traditional small-scale thrusters: reduced propellant storage pressure, abatement of catastrophic nozzle plugging, ease and flexibility of construction, and reduced valve actuation requirements for small impulse bits.

An iterated design concept has been developed for the FMMR using parametric studies with the DSMC numerical technique. The nominal design calls for a 100-μm-wide and 8-mm-long slot with an expansion angle of 54.74 deg. The heating element optimum width is found to be approximately three times the slot width (300 μm) with a surface temperature between 600 and 1200 K.

Calculations using an argon propellant at a stagnation temperature of 600 K show that a thrust level near 0.25 mN at a specific impulse of approximately 45 s can be achieved for a thruster configuration utilizing 10 expansion slots and associated heating elements. For higher thrust requirements, stagnation pressure, slot width, slot length, and number of slots can be changed to allow for higher mass flows from the thruster while still maintaining the free molecule condition. For propellant storage ease, future thruster designs using ammonia or water propellants will be investigated. For these cases, the gas–surface interaction models used in the DSMC code are expected to dominate the results due to coupling with internal energy modes.

Acknowledgment

The Navier–Stokes calculations were made possible by a grant of computer time and resources from the HPCMP Air Force ASC Major Shared Resource Center.

References

[1]Mueller, J., Tang, W., Wallace, A., Li, W., Bame, D., Chakraborty, I., and Lawton, R., "Design, Analysis, and Fabrication of a Vaporizing Liquid Micro-Thruster," AIAA Paper 97-3054, Seattle, WA, July 1997.

[2]Janson, S., "Batch-Fabricated Resistojets: Initial Results," International Electric Propulsion Conf., Paper 97-070, Cleveland, OH, July 1997.

[3]Muntz, E. P., and Ketsdever, A., "Microspacecraft Exhaust Plumes, Thrust Generation and Envelope Expansion to Lower Altitudes," Presented at the AFOSR Micropropulsion Special Session and Workshop, San Diego, CA, July 1997.

[4]Sutton, G., *Rocket Propulsion Elements*, 6th ed., Wiley, New York, 1992.

[5]Bird, G., *Molecular Gas Dynamics and the Direct Simulation of Gas Flows*, Clarendon Press, Oxford, England, U.K., 1994.

[6]Wadsworth, D., Erwin, D., and Muntz, E. P., "Transient Motion of a Confined Rarefied Gas Due to Wall Heating or Cooling," *Journal of Fluid Mechanics*, Vol. 248, 1993, pp. 219–235.

[7]Wadsworth, D., and Muntz, E. P., "A Computational Study of Radiometric Phenomena for Powering Microactuators with Unlimited Displacements and Large Available Forces," *Journal of Microelectromechanical Systems*, Vol. 5, 1995, pp. 59–65.

[8]Wadsworth, D., Muntz, E. P., Pham-Van-Diep, G., and Keeley, P., "Crookes' Radiometer and Micromechanical Actuators," *Rarefied Gas Dynamics, Proceedings of the 19th International Symposium*, edited by J. Harvey and G. Lord, Oxford Univ. Press, Oxford, England, U.K., 1995, pp. 708–714.

[9]Wadsworth, D., "Slip Effects in a Confined Rarefied Gas, {I}: Temperature Slip," *Physics of Fluids A*, Vol. 5, 1993, pp. 1831–1839.

[10]Ketsdever, A., Wadsworth, D., Vargo, S., and Muntz, E. P., "Flow Properties of a Free Molecule Micro-Resistojet for Small Spacecraft Applications," *Rarefied Gas Dynamics, Proceedings of the 21st International Symposium*, edited by R. Brun, R. Campargue, R. Gatignol, and J.-C. Lengrand, Cepadues-Editions, Paris, 1999, pp. 601–606.

[11]Rae, W., "Some Numerical Results on Viscous Low-Density Nozzle Flows in the Slender-Channel Approximation," *AIAA Journal*, Vol. 9, No. 5, 1971, pp. 811–817.

[12]Ivanov, M., Markelov, G., Ketsdever, A., and Wadsworth, D., "Numerical Study of Cold Gas Micronozzle Flows," AIAA Paper 99-0166, Reno, NV, Jan. 1999.

[13]Janson, S., and Helvajian, H., "Batch-Fabricated Microthrusters: Initial Results," AIAA Paper 96-2988, Buena Vista, FL, July 1996.

[14]Stevens, K., *Statics and Strength of Materials*, Prentice–Hall, New York, 1979.

[15]Humble, R., Henry, G., and Larson, W., *Space Propulsion Analysis and Design*, Space Technology Series, McGraw–Hill, New York, 1995.

Chapter 6

Study of Very Low-Power Arcjets

Hideyuki Horisawa*
Tokai University, Hiratsuka-shi, Kanagawa, Japan
and
Itsuro Kimura†
University of Tokyo, Yokohama, Japan

Nomenclature

C = convergent nozzle
D = convergent–divergent nozzle
d_{con} = constrictor diameter
I = discharge current
I_{sp} = specific impulse
l_{ac} = distance between anode and cathode
l_{con} = constrictor length
$\dot{m}$ = propellant mass flow rate
P_{in} = electrical input power
P_{sp} = specific power
T = total thrust
T_c = thrust of cold-gas jet
T_{rot} = rotational temperature
T_{vib} = vibrational temperature
V = discharge voltage
η = thrust efficiency

I. Introduction

THE current trend toward smaller spacecraft, which is not only mass limited but also power limited, has produced a strong interest in the development of micropropulsion devices.[1–3] The significance of reducing launch masses has attracted growing interest with regard to reducing mission costs and increasing launch rates.

Copyright © 2000 by the American Institute of Aeronautics and Astronautics, Inc. All rights reserved.
*Assistant Professor, Department of Precision Mechanics, School of Engineering.
†Professor Emeritus, University of Tokyo, and Nara Textile Inc.

Although, in the past, many very small spacecraft have lacked propulsion systems altogether, future microspacecraft will require significant propulsion capability to provide a high degree of maneuverability and capability. The benefit of using electric propulsion for the reduction of spacecraft mass will likely be even more significant for mass-limited microspacecraft missions.[2] Feasibility studies of microspacecraft are currently under development for a mass of less than 100 kg with an available power level for propulsion of less than 100 W.[2–20] Various potential propulsion systems for microspacecraft applications, such as ion thrusters,[4–7] field emission thrusters,[8,9] PPT,[10,11] vaporizing liquid thrusters,[12,13] resistojets,[14–16] microwave arcjets,[17] and pulsed arcjets,[18,19] have been proposed and are under significant development for primary and attitude control applications.

As for low-power DC arcjets operational at power levels down to about 300 W, several investigations have been conducted on their use for north–south stationkeeping (NSSK) on geosynchronous satellites.[21–32] For the reduction, or minimization, of arcjet input power, in a previous study, operation with throttleability from 100 to 300 W was demonstrated, in which steady operation was achieved down to 40 W, but with considerable voltage fluctuations.[32] Also, there have been various reports on the effects of constrictor sizes, or constrictor diameter and length, and electrode gap on thrust performance, i.e., thrust, specific impulse, or thrust efficiency, of low-power arcjets operational at power levels down to 300 W.[20–25,27,29,30] In previous studies, it was reported that a reduction of constrictor diameter contributes to thrust performance improvement.[23,25,27,29] As for the electrode gap, it was reported that changes in the gap do not significantly affect performance.[23,25,29] On the other hand, performance improvement was confirmed with a reduction of the gap, in which the operation became more stable.[27] However, there have been no investigations on these geometric effects for the performance improvement of very low-power DC arcjets operational below 30 W[20] for microspacecraft propulsion devices, relating not only to the thrust performance but also to the fundamentals of very low-power DC discharges.[33] The structural simplicity of an arcjet may be favorable for both size and mass reduction of the thruster; also, further reduction of the input electrical power, to less than 100 W, for example, may be effective for reducing the mass of the power supplies. In addition, operation of the arcjets at reduced specific power levels with a lower temperature of the propellant, which is heated through the discharge, will elongate the life of electrodes and reduce frozen flow losses and electrode losses, through the reduction of heat transfer from the heated propellant gas.[34] Although the specific impulse achievable during operation will be reduced at low specific power levels, it will be recovered to some extent through the achievement of loss reduction.

The objective of this study is to investigate the fundamentals of discharge characteristics and the thrust performance of very low-power DC arcjets with electrical input power levels ranging from approximately 5 to 30 W to ascertain the effective operational condition that possibly results in a higher thrust performance. In this study, the conditions for stable operation and thrust performance, such as thrust, specific impulse, and thrust efficiency, of very low-power arcjets are evaluated for nozzles of various dimensions made from different materials. Diagnostics of the internal flow of the arcjets and temperature and heat content measurements at the constrictor exit are also conducted, using a nozzle without any divergent parts. Then the performances of these various nozzles are evaluated to investigate the

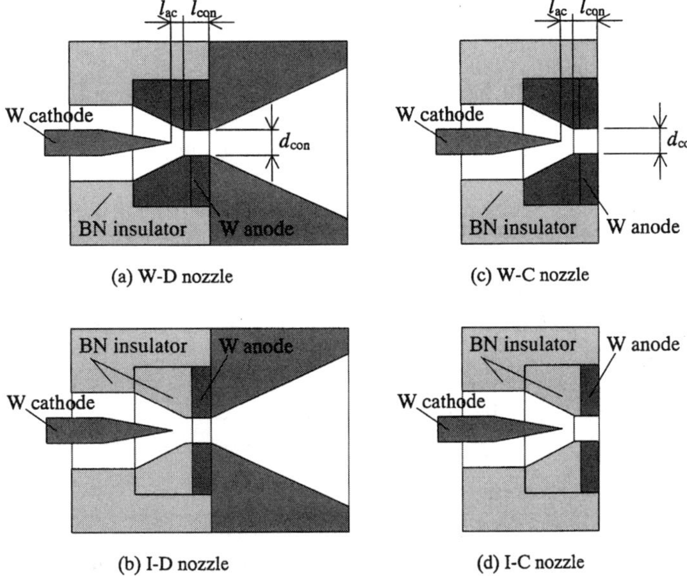

Fig. 1 Schematics of arcjet nozzles tested in the experiment: W-, tungsten nozzle; I-, insulator inserted nozzle; C, convergent nozzle; D, divergent nozzle.

most favorable structural and operational condition that results in a higher thrust performance.

II. Experiment

A. Arcjet Thruster

Cross-sectional schematics of the four types of nozzles used in this study are shown in Fig. 1, and their dimensions are given in Table 1. In general, an arcjet nozzle consists of a metallic material and serves a dual function, as an anode and an arc column constrictor, except for a divergent section. In this study, a ceramic material with a low heat conductivity for a convergent section and the following part of the constrictor in a nozzle was used to reduce electrode losses. Using modified or partially insulated nozzles (I-nozzles) and conventional tungsten nozzles (W-nozzles), the effects of the constrictor material and dimensions on the characteristics of very low-power DC discharge, the propulsive performance, and the thermal characteristics of arcjets were evaluated. For evaluation of the propulsive

Table 1 Dimensions of tested arcjets

Type	W-C, W-D, I-C, I-D
d_{con}, mm	0.3, 0.5, 0.7
l_{con}, mm	0.5, 1.0, 2.0
l_{ac}, mm	0.25, 0.5, 0.75

performance, convergent–divergent nozzles (D-nozzles) were utilized. To diagnose the thermal characteristics of very low-power discharges established in the constrictors, investigations using nozzles without any divergent parts or convergent nozzles (C-nozzles) were also performed. As illustrated in Fig. 1, conventional nozzles are denoted W-, (a) W-D or (c) W-C, and consist of an assembly of pure tungsten nozzle parts, while modified nozzles are denoted I-, (b) I-D or (d) I-C, and consist of an assembly of an insulator and a tungsten anode. For the ceramic material, high-purity boron–nitride (BN) was used for part of the constrictor to allow the arc column to penetrate farther downstream of the constrictor or to maintain the high-voltage mode discharges and, possibly, reduce electrode losses. The cathode used in the tests was made from a tungsten rod 1 mm in diameter with a conical tip angle of 15 deg. Nitrogen gas was used as the propellant, and the feed pressure was measured upstream of the plenum. In this study, to establish stable discharges at very low current levels, ranging from 10 to 150 mA, a high-voltage power supply with a high ballast resistance was used.

B. Propulsive Performance Tests

A thrust performance test was conducted for each nozzle in a vacuum vessel. The background pressure during most of the experiments was maintained at below 4 Pa. A calibrated pendulum-type thrust stand was used for the measurements.

The specific impulse I_{sp} and the thrust efficiency η are calculated as follows:

$$I_{sp} = \frac{T}{\dot{m} g} \tag{1}$$

and

$$\eta = \frac{T^2 - T_c^2}{2\dot{m} I V} \tag{2}$$

where T is the total thrust and T_c is the thrust of the cold-gas jet.

It is necessary to apply the definition of Eq. (2) to very low-power thrusters in which the thrust increment by arc augmentation is relatively small compared with the cold-gas thrust.[25] Here, typical values of T_c measured in the experiment were $T_c = 6.1$ mN for the I-D or W-D nozzle for $d_{con} = 0.3$, and $T_c = 4.3$ mN for the I-D or W-D nozzle for $d_{con} = 0.5$ mm, at $\dot{m} = 5$ mg/s.

C. Thermal Efficiency and Gas Temperature Diagnostics

The net thermal power output of the arcjet was measured directly using a flow calorimeter at the constrictor exit of the convergent nozzles (W-C, I-C) under atmospheric pressure. The calorimeter consists of a coil of 2-mm-diameter copper tube wound against the inside wall of a 15-mm-internal diameter copper tube. The outside wall is insulated. The calorimeter has a 180-deg bend near the exit to minimize energy loss by radiation. A thin ceramic insulator electrically isolates the calorimeter from the anode. The temperature rise of the cooling water that flows through the calorimeter, or the copper tube coil, at the inlet and outlet and the gas temperature at the exit of the calorimeter were measured using copper–constantan thermocouples. The power absorbed by the calorimeter was computed as the product of the temperature rise and mass flow rate of the cooling water and the specific heat of the water. The thermal efficiency of the exhaust propellant gas from the arcjet constrictor was computed as the ratio of net thermal power output to arcjet input power or the product of arc current and discharge voltage.

Spectroscopic measurements were performed to evaluate the effects of variation in the constrictor material, dimensions, mass flow rate, and discharge current on the heavy particle temperature of the heated propellant at the constrictor exit under atmospheric pressure. To evaluate the gas temperature at the constrictor exit of the convergent nozzles (W-C, I-C), the wavelength (300–800 nm) and spectrum intensity of spontaneous emission of the gas 0.5 mm downstream from the constrictor exit on the central axis of the nozzle were measured using a multichannel spectrum analyzer. Based on spectroscopic theories,[35] the spontaneous emission spectrum resulting from the electronic transitions of nitrogen molecules (first positive system, $B^3\Pi_g$–$A^3\sum_u^+$)[36] was calculated for a given set of conditions. Here, the parameters assumed for computing each theoretical spectrum were the number density of the gas, the vibrational temperature (T_{vib}), and the rotational temperature (T_{rot}). The temperatures (T_{vib} and T_{rot}) of the gas at the constrictor exit were found by selecting a set of parameters that makes the calculated spectrum fit the measured spectrum well.[33]

III. Results and Discussion

A. Propulsive Performance of Very Low-Power Arcjet Thrusters

1. Current–Voltage Characteristics

The discharge current–voltage characteristics observed for W- or I-nozzles with constrictor diameters $d_{con} = 0.3$ and 0.5 mm and a constrictor length $l_{con} = 0.5$ mm are shown in Figs. 2 and 3. For clarity, only the data taken at flow rates of 10 and 5 mg/s are shown. Similar trends were observed for other flow rates. It is shown that the discharge voltage for all cases decreases as the current rises over the range of current $I = 10$–75 mA, except in the W-nozzle case at 5 mg/s with $d_{con} = 0.5$ mm.

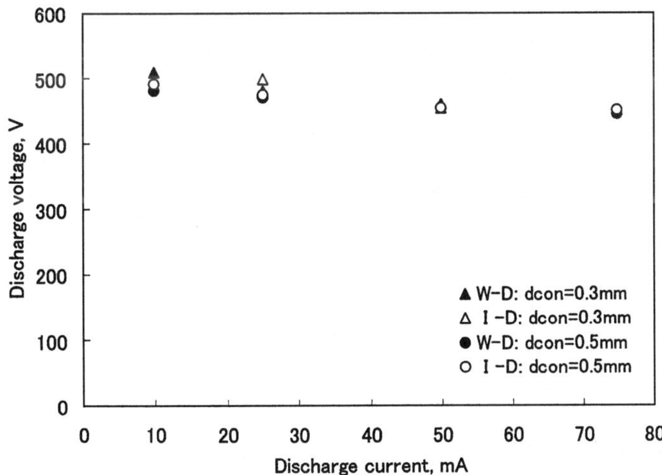

Fig. 2 Discharge current–voltage characteristics. Propellant mass flow rate: 10 mg/s.

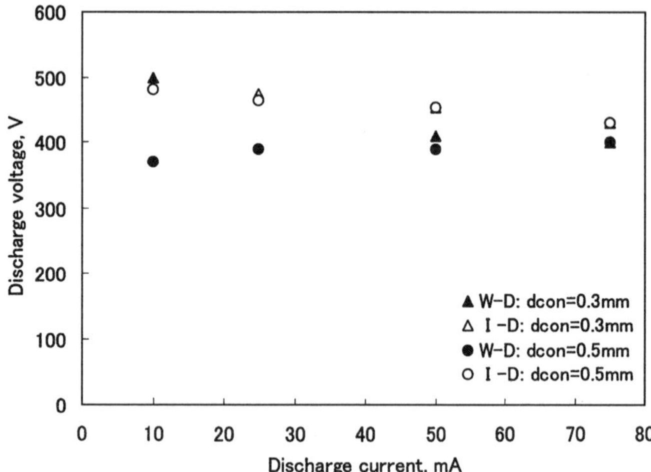

Fig. 3 Discharge current–voltage characteristics. Propellant mass flow rate: 5 mg/s.

This tendency is a typical electrical characteristic of an arc discharge,[37] and it was confirmed that the arc discharge was established even at this very low current level. However, in the W-nozzle case at 5 mg/s with $d_{\text{con}} = 0.5$ mm, or plenum pressure below about 30 kPa, the discharge voltage gradually decreases with decreasing current. This current–voltage trend is a typical characteristic of a glow discharge,[37] although this arc–glow transition might be inevitable at this very low current level under the low plenum pressure.

In all cases shown in Figs. 2 and 3, it is clear that the discharge voltages of I-nozzles are slightly higher than those of W-nozzles. This tendency may be due to the fact that in the I-nozzle cases, compared to the W-nozzle cases, the electric field around the anode may change, or the arc column may attach farther from the cathode due to the insulator inserted between the cathode and the anode, resulting in the voltage rise. Also, discharge voltages at lower mass flow rates are slightly lower than those at higher mass flow rates in all cases.

Although not plotted, it was observed that the plenum pressure increases with increasing arc current. The pressure increases with arc current due to the flow blockage induced by a bigger arc column with a higher arc current.[23] The plenum pressure of I-nozzles was also higher than that of W-nozzles. It is hypothesized that in the I-nozzle cases, compared with the W-nozzle cases, the arc anode attachment is farther from the cathode tip on the throat or the divergent section of the anode nozzle due to the insulator inserted between the anode and the cathode. This results in a decrease in the effective diameter of the throat due to the flow blockage by the arc column and the increase in the plenum pressure as the arc column penetrates farther into the throat. As shown in Figs. 2 and 3, to obtain operation at a lower power under the conditions of this study, it is necessary to reduce both the arc current and the mass flow rate.

2. *Propulsive Performance*

Figures 4 and 5 show plots of specific impulse vs arcjet input power for propellant mass flow rates of 10 and 5 mg/s, respectively. The accuracy of the specific impulse evaluated here was about 4%. It must be noted that the specific impulse at each

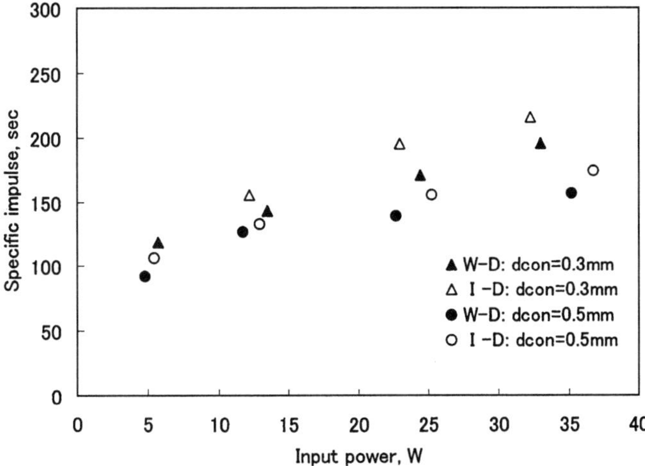

Fig. 4 Specific impulse vs arcjet input power. Propellant mass flow rate: 10 mg/s.

mass flow rate shows a linear increase and arrives at ~280 s ($d_{con} = 0.3$ mm; mass flow rate, 5 mg/s) even at very low power levels of ~32 W. In all cases, I-nozzles give a significantly higher specific impulse, or higher propulsive performance, compared with W-nozzles over a wide range of input power. There is a significant decrease in the specific impulse for W-nozzles with $d_{con} = 0.5$ mm at a mass flow rate of 5 mg/s, in which the discharge type changes into a glow discharge. In this case, it is shown that little improvement of propulsive performance with increasing input power up to 30 W is obtainable with the use of a DC glow discharge or a glow jet.

Figures 6 and 7 show replots of Figs. 4 and 5 where the specific power is plotted along the abscissa. The specific impulse shows a linear increase with specific power, as with the input power. It is shown that, at higher specific powers, the specific impulse at a given specific power is relatively independent of the mass

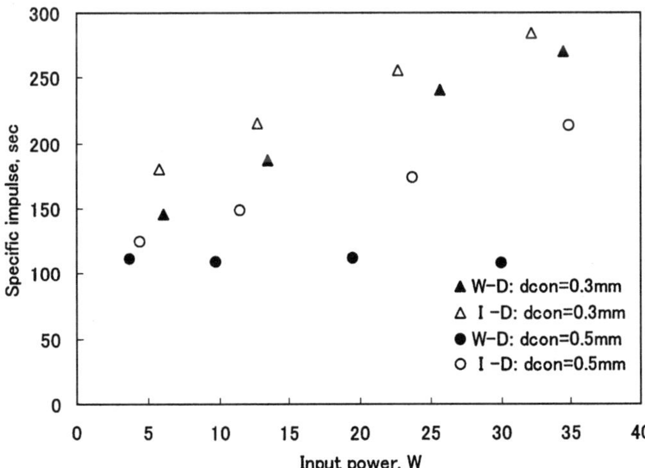

Fig. 5 Specific impulse vs arcjet input power. Propellant mass flow rate: 5 mg/s.

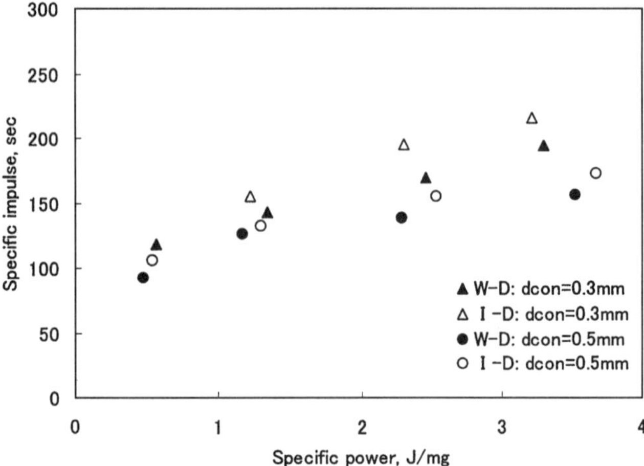

Fig. 6 Specific impulse vs specific power. Propellant mass flow rate: 10 mg/s.

flow rate. At lower specific powers, the curve for the lower mass flow rate is slightly above that for the higher mass flow rate. In all cases in Figs. 6 and 7, it must be emphasized that the specific impulse is significantly higher with the utilization of I-nozzles compared to W-nozzles. This fact indicates that I-nozzles may efficiently contribute to the reduction of heat transfer to the constrictor wall or electrode losses. Also, it was observed that a reduction of the constrictor diameter and constrictor length results in a higher specific impulse. It must be noted again that the specific impulse with glow discharges (W-D nozzle, $d_{con} = 0.5$ mm, mass flow rate = 5 mg/s) is significantly decreased compared with the identical values of specific power with arc discharges.

Figure 8 shows the relationship between specific impulse and thrust efficiency for each nozzle with $d_{con} = 0.3$ mm. The estimated accuracy of the thrust efficiency was approximately 8%. It is expected, at reduced specific power levels as in this

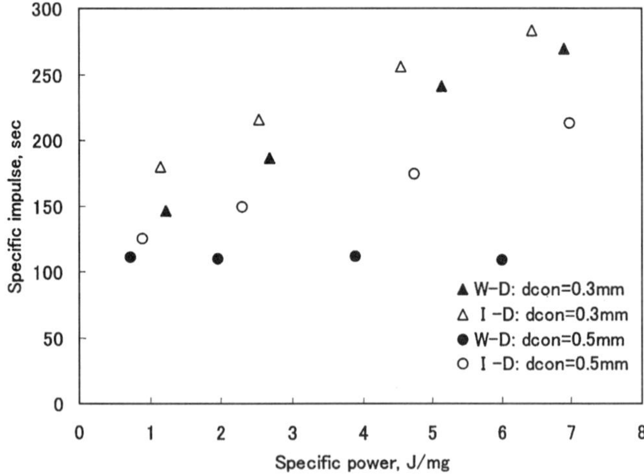

Fig. 7 Specific impulse vs specific power. Propellant mass flow rate: 5 mg/s.

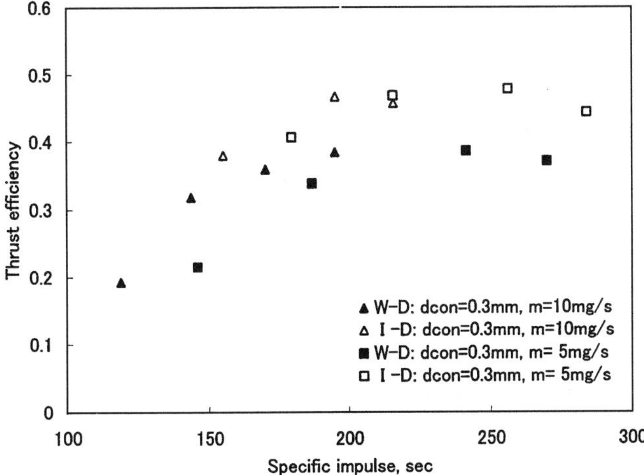

Fig. 8 Specific impulse vs thrust efficiency. Constrictor diameter, d_{con}: 0.3 mm.

experiment, that electrode loss and frozen flow loss are decreased to some extent,[34] while local heat losses of the electrodes caused by near-electrode voltage drops may be large under the applied discharges of low current and high voltage.[31] Since a detailed analysis of the mechanism of the losses[21,26,30,31,34] is not included in this study, further study is needed to elucidate these issues. In Fig. 8, it is shown that the thrust efficiency increases with specific impulse, and gradually decreases in the range of larger specific impulse, or specific power. A similar tendency was observed in some of the results obtained in previous reports on low-power arcjets with higher specific power.[26,30] In all cases, it is shown that I-nozzles give a significantly higher thrust efficiency than W-nozzles. Also, in most cases, a higher thrust efficiency is obtainable using nozzles with a smaller constrictor diameter. Though it is not plotted, the thrust efficiency with the glow discharge was significantly below those with arc discharges, i.e., by about 1%. In this case, because of the very low current under low plenum pressure, the current–voltage characteristics observed exhibit the glow discharge trend, as shown in Fig. 3, and the specific impulse and thrust efficiency are lowered significantly. These are considered to be due to the significant reduction of Joule heating, or the increase in frozen flow losses that accompany the glow discharges, resulting in significant decreases in the specific impulse and thrust efficiency. From the results, it was reconfirmed that little improvement in the propulsive performance of the very low-power arcjet thrusters is expected when the discharge type turns into a glow discharge.

B. Diagnostics of Gas Temperature and Thermal Efficiency of Very Low-Power Arcjets

In this study, heavy particle temperature and heat content measurements at the constrictor exit were conducted to assess the qualitative differences in the temperature and thermal efficiency with changes in the constrictor diameter, length, material, and electrode gap, using convergent nozzles for a propellant mass flow rate of 60 mg/s and a static pressure at the nozzle exit of ~0.1 MPa. Here, the mass flow rate and the input power were set higher than those in the thrust measurement

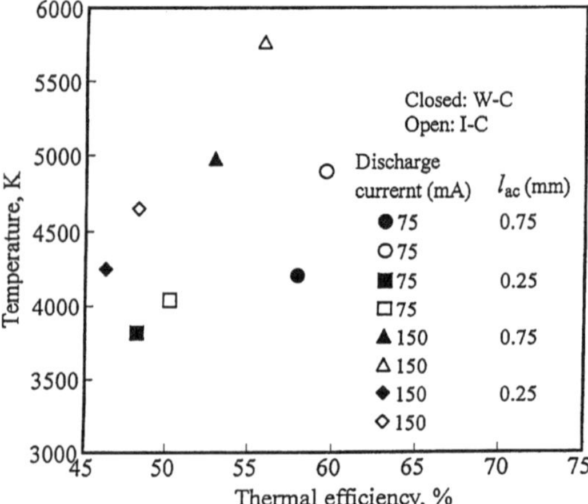

Fig. 9 Gas temperature vs thermal efficiency. d_{con}, 0.5 mm; l_{con}, 0.5 mm; propellant mass flow rate, 60 mg/s.

so that both the thermal and the spectroscopic data became obtainable within acceptable errors under atmospheric pressure. However, under the condition of this experiment, similar trends of the geometric effects were observed for other mass flow rates. The repeatability of both the heat and the temperature measurements was estimated to be within about 5%.

Under the experimental conditions, it was observed that the rotational temperature (T_{rot}) and the vibrational temperature (T_{vib}) of the heated propellant gas at the constrictor exit estimated through the spectroscopic measurements were nearly in equilibrium. Figures 9 and 10 show the relationships between the thermal efficiency and the heavy particle temperature at the core, about 20 μm wide, of the exhaust

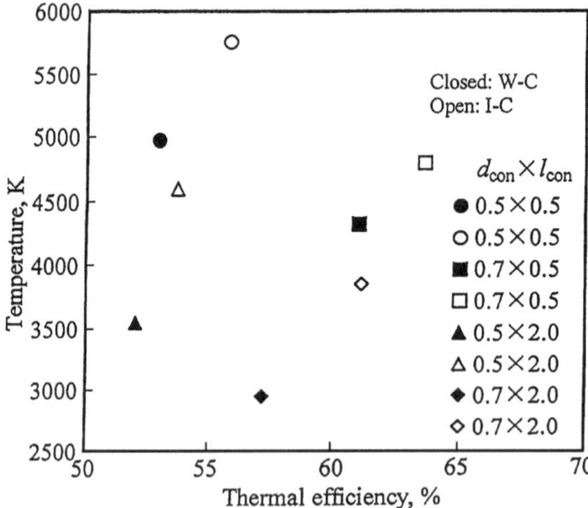

Fig. 10 Gas temperature vs thermal efficiency. Current, I, 150 mA; l_{ac}, 0.75 mm; propellant mass flow rate, 60 mg/s.

propellant gas measured downstream from the constrictor exit. In Fig. 9, it is shown that the temperature is elevated, to roughly 3000–6000 K, and rises with increasing arc current, while the thermal efficiency of the exhaust propellant decreases. Also, it is shown that both the temperature and the thermal efficiency increase as the electrode gap increases. This increase in the electrode gap is preferable to improve the performance of arcjet thrusters.

In all cases in Figs. 9 and 10, it must be emphasized that both the temperature and the thermal efficiency are higher with I-nozzles than with W-nozzles. This is related to the reduction of the heat transfer rate to the nozzle wall and/or the increase in the heating rate of the propellant followed by the further penetration of the arc column toward the constrictor exit with the utilization of a ceramic material of low heat conductivity. In Fig. 10, it is shown that, for all cases, the thermal efficiency increases with reducing constrictor length and with increasing constrictor diameter. It is also shown that a reduction of the constrictor diameter and constrictor length results in a higher temperature of the heated propellant gas. It is considered that the increase in the thermal efficiency with decreasing constrictor length is caused mainly by the decrease in the area of the constrictor wall to which heat is transferred. It must be noted that with decreasing constrictor diameter, the temperature rises through intensification of the thermal pinch effect, while the thermal efficiency decreases with the increase in heat transfer rate to the constrictor wall due to the reduced distance between the arc column surface and the wall.

IV. Conclusions

A study of the discharge characteristics and thrust performance of very low-power DC arcjets with electrical input power levels ranging from 5 to 30 W was carried out. To clarify the potential of the effective operational condition that possibly results in a higher thrust performance within these power levels, the effects of the constrictor material and the dimensions on the characteristics of the discharges and the propulsive performance were investigated. Additional diagnostics of the thermal characteristics of the internal flow were also performed, and the following results were obtained.

1) Stable arcjet thruster operation with specific impulse levels up to ~280 s at very low power levels, ranging from 5 to 30 W, with a constrictor diameter of 0.3 and 0.5 mm were confirmed, except in a singular case involving a DC glow discharge or a glow jet, in which little improvement in propulsive performance with an input power up to ~30 W was observed.

2) At higher specific powers, the specific impulse was relatively independent of the mass flow rate. At lower specific powers, the specific impulse for a lower mass flow rate was slightly higher than that for a higher mass flow rate.

3) In thrusters with partially insulated nozzles the specific impulse and thrust efficiency were significantly higher than those for conventional nozzles. Also, both the heavy particle temperature and thermal efficiency at the constrictor exit were higher with partially insulated nozzles than with conventional nozzles.

4) Both the temperature and the thermal efficiency increase as the electrode gap increases. In all cases, the thermal efficiency increases with decreasing constrictor length and with increasing constrictor diameter. In addition, a reduction of the constrictor diameter and length results in a higher temperature of the heated propellant gas.

References

[1]Myers, R. M., Oleson, S. R., Curren, F. M., and Schneider, S. J., "Small Satellite Propulsion Options," AIAA Paper 94-2997, June 1994.

[2]Mueller, J., "Thruster Options for Microspacecraft: A Review and Evaluation of Existing Hardware and Emerging Technologies," AIAA Paper 97-3058, July 1997.

[3]Leifer, S., "Overview of NASA's Advanced Propulsion Concepts Activities," AIAA Paper 98-3183, July 1998.

[4]Young, M., Muntz, E. P., and Ketsdever, A. D., "Investigation of a Candidate Nonmagnetic Ion Micro-Thruster for Small Spacecraft Applications," AIAA Paper 98-3917, July 1998.

[5]Mueller, J., Pyle, D., Chakraborty, I., Ruiz, R., Tang, W., and Lawton, R., "Microfabricated Ion Accelerator Grid Design Issues: Electric Breakdown Characteristics of Silicon Dioxide Insulator Material," AIAA Paper 98-3923, July 1998.

[6]Yashko, G., Giffin, G., and Hastings, D., "Design Considerations for Ion Microthrusters," IEPC 97-072, Aug. 1997.

[7]Gorshkov, O. A., "Low-Power Hall Type and Ion Electric Propulsion for the Small Sized Spacecraft," AIAA Paper 98-3929, July 1998.

[8]Marcuccio, S., Genovese, A., and Andrenucci, M., "Experimental Performance of Field Emission Microthrusters," *Journal of Propulsion and Power*, Vol. 14, No. 5, 1998, pp. 774–781.

[9]Fehringer, M., Rudenauer, F., and Steiger, W., "Space-Proven Indium Metal Field Ion Emitters for Microthruster Applications," AIAA Paper 97-3057, July 1997.

[10]Guman, W. J., and Peko, P. E., "Solid-Propellant Pulsed Plasma Microthruster Studies," *Journal of Spacecraft and Rockets*, Vol. 5, No. 6, 1968, pp. 732–733.

[11]Turchi, P. J., "An Electric Propulsion Development Strategy Based on the Pulsed Plasma Microthruster," AIAA Paper 82-1901, Nov. 1982.

[12]Mueller, J., Tang, W. C., Wallace, A. P., Li, W., Bame, D., Chakraborty, I., and Lawton, R., "Design, Analysis, and Fabrication of a Vaporizing Liquid Micro-Thruster," AIAA Paper 97-3054, July 1997.

[13]Mueller, J., Chakraborty, I., Bame, D., Tang, W. C., Lawton, R., and Wallace, A. P., "Proof-of-Concept Demonstration of a Vaporizing Liquid Micro-Thruster," AIAA Paper 98-3924, July 1998.

[14]Jankovsky, R., Sankovic, J., and Oleson, S., "Performance of a FAKEL K10K Resistojet," AIAA Paper 97-3059, July 1997.

[15]Ketsdever, A., Wadsworth, D. C., Vargo, S., and Muntz, E. P., "The Free Molecule Micro-Resistojet: An Interesting Alternative to Nozzle Expansion," AIAA Paper 98-3918, July 1998.

[16]Lawrence, T. J., Sweeting, M., Paul, M., Sellers, J. J., LeDuc, J. R., Malak, J. B., Spanjers, G. G., Spores, R. A., and Schilling, J., "Performance Testing of a Resistojet Thruster for Small Satellite Applications," AIAA Paper 98-3933, July 1998.

[17]Nordling, D., Souliez, F., and Micci, M. M., "Low-Power Microwave Arcjet Testing," AIAA Paper 98-3499, July 1998.

[18]Willmes, G. F., and Burton, R. L., "Thrust Performance of a Very Low Power Pulsed Arcjet," AIAA Paper 94-3125, June 1994.

[19]Willmes, G. F., and Burton, R. L., "Performance Measurements and Energy Losses in a 100 Watt Pulsed Arcjet," AIAA Paper 96-2966, July 1996.

[20]Horisawa, H., and Kimura, I., "Influence of Constrictor Size on Thrust Performance of a Very Low Power Arcjet," AIAA Paper 98-3633, July 1998.

[21]Curran, F. M., and Sarmiento, C. J., "Low Power Arcjet Performance Characterization," AIAA Paper 90-2578, July 1990.

[22]Andrenucci, M., Saccoccia, G., Scortecci, F., Panattoni, N., Schulz, U., and Deininger, W. D., "Performance Study of a Laboratory Model of Low Power Arcjet," International Electric Propulsion Conf. Paper 91-045, Oct. 1991.

[23]Capecchi, G., Scortecci, F., Repola, F., and Andrenucci, M., "Parametric Test Results of a Low Power Arcjet," International Electric Propulsion Conf. Paper 93-213, Sept. 1993.

[24]Macfall, K. A., Tilley, D. L., and Gulczinski, F. S., III, "Low Power Arcjet Performance Evaluation," International Electric Propulsion Conf. Paper 95-18, Sept. 1995.

[25]Ogiwara, K., Hosoda, S., Suzuki, T., Toki, K., Kuriki, K., Matsuo, S., Nanri, H., and Nagano, H., "Development and Testing of a 300 W-Class Arcjet," International Electric Propulsion Conf. Paper 95-017, Sept. 1995.

[26]Sankovic, J. M., and Jacobson, D. T., "Performance of a Miniaturized Arcjet," AIAA Paper 95-2822, July 1995.

[27]Izumisawa, H., Yukutake, T., Andoh, Y., Onoe, K., Tahara, H., Yoshikawa, T., Ueno, F., and Ishii, M., "Operational Condition and Thrust Performance of a Low Power Arcjet Thruster," 20th ISTS Paper 96-a-3-23p, Gifu, Japan, May 1996.

[28]Birkan, M. A., "Arcjets and Arc Heaters: An Overview of Research Status and Needs," *Journal of Propulsion and Power*, Vol. 12, No. 6, 1996, pp. 1011–1017.

[29]Auwerter-Kurtz, M., Glocker, B., Golz, T., Kurtz, H. L., Masserschmid, E. W., Riehle, M., and Zube, D. M., "Arcjet Thruster Development," *Journal of Propulsion and Power*, Vol. 12, No. 6, 1996, pp. 1077–1083.

[30]Sankovic, J. M., and Hopkins, J., "Miniaturized Arcjet Performance Improvement," AIAA Paper 96-2962, July 1996.

[31]Martinez-Sanchez, M., and Pollard, J. E., "Spacecraft Electric Propulsion—An Overview," *Journal of Propulsion and Power*, Vol. 14, No. 5, 1998, pp. 688–699.

[32]Sankovic, J. M., "Ultra-Low-Power Arcjet Thruster Performance," *Proceedings of the 1993 JANNAF Propulsion Meeting*, Vol. V, Chemical Propulsion Information Agency, Nov. 1993, pp. 371–386.

[33]Horisawa, H., and Kimura, I., "Optimization of Arc Constrictor Sizes in Low Power Arcjet Thrusters," AIAA Paper 97-3202, July 1997.

[34]Butler, G. W., and Cassady, R. J., "Directions for Arcjet Technology Development," *Journal of Propulsion and Power*, Vol. 12, No. 6, 1996, pp. 1026–1034.

[35]Arnold, J. O., Whiting, E. E., and Lyle, G. C., "Line by Line Calculation of Spectra from Diatomic Molecules and Atoms Assuming a Voigt Line Profile," *Journal of Quantitative Spectroscopy and Radiative Transfer*, Vol. 9, 1969, pp. 775–798.

[36]Pearse, R. W. B., and Gaydon, A. G., *The Identification of Molecular Spectra*, 4th ed., Chapman and Hall, 1976, pp. 217–219.

[37]von Engel, A., *Ionized Gases*, 2nd ed., Oxford Univ. Press, Oxford, England, U.K., 1965, pp. 217–287.

Chapter 7

Low-Power Microwave Arcjet Testing: Plasma and Plume Diagnostics and Performance Evaluation

F. J. Souliez,* S. G. Chianese,* G. H. Dizac,* and M. M. Micci†
Pennsylvania State University, University Park, Pennsylvania

Nomenclature

c	= speed of light
f_r	= resonant frequency
I_{sp}	= specific impulse
P_{cavity}	= power absorbed by cavity
$P_{forward}$	= power forwarded to cavity
$P_{reflected}$	= power reflected by cavity
P_0	= mean chamber pressure
SP	= specific power
T_e	= electron temperature
T_0	= mean chamber temperature
λ	= emission wavelength
$\Delta\lambda$	= Doppler shift
ν	= emission frequency
γ	= specific heat ratio
η	= coupling efficiency

I. Introduction

ALTHOUGH propulsion systems on the majority of space vehicles to date have consisted of chemical thrusters, an increasing number of spacecraft launched recently use electric propulsion devices. Many of these systems consist of either ion thrusters, manufactured by Hughes for use on their satellites that have been under development for almost 35 years,[1] or arcjet thrusters manufactured by

Copyright © 2000 by the American Institute of Aeronautics and Astronautics, Inc. All rights reserved.

*Graduate Research Assistant, Aerospace Engineering Department.
†Professor, Aerospace Engineering Department. Associate Fellow AIAA.

Primex Aerospace[2] and now offered by Lockheed Martin, as a high-performance alternative for geosynchronous satellite north–south station keeping (NSSK).

The increasing interest in electric propulsion derives from much higher specific impulses relative to chemical thrusters, coupled with technological advances in power subsystem capabilities.[3] However, some of these systems do not fit all types of missions. Ion engines exhibit high efficiencies in specific impulses between 3000 and 9000 s. These values result in thrust levels that are too low for near-Earth missions at typical power levels of 1 to 10 kW. These systems generate very low thrust levels that are inefficient for drag makeup operations or require too much time for orbit insertion maneuvers.[4] The optimal specific impulses for orbit-raising and in-orbit maneuvering are in the 1000- to 2000-s range, in which ion thrusters are quite inefficient.[5]

Conventional arcjets suffer from cathode erosion problems[6] as well as decreased efficiencies when operating in lower power ranges.[7] Operation at 250 W on simulated ammonia yielded specific impulse levels from 360 to 470 s, with corresponding efficiencies of between 0.28 and 0.36. Arcjets also cannot operate efficiently in a pulsed mode required for attitude control. Resistojets have a material-based limitation that the propellant gas temperature cannot exceed the maximum allowable temperature of the heating element or any other propellant-wetted surface.

Outside the class of electrothermal thrusters, Hall thrusters operating at 300 W have given specific impulses up to 1160 s with hydrazine and a corresponding efficiency of only 0.32.[8] Performance measurements have also been taken with a Hall thruster operating at power levels from 250 down to 90 W using xenon propellant: the corresponding efficiencies and specific impulses went from 0.31 and 1230 s down to 0.14 and 521 s, respectively.[9]

Microwave thrusters are electrothermal thrusters that are electrodeless and therefore do not suffer life limitations of electrode erosion. A systematic illustration is given in Fig. 1, showing the major components of a microwave propulsion system.

Microwave resonant cavity thrusters can be distinguished from other electrothermal engines such as arcjets and resistojets by their method of heating the propellant gas. Arcjets use an electric-arc discharge to heat the propellant, whereas resistojets use electrical heating through a wall to increase the propellant stagnation temperature. In the case of the microwave thruster, the energy of standing

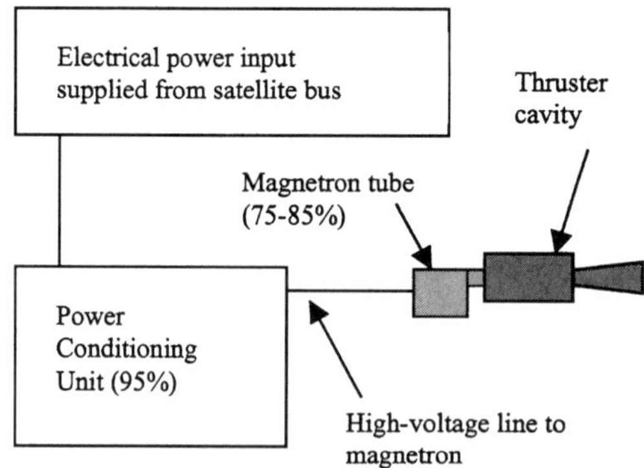

Fig. 1 System illustration with component efficiencies.

microwaves inside a resonant cavity is absorbed by the incoming propellant without contacting any of the engine's elements.

Previous research on microwave heated propulsion utilized much higher power levels (up to 2200-W CW) with plasmas confined in one single type of resonant cavity at 2.45 GHz. Over the past three years, the program has shifted from 2-kW-class thrusters to the development of lower-power microwave thrusters operating at power levels down to 70 W.

II. Experiment

The system introduces the microwaves into the engine as illustrated in Fig. 2. As can be seen, the cavity is partitioned in two halves separated by a dielectric quartz plate. The propellant is swirl-injected tangentially into the nozzle side of the cavity (plasma chamber). This is done both for cooling of the chamber interior walls and for radial stability of the plasma. The other side near the antenna is kept pressurized to ensure that the plasma formation takes place only in the plasma chamber, where the propellant is fed in at a lower pressure and brought slowly up to the desired chamber pressure.

The plasma is created by the region of high electric field strength formed on the axis of the cavity near the nozzle. The propellant gas is heated by being forced to flow in close contact to the plasma as it expands through the nozzle, converting thermal energy to directed kinetic energy, creating thrust.

This design has been tested at two resonant frequencies and proven to be effective with propellants such as gaseous nitrogen, helium, hydrogen, ammonia, and water vapor.[10] At low power levels (<70 W), vacuum starts have been successful at both frequencies (2.45- and 7.5-GHz engine) using helium, nitrogen, or ammonia as propellant. A vacuum start is a procedure that brings the resonant cavity to a pressure low enough for the microwave breakdown to occur and the plasma to form, thus simulating realistic start-up conditions in space. Plasma ignition is instantaneous and does not result in the erosion of any thruster components.

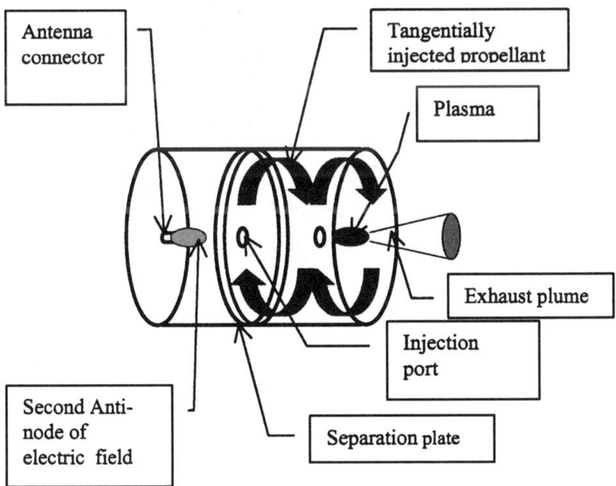

Fig. 2 Schematic of microwave thruster.

Coupling efficiencies between the injected gas and the incoming microwave energy of up to 99% have been measured.

III. Propellant Testing

The goals of this experiment were to obtain values for plasma chamber stagnation pressure, specific powers, and cavity efficiencies. Each of these values could be obtained from pressure, power, and mass flow rate measurements. Equations (1–3) give the formulas for specific power and cavity coupling efficiency:

$$P_{\text{cavity}} = P_{\text{forward}} - P_{\text{reflected}} \tag{1}$$

$$SP = \frac{P_{\text{cavity}}}{\dot{m}} \tag{2}$$

$$\eta = \frac{P_{\text{cavity}}}{P_{\text{forward}}} \tag{3}$$

The other objective was to test the 7.5-GHz thruster under vacuum conditions to characterize the operation of the thruster with respect to its start-up behavior and achievement of steady-state operation at low power levels using various propellant gases (helium, nitrogen, and ammonia). During these tests, the vacuum tank was pumped using only a Stokes mechanical pump. The experimental setup is illustrated in Fig. 3.

This testing showed that the low-power thruster could be operated autonomously under realistic conditions and for extended periods without any damage to any

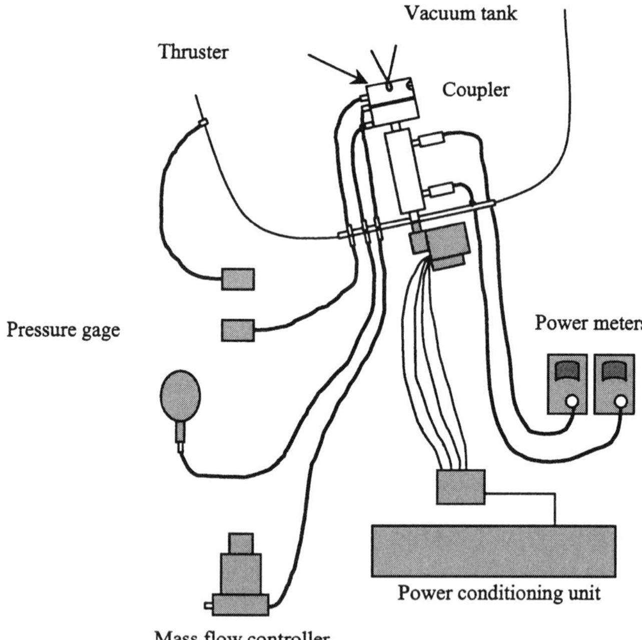

Fig. 3 Schematic of assembly used during vacuum tank testing.

Table 1 Helium hot-firing data in vacuum

$\dot{m}$, mg/s	P_{cavity}, W	P_0, psia	η, %	SP, MJ/kg
2.16	34.5	8.85	55.0	15.97
4.32	47.9	21	64.3	11.08
6.48	65.6	30.25	98.5	10.12
12.37	67.1	45.28	98.5	5.42
14.41	77.0	54.00	98.7	5.34

component. Most of the testing has concentrated on the 7.5-GHz cavity using all three propellants since no significant thruster chamber pressure level could be reached using the 2.45-GHz engine before the plasma would extinguish.

Tables 1–3 present data that were collected using all three propellants including mass flow rate, microwave power input to plasma, chamber stagnation pressure, coupling efficiency, η, between the input microwave power and the plasma, and specific power from initial start-up to steady-state conditions.

The data show that the plasma chamber pressure could be brought up to 45 psia with a stable helium plasma in place absorbing a power of slightly less than 70 W. A higher pressure of 54 psia could be achieved by increasing the power inside the cavity up to 77 W, with a corresponding coupling efficiency of 98.5% or more. Stable nitrogen plasmas could be maintained at pressure levels up to 55 psia with coupling efficiencies between the gas and the incoming microwave power above 98%. The orientation of the thruster within the vacuum tank combined with buoyancy forces resulted in nitrogen plasmas being generally less stable than helium plasmas. Data with ammonia plasmas could be obtained with chamber pressures up to 7 psia with coupling efficiencies above 86%.

The different plasma pressure levels that could be attained are of the same order of magnitude as those encountered during previous investigations at much higher power levels.[11,12] The difference in behavior between helium plasmas and other molecular propellants, in that for a given microwave power a helium plasma can be sustained at higher pressures, had also been observed at higher power levels.[13] The most probable explanation is that helium is monatomic, whereas the other propellants are diatomic or polyatomic molecules. The helium atom has few modes of internal energy storage; because of this, inelastic collisions with high-energy electrons are highly effective in liberating free electrons and creating helium ions. On the other hand, the polyatomic propellants have a relatively large

Table 2 Nitrogen hot-firing data in vacuum

$\dot{m}$, mg/s	P_{cavity}, W	P_0, psia	η, %	SP, MJ/kg
2.60	76.4	4.71	76.5	29.38
7.81	67.9	11.0	98.5	8.69
10.42	67.9	24.6	98.5	6.52
14.59	67.9	32.7	98.5	4.65
20.32	86.0	50.0	99.0	3.66
26.05	86.0	55.0	99.0	3.30

Table 3 Ammonia hot-firing data in vacuum

$\dot{m}$, mg/s	P_{cavity}, W	P_0, psia	η, %	SP, MJ/kg
0.65	72.5	4.18	98.6	113.54
0.83	65.4	4.86	98.2	78.79
0.97	58.5	5.45	94.4	60.31
1.07	57.3	5.95	89.1	53.55
1.28	86.0	7.13	86.0	67.19

number of internal modes of energy storage compared with the helium atom. Inelastic collisions that would ionize a helium atom may act only simply to excite an internal mode of the polyatomic molecule.

Another argument to explain the premature loss of molecular plasmas is the difference in thermal conductivities of helium and other polyatomic propellants at high temperatures: 0.25 W/(mK) for helium, 0.067 W/(mK) for ammonia gas, and 0.044 W/(mK) for nitrogen.[14] The reduced values of the thermal conductivities of nitrogen and ammonia result in a decrease in the effective volumetric area the thermal energy of the plasma core can influence, thus limiting the amount of heat transferred from the plasma to the swirling cold propellant gas. Such a limitation on the maximum attainable enthalpy rise restrains the thruster chamber pressure from reaching pressure levels of the order of magnitude of that of helium propellant when using the same amount of microwave power.

A reliable manner to determine the temperature rise once the plasma forms inside the cavity is to compare the pressure values for a given mass flow rate between cold flows and hot-firings. As shown in Eq. (4), forming their ratio allows us to eliminate the throat area A, which is known with a very poor precision due to its small size, 0.25-mm ID:

$$\frac{T_{0_{\text{hot}}}}{T_{0_{\text{cold}}}} = \left(\frac{P_{0_{\text{hot}}} \sqrt{\gamma_{\text{hot}}} \left(\frac{2}{\gamma+1} \right)_{\text{hot}}^{\left(\frac{\gamma+1}{2(\gamma-1)} \right)_{\text{hot}}}}{P_{0_{\text{cold}}} \sqrt{\gamma_{\text{cold}}} \left(\frac{2}{\gamma+1} \right)_{\text{cold}}^{\left(\frac{\gamma+1}{2(\gamma-1)} \right)_{\text{cold}}}} \right)^2 \tag{4}$$

The procedure requires several iterations: an initial inlet temperature has to be assumed in order to have the corresponding γ value, then the chamber stagnation temperature can be evaluated using Eq. (4) and compared with the initial guess. This is done until both values coincide. This gives a set of mean chamber stagnation temperature values that are plotted in Figs. 4a–c for helium, nitrogen, and ammonia propellants, respectively.

As expected, the inlet temperature for ammonia propellant is relatively low, since some of the microwave energy absorbed by the ammonia gas goes into various energy storage modes such as vibrational or rotational modes instead of raising the translational energy of the propellant, i.e., its temperature. The variations of temperature are dominated by the location of the plasma. At high mass flow rates, the resonance frequency is such that the plasma is right next to the nozzle inlet, whereas at low flow rates, it tends to stay away from the nozzle. This determines how heat is transferred to the cold propellant gas before exiting the chamber.

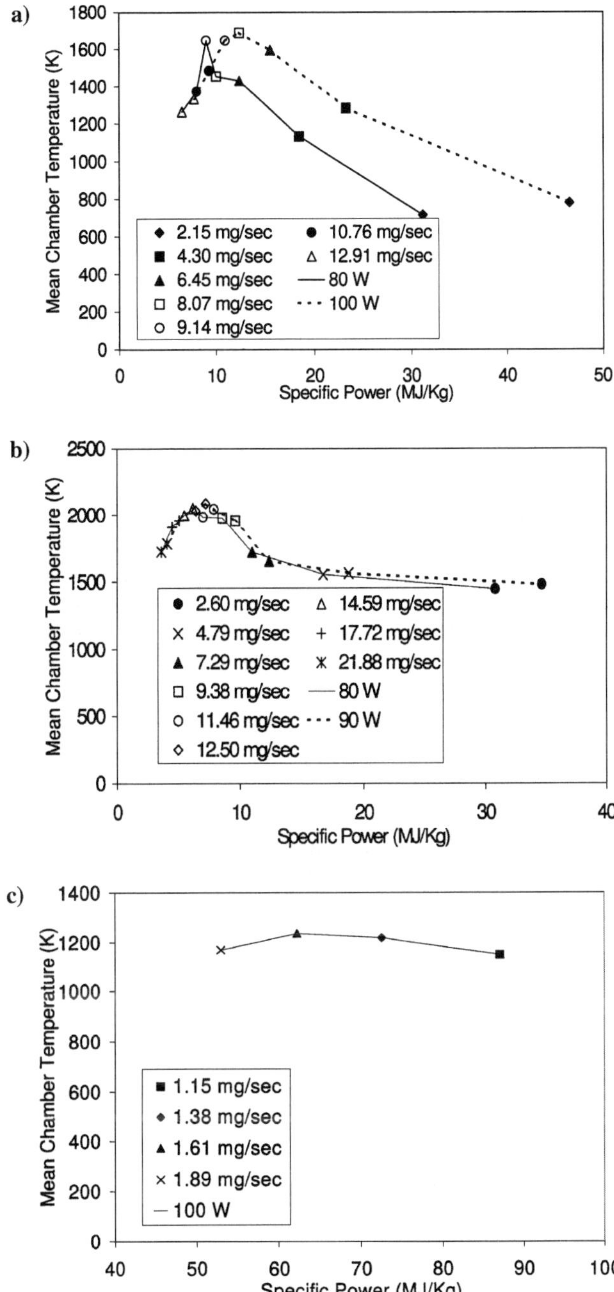

Fig. 4 Measured mean chamber stagnation temperature for a) helium, b) nitrogen, and c) ammonia propellants at various power levels.

Testing so far of the larger-cavity 2.45-GHz thruster has produced stable plasma discharges using helium, nitrogen, and ammonia. Plasma chamber pressure levels up to 16 psia have been obtained using helium propellant. The 2.45-GHz thruster chamber pressure has been brought up to 5 psia using nitrogen or ammonia.

IV. Electron Temperature Experiment

Emission spectroscopy provides a nonintrusive method of determining the temperature of gaseous discharges. It does not require the use of probes exposed to the plasma that might change its properties. The present work focused on the measurement of the free electron temperature T_e. To determine that quantity accurately, possible deviations from thermal equilibrium behavior within the plasma had to be evaluated, as well as their impact on the selected spectroscopic method.

Most of the data were taken well above atmospheric pressure levels. Previous investigations in argon plasmas by Eddy[15] indicate that the heavy particle temperature approaches the electron temperature at this level of pressure. Many techniques have been considered, in particular, absolute continuum spectroscopy and relative line intensity, the latter being the one chosen for use. In all cases, because of the local thermodynamic equilibrium (LTE) assumption involved, the obtained temperature values must be interpreted very carefully. This method is the most commonly used spectroscopic diagnostic to determine the electron temperature inside laboratory plasmas. It consists in determining the relative intensities of two spectral lines. It may be shown that for two emission lines at the wavelengths λ_A and λ_B,

$$\frac{I_A}{I_B} = \frac{(\nu g_2 A_{21})_A}{(\nu g_2 A_{21})_B} e^{-(E_{2_A} - E_{2_B})/kT_e} \tag{5}$$

where I_A and I_B are the intensities measured at the wavelengths λ_A and λ_B, respectively. The frequency ν, the transition probability A, the degeneracy g, and the upper energy level E are known constants. The electron temperature T_e can then be evaluated from the following expression:

$$T_e = \frac{1}{k} \frac{E_{2_B} - E_{2_A}}{\ln(I_A(\nu g_2 A_{21})_B / I_B(\nu g_2 A_{21})_A)} \tag{6}$$

where the energy gap $E_{2_A} - E_{2_B}$ has to be as large as possible for this method to be temperature sensitive.[16] Another condition is that the plasma pressure must be as high as possible to minimize deviation from LTE assumptions, as described in the following equation:

$$\frac{\Delta T}{T_e} = \frac{T_e - T_g}{T_e} = \frac{(\lambda_e E)^2}{\left(\frac{3}{2}kT_e\right)^2 8m_e} \tag{7}$$

The temperature difference $T_e - T_g$ between electrons and heavy particles, and thus the degree of nonequilibrium, can be decreased only at high pressures (small mean free path length λ_e) and low power levels (small electric field strength E). Another issue is that the transitions must correspond to the highest states that can be detected by the optical setup for the LTE assumptions to be valid. High collision rates are the predominant factor that can bring the plasma to equilibrium. Large upper quantum state numbers, n, correspond to larger collision cross sections σ

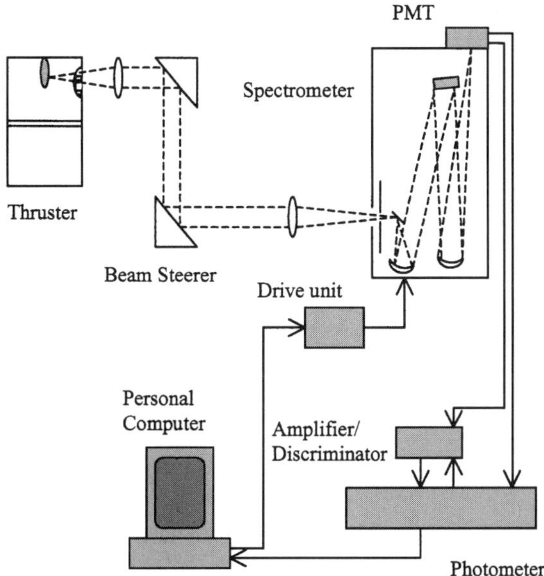

Fig. 5 Schematic of optical setup used in the electron temperature experiment.

(e.g., σ is proportional to n^4 for excited hydrogen atoms). As shown in the equation below, the time t between two consecutive collisions is short when the collision cross section is large, and hence departures from LTE are small[17]:

$$t = \frac{1}{N_e \sigma V} \tag{8}$$

The spectroscopic measurement is performed by collecting the light emitted by the plasma through the engine viewing window. The spectroscopic system that is used throughout the different experiments is shown in Fig. 5.

The collected light is focused on the entrance slit of the spectrometer, which is a 0.5-m, f/6.9 Czerney–Turner system. A 1200-line/mm Bausch & Lomb ruled grating is used. This spectrometer/grating combination imposes an upper limit to the wavelengths that may be detected of 660 nm. Connected to the output of the spectrometer is a Hamamatsu 1P28A photomultiplier tube (PMT) with a detection range from 185 to 700 nm and a maximal spectral response at 450 nm. The quantum efficiency of the PMT over the visible range had to be calibrated by its manufacturer, as well as the diffraction grating efficiency of the spectrometer. This is done to obtain the true intensity from the measured intensities correcting for the various instrumental sensitivities. The PMT serves to amplify the signal, which is then further amplified and converted into an electric current analog signal by a Pacific Instruments Model 126 photometer. The spectrometer control and data acquisition were performed by a personal computer with a data translation series A/D board driven by a BASIC program.

A set of four lines of helium $\lambda = 5876$ Å (upper quantum level $n = 3$), $\lambda = 5016$ Å ($n = 3$), $\lambda = 4922$ Å ($n = 4$), and $\lambda = 4471$ Å ($n = 4$) was scanned twice for each set of operating conditions. This was meant to check for the repeatability of the measurements in time. The lines corresponding to higher quantum numbers were either out of range of the spectrometer or of an intensity too low to be detected by the instruments.

For the reasons already mentioned, the data were collected at increasing pressure levels and the commonly made assumption of LTE was examined. The lowest operating plasma chamber pressure was 28 psia: this corresponds to twice the minimum pressure value for which the LTE assumption is usually made, i.e., 1 atm. The highest testing pressure was 50 psia: this is due mainly to the injection pressure limitation across the digital mass flow controller that cannot exceed 50 psia, as well as the design of the engine, since the pressure differential across the separation plate cannot exceed 50 psia.

Under these conditions, the resultant line broadening mechanism is dominated by the pressure broadening effect (or Stark effect), so that the line intensities were corrected assuming a Lorentzian shape profile out into the far wings. Huddlestone[18] gave the two following equations to evaluate the whole energy associated with each peak:

$$\frac{\Delta\lambda_{\mathrm{B}}^{*}}{\Delta\lambda_{\frac{1}{2}}^{t}} = \left[\left(\frac{\Delta\lambda_{\mathrm{B}}^{*}}{\Delta\lambda_{\frac{1}{2}}^{*}}\right)^{2} - 2\right]^{\frac{1}{2}} \tag{9}$$

$$\frac{I^{t}}{I^{*}} = \frac{\pi}{2}\left[\tan^{-1}\left(\frac{\Delta\lambda_{\mathrm{B}}^{*}}{\Delta\lambda_{\frac{1}{2}}^{t}}\right) - \left(\frac{\Delta\lambda_{\mathrm{B}}^{*}}{\Delta\lambda_{\frac{1}{2}}^{t}} + \frac{\Delta\lambda_{\frac{1}{2}}^{t}}{\Delta\lambda_{\mathrm{B}}^{*}}\right)\right] \tag{10}$$

where I^{t} is the true line intensity and $\Delta\lambda_{\frac{1}{2}}^{t}$ the true half-width at full maximum, both recovered from the measured values I^{*}, $\Delta\lambda_{\frac{1}{2}}^{t}$, and $\Delta\lambda_{\mathrm{B}}^{*}$ (half-width of the truncated base of the peak). Only the data satisfying a 15% repeatability between each scan were used to estimate the electron temperature.

Using Eq. (6) combined with the reading of different scans, electron temperature values could be determined for various pressure conditions. The frequency ν, the transition probability A, the degeneracy g, and the upper energy level E needed to evaluate T_e were from 5 to 15% accurate.[19] The different temperature values for each emission line and each set of operating conditions are listed in Table 4.

These values are much lower than those obtained by Mueller and Micci[20] or Balaam and Micci[21,22] during previous spectroscopic investigations. This can be explained by several factors: their data were collected at a much higher power level, and the absolute continuum technique was used. Generally, the electron temperature values obtained with their method ranged from 11,000 K up to 12,500 K.

Table 4 Electron temperature data

P_0, psia	$T\left(\frac{I_{\lambda=5876}}{I_{\lambda=4922}}\right)$, K	$T\left(\frac{I_{\lambda=5876}}{I_{\lambda=4771}}\right)$, K	$T\left(\frac{I_{\lambda=5016}}{I_{\lambda=4922}}\right)$, K	$T\left(\frac{I_{\lambda=5016}}{I_{\lambda=4471}}\right)$, K	$\frac{\sum T}{4} \pm \%$
28	10,078	5,893	3,946	3,077	5,749 ± 75%
37	5,892	5,635	3,166	2,982	4,419 ± 33%
50	4,865	4,147	3,726	3,283	4,005 ± 18%

These values show that, at 28 psia, the plasma is far from LTE conditions, and each line ratio would yield the same value. As expected, the temperature values tend to converge toward one single value around 4000 K as the plasma pressure increases to 50 psia. The dispersion of the average values goes from 75% down to 18%. The measured temperatures for the line ratios at 50 psia can be assumed to be very close, considering that the calculation of the temperature is subject to both experimental errors and errors arising from uncertainties in the spectroscopic constants that are employed. The relative error in the temperature is related to any error through Eqs. (11) and (12),

$$\left|\frac{\Delta T}{T}\right| = \frac{kT}{E_2 - E_1}\left|\frac{\Delta X}{X}\right| \tag{11}$$

$$X = \frac{(\nu g A)_2 I_1}{(\nu g A)_1 I_2} \tag{12}$$

where experimental errors in $\Delta X/X$ are rarely below 0.1, and the prefactor of this quantity is typically of order 1. This suggests a $\pm 10\%$ error in electron temperatures from such line ratios, even under favorable conditions concerning the validity of LTE conditions and the availability of accurate A values. Those results were encouraging enough so that no attempt was made to use the absolute continuum technique.

V. Doppler Shift Experiment

If light is emitted from a particle moving toward a detector, the frequency of that light will be shifted. If this shift can be measured, then the velocity u at which the molecule was moving toward the detector can be calculated as

$$u = \frac{c\Delta\nu}{\nu} \tag{13}$$

where ν is the frequency of a reference signal, c the speed of light, and $\Delta\nu$ the Doppler shift between the moving particle and the reference signal. This application assumes that the gas exiting the nozzle either is self-radiant (i.e., a high-temperature plasma exhaust) or can be induced to fluoresce using some type of excitation process (i.e., laser-induced fluorescence). This technique as applied to electrothermal thrusters has concentrated primarily on resolving the velocity profile of the atomic hydrogen species within the exhaust plume. In our case, this has been used mainly to characterize the exhaust plume of helium plasmas. A picture of a helium plasma plume is shown in Fig. 6. The nozzle area ratio is 52 : 1.

In practice, the lens that is supposed to collect the Doppler-shifted signal is slightly off-axis so that the unshifted pressure-broadened signal from the plasma inside the cavity does not override through the nozzle the shifted emission from the plume. A picture of the collection optics is shown in Fig. 7.

The instruments needed for this experiment are very similar to those required for the electron temperature measurement. Some critical elements had to be added to be able to measure the Doppler shift, including a high-spectral resolution Burleigh TL-15 Fabry–Perot interferometer and two high-transmittance fiber-optic cables mounted with special collimators.

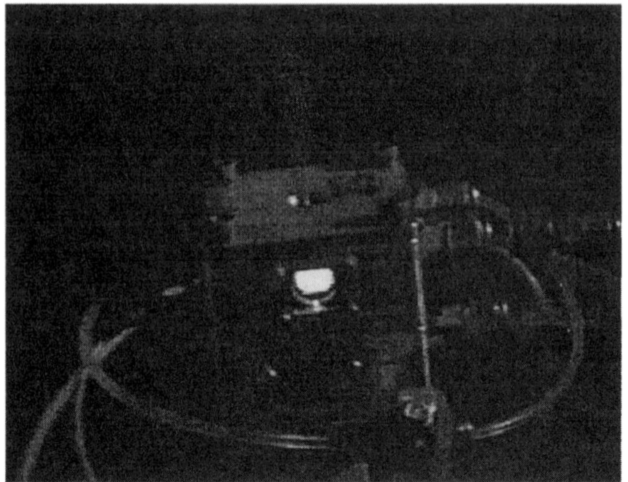

Fig. 6 A 7.5-GHz thruster helium firing in vacuum.

Centerline specific impulses have been measured for specific powers ranging from 15 up to 30 MJ/kg using helium propellant with an input power of 80 W. The Doppler shift was evaluated using the emission line corresponding to a wavelength of 5876 Å. At lower specific power levels (corresponding to higher mass flow rates), the lines were pressure-broadened so that the shift could not be resolved for pressures above 31 psia. Figure 8 clearly shows that the centerline shift can be accurately measured at high specific power values.

For verification purposes, several scans of the plume seen from the same angle in the same operating conditions were compared to check that no significant drift occurs. Figure 9 is a summary of all the data points collected between 15 and 30 MJ/kg.

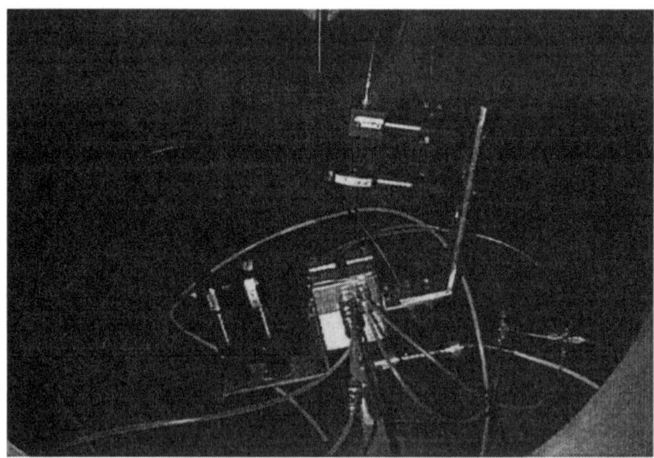

Fig. 7 Optical assembly mounted inside the vacuum facility.

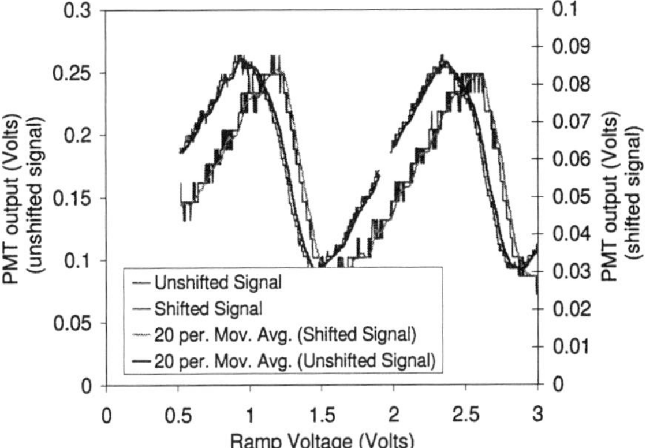

Fig. 8 Trace of Fabry–Perot etalon output at a high specific power.

As expected these plots show that specific impulse increases with specific power. The error bar is bigger for low specific power results since the plume was pressure-broadened and the expected specific impulse was lower, thus making the discrimination between two closely spaced lines more difficult. No result could be obtained at higher specific power levels, mainly due to the low light intensity of the plume at low pressure, thus limiting the maximum specific impulse that could be experimentally measured.

VI. Thrust Measurement

Thrust measurements are required to determine the performance of the microwave arcjet. Previously, Nordling[23] employed an inverted pendulum stand for use with the thruster operating in a horizontal configuration. Thrust measurements were taken under ambient conditions. However, buoyancy effects caused plasma instability and a loss of measured thrust. To eliminate these losses, a stand that allows the thruster to operate in the vertical position was developed.

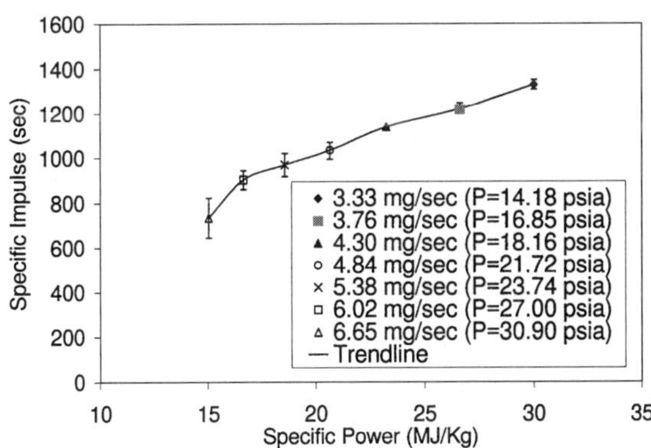

Fig. 9 Specific impulse vs specific power for helium propellant at 80 W.

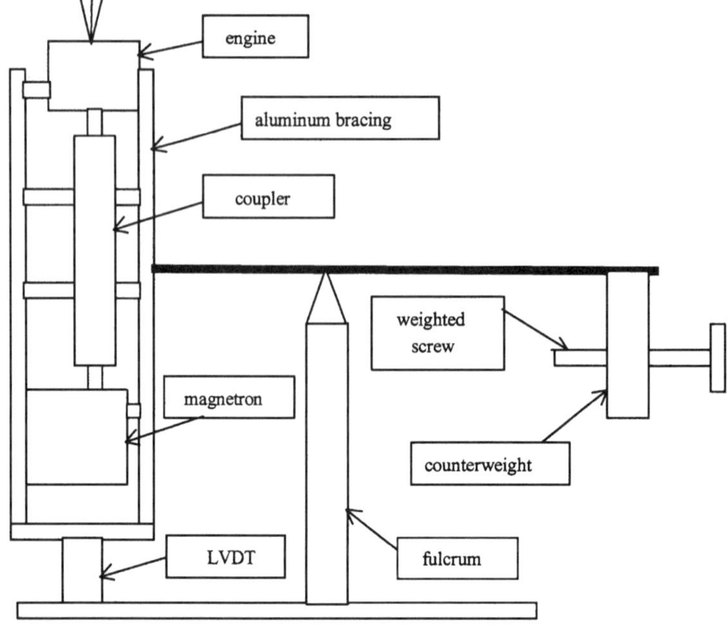

Fig. 10 Test stand schematic (not to scale).

A resolution of 1 mN or better is desired. To achieve this, the 5-kg mass of the thruster system must be counterbalanced. This is done with a lead counterweight attached to a knife-edge fulcrum system. Figure 10 shows a schematic of the system.

The thruster system has a moment arm of 10 cm, and the counterweight has a moment arm of 28.5 cm. The lever arm is attached to the thruster system and counterweight above their respective centers of mass, to provide inherent rotational stability. To balance the system finely, a fully threaded 6-in. screw can be adjusted. When the engine is fired, the thruster system displaces downward. For small angular rotations the motion is proportional to the applied thrust. A force-sensing linear variable differential transducer (LVDT) is used to measure the thrust.

For analysis, the fulcrum is assumed to be frictionless. The largest thrust expected, approximately 50 mN, causes a vertical deflection of no more than 0.05 mm. The resulting angular deflection is no more than 0.035 deg. This allows the restoring force of gravity to be neglected. The sensitivity of the system is dominated by the limitations in the electronics. The LVDT, and thus the test stand, has a resolution of approximately 0.5 mN, which is within the desired range.

The test stand is used while the thruster is operating in the vacuum tank. During firings the vacuum pump is not operated because the vibrations cause too much noise in the system. Because the power not absorbed by the propellant is reflected back to the magnetron, the microwave generating device must be cooled. This is done by flowing two jets of helium gas against the sides of the magnetron after each test. The vacuum tank is evacuated during and after this process. The stand is calibrated by placing known masses on the thruster and recording the response before the vacuum pump is operated for the first time each testing session.

The test stand will be used to determine the thrust and specific impulse produced by the microwave arcjet with helium, nitrogen, and ammonia as propellants for several power levels and mass flow rates.

VII. Conclusions

The ability to create and maintain plasmas at low microwave power levels under realistic space conditions has been demonstrated with propellants such as helium, nitrogen, and ammonia. Mean chamber stagnation temperatures could be measured, employing an iterative process providing inlet temperatures as high as 1700 K for helium propellant, 2100 K for nitrogen propellant, and 1240 K for ammonia propellant. As expected, a large fraction of microwave power is lost in various energy storage modes of ammonia molecules, hence reducing its temperature rise.

Spectroscopic measurements using the relative line intensity method have yielded electron temperature values for helium converging to an average value of 4000 K with less than 100 W of input power. The LTE assumption has been validated at pressure levels approaching 50 psia. The Doppler shift experiment has been successful and has produced centerline specific impulse values ranging from 730 up to 1330 s for specific powers between 15 and 30 MJ/kg with helium propellant.

Acknowledgment

This work was supported by Air Force Office of Scientific Research Grant F49620-97-1-0117.

References

[1]Schreib, R., "Utility of Xenon Ion Statiokeeping," AIAA Paper 86-1849, June 1986.

[2]Smith, R. D., Yano, S. E., Armbruster, K., Roberts, C. R., Lichtin, D. A., and Beck, J. W., "Flight Qualification of a 1.8 kW Hydrazine Arcjet System," IEPC Paper 93-007, 23rd International Electric Propulsion Conf., Sept. 1993.

[3]Bennett, G. L., Brandhorst, H. W., Jr., Bankston, C. P., and Sovie, R. J., "An Overview of Electric Power: A Key Technology for Electric Propulsion," IEPC 97-006, 25th International Electric Propulsion Conf., Aug. 1997.

[4]Keiser, T. L., and Micci, M. M., "Application of Microwave Thrusters for Stationkeeping and Orbit Raising on the INTELSAT Series Communication Satellites," IAF 90-228, 41st Congress of the International Astronautical Federation, Oct. 1990.

[5]Hill, P. G., and Peterson, C. R., *Mechanics and Thermodynamics of Propulsion*, Addison–Wesley, Reading, MA, 1965.

[6]Nordling, D., and Micci, M. M., "Low Power Microwave Arcjet Development," IEPC 97-089, 25th International Electric Propulsion Conf., Aug. 1997.

[7]Sankovic, J., and Hopkins, J., "Miniaturized Arcjet Performance Improvement," AIAA 96-2962, 32nd AIAA/ASME/SAE/ASEE Joint Propulsion Conf., Lake Buena Vista, FL, July 1996.

[8]Manzella, D., Oleson, S., Sankovic, J., Haag, T., Semenkin, A., and Kim, V., "Evaluation of Low Power Hall Thruster Propulsion," AIAA 96-2736, 32nd AIAA/ASME/SAE/ASEE Joint Propulsion Conf., Lake Buena Vista, FL, July 1996.

[9]Jacobson, D., and Jankovsky, R., "Test Results of a 200 W Class Hall Effect Thruster," AIAA 98-3792, 34th AIAA/ASME/SAE/ASEE Joint Propulsion Conf., Cleveland, OH, July 1998.

[10]Sullivan, D. J., *Development and Performance Characterization of a Microwave Electrothermal Thruster Prototype*, Ph.D. Dissertation, Aerospace Engineering Dept., Pennsylvania State Univ., University Park, PA, Jan. 1996.

[11]Sullivan, D. J., Kline, J., Philippe, C., and Micci, M. M., "Current Status of the Microwave Arcjet Thruster," AIAA 95-3065, 31st AIAA/ASME/SAE/ASEE Joint Propulsion Conf. and Exhibit, San Diego, CA, July 1995.

[12]Balaam, P., Maul, W., and Micci, M. M., "Characterization of Free-Floating Nitrogen and Helium Plasmas Generated in a Microwave Resonant Cavity," IEPC 88-099, 20th DGLR/AIAA/JSASS International Electric Propulsion Conf., Garmisch-Partenkirchen, Germany, Oct. 1988.

[13]Sullivan, D. J., and Micci, M. M., "The Effects of Molecular Propellants on the Performance of a Resonant Cavity Electrothermal Thruster," IEPC 91-034, 22nd AIDAA/AIAA/DGLR/JSASS International Electric Propulsion Conf., Viareggio, Italy, Oct. 1991.

[14]Lide, D. R., *Handbook of Chemistry and Physics*, 74th ed., CRC Press LLC, Boca Raton, 1993–1994.

[15]Eddy, T. L., "Low Pressure Plasma Diagnostic Methods," AIAA 89-2830, 25th Joint Propulsion Conf., Monterey, CA, July 1989.

[16]Griem, *Principles of Plasma Spectroscopy*, Cambridge Univ. Press, New York, 1997.

[17]Dalgarno, A., and Layzer, D., *Spectroscopy of Astrophysical Plasmas*, Cambridge Univ. Press, New York, 1987.

[18]Huddlestone, R. H., and Leonard, S. L., *Plasma Diagnostic Techniques*, Academic Press, New York, 1965.

[19]Wiese, W. L., Smith, M. W., and Glennon, B. M., *Atomic Transition Probabilities (a Critical Data Compilation), Vol. I. Elements Hydrogen Through Neon*, U.S. Dept. of Commerce, National Bureau of Standards, Washington, DC, 1966.

[20]Mueller, J., and Micci, M. M., "Microwave Waveguide Helium Plasmas for Electrothermal Propulsion," *Journal of Propulsion and Power*, Vol. 8, No. 5, 1992, pp. 1017–1022.

[21]Balaam, P., and Micci, M. M., "Investigation of Free-Floating Resonant Cavity Microwave Plasmas for Propulsion," *Journal of Propulsion and Power*, Vol. 8, No. 1, 1992, pp. 103–109.

[22]Balaam, P., and Micci, M. M., "Investigation of Stabilized Resonant Cavity Microwave Plasmas for Propulsion," *Journal of Propulsion and Power*, Vol. 11, No. 5, 1995, pp. 1021–1027.

[23]Nordling, D. A., "High-Frequency Low-Power Microwave Arcjet Thruster Development," M.S. thesis, Aerospace Engineering Dept. Pennsylvania State Univ., University Park, PA, Aug. 1998.

Chapter 8

Vaporizing Liquid Microthruster Concept: Preliminary Results of Initial Feasibility Studies

Juergen Mueller,* Indrani Chakraborty,† David Bame,‡ and William Tang§
Jet Propulsion Laboratory, California Institute of Technology
Pasadena, California

I. Introduction

RECENTLY, a strong interest in micropropulsion devices has arisen within the space community. Such devices, capable of delivering very low thrust values, in the milli-Newton range and below, and impulse bit values as low as a few micronewton-seconds and being orders of magnitude smaller in size and mass than available state-of-the-art technologies,[1] would be needed for microspacecraft designs. This type of spacecraft, typically viewed as having wet masses of a few tens of kilogram or less, has gained attention within NASA as well as the U.S. Air Force.[2] This is because these craft will allow new and unique mission profiles to be flown, involving constellations of microspacecraft charting entire regions of space simultaneously and cost efficiently, or increase mission reliablility by offloading instruments from a large single spacecraft onto a fleet of microspacecraft. Thus, the loss of a single or even a few microspacecraft will not jeopardize the entire mission. In military applications, constellations of microspacecraft are being envisioned to deploy phased antenna arrays for high-resolution radar observation of objects of military interest on Earth. Reduced spacecraft mass will result in decreased launch costs. Typically, launch costs may contribute as much as 30% to the total mission cost and are, to a large extent, determined by spacecraft mass. In particular, in the case of large spacecraft constellations, costs for these missions may be reduced significantly by resorting to microspacecraft designs.

Micropropulsion devices may also find applications on larger spacecraft. Interferometry missions, for example, may not necessarily be based on microspacecraft

Copyright © 2000 by the American Institute of Aeronautics and Astronautics, Inc. The U.S. Government has a royalty-free license to exercise all rights under the copyright claimed herein for Governmental purposes. All other rights are reserved by the copyright owner.

*Advanced Propulsion Technology Group.
†MEMS Group, Microdevices Laboratory.
‡Propulsion Flight Systems Group.
§Formerly Supervisor, MEMS Group, Microdevices Laboratory; currently at DARPA.

designs, yet have a need for very low impulse bit propulsion systems for fine-positioning and pointing.[3] Some of these types of missions will target the search for planets and possibly even the evidence of life around other stars and are anticipated to play a central role in future NASA activities.

One of the most challenging aspects of both mission types, microspacecraft and interferometry, is attitude control. Due to either the low mass of the microspacecraft or the stringent pointing and positioning requirements, very small impulse bits will be required, which could reach into the micronewton-second range.[1] Furthermore, since multiple thrusters will be required for attitude control, thruster units will have to be extremely small and lightweight, in particular for microspacecraft applications, to meet the severe mass and volume design constraints. Similarly, power constraints will need to be adhered to. Total microspacecraft power levels may not exceed a few tens of watts, possibly considerably less.[1]

A new micropropulsion concept, termed the vaporizing liquid microthruster[4,5] (VLM), has recently been introduced, aimed at providing attitude control for microspacecraft. This microfabricated thruster concept is targeted to provide very low thrust values, in the range of 0.1–1 mN, and impulse bits in the 10^{-7} to 10^{-5} Ns range (depending on the valve used) at an extremely low thruster weight and size, able to be placed onto a silicon chip. In this thruster concept, propellant will be vaporized on demand, generating thrust by thermally expanding propellant vapor through a nozzle. The propellant can be stored compactly in its liquid phase using lightweight, low-pressure tanks, and leakage concerns can be significantly reduced or eliminated due to liquid storage. In principle, this thruster may use any propellant that can be vaporized; initial testing, however, is limited to water propellant. Other propellants under consideration are ammonia, with a heat of vaporization about half that of water, or possibly hydrazine due to flight heritage.

Development of the VLM is in its earliest stages, as characterized by the work presented in this chapter. The initial focus of VLM-related work is the determination of the feasibility of this thruster concept, to be followed up by the determination of its performance and operating characteristics in future studies. This chapter, after introducing the VLM design concept, discusses the design and fabrication of the concept, initial tests determined to characterize heater performance, and preliminary results obtained during propellant (water) vaporization tests. Results are limited due to the unavailability of appropriate diagnostic equipment at the time of the experiments, in particular, also with respect to low-liquid flow rate measurements. Development efforts and searches for appropriate equipment are currently under way to remedy this limitation.

II. Chip Design and Fabrication

The VLM device is a microfabricated device using silicon-based MEMS technologies. The thruster schematic and an actual device are shown in Figs. 1 and 2. The VLM is constructed of a laminate of three chips, as shown in Fig. 3. The top and bottom chips contain the vapor-deposited thin-film heaters, as well as the nozzle and inlet. Also featured on these chips are two vias (through-holes) to contact the heater elements electrically. These wafers are bonded into a stack via a spacer, or channel, chip. This channel chip features a cutout that forms the sidewalls of the flow channel as well as vias needed to contact the lower heater element, since electrical contacts are made from only one side of the chip. The

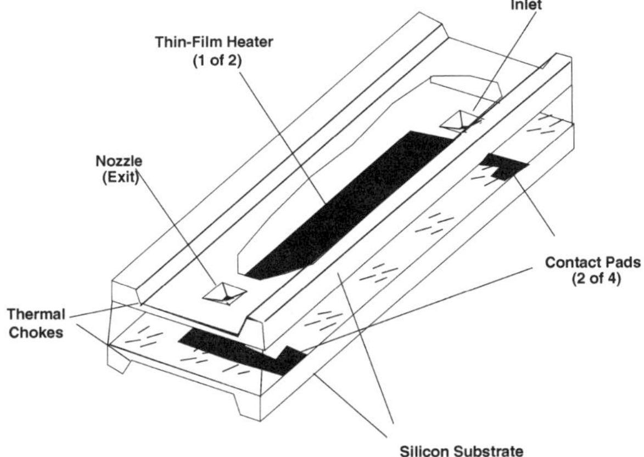

Fig. 1 Concept of the vaporizing liquid microthruster.

cutout forming the flow channel walls was fabricated using a state-of-the-art deep trench reactive ion etching (DRIE) technique that allows straight channel walls to be formed, in contrast to the anisotropically etched, angled nozzle and via walls. Liquid propellant enters the chip through an orifice in the bottom of the stack, flows along the heaters through the microchannel formed by the spacer chip and the heaters, and then exits the chip through a nozzle in a gaseous state.

The channel has a width of 0.95 mm. The channel height is determined by the thickness of the spacer wafer. In some cases, a full-thickness spacer wafer was used, resulting in a channel height of 0.6 mm. In others, the spacer wafer was etched back to a thickness of 0.3 mm, leading to an accordingly reduced channel height. As will be seen below, the flatter channel profile ensures improved heat conduction into the liquid due to a higher surface-to-volume ratio of the microchannel, resulting

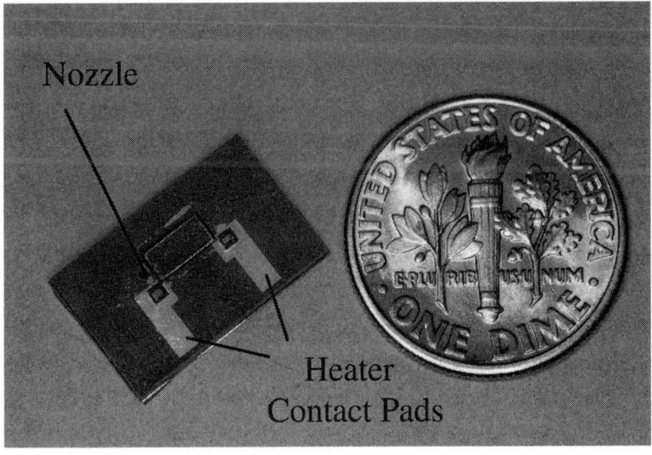

Fig. 2 Vaporizing liquid microthruster chip.

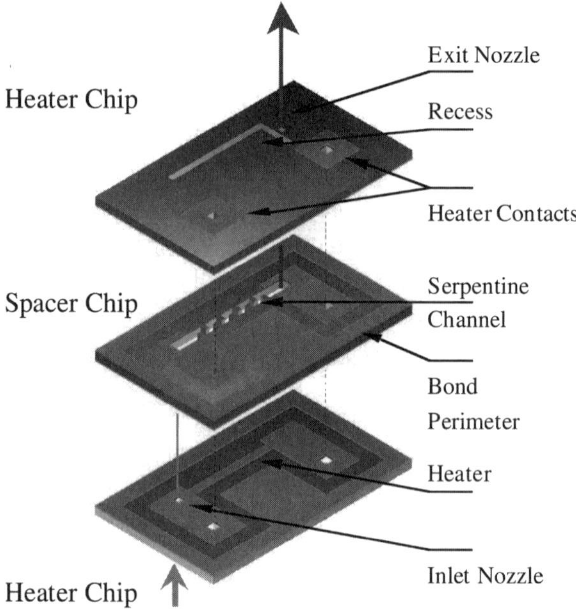

Fig. 3 VLM chip design components.

in more complete vaporization of the propellant. Just opposite the heater strip a recess is machined into the silicon substrate, thinning the substrate material at this location (compare with Figs. 1 and 3). The purpose of this design feature was to provide a thermal choke to reduce heat conduction from the heater surface to the remainder of the chip.

Channel inlet and nozzle are square-shaped, following the anisotropic etch patterns of 100-silicon wafers, and have throat dimensions of $50 \times 50\ \mu m^2$. The nozzle is symmetric with respect to its converging and diverging sections, and since the silicon wafer into which the nozzle was machined is 0.6 mm thick, the length of the diverging (and converging) nozzle section is 0.3 mm. It should be noted that this current nozzle shape is merely a place holder for more optimized nozzle contours to be integrated into future versions of the VLM. Improved nozzle contours have been investigated at MIT.[6,7] The chip weight is a few grams, and current chip sizes are about $0.9 \times 1.5 \times 0.1\ cm^3$, with a slightly smaller version of $1 \times 1.1 \times 0.1\ cm^3$ (not shown) recently completed.

The three wafers making up the VLM chip are bonded via a thin gold layer through a metal-to-metal thermal compression process. This has the advantage that the bonding medium as well as the heater elements, also formed through the deposition of a gold layer, can be processed in the same fabrication step. Gold is being used as the heater material since its low resistance will result in low voltage requirements for the thruster. One set of chips was fabricated featuring polysilicon heaters for comparison. Polysilicon heaters would be required if, in future versions of the VLM, the silicon wafers were to be fusion bonded, rather than thermal compression bonded. Fusion bonds are believed to be much stronger than thermal gold compression bonds. However, fusion bonding does require a high-temperature (900–1000°C) annealing step, which would not allow gold heaters to be used.

In past design iterations, Pyrex material was intended for use as the spacer wafer material, to be bonded anodically to the silicon nozzle and inlet wafers.[4,5] This Pyrex–silicon anodic bond also features high bond strengths,[4] however, it does not require a high-temperature anneal as in the case of a fusion bond. Unfortunately, it was noted during the fabrication of the Pyrex spacer, requiring ultrasonic machining processes to form the channel and vias, that unacceptable surface roughnesses had appeared on one side of the wafer, preventing successful bonding. The surface roughness was not found to be homogeneous across the wafer surface, but appeared only in certain regions of the wafer, distributed in a regular pattern. It appeared that an abrasive slurry, used during the ultrasonic machining process, had caused abrasions in the antinode regions of the vibrating spacer, with the spacer vibrations being caused by the ultrasonic tool. Subsequent polishing attempts of the Pyrex spacer failed and led to repeated wafer breakage, likely due to internal stresses caused by a combination of drilling a multitude of holes per wafer (more than a dozen VLM spacers were fabricated per 3-in.-diam Pyrex wafer) and the thinness of the wafer (0.5 mm). Repeated attempts by the vendor to deliver a satisfactory product failed, and, finally, this technique was abandoned in favor of the gold compression bonding technique using a silicon spacer.

The chips were packaged by placing them into a ceramic (alumina) hybrid chip carrier with a port drilled into the bed for access to the chip inlet. This carrier provided both electrical and propellant interfaces. Bonding of the chip to the carrier was facilitated by a high-temperature epoxy. A threaded nut (aluminum or Vespel) was bonded to the bottom of the chip carrier, with the through-hole overlapping the port drilled into the carrier and the chip inlet. This nut will allow the packaged chip to be plumbed to a feed system providing the propellant. It should be noted that this packaging scheme serves initial bench-top tests aimed at preliminary characterization of thruster chip performances only. It has the advantage of being cheap and consists of readily available commercial components. Thruster packages more closely resembling flight hardware will require customized packaging.

In the following, a first set of VLM bench-top experiments is described. These tests were very preliminary in nature, with the intended goal of demonstrating the functionality of the VLM concept. More detailed performance measurements will follow in future studies. Included in these initial tests are heater performance experiments, aimed at determining power and voltage requirements versus heater temperature, and a functionality test of the VLM using water propellant, demonstrating, for the first time, that complete propellant vaporization is possible with a VLM. Vaporization of propellant (water) has been achieved at power levels of about 2 W and voltages of slightly under 2 V.

III. Heater Characterization

A. Description of Experiment

The goal of the heater experiments was to determine power and voltage requirements to achieve a given heater temperature. This thermal characterization of the heater is an important evaluation criterion for the VLM, since the available power onboard a microspacecraft may be severely limited, potentially not exceeding a few tens of watts of total spacecraft power. Similarly, bus voltages onboard microspacecraft may be significantly lower than are common on spacecraft today, possibly as low as 5 V, and any microthruster considered for microspacecraft use should be able to meet these requirements.[1] Consequently, initial tests were performed measuring required thruster power levels and voltages vs chip temperature.

To perform the temperature measurements the chip was placed under an infrared (IR) camera and power supplied to the chip was measured for a given chip temperature. The thruster temperature could be measured only on the outside walls of the chip, and a position in the recess area, just opposite one of the heater elements (compare with Fig. 1), was chosen as the location for the IR camera to lock on. A small dot of black graphite-based paint was applied to the chip at this location to provide a surface of well-characterized emissivity for the IR camera to lock on. Due to the location of temperature measurement, actual heater temperatures may have been slightly higher, as a temperature drop may have occurred across the 300-μm-thick silicon membrane separating the heater element from the actual measurement position. However, given the rather small silicon substrate thickness and high silicon heat conductivity, differences are believed to be small.

Tests were performed with a chip featuring a 4-mm-long gold heater, packaged as described above. Although an aluminum or a Vespel nut was bonded to the chip carrier in these measurements, the nut was not connected to a feed system. All initial tests were performed without water vaporization occurring inside the chip to check the thermal design of the chip independent of the vaporization processes and power requirements associated with them.

B. Results

Results of the thermal characterization of the packaged chip are shown in Figs. 4 and 5, representing electric input power and voltage vs heater temperature. Initial tests were performed using an aluminum nut connected to the chip carrier. During

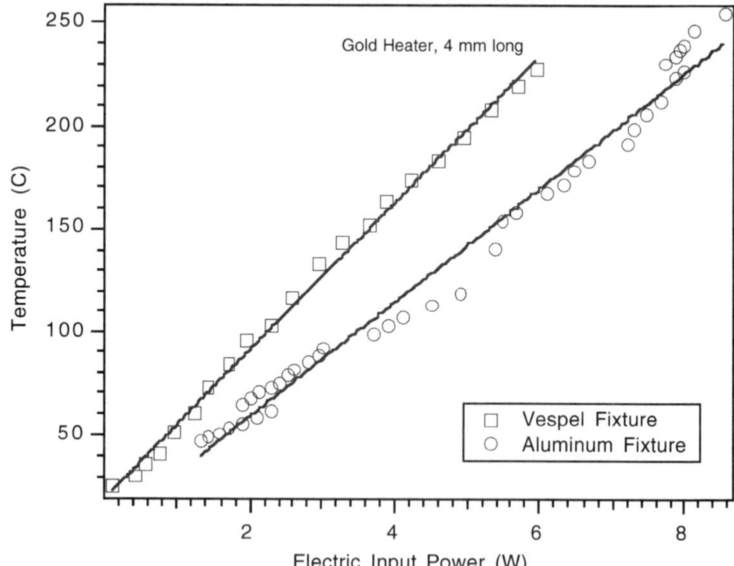

Fig. 4 VLM chip temperature vs electric input power for different packaging approaches.

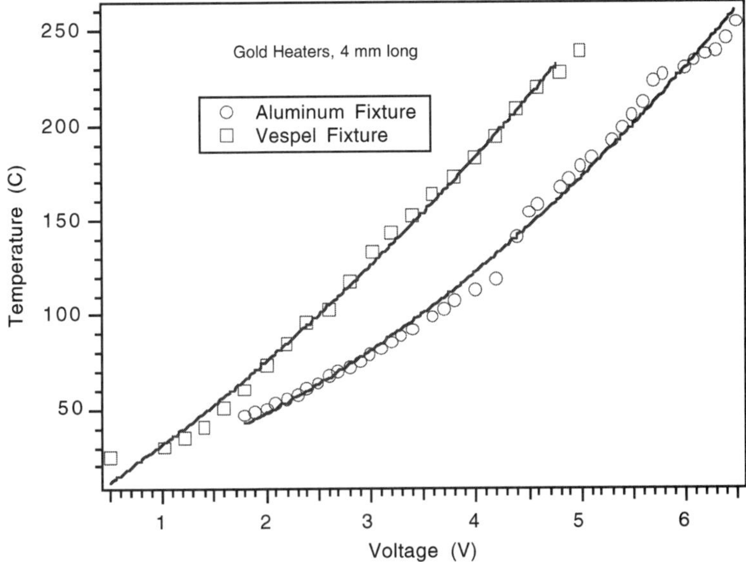

Fig. 5 Chip temperature vs voltage for different chip packaging approaches.

the tests it was noted that, despite the ceramic chip carrier being placed between the chip and the aluminum nut, insulation was poor and the aluminum nut got very hot, obviously acting as a heat sink for a considerable amount of the energy provided to the chip. Thus, a nut made from Vespel was bonded to the chip carrier to investigate how such a relatively simple design change would impact chip performance.

As Fig. 4 indicates, the required chip power levels, and thus voltages, could be lowered significantly using the Vespel fixture. At 100°C, the required power level and voltage were 2.3 W and 2.6 V, respectively. The corresponding values were 3.6 W and 3.4 V at 150°C and 5 W and about 4.3 V at 200°C. Required voltages could be kept well below 5 V for the Vespel package, even in the cases of the highest temperatures.

Further design changes are anticipated to improve the obtained results. During the tests described above it was noted that, by placing multiple graphite paint dots onto the chip surface at various locations to measure the temperature there, the temperature profile across the chip surface was rather uniform. Ideally, a peaked temperature distribution is desired, with maximum temperatures occurring at the location of the heater elements in the center of the chip. A thermal choke created by etching a recess underneath the heater elements (see Section II above) was supposed to ensure such a peaked temperature distribution. Obviously, the recess in its current design layout, 300 μm deep and leaving a 300-μm-thick silicon membrane underneath the heater element, was rather ineffective. A new chip is currently under construction with a thinner silicon membrane (100 μm), which should improve the thermal choke and lower the temperatures of the chip near its perimeter where it bonds to the chip carrier, thus reducing heat losses. This new generation of VLM chips will also feature a smaller footprint, aimed at reducing heat conduction losses further by reducing the bond area between chip and carrier by another 20%.

IV. Propellant Vaporization: Initial Studies

A. Description of Experiment

It is crucial for the success of the VLM concept to achieve complete vaporization inside the thruster and avoid the ejection of droplets. Droplet nozzle exit velocities are slow in comparison with gaseous ejections and would thus significantly lower the specific impulse and result in inefficient use of propellant. Achieving this goal with the VLM poses a major challenge since vaporization has to be accomplished over very short heater lengths (of the order of a few millimeters) to be compatible with typical chip dimensions, while, at the same time, keeping power requirements low. Power values of 5 W or less were targeted. Achieving these goals was by no means certain at the outset of this project, and a successful demonstration of complete vaporization was thus considered a critical first milestone in the development of the VLM. A simple bench-top experiment was performed to gain initial experience with VLM operation and to provide feedback to guide future design improvements.

This initial set of experiments was seriously hampered by the unavailability of flow meters able to measure the very low liquid flow rates. As a consequence, the results obtained are largely qualitative. However, leads have been developed possibly providing adequate flow diagnostic equipment for future experiments. The set of experiments described in this study will form the basis for a more quantitative investigation in the future. Despite these limitations, however, the set of experiments described here represents an important, although very preliminary, milestone in the development of the VLM, testing the functionality of the concept, demonstrating vaporization on chip scales, allowing qualitative comparisons between different chip designs, and, thus, narrowing the field of designs needed to be tested in more sophisticated future experiments.

1. *Test Chips*

Different chip configurations were tested, listed in Table 1. They included two chips featuring 4-mm heater lengths with a channel cross section of $950 \times 600\ \mu m^2$, where the first number represents the width of the channel and the second number the channel height. One of these chips was integrated into an assembly featuring an aluminum nut, the other a Vespel nut, connecting the chip to the water tank.

Table 1 VLM chip test articles

Designation	Heater length, mm	Channel cross section, μm^2	Channel type	Fixture material
SC-4-600-Al	4	950×600	Straight	Aluminum
SC-4-600-V	4	950×600	Straight	Vespel
SC-5-300-V	5	950×300	Straight	Vespel
SC-5-300-V/Poly[a]	5 (PolySi)	950×300	Straight	Vespel
SC-6-300-V	6	950×300	Straight	Vespel
MC-12–300-V	12.16	400×300	Meandering	Vespel

[a]Featuring a heater made from polysilicon. All other heaters made from gold.

The 600-μm channel height was determined by the spacer wafer thickness, and the channel width was chosen rather arbitrarily, close to 1 mm, as a starting point for VLM investigations. As shown below, this type of microchannel led to poor and incomplete vaporization even for relatively high power levels. Therefore additional sets of chips were fabricated featuring different channel dimensions aimed at improving heat transfer into the liquid.

The second type of chip featured the shallower channel cross section of $950 \times 300\ \mu m^2$, which was obtained by thinning the spacer wafer to one-half its original thickness through etch-back. By lowering the channel height, the surface-to-volume ratio of the heater channel was increased, increasing the heat transfer per unit volume for this channel. This type of chip also featured different channel lengths of 4, 5, and 6 mm, respectively, to study the effect of channel lengths. Only a 5- and a 6-mm channel were available for testing.

All of these chips were packaged into assemblies featuring Vespel nuts. Heater widths in all cases were 650 μm, i.e., slightly narrower than the channel width. This is necessary to avoid electrical contact between the gold heater and the gold layer acting as bonding agent between the various wafers making up the VLM chip. One chip of this type featured a polysilicon heater to study differences between this heater material and the thin-film deposited gold material used in the cases of all other chips tested.

Finally, a third type of chip, featuring a meandering channel layout with a channel height of 300 μm and a channel width of 400 μm, as shown in Fig. 6, was fabricated. The motivation for this design was to increase the heated surfaces exposed to the liquid flow further by inserting fins into the original straight channel section (compare with Fig. 6). The silicon fins are equidistantly spaced and are 100 μm wide, protruding 550 μm from the side-walls. The heater elements in the case of this particular chip are 5 mm long and 650 μm wide, i.e., remain rectangular in shape and extend below the fins. As in the case of most other chips, these heater elements are made of thin-film deposited gold material. The fins are being heated through direct contact with the heater elements and, since silicon is an excellent heat conductor, serve as additional heating surfaces for the flow entering the heater channel. As a result of the insertion of the fins, a meandering flow path has been

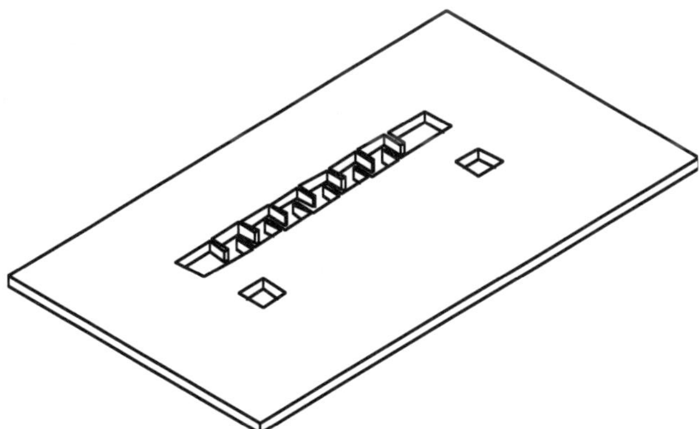

Fig. 6 Channel (spacer) chip with meandering flow path design.

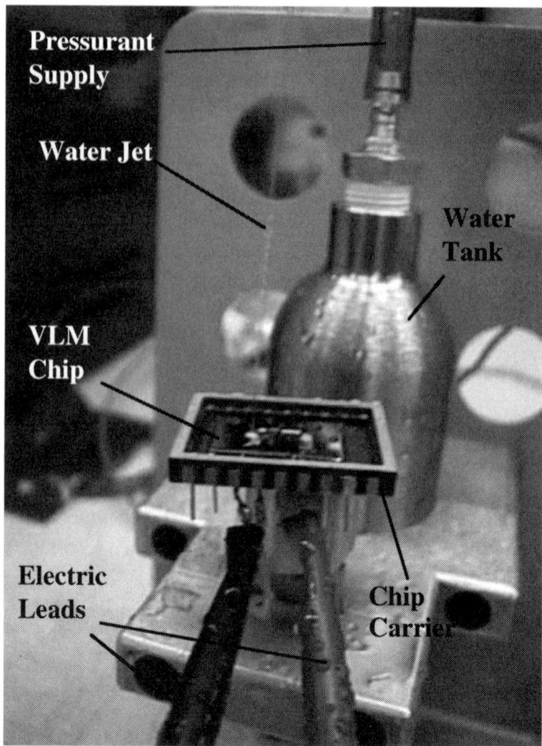

Fig. 7 VLM chip on chip carrier attached to water tank. Note the water jet exciting nozzle. No heat input to chip.

created with a total channel length of about 12 mm. (A chip with a 4-mm straight channel and a second chip with a meandering channel, but more closely spaced fins, were also machined, however, they were not able to flow any water. This may have been due to contamination inside the channels, blocking the flow, acquired either during handling of the chips during testing or during the epoxy-bonding process when packaging the chips. It had been noted prior in a related project that epoxy, in a liquid phase during the bonding process, may flow into narrow channel sections due to capillary forces. Bonding procedures were subsequently adjusted to prevent epoxy capillary flow, but contamination as a result of this observed epoxy behavior may not be entirely excluded.)

2. *Experimental Setup*

The packaged test chips (including chip carrier and nut) were mounted onto a test rig consisting of a water tank, a 2-μm filter placed at the tank outlet, and a small commercial solenoid valve manufactured by the Lee Company. The packaged chip assembly was mounted onto a port via the already-mentioned threaded aluminum or Vespel nuts (see Fig. 7). A pressurant supply was connected to the water tank. By pressurizing the tank, water is forced through the filter and valve and into the chip. In early tests, feed pressures had to be lowered to below 1 psig to adjust water flow to levels low enough to achieve vaporization. In these cases, the pressurant supply was replaced by an additional water tank placed at an elevated position above the chip assembly and the propellant was gravity-fed.

3. *Test Procedure*

A simple test procedure was followed in this first set of preliminary bench-top experiments. The feed pressure was adjusted to a given level, and water was fed into the chip. (Since, as mentioned above, direct flow measurements were not yet possible given the very low liquid flow rates through the VLM chips, feed pressure was the only parameter having some bearing on the flow rate that could be measured. Given that different channel designs will result in different flow rates for a given feed pressure, the limitations of this first set of experiments are evident.)

The exit flow out of the chip was observed visually. Without heater power, depending on the flow rate (not yet measurable), either a water jet (see Fig. 7), or a droplet forming at the nozzle exit was observed. When adding and increasing the heater power, first an increase in water jet temperature was noted. At somewhat higher power levels intermittent vapor and liquid water ejections were observed, accompanied by a sputtering noise. Increasing the chip power further led finally to the ejection of pure steam that could not be observed directly and could be evidenced only by observing its condensates on a mirror or glass slide placed in the vapor jet. At power levels still not quite high enough, this vapor emission lasted for only a few seconds. Cooling of the chip due to forced convection as a result of the propellant flow then led again to the emission of liquid water droplets. Cooling of the chip was evidenced by observing a shift toward lower voltage and higher current levels during propellant flow, indicating a decrease in resistance with lower heater temperatures, as expected for metallic (gold) heaters. Finally, on increasing the power level further still, continuous vapor ejection was observed. The tests were then repeated at different feed pressures, and the power required for vaporization for different feed pressures (flow rates) was recorded. For the same chip, this will allow for a qualitative comparison of chip behavior and heater effectiveness. Comparing chips of different channel designs is not possible with this method since the same feed pressure may result in different flow rates for different chips, depending on the channel dimensions.

B. Preliminary Results

1. *Initial Testing*

Preliminary results obtained with the chips listed in Table 1 are shown in Figs. 8 and 9. Data are plotted vs feed pressure for reasons given above. The first tests, conducted using chips featuring a 600-μm channel height and packaged using the aluminum nut, resulted in relatively high required power levels at extremely low feed pressures to achieve vaporization. The VLM required 7 W at about 0.25-psig feed pressure.[5] This data point is shown in the upper left-hand corner of Fig. 8. As mentioned, this chip featured an aluminum nut, which had been identified as a major heat sink due to the poor insulation provided by the ceramic carrier package. Replacing the aluminum nut with one machined out of Vespel material reduced required the power levels dramatically, by about 30%, to just under 5 W at the same feed pressure. Using this chip assembly, vaporization was still possible

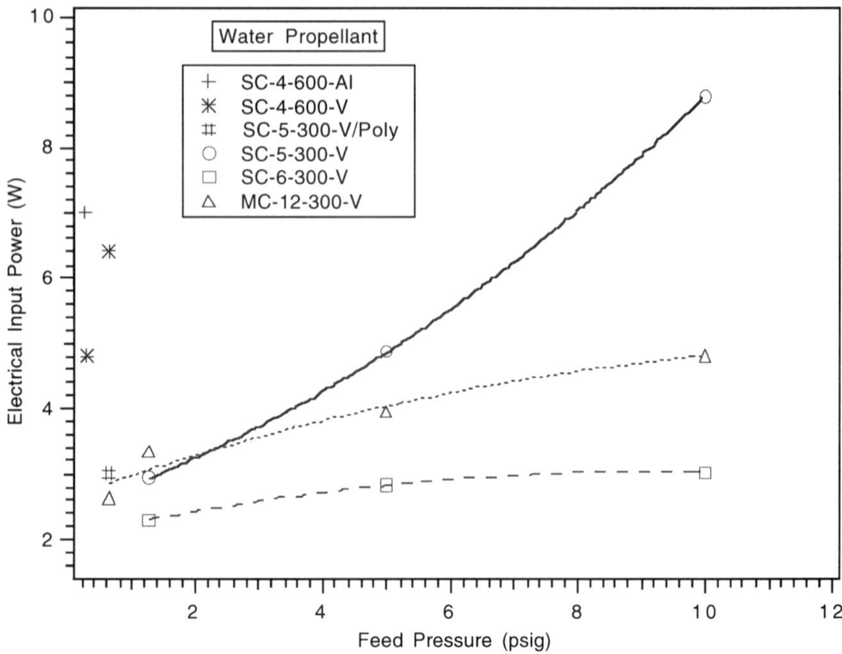

Fig. 8 VLM vaporization tests (power vs feed pressure).

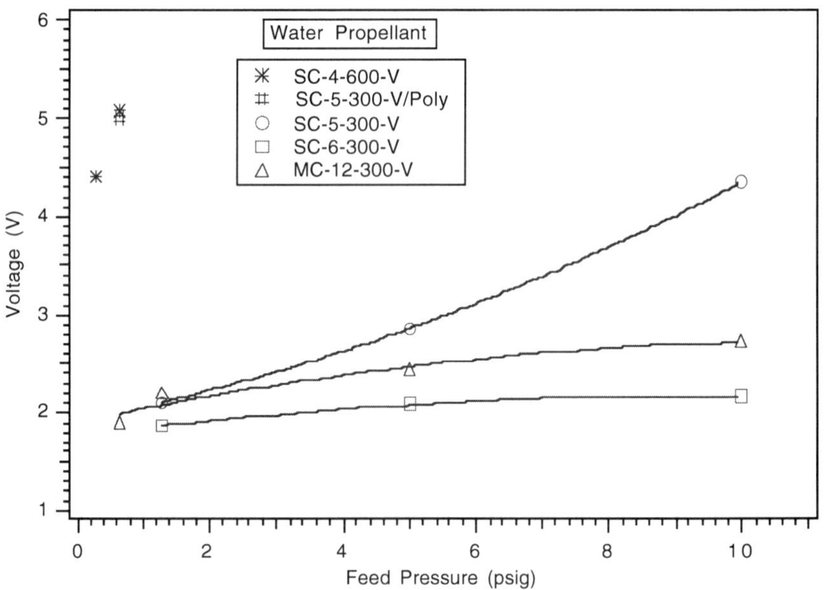

Fig. 9 VLM vaporization tests (voltage vs feed pressure).

at a feed pressure more than twice that value (0.64 psig) at about 6.5 W. These power levels, however, were still approaching or exceeding the maximum targeted 5-W value, and, under vaporization conditions, only very low feed pressures could be maintained. In addition, vaporization was noted to be poor, and certainly not complete, as evidenced by a continuous sputtering noise and frequent visible liquid droplet ejections, in addition to the steam otherwise generated.

The low required feed pressures were not expected. A major concern in the development of the VLM had been anticipated high viscous losses, in particular, in the case of liquid propellant flow. Consequently large channel cross had been conservatively chosen to accommodate the flow without encountering excessive flow resistances. This concern could clearly be dismissed after this first set of tests. Thus, new VLM designs featuring shallower heater channels, as described above, were fabricated and tested. The goal was to bring more liquid into more immediate contact with the heaters and thus effect a more efficient heat transfer.

2. *Effects of Heater Channel Dimensions*

Using this second type of chip, a significant improvement in performance was noted. In contrast to tests conducted with chips featuring 600-μm channel heights, vaporization was complete once a critical power level had been attained for a given feed pressure. Exhaust was invisible but could be evidenced by condensates on a cool glass slide placed in the vapor jet. No sputtering noises were observed. The critical power levels required to attain these conditions were also significantly lower than in earlier tests. For a chip featuring a 6-mm-long heater, the required power levels were only 2.3 W at an ~1.3-psig feed pressure, increasing to 2.8 W at a 5-psig feed pressure and 3 W at a 10-psig feed pressure. These feed pressure values are within the realm of practical applicability, even for use in space systems, and the power levels are well below the initially targeted value. In addition, the voltage requirements in this case range just below or above 2 V (see Fig. 9), depending on the feed pressure, and are thus well within the capability of future microspacecraft. Note, however, that, unless the flow rate is known, these data do not allow for a complete evaluation of the VLM concept since the efficiency remains unknown, as well as specific impulse.

In the case of a chip of the same type but featuring a 5-mm-long heater, the power levels were found to be somewhat higher but are still well within the design targets, with the exception of one data point at a 10-psig feed pressure, in which case the power requirements exceeded 8 W. This rather dramatic increase in power requirements with increased feed pressure (thus flow rate) for chips with shorter heater elements is reasonable, since less heat transfer to the liquid can occur over shorter heater lengths, requiring higher heater temperatures and power levels to achieve complete vaporization. However, even for this chip a voltage of no more than 5 V was required.

3. *Effects of Meandering Channel Design*

In the case of the meandering channel design, superimposed on a 5-mm-long heater element, the required power levels were higher than in the 6-mm-long heater design even though the flow path in the meandering channel design is less than half as wide and twice as long as in the 6-mm-long straight channel design. This appears to be surprising and inconsistent, however, as noted above, due to the lack of suitable flow rate measurements, direct quantitative comparison between the different chip designs is not yet possible. Despite this lack of accurate flow measurements, it was noted that the meandering channel design flowed considerably more liquid than the straight channel sections. While all other chips, even at the highest feed pressures, formed a liquid droplet at the nozzle outlet in the unheated state, the meandering chip design emitted a water jet at the same pressures, similar to the one shown in Fig. 7, thus indicating a higher flow rate. While this in itself is surprising, since higher flow resistance should have been expected in the meandering chip design, it does explain—at least qualitatively—the observed difference in required heater powers.

A possible explanation for these observed qualitative differences in flow behavior for the different chips is the potential of contamination located inside the narrow flow channels, accumulated either during handling and testing of the chip or through the solidification of capillary-fed epoxy flow into the channel sections during bonding procedures. Poor alignment between the various vias in the different chip laminates and the ports machined into the carriers and connecting nuts in some cases may also have led to local flow restrictions. In all cases, actual flow rates may be reduced due to locally decreased flow cross sections. So far, evidence of contamination during handling and testing may have been found in the case of one chip that seized conducting liquid during testing. The same chip featured a misalignment between the port machined into the chip carrier and the chip inlet hole, with the chip carrier surface partly covering the latter. In addition, one chip did not flow liquid at all, and contamination, if any, may have occurred prior to testing, possibly at some stage in the packaging phase, where chips leave the cleanroom environment. Once packaging is complete, chips are sealed in containers and not exposed to the environment prior to testing. Neither one of the failed chips is listed in Table 1 or shown in Figs. 8 and 9. These experiences clearly demonstrate the need for strict contamination control during testing.

4. *Effect of Heater Material (Polysilicon vs Gold)*

One chip featuring a polysilicon heater was tested and its performance compared with that of a similar chip featuring a gold heater. The polysilicon chip again featured a straight channel and a 5-mm-long heater. Comparing the power requirements of this chip with those of a chip featuring a 5-mm gold heater, they are found to be very similar for nearly identical feed pressures, as would be expected. However, examining Fig. 9, it can be seen that the polysilicon chip has voltage requirements well exceeding those of the gold heater, by more than a factor of two, at approximately the same power levels, i.e., about 5 vs 2 V in the case of the gold heater. This difference in voltage requirements is due to the higher resistance of polysilicon compared with gold and may be critical in view of microspacecraft applications.

V. Preliminary Conclusions and Future Work

Preliminary tests with a microfabricated VLM were conducted. The VLM concept is in its early stages of development, and results obtained to date are very limited. The thruster chip, about $0.9 \times 1.5 \times 0.1\,\mathrm{cm}^3$ in size and weighing but a few grams, was able to vaporize water propellant completely at power levels as low as 2 W and required voltages of merely 2 V for a 6-mm-long heater at feed pressures of about 1.3 psig. At 10-psi feed pressures, the same chip achieved complete vaporization at 3 W and an only slightly increased voltage of 2.2 V.

Different chip configurations were tested. Shorter heaters resulted in higher power requirements to achieve vaporization, however, the power levels remained below the 5-W level in most cases and the voltage requirements stayed below 3 V in most cases. However, the data obtained are very preliminary and do not allow for a quantitative comparison between different chips due to nonexisting measuring capabilities for very low liquid flow rates, as required for VLM characterization. Chip designs featuring shallower heater channel geometries, placing a larger portion of the liquid into more immediate contact with the heater elements, resulted in clearly improved vaporization at reduced power levels over chips featuring larger channel heights. Thermal characterization of the thruster revealed that approximately 2.6 W is required to achieve a heater temperature of 100°C for a 4-mm-long heater, 3.4 W to achieve 150°C, and about 4.3 W to reach 200°C.

Despite this very preliminary data set, several lessons could be learned from these tests that will benefit future work.

1) It could be successfully shown that vaporization over chip length scales is possible within the power constraints and bus voltage levels expected to be found on microspacecraft.

2) The value of shallower channel profiles, allowing for improved heat transfer between the heater and the liquid propellant, was demonstrated.

3) Viscous flow losses for water in microchannels featuring cross sections of $950 \times 300\,\mu m^2$ over channel lengths of several millimeters and flow orifices of $50 \times 50\,\mu m^2$ are not very significant.

4) Gold heaters are preferred over polysilicon heaters due to their lower voltage requirements. For use on larger spacecraft, this may not be a critical design issue.

5) There exists a need for improved contamination control during the test phase. Although a 2-μm filter was placed in the feed line to the VLM, there appears to be some evidence of flow passage blockage.

6) Since proper packaging can significantly reduce heat losses from the chip, designs may be envisioned where the entire chip is allowed to "float" thermally at heater temperatures while remaining highly insulated from its propellant and structural interfaces to the remainder of the feed system or spacecraft.

Significant future work is still required in this early stage of VLM development. Methods have to be devised to measure extremely low liquid flow rates. Vacuum testing will need to be performed. Exposure to vacuum will likely reduce power requirements to achieve complete vaporization. Pending suitable flow diagnostics, and using existing thrust stand hardware, detailed performance measurements will also need to be performed at this stage. Improved understanding of microchannel two-phase flow phenomena as encountered inside the chip is highly desirable. Currently, the state of knowledge in this area is poor. The existing literature in this area addresses some special cases, however, in many cases two-phase flow phenomena are not considered at all. An improved understanding of microchannel two-phase flow physics is essential in designing improved VLM heater designs, offering the potential of designing for lower power requirements or shortened vaporization channels, further decreasing chip sizes. Detailed propellant compatibility studies will be required, and, if needed, appropriate chip coating techniques need to be explored to avoid chip erosion. Concerns related to the dribble volume, formed by the heater channel volume downstream of the thruster valve, which may broaden the thruster impulse bit, will need to be addressed. Although longer channels may be beneficial for vaporization reasons, they will increase the dribble volume unless channel cross sections are constrained. Finally, assuming that satisfactory performance of the VLM can be demonstrated, chip-based thruster integration with other feed system components as well as chip-based control electronics will need to be addressed aggressively to take full advantage of chip-based propulsion designs by arriving at highly compact MEMS-based propulsion modules.[1]

Acknowledgments

The authors would like to thank Larry Rupel for his help in performing the tests, Sandee Chavez and John Rice for their work in packaging the chips, and Ronald Ruiz for his assistance in performing the thermal characterization. We would also like to thank many past members of the team who have gone on to other assignments but who have contributed significantly to the project in the past, namely, Andrew Wallace, for performing thermal finite element calculations and improving chip design and packaging, Lilac Muller and Thomas George, for formulating the first microfabrication approaches and contributing many design ideas, and Stephanie Leifer, coinventor of the VLM. Finally, we would like to thank Ross Jones for his vision in providing the initial funding to this project and Virendra Sarohia, Christopher Salvo, Barry Hebert, John Stocky, and Elizabeth Kolawa for their continued support.

The research described in this chapter was carried out by the Jet Propulsion Laboratory, California Institute of Technology, under a contract with NASA.

References

[1]Mueller, J., "Thruster Options for Microspacecraft: A Review and Evaluation of State-of-the-Art and Emerging Technologies," *Micropropulsion for Small Spacecraft*, Progress in Astronautics and Aeronautics, Vol. 187, edited by M. Micci and A. Ketsdever, AIAA, Reston, VA, 2000, Chap. 3 (this volume).

[2]Collins, D., Kukkonen, C., and Venneri, S., "Miniature, Low-Cost Highly Autonomous Spacecraft—A Focus for the New Millennium," IAF Paper 95-U.2.06, Oslo, Norway, Oct. 1995.

[3]Blandino, J., Cassady, R., and Sankovic, J., "Propulsion Requirements and Options for the New Millennium Interferometer (DS-3) Mission," AIAA Paper 98-3331, 34th Joint Propulsion Conf., Cleveland, OH, July 1998.

[4]Mueller, J., Tang, W., Wallace, A., Li, W., Bame, D., Chakraborty, I., and Lawton, R., "Design, Analysis, and Fabrication of a Vaporizing Liquid Micro-Thruster," AIAA Paper 97-3054, 33rd Joint Propulsion Conf., Seattle, WA, July 1997.

[5]Mueller, J., Chakraborty, I., Bame, D., Tang, W., and Wallace, A., "Proof-of-Concept Demonstration of a Vaporizing Liquid Micro-Thruster," AIAA Paper 98-3924, 34th Joint Propulsion Conf., Cleveland, OH, July 1998.

[6]Bayt, R., Ayaon, A., and Breuer, K., "A Performance Evaluation of MEMS-Based Micronozzles," AIAA Paper 97-3169, 33rd Joint Propulsion Conf., Seattle, WA, July 1997.

[7]Breuer, K., and Bayt, R., "Viscous Effects in Supersonic MEMS-Fabricated Micronozzles," Proceedings, Formation Flying and Micro-Propulsion Workshop, Air Force Research Lab., Lancaster, CA, Oct. 1998.

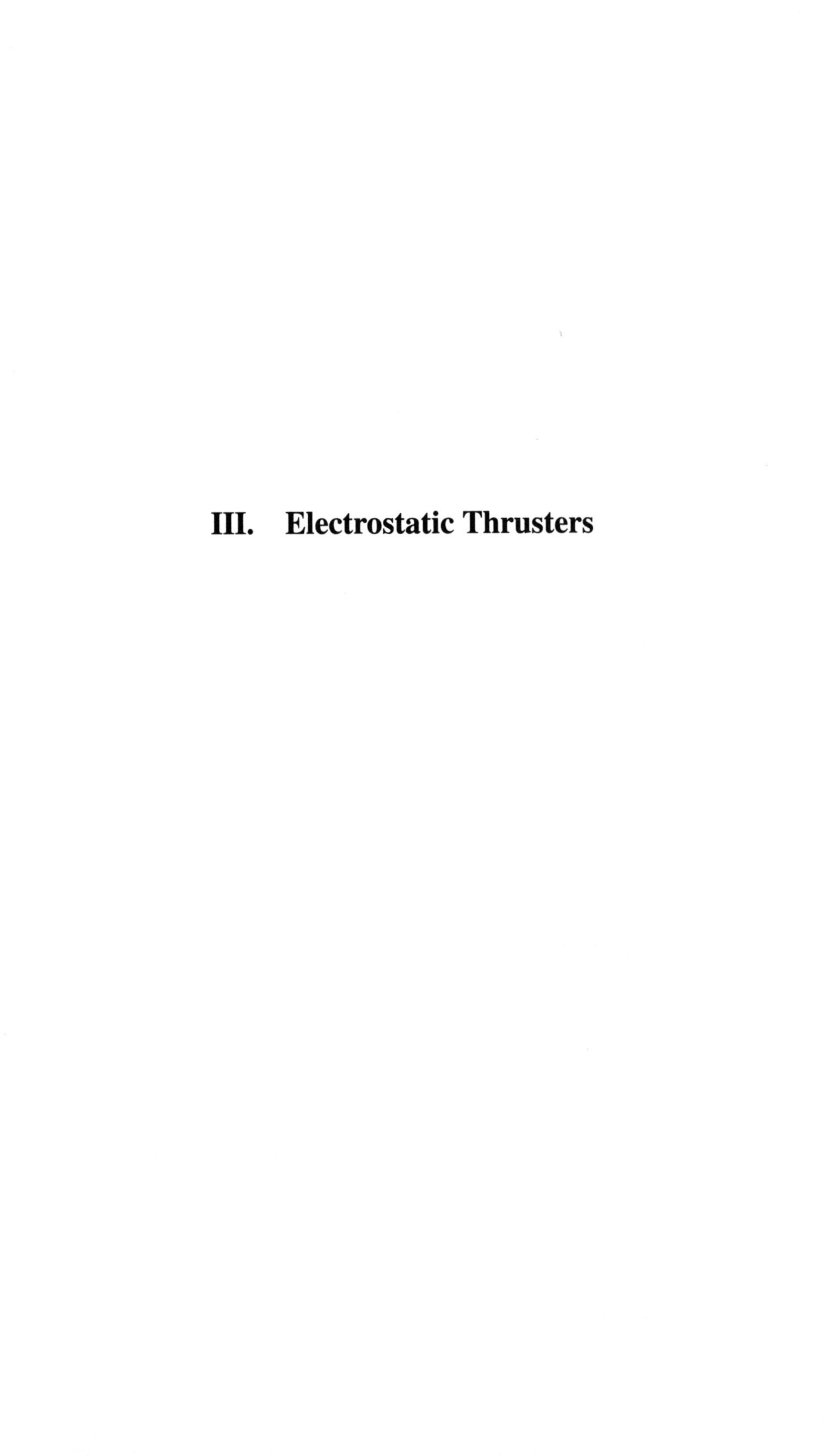

III. Electrostatic Thrusters

Chapter 9

Fifty-Watt Hall Thruster for Microsatellites

V. Khayms* and M. Martinez-Sanchez†
Massachusetts Institute of Technology, Cambridge, Massachusetts

I. Introduction

HIGH-PERFORMANCE electric propulsion systems are now being adopted to a number of spacecraft designs in the power range between 1 and 10 kW. In spite of this recent development, with the exception of currently existing low-efficiency PPTs, no equivalent propulsion technology exists for very small satellites.[1] In view of the need to develop a miniature propulsion system for potential use on microsatellites, an attempt was made to miniaturize a Hall thruster. This device has been chosen as one of the better-developed and more extensively tested thrusters in its class.

A simplified scaling model was developed to highlight potential practical limitations and to provide guidance in the design of the microthruster.[2] The scaling model was based on the physics of the existing larger devices and was used to evaluate various thruster configurations. It aided in the selection and sizing of the thruster components as well as in specifying the desired performance characteristics. The thruster has been configured to operate at 50 W. It was designed and manufactured at the Space Systems Laboratory of the Massachusetts Institute of Technology. Preliminary vacuum tank tests were initially performed to demonstrate thruster operation,[3] although, because of the lack of a sensitive microbalance at the time, no definitive conclusions could be made. More recently, a horizontal arm balance at the EPPDy Lab at Princeton University was used to perform detailed thrust measurements. A summary and analysis of the test results as well as the conclusions and recommendations are presented in Secs. VI and VII of this chapter.

II. Hall Thruster Operation

A conventional Hall thruster has the shape of an annular shell with cylindrical symmetry as shown in Fig. 1. During operation, propellant is injected through the

Copyright © 2000 by the American Institute of Aeronautics and Astronautics, Inc. All rights reserved.

*Graduate Student, Department of Aeronautics and Astronautics. Student Member AIAA.

†Professor, Department of Aeronautics and Astronautics. Senior Member AIAA.

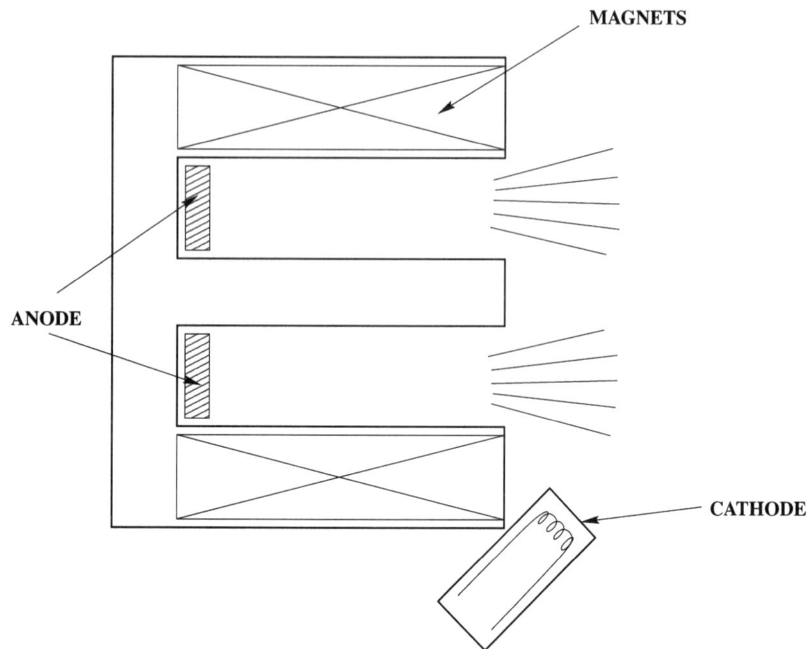

Fig. 1 SPT schematic (Dexter Magnetic Materials Division).

back side into the main discharge chamber. The cathode located in front of the channel is heated by an external power source to achieve the required neutralizing current by thermionic emission. A radial magnetic field established across the accelerator channel prevents the electrons emitted by the cathode from streaming directly into the anode. Instead, they spiral around the magnetic field lines and drift azimuthally between ionizing collisions with neutrals. The magnetic field strength is chosen so that the gyromagnetic radius of the electrons is much smaller than the characteristic dimension of the thruster. Through collisions and turbulent scattering, both primary and secondary electrons can penetrate the magnetic field toward the anode, sustaining the discharge in the channel. A potential difference externally applied between the anode and the cathode produces an axial electric field. The two fields set in mutually perpendicular directions result in an electron $\boldsymbol{E} \times \boldsymbol{B}$ drift motion in the azimuthal direction. The $q\boldsymbol{v} \times \boldsymbol{B}$ force on the electrons due to this drift balances the electrostatic force on them, and the corresponding reaction on the magnetic assembly transmits the thrust to the body. This is the origin of the generic "Hall thruster" designation for this device.

The ions are produced by numerous collisions between the neutrals and the electrons trapped by the magnetic field. These ions are accelerated in the electric field along the annular channel to exhaust velocities of the order of 10,000–20,000 m/s. The ion beam is neutralized upon its exit from the channel by additional electrons drawn from the heated cathode.

III. Scaling Model

As a first step in the design of a miniaturized thruster it is necessary to estimate some of the critical parameters, such as physical dimensions, magnetic field strength, electric potential, mass flow rate, and power, based on our current understanding of the physics of the existing larger models of the Hall thruster. The estimates are based on scaling laws that relate various parameters to changes in the characteristic length scale.

Our goal is to achieve a reduction in the length scale of the device while preserving both the efficiency of the thruster and its specific impulse I_{sp}. The specific impulse is directly related to the speed of ions exiting the channel. The exhaust speed, in turn, is controlled entirely by an electric potential applied between the anode and the cathode. Therefore fixing the applied voltage will ensure the invariance of I_{sp} under scaling. An additional assumption that is justified in this section is that the electron temperature must remain constant as well.

We would like to determine the relationship that governs the dependence of particle density on the length scale. This can be done by insisting that the ratio of the mean free path of each species for all types of collisions to the characteristic dimension be preserved:

$$\frac{\lambda}{L} = \frac{1}{nQL} = \text{const} \tag{1}$$

Here we have assumed that the collision cross section Q, being a function of the electron temperature, does not change because of the invariance of T_e. Thus particle densities (both n_e and n_n) are inversely proportional to the characteristic dimension:

$$n_e, n_n \sim \frac{1}{L} \tag{2}$$

Invariance of T_e also yields a similar result for the pressure p:

$$p \sim \frac{1}{L} \tag{3}$$

Mass flow rate, which scales as the product of the particle density with the area (square of the characteristic dimension), scales linearly with L:

$$\dot{m} \sim L \tag{4}$$

The dependence of the total current on the length scale can be deduced from a generalized form of Ohm's law: $j = \sigma(E + (1/en)(\mathrm{d}P_e/\mathrm{d}x))$, where E is the electric field along the channel and σ is the electrical conductivity. Since the electron density and the scattering frequency both scale inversely with length, σ, which scales as the ratio of the two, must remain constant. The field is inversely proportional to L under the assumption that the electric potential remains constant. Therefore the electron current density scales as $1/L$, and the total current as

$$I_e \sim L \tag{5}$$

The ion current density is governed by the distribution of ionization rates ($\nabla \cdot \mathbf{j_i} = e\dot{n}_e$), and since the ionization mean free path remains the same fraction of length, $\dot{n}_e$ will be distributed the same way, and its magnitude will scale as $n_e v_i / L \sim 1/L^2$, giving

$$j_i \sim 1/L, \qquad I_i \sim L \tag{6}$$

Comparing Eqs. (4) and (6) shows preservation of the utilization efficiency $\eta_u = m_i I_i / e\dot{m}$. Finally, we would like to obtain a relationship for the total power P

required for the thruster, which scales as the applied voltage times the ion current. Hence,

$$P \sim L \tag{7}$$

The wall loss can be estimated as the current ($v_{\text{Bohm}} n_e e A$) into the wall times the ionization potential. Hence,

$$P_{\text{wall}} \sim L \tag{8}$$

Clearly, the overall efficiency, which depends mainly on the wall recombination losses (P_{wall}/P), the degree to which ions are ionized at the highest potential, the ionization–radiation losses, and neutral losses (η_u), will remain invariant, as desired.

The scaling relationship for the magnetic field can be deduced from the fact that the ratio of the electron Larmor radius to the characteristic dimension must remain constant. This condition yields

$$B \sim \frac{1}{L} \tag{9}$$

Although most of the parameters considered above scale with the thruster dimension as desired, the sheath thickness (Debye length), which goes as $\sqrt{T_e/n_e}$, scales as $\sqrt{L}$, hence violating the general trend. As a result, sheaths will become proportionally thicker as the characteristic dimension is reduced. It is believed, however, that this does not have a strong impact on the general scale invariance as long as the sheaths remain much thinner than the channel width.

Finally, an electron energy equation can be used as a check on the assumption of the electron temperature invariance under scaling. Neglecting for the moment radiation and elastic losses, the electron energy equation reads

$$\nabla \cdot \left(-\frac{\boldsymbol{j}_e}{e} h_e - K_e \nabla T_e \right) = \boldsymbol{E} \cdot \boldsymbol{j}_e \tag{10}$$

where h_e is the electron enthalpy (including ionization), and is invariant if T_e is constant, and K_e is the electron thermal conductivity. The general Wiedemann–Franz law states that K_e/σ is proportional to T_e, the proportionality constant being $\sim(k_B/e^2)$ for a plasma; since σ and T_e are invariant, so is K_e. It can therefore be verified that each term in the electron energy equation scales as $1/L^2$, and the conservation law is satisfied by the same T_e at all scales, as assumed. Elastic power loss to heavy particles also scales (per unit volume) as $1/L^2$, as can be easily verified.

Radiative losses deserve special consideration. Due to the preservation of all mean-free-path ratios, the gas opacity will be preserved, and, since radiation is volumetric (optically thin limit) in existing Hall devices, it will remain so upon scaling. The radiative power loss per unit volume will then scale as the ionization rate, $1/L^2$, since both ionization and excitation of atoms are driven by electron–atom collisions, and photon emission is prompt. Thus, the common practice of accounting for radiative losses by using in the electron energy balance an "effective" ionization energy of two to three times eV_i will remain valid, with the same factor at all scales. Note that for devices or components where radiation is from the surface only (optically thick limit), radiative power losses per unit volume will scale as $1/L$ and will become weaker than other terms in the energy balance when L is reduced.

Table 1 Scaled design parameters

Parameter	SPT-100	Mini-SPT
Power, W	1350	50
Thrust, mN	83	3
Specific impulse, s	1600	1600
Efficiency	50	50
Channel diameter, mm	100.0	3.7
Flow rate, mg/s	5.0	0.2
Magnetic field, T	0.02	0.5

Finally, a reduction of the characteristic dimension along with an increase in the particle flux both result in a significant reduction of the lifetime ($\sim L^2$). The lifetime must be the subject for further studies in selecting better sputter-resistant materials for this application.

Although the eventual detailed design of the thruster evolved toward a different configuration, the initial approximate design parameters were obtained by scaling from a ceramic-based 1.35-kW SPT-100 thruster, which was selected because of the readily available optimized performance data. The scaling factor was chosen to be roughly 27. This corresponds to the power requirement reduction from 1350 W for SPT-100 to about 50 W, which is ideal for many potential microsatellite applications. A summary of the major results is presented in Table 1.

IV. Thruster Design

A. General Considerations

A Hall thruster in its standard configuration consists of a magnet, an insulated plasma channel, an anode, a propellant feed system, and a hollow cathode. Often the plasma channel is thermally insulated from the surroundings by a layer of high-temperature ceramic that protects the magnets and other components from excessive heating. However, because of the larger heat fluxes involved, internal temperatures may exceed those imposed by the material limitations. By leaving a vacuum gap between the insulator and the body of the thruster, heat fluxes from the channel can be significantly reduced and limited to those due mainly to thermal radiation. Although the latter configuration seems to resolve the heating problem, it raises another point of concern. The manufacturing of the tiny ceramic channel, which would have to withstand high temperatures and be resistant to thermal shock, may prove to be extremely complicated. For this reason we have decided to abandon the idea of using an insulator but, rather, extend the anode farther into the channel and conduct all of the heat radiated by the anode through the thruster body out to the surroundings.

This Hall thruster configuration with an extended anode and bare metallic walls is known as TAL (thruster with anode layer). The walls are usually kept at the cathode potential, thus repelling electrons trapped by the magnetic field near the exit of the channel. Although the physics of near-wall processes in the TAL is somewhat different from that of the SPT, the reported performance characteristics of the two thrusters are almost identical.

B. Magnetic Circuit Design

If the electric coils are to be used to generate the required magnetic field, the current externally supplied to them would be estimated from the following relationship:

$$B = \frac{\mu IN}{g} \tag{11}$$

where g is the gap size. Since the B-field scales inversely with length, this implies that the number of ampere-turns (IN) for the coils must remain constant. Power dissipated in the coils scales as the square of the current times the resistance of the wire. Assuming that the current driving the solenoid scales directly with length [see Eq. (5)], the number of turns required would be inversely proportional to L. The wire resistance scales as the ratio of its length NL to the cross-sectional area A. Assuming that the solenoid wire is wrapped in layers such as to preserve its aspect ratio, the cross-sectional area A would scale as $L^2/N \sim L^3$. The overall wire resistance can then be written

$$R = \frac{\rho l}{A} \sim \frac{1}{L^3} \tag{12}$$

Hence, the power dissipated in the coils scales inversely with the square of the characteristic dimension. Since both the effective conduction and the radiation areas are simultaneously scaled by a factor of L^2, the heat fluxes both scale as $1/L^4$. Hence,

$$\Delta T_{\text{cond}} \sim \frac{1}{L^2} \tag{13}$$

$$T_{\text{rad}} \sim 1 \tag{14}$$

Since conduction is the dominant heat transfer mechanism, Eq. (13) yields extremely high temperatures on the coil surfaces. Such high temperatures can result in severe damage to the insulation, short-circuiting between the individual turns, and, finally, complete loss of the solenoid.

A different approach to the problem is to abandon the idea of using electric coils in favor of permanent magnets. Although a number of advantages make them ideal for this application, there is one drawback. Scaling of the thruster dimension leads to a significant reduction in the overall radiative surface area, causing larger heat fluxes and, as a result, potentially higher temperatures (see Sec. IV.C). The maximum allowable temperatures, however, are limited due to the possibility of demagnetization of the permanent magnets. Therefore it becomes necessary to provide for adequate heat escape paths, which may substantially complicate both the design and the manufacturing process. Startup problems are possible due to the existence of a strong magnetic field that may prevent the electrons from entering the channel to initiate ionization of the neutrals during ignition. An increase in the required magnetic field poses a question of feasibility of using permanent magnets of reasonable sizes for this application.

A single shell of permanently magnetized material was used to achieve the required B-field, which was estimated to range from 0.3 to 0.5 T within the gap (see Table 1). A simplified analysis was done for preliminary assessment and comparison of the possible circuit geometries and materials to be used for construction. Assuming a constant cross section and infinite permeability of the iron core, Ampere's law around the loop yields

$$H_m d + \frac{B_m g}{\mu_0} = 0 \tag{15}$$

where $H_m < 0$ is the magnetic field and B_m is the magnetic induction within the magnet. A linear approximation to the magnetization curve for a given magnetic material can be written in the form

$$B_m = \frac{B_0}{H_0} H_m + B_0 \tag{16}$$

where $H_0 > 0$ and B_0 are the coercive force and the magnetic remanence, respectively. Eliminating H_m, the two equations yield a rough estimate of the resultant magnetic field within the gap:

$$\frac{B_m}{B_0} = \frac{1}{1 + \mu_0(g/d)(B_0/H_0)} \tag{17}$$

The gap width is limited by the channel opening and the ratio g/d is a fixed parameter. Thus, to achieve the required B-field, the material of choice would have to meet the following specifications:

$B_0 \sim 0.7$–1.0 T
Low B_0/H_0, hence, high coercivity (H_0)

In addition, the material would be required to withstand elevated temperatures without significant loss of magnetization. Only three materials were identified to be applicable to a wide selection of magnetic alloys currently available in the market. Both NdFeB and Alnico magnetic alloys possess excellent magnetic properties. However, they have a low Curie temperature (350°C) and low coercivity, respectively. SmCo alloys, on the other hand, with their high coercivity (9000 Oe) and high maximum operating temperature of 275°C (Curie point of 750°C) were found to be optimal for our application. Attention was paid in the design to the magnetic field profile in the gap. The goal was to prevent any electrons emitted by the cathode from entering the channel and traveling toward the anode. This imposed a constraint on the field geometry such that the adjacent field lines be everywhere tangent to the anode frontal surface. This configuration would also prevent the ion beam from deflecting either away from or toward the center of the channel. Such a deflection may be a result of nonuniformities in the electric field caused by the presence of electrons trapped in the magnetic field.

Two-dimensional numerical magnetic computations were run for several variations of the proposed geometry that is shown schematically in Fig. 2 (simulation is courtesy of Dexter Magnetic Materials Division). Two SmCo permanent magnets are used in the design of the magnetic circuit. One of the magnets is designed in the shape of an axially polarized cylindrical shell. It is the main "driving magnet." The second one, consisting of eight arc segments, is polarized radially in the direction perpendicular to the polarization of the main magnet and is used to force the flux into the gap and to shape the field lines to meet our profile specifications. Propellant and power feed lines can be inserted into the channel via four circular holes in the back side of the iron core. The iron return path is designed to extend all the way around the main magnet to reduce the flux leakage out of the circuit and to help direct the flux into the segmented magnet (see Fig. 3). The attained values of the magnetic field strength in the gap for this geometry were numerically estimated to range from 0.4 to 0.6 T at the designed maximum operating temperature (250°C), thus meeting our specifications.

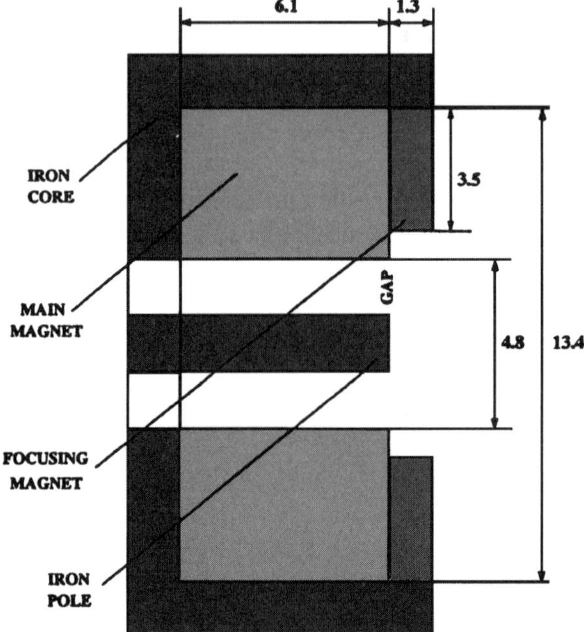

Fig. 2 Magnetic circuit geometry (all dimensions in millimeters).

C. Thermal Design/Material Selection

The two mechanisms for evacuating heat from solid components (anode, magnets, etc.) are radiation and conduction, and the resulting equilibrium temperatures depend on which of the two predominates. The heat deposition rate per unit area scales as power/area $\sim 1/L$. The radiative cooling rate per unit area is $\epsilon(T)\sigma T^4$, independent of scale, and so, if radiation dominates, the wall temperature would

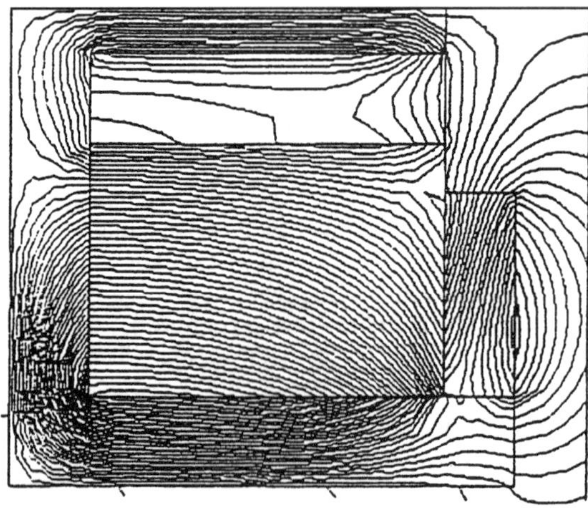

Fig. 3 Magnetic circuit simulation (Dexter Magnetic Materials Division).

increase roughly as $1/L^{\frac{1}{4}}$ (a significant problem for small L, as contemplated here). The conductive cooling rate per unit area is $k\Delta T/l$, where l is some thickness, and so this cooling rate scales as $1/L$, the same as the deposition rate; thus, if conduction is dominant, wall temperatures can be kept invariant upon scaling. Therefore high temperatures within the channel and at the anode can be avoided by designing adequate heat conduction paths from the channel out to the surroundings. Since the anode is kept at a positive potential with respect to ground, it is virtually impossible to maintain direct contact of the anode with an external heat sink. Instead, the anode, along with its electrical leads can be designed as a dual-purpose device. The propellant feed lines that normally supply gas to the anode can be used to conduct heat generated inside the channel toward the back of the thruster. Using for the back support structure a material that is a good electrical insulator as well as a good thermal conductor, the heat can be transferred to the surrounding metal, maintained near ambient temperature. From the materials viewpoint, a heat-conducting ceramic such as BN or AlN would be most appropriate. Although AlN has a higher thermal conductivity, it is harder to machine, hence BN was selected.

Another important aspect from the materials standpoint is the anode design. A simple one-dimensional thermal model has shown that the anode tip temperatures would not exceed 1300°C. The analysis was carried out under the assumption that all of the heat generated at the anode is allowed either to be radiated to the magnets or to be conducted through the gas-feed lines to the back of the thruster, where it is rejected to the surroundings. Since the heat generated inside the channel arises due mainly to the thruster inefficiency, roughly 25 W of the total power would be dissipated as heat to the walls. Localized heating due to the impinging ions was neglected in this simple model.

The analysis has shown that conduction through the feed lines is a dominant process in the heat rejection mechanism. In addition, because of the large contact area at the back, the temperature of the magnet surrounding the anode remains close to ambient. Therefore, the material of choice for both the anode and the feed lines would need to have the following characteristics:

1) High thermal conductivity to ensure adequate heat rejection rate through conduction.

2) High surface emissivity to enhance heat rejection by radiation.

3) Adequate melting point ($T_m > 1700°C$), compatible with the limiting values predicted by the model.

4) Easy machinability.

Although refractory materials, such as W, Ta, and Mo, seem to fit this category best, the intricate shapes of such small dimensions are extremely hard to manufacture out of pure metals. Alloys of these refractories with copper, nickel, or iron, although more ductile, have melting temperatures that are below the acceptable limits and therefore would not be appropriate for this application. Finally, the preference was given to molybdenum, as it is an excellent heat conductor and has a melting point of 2700°C, which is more than adequate for this application.

D. Cathode Design

Little attention has been given to the design of a new cathode. Although the new generation of field-effect (cold) cathodes seems to be promising for this application, no state-of-the-art technology exists at the time for their successful implementation.

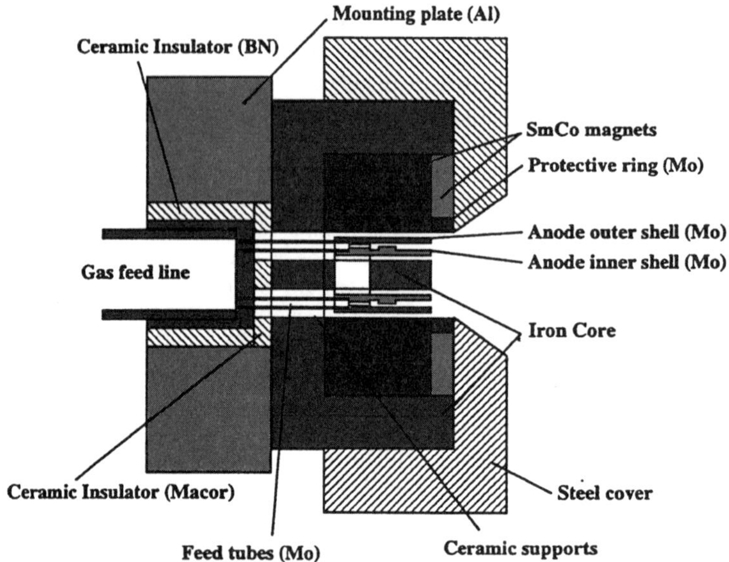

Fig. 4 Final design schematic.

Hollow cathodes, conventionally employed in SPTs, do not easily conform to the photographic scaling, yet they do have some potential for being miniaturized for future applications. Thoriated tungsten filaments were used in the preliminary tests because of their ease in implementation. The main disadvantage of such filaments is that they consume a great deal of power (50 W/cm^2) because of their high-temperature operation. Materials that have a lower work-function such as LaB_6 would be more suitable for this application, however, they are somewhat difficult in handling and operation. A hollow cathode was utilized in the most recent experiments. Refer to Sec. 5.3 for details. The optimal choice of a cathode for this thruster is a subject for further studies and detailed trade-off analysis.

E. Final Design

The final version of the miniaturized Hall thruster is shown schematically in Fig. 4. The magnetic assembly is identical to that shown in Fig. 2. The anode is designed as a hollow concentric channel with a flow buffer to achieve a uniform azimuthal jet distribution. The four capillaries attached to the anode deliver the propellant from a gas distributor region located in the back of the assembly to the plasma channel. The frontal cap (made of stainless steel) is a structural component that presses the magnets and a protective molybdenum ring into the iron assembly and holds them in place. Ceramic inserts were used to support the anode and improve its alignment with respect to the iron core.

V. Testing Facility

After a number of preliminary trials, conclusive performance tests were conducted in a vacuum tank facility at the Plasma Propulsion Laboratory of Princeton

University. Brief descriptions of the facility and the diagnostics equipment are provided below.

A. Vacuum Tank

The vacuum tank was equipped with two mechanical roughing pumps and a single diffusion pump. In addition, a baffle cooled with liquid nitrogen was operated for the duration of the test runs. At the nominal Xe flow rate of 2 sccm and a cathode flow rate of 1.4 sccm, the steady-state chamber pressure did not exceed 5×10^{-5} Torr. Although such background pressures may be considered marginal for testing larger devices, favorable scaling of particle densities at small sizes make these pressures more than adequate for microthruster testing. The chamber was equipped with a number of gas and electrical ports, thus allowing all of the monitoring and control circuitry for the thrust balance, power supplies, and flow controller to be positioned outside the vacuum tank. Main power to the thruster was provided by a 300-V, 5-A DC power amplifier. Two additional supplies were used to power the cathode.

B. Thrust Balance, Calibration, and Data Aquisition

The microbalance used in the experiments was a horizontal arm design with a thruster mounted at its tip. The propellant supply line and a flexible power cable were both attached to two fixed cable clamps, one mounted on the arm and the other mounted on the fixed platform to avoid variations in stiffness or dry friction. To avoid imposing additional stiffness, the cathode together with its power cables and the gas feed were mounted to the fixed platform in close proximity to the thruster. As the balance arm failed to provide sufficient heat capacity to accommodate the heat dissipated by the thruster, a cooling jacket (Fig. 5) was mounted to the thruster walls. Cooling lines were attached to the arm by the cable clamps and diverted to the exterior water supply.

An LVDT (linear variable differential transformer) sensor was used to detect movements of the arm away from its equilibrium position. The signal was fed into a derivative amplifier, which in turn activated the damper coil. Two motors were used to control the longitudinal and lateral positions of the platfrom. Prior to operation of the balance the motors were adjusted so as to bring the arm to its equilibrium position. The LVDT signal was sampled by a digital oscilloscope and stored into a data file for further analysis.

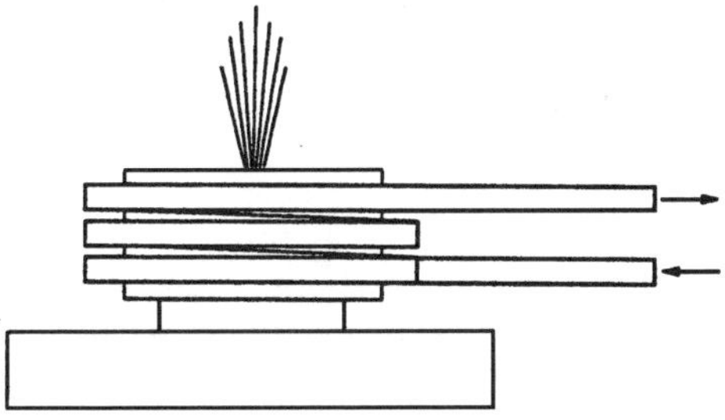

Fig. 5 Cooling jacket schematic.

The thrust balance was calibrated in air by determining the effective spring constant of the assembly. This was done by applying a known impulse to the arm at the point of thrust application and measuring the resultant initial linear velocity of the arm at that location. The impulse was determined by integrating the signal from the calibrated force transducer and the effective mass of the arm was computed knowing the impulse and the linear velocity. The spring constant was finally obtained by measuring the undamped natural frequency of the arm, which was estimated from the samples of the numerous LVDT traces.

C. Cathode

A hollow cathode manufactured by Busek Co. was used in the experiments. The cathode was supplied with Xe at a flow rate of 1.4 sccm. Since the cathode was originally designed to operate with larger thrusters in a self-heating mode, the lower currents drawn by this thruster were not sufficient to maintain stable cathode operation. For this reason, once in a standby mode, the heater current was reduced from the nominal 6.5 to 3 A while the keeper current was maintained at 0.5 A.

D. Flow System

Xenon was supplied to the vacuum tank by means of a digital MKS flow controller calibrated for flow rates between 1 and 10 sccm (0.098 to 0.98 mg/s). The upstream pressure in the Xe tank was kept at 50 psi. The flow rate was calibrated before and after the test runs for each of the flow rate settings by passing xenon bubbles through a beaker immersed upside down in a container with water, thus maintaining nearly atmospheric pressure. The cathode flow was supplied from the same Xe tank via a needle valve calibrated to deliver 1.4 sccm.

VI. Experimental Results

A series of vacuum tank tests was performed to assess the overall performance of the micro-Hall thruster. The thrust balance arm was calibrated prior to evacuating the tank while the cooling water was allowed to flow in the lines. Additional calibration was performed at the end of the testing procedure once the tank was vented to the atmosphere. The calibration constant remained unchanged under vacuum as indicated by the arm's natural frequency measured before and after the tank was evacuated. Prior to each test sequence the main flow was turned on and the cathode was set in a standby mode until all thermal and mechanical transients in the thrust signal could no longer be observed. Both the anode voltage and the discharge current were monitored at all times, while the thrust signal was visually observed on the oscilloscope screen and logged by the data acquisition system upon completion of each test sequence.

Preliminary test trials have shown a significant thermal drift in the arm's reference position throughout the duration of the firings. After the cooling jacket was mounted to the thruster, however, thermal drifts were almost entirely eliminated. As part of the preliminary validation process for the balance, the arm was allowed to rest at one of its stops while the LVDT signal was monitored with the thruster operating at its nominal conditions. This was done to ensure that the LVDT or any other auxiliary cables that run to the balance and are in close proximity to the plasma do not pick up any spurious signals interfering with the thrust measurement. No deviations in the signal were observed.

The test runs were performed at three voltages—200, 250, and 300 V—and three flow rates—0.100, 0.168, and 0.215 mg/s. For each test run sufficient time was alloted for the thrust signal and the discharge current to stabilize. Each run lasted approximately 7–8 min and included 1 min before and after each firing to obtain a stable signal reference. The thruster was run twice at each operating condition to ensure that the measurements were repeatable and were not affected by any thermal drifts in the thruster or the balance arm assembly. Thruster efficiencies and specific impulse values were computed from the measured current, voltage, flow rate, and thrust using the following relations:

$$\eta = \frac{T^2}{2IV\dot{m}} \tag{18}$$

$$I_{\text{sp}} = \frac{T}{\dot{m}g} \tag{19}$$

where T is the thrust, $\dot{m}$ is the flow rate, and g is the acceleration due to gravity.

Results of these test runs are conveniently summarized in Fig. 6, showing the efficiency vs the specific impulse at different flow rates. Additional test data are listed in Table 2. Analysis of the data shows that at the near-nominal flow rate of 0.215 mg/s the measured thrust levels were about a factor of two lower than expected, while the discharge current was twice its nominal design value. The resulting efficiencies were, therefore, unexpectedly low, at only one-eighth of the nominal 50%. Assuming that most of the voltage drop develops downstream of

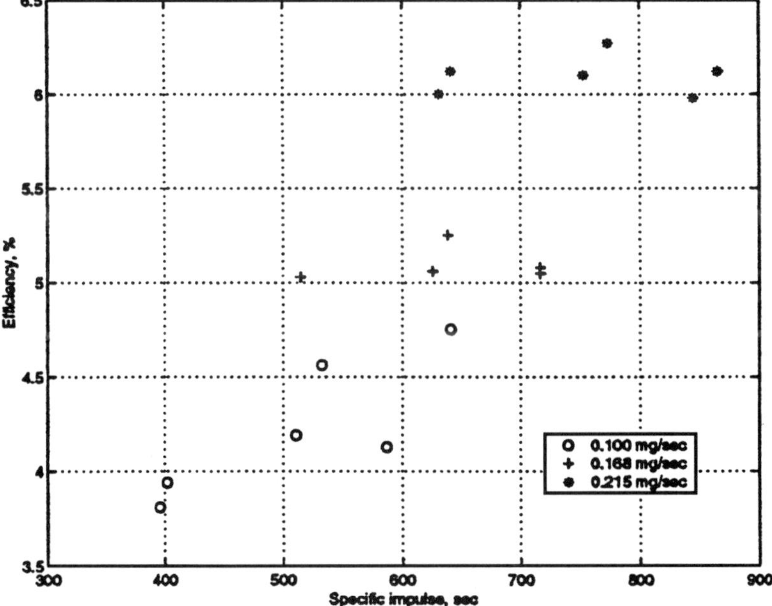

Fig. 6 Fifty-watt Hall thruster performance.

Table 2 Performance data

Flow, kg/s	Voltage, V	Current, A	Thrust, N	I_{sp}, s	Efficiency, %	I_b/I_a, %	Util. effic. %
1.005e–07	250	0.120	5.029e–04	510.6	4.19	16.04	26.14
1.005e–07	300	0.134	5.778e–04	586.7	4.13	15.07	27.42
1.005e–07	250	0.120	5.243e–04	532.3	4.56	16.73	27.25
1.005e–07	300	0.139	6.313e–04	641.0	4.75	15.87	29.96
1.005e–07	200	0.099	3.959e–04	402.0	3.94	17.12	23.01
1.005e–07	200	0.099	3.895e–04	395.5	3.81	16.84	22.64
1.676e–07	200	0.212	8.453e–04	514.6	5.03	17.07	29.46
1.676e–07	200	0.212	8.453e–04	514.6	5.03	17.07	29.46
1.676e–07	250	0.250	1.049e–03	638.4	5.25	16.06	32.69
1.676e–07	250	0.249	1.027e–03	625.4	5.06	15.79	32.02
1.676e–07	300	0.273	1.177e–03	716.6	5.05	15.07	33.49
1.676e–07	300	0.271	1.177e–03	716.6	5.08	15.18	33.49
2.146e–07	200	0.346	1.348e–03	641.1	6.12	16.68	36.69
2.146e–07	200	0.342	1.327e–03	630.9	6.00	16.61	36.11
2.146e–07	250	0.393	1.626e–03	773.3	6.27	15.84	39.59
2.146e–07	250	0.383	1.584e–03	753.0	6.10	15.83	38.55
2.146e–07	300	0.420	1.819e–03	864.9	6.12	15.14	40.42
2.146e–07	300	0.410	1.776e–03	844.6	5.98	15.14	39.47

the ionization zone, the total efficiency can be broken down into the product of the utilization and the acceleration efficiencies:

$$\eta = \eta_u \eta_a \tag{20}$$

where η_u is given by

$$\eta_u = \frac{I_b m_i}{\dot{m} e} = \eta_a \frac{I_a m_i}{\dot{m} e} \tag{21}$$

and $\eta_a = I_b/I_a$. Since the overall efficiency η can be obtained from the experimentally measured quantities only, both η_a and η_u could be determined from

$$\eta_a = \sqrt{\eta \frac{\dot{m} e}{I_a m_i}} \tag{22}$$

$$\eta_u = \frac{\eta}{\eta_a} \tag{23}$$

The resulting utilization efficiency ranged anywhere between 22 and 40%, values that are unusually low for these types of thrusters. Additionally, measured values of I_{sp}, even at high voltages (300 V), have not exceeded 850 s. All of these observations point to one plausible explanation: a low degree of ionization in the plasma results in poor utilization of neutrals, hence, a low thrust for a given flow rate. Poor ionization can be a result of partial degradation or loss of magnetic confinement of electrons. The unusually high values of the ratio of the anode current to the mass flow rate at high voltages, of the order of 2.6–2.7, indicate excessive leakage of electrons and support the hypothesis. In the search for possible clues, two mechanisms consistent with the observed behavior were identified as most likely to explain the loss of magnetic confinement.

1) Degradation and nonreversible loss of the magnetic field due to excessive heating and damage to permanent magnets may be a possible cause of electron leakage. Although this mechanism is plausible, thermocouple measurements throughout the tests have indicated that the temperature of the thruster assembly has never exceeded 42°C. Since the path between permanent magnets and the thruster mount has a relatively low thermal impedance, it is unlikely that the magnets ever reached their maximum allowable temperature of 275°C. Subsequent measurements of the magnetic field strength using a miniature Gauss probe indicated that the mean field strength in the vicinity of the pole averaged at about 0.17 T with a maximum of 0.25 T. Because the diameter of the axial probe used in the measurements was comparable to that of the central pole piece, and because the field strength falls off rapidly from the pole tip, it is very likely that the measured values of the average field strength over the volume of the probe are a factor of two to three lower than the expected value within the gap. Since the nominal field strength should be on the order of 0.4–0.5 T, it can be concluded with a certain degree of confidence that the permanent magnets have experienced at worst minimal damage during the course of the experiments.

2) The center pole piece is not properly cooled due to its small cross-sectional area and, as a result, may have locally reached temperatures at which the magnetic permeability of iron drops sharply to zero. Such a dramatic reduction of permeability would have caused increased fringing of the magnetic field inside the thruster channel, possibly resulting in the loss of field strength at the channel exit and forcing some of the field lines to cross the anode. If that were the case, electrons emitted by the cathode would follow the field lines and easily leak into the anode without fully ionizing the propellant. To validate this hypothesis further, one can estimate the temperature gradient across the magnetic pole subjected to external heating by the plasma and the anode. Although the anode temperatures were never measured directly, prior designs that utilized platinum as the anode material showed slight signs of melting at the tip. Assuming that the tip of the anode gets at least as hot as the melting point of platinum (~2000 K) and radiates heat from a section at the tip (~3 mm deep), the amount of heat deposited into the iron pole just from the anode is of the order of 6 W. Accurately accounting for the additional heat due to the impinging ions and electrons is more difficult, however, it is safe to assume that, including the heat radiated from the anode, a total of 10 W is deposited into the iron pole. In steady state, this amounts to a tip temperature differential of 950 K with respect to the iron base. Since the Curie temperature for iron is about 1061 K and the base temperature is no lower than 350 K, it is very likely that the magnetic flux does not penetrate all the way to the tip of the iron pole, thus modifying the field profile and reducing its strength within the gap. As a result, magnetic field lines can intersect the anode and collect excessive electron current without allowing sufficient residence time for the electrons to ionize the gas fully.

VII. Alternative Scaling Scenarios: Universal Scaling

In view of the numerous difficulties associated with the manufacturing and operation of a small-scale device designed to satisfy the strict scaling relations outlined in Sec. III, alternative scaling models have been considered in the hope of alleviating some of these difficulties while preserving the superior performance characteristics achieved for the larger devices.

Consider an arbitrary scaling scenario in which both the power and the length scale reduction are independent. Then, the power and the flow rate both scale in proportion to the product of the plasma density and the characteristic area of the device:

$$P \sim nL^2 \tag{24}$$

$$\dot{m} \sim nL^2 \tag{25}$$

The ratio of the mean free path to the length scale, $h = \lambda/L$, is no longer fixed as an invariant quantity but, instead, is allowed arbitrary variations with the plasma density and the length scale:

$$h = \left(\frac{v_n}{c_e Q}\right)\frac{1}{nL} \sim \frac{1}{nL} \tag{26}$$

Suppose a nominal device with the operating/geometrical characteristics given by $n_0, P_0, L_0, h_0, \dot{m}_0$, and η_0 is to be scaled down in power. Introducing nondimensional parameters scaled to the nominal:

$$\tilde{n} = \frac{n}{n_0} \tag{27}$$

$$\tilde{P} = \frac{P}{P_0} \tag{28}$$

$$\tilde{L} = \frac{L}{L_0} \tag{29}$$

$$\tilde{h} = \frac{h}{h_0} \tag{30}$$

$$\tilde{\dot{m}} = \frac{\dot{m}}{\dot{m}_0} \tag{31}$$

$$\tilde{\eta} = \frac{\eta}{\eta_0} \tag{32}$$

relations in Eqs. (24–26) can be written as

$$\tilde{P} = \tilde{n}\tilde{L}^2 \tag{33}$$

$$\tilde{\dot{m}} = \tilde{n}\tilde{L}^2 \tag{34}$$

$$\tilde{h} = \frac{1}{\tilde{n}\tilde{L}} \tag{35}$$

Substituting Eq. (35) into Eq. (34) yields an expression for the power scaling as a function of the characteristic parameter $\tilde{h}$:

$$\tilde{P} = \frac{\tilde{L}}{\tilde{h}} \tag{36}$$

The disproportionate increase of the mean free path in comparison to the size of the device is what is thought to produce the drop in the utilization efficiency. Therefore, obtaining a relationship between $\tilde{h}$ and $\tilde{\eta}$ would allow one to predict the degradation in performance under arbitrary scaling conditions.

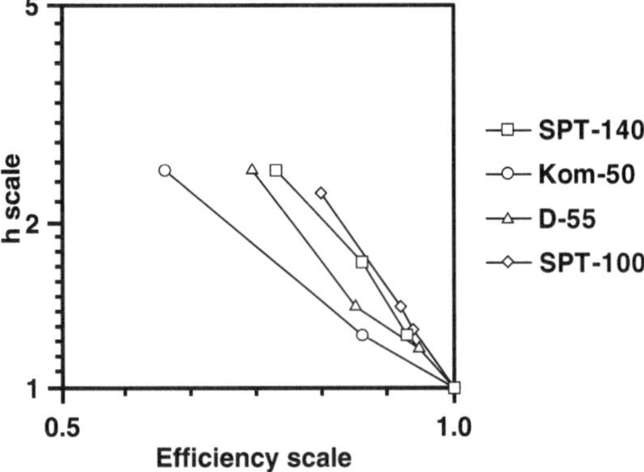

Fig. 7 Dependence of the efficiency on the mean free path parameter $\tilde{h}$.

It was already noted that an increase in the quantity h can be caused by either a disproportionately small reduction in size for a given decrease in the flow rate or, alternatively, a corresponding decrease in the flow rate for a device with a fixed geometry. Several existing thruster models for which the efficiency data are readily available have been operated at off-design conditions, specifically at reduced flow rates. The empirical dependence of $\tilde{\eta}$ vs $\tilde{h}$ for several commercial thrusters in the SPT series: SPT-100, SPT-140, a TAL model D-55, and a 50-mm laboratory version by Komurasaki was extracted from the data and is graphically summarized in Fig. 7. The experimental quantities plotted in Fig. 7 were scaled to the operating conditions corresponding to the maximum obtained efficiency for a given thruster. Figure 7 clearly indicates a drop in the overall efficiency as the mean free path becomes disproportionately larger in comparison to the size of the device. Moreover, as the data suggest, the dependence of $\tilde{h}$ on $\tilde{\eta}$ can be modeled simply as an inverse power law that, surprisingly, is very weakly dependent on the dimension of the device and can be extrapolated to predict the efficiency of smaller thrusters with an equivalent deviation in $\tilde{h}$. Extracting the empirical dependence of $\tilde{h}$ on $\tilde{\eta}$ for the SPT series as approximately given by

$$\tilde{h} \approx \frac{1}{\tilde{\eta}^3} \tag{37}$$

the contours of constant efficiency can be determined by substituting Eq. (37) into Eq. (36):

$$\tilde{\eta}^3 = \frac{\tilde{P}}{\tilde{L}} \tag{38}$$

The contours of constant efficiency are plotted as dashed lines in the $\tilde{P}$–$\tilde{L}$ scale space as shown in the accompanying Fig. 8. Note the strict scaling for which the efficiency remains constant and for which $P \sim L$ is easily recovered as the curve $\tilde{\eta} = 1$. The lines of constant operating life referenced to the nominal can be determined as the ratio of the length scale to the plasma density:

$$\tilde{t}_l = \frac{\tilde{L}}{\tilde{n}} = \frac{\tilde{L}^3}{\tilde{P}} \tag{39}$$

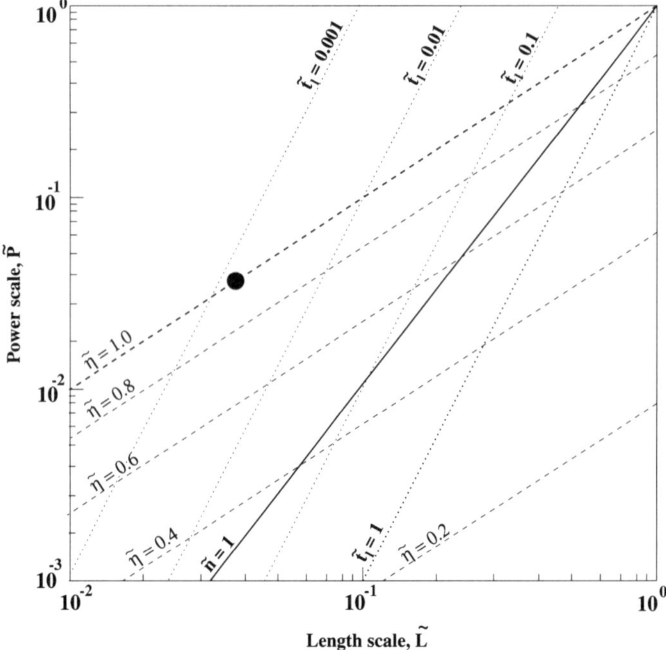

Fig. 8 Universal scaling.

These lines can be overlayed on top and are shown as dotted lines in Fig. 8. Photographic scaling can also be shown as the line of constant plasma density (solid line):

$$\tilde{n} = \frac{\tilde{P}}{\tilde{L}^2} \tag{40}$$

The scaling alternatives conveniently summarized in Fig. 8 can be easily interpreted. Following strict scaling (solid line $\tilde{\eta} = 1$) from the upper right corner, the reduction of power from 1350 W to the desired 50-W level ($\tilde{P} = 0.037$) corresponds to a reduction in size by a factor of 27, or $\tilde{L} = 0.037$. The location of this operating point is marked on the diagram. Although the efficiency has remained constant, the lifetime is reduced by almost a factor of 700 from the nominal. Alternatively, following the photographic scaling strategy (solid line) to the power level corresponding to $\tilde{P} = 0.037$, the reduction in size corresponds to only $\tilde{L} = 0.2$. At that location, the lifetime is reduced by only a factor of 5, however, the efficiency is down by a factor of 0.6 from the nominal. The plot also indicates that the two scaling strategies presented earlier are not special and are just examples of limiting behavior ($\tilde{\eta} = 1$ or $\tilde{n} = 1$), so that for a given reduction in power (line $\tilde{P} = 0.037$) there is an unlimited number of scaling alternatives for which the drop in efficiency can be traded with the loss in operating life. For a specified minimum operating life and a minimum tolerable eficiency, there is a maximum reduction of power for which these specifications can be met with a single scaling strategy (Fig. 9a). If the desired power level reduction is less than the maximum, there

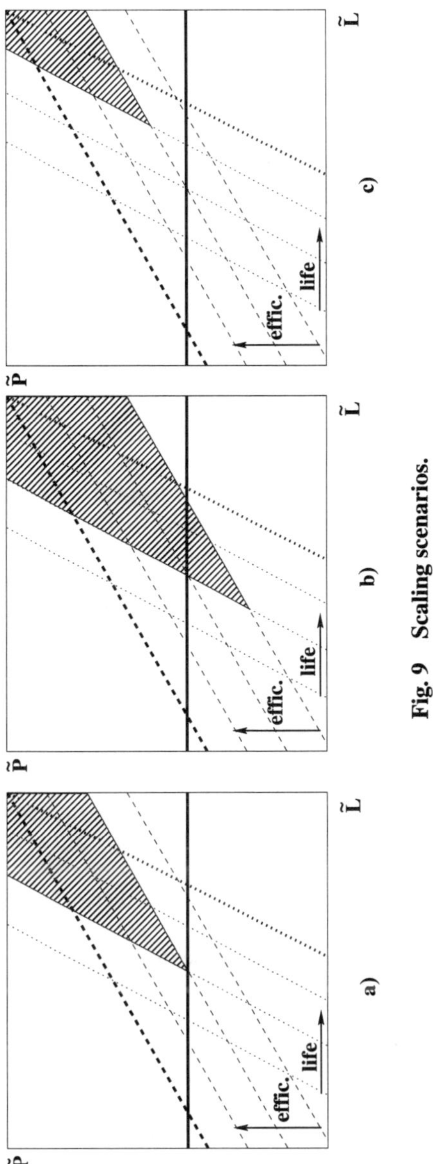

Fig. 9 Scaling scenarios.

exists an unlimited number of alternatives (Fig. 9b). However, if the reduction of power exceeds the maximum allowable, no scaling solutions are available to meet the specifications (Fig. 9c). The use of alternative scaling strategies presents a viable solution for overcoming or offsetting the intrinsic lifetime limitations that most plasma devices experience at small scale by partially sacrificing their operating efficiencies. The applicability of these devices to missions of future interest, comparison to other competing technologies, and possible implementation of these alternative scaling strategies are important avenues for further exploration.

VIII. Conclusions and Recommendations

Despite the disappointingly low performance, the effort to miniaturize a Hall thruster has provided a number of useful insights for any such attempts in the future. Most important, this work has highlighted the generic difficulty, common to all plasma thrusters, associated with the increase of the plasma density as the scale of the device is reduced. The consequences of such scaling, most notably the higher particle fluxes that cause an increase in the erosion rates and significant loss of operating life at a small scale, create a strong incentive to search for propulsion schemes that avoid ionization by electron bombardment.

Another important aspect of the higher plasma density is the increased heat flux into the wall. It was found that with an appropriate choice of materials and component configurations, the intrinsically stronger thermal gradients could be exploited to shunt heat to noncritical areas for radiation. The final dimensions of the iron center pole, however, were such that sufficient heat conduction was not achieved. This resulted in the partial loss of magnetic permeability near the tip and a decrease of the magnetic field strength within the gap. In retrospect, the situation could have been improved if the center pole had been designed somewhat thicker with a conical taper and a thicker base. Even if attempted, however, tight tolerances, as well as the alignment and geometrical constraints dictated by scaling, make these alterations extremely difficult, if not impossible.

Another lesson learned was the need to use permanent magnets to achieve the higher magnetic fields dictated by scaling. Although the use of permanent magnets had extended the available scaling range to power levels beyond what is normally possible with electric coils, intrinsic limitations of the existing magnetic alloys do not permit scaling below power levels of about 40–45 W. In addition, the sensitivity of these magnets to temperature makes thermal design of these and other components especially important. The current design appears to be satisfactory in this respect, except that, because of the unexpectedly low efficiency and, hence, additional unforeseen heat dissipation, it is possible that some reversible field weakening may have occurred during operation. As suggested by the subsequent measurements of the magnetic field strength, it is unlikely that excessive heating had caused any permanent damage to the magnets.

The very small physical dimensions of the 50-W Hall thruster have posed a number of practical difficulties. Some of the more significant and severe ones are outlined here:

1) Measuring and mapping the profile of the magnetic field strength within the narrow 1.6-mm gap to even moderate accuracy requires the use of magnetic field sensors that are less than 50–100 μm in size. The lack of adequate instrumentation has precluded any detailed measurements from being made to validate the nominal

field strength against the results of a numerical simulation, or to examine the changes, if any, in the field strength that may have occurred over the course of the experiments. A large 1.5-mm Gauss probe was used instead to provide a rough indication of the mean magnetic field strength in the vicinity of the central pole piece.

2) Following the configurations of the typical TAL designs, the anode of the miniature thruster was extended toward the exit of the channel where the field profile is tailored so as to be tangent to the frontal face of the anode. Hence, even minimal misalignment of the anode with respect to the thruster center line may force the field lines to intersect the anode and cause excessive electron leakage. Attaining relative tolerances on centering and alignment comparable to those in larger devices is generally more difficult with miniature components.

3) Rough measurements of the magnetic field strength using a 1.5-mm Gauss probe have also indicated a nonuniformity of the field strength in the azimuthal direction. Significant variations of the field strength along the circumference of the channel, at certain locations by as much as a factor of two, can be attributed to the variability in the shapes of the segmented magnets. Crumbling and the lack of axial symmetry due to manufacturing and assembly imperfections may have contributed to the field variations as well. Such strong azimuthal gradients suggest that, even before thruster operation, the nominal field profile may not have been properly shaped to prevent intersections of the field lines with the anode.

4) Reduction in component dimensions made the manufacturing of the critical parts, such as the anode and the magnetic circuit assembly, more difficult. The choice of a TAL configuration with an extended anode was primarily dictated by the desire to avoid the use of a thin and fragile ceramic insulator but, in turn, it introduced an unwanted sensitivity to alignment.

5) Small magnetic particles had a tendency to lodge in the narrow (0.3 mm) gap between the anode and the main magnet, creating occasional shorting. Although some of these particles were metallic dust collected during handling and installation, the majority of them were debris from the SmCo magnet. The magnets were ultimately encapsulated in nickel to reduce crumbling.

6) Removal of the trapped magnetic particles from the gap between the anode and the magnetic assembly prior to thruster installation required the ability to access the interior of the channel. Because of the need to provide for assembly and disassembly of the anode supported by its metallic feed tubes, press fits were used at one of their ends. The problem of ensuring gas tightness at that location was not fully resolved and gas leaks could not be completely ruled out. Improvements are needed in this regard.

7) The most severe limitation, however, is the loss of operating life resulting from the increased particle fluxes and erosion rates at small scale. It was shown in the earlier sections that the expected lifetime of the micro-Hall thruster is reduced by almost a factor of a thousand from the nominal life of 7000 h experimentally obtained for a larger 1.35-kW thruster (SPT-100). Although no specific erosion rate measurements were performed, the signs of accelerated erosion can be clearly seen with a naked eye after about 8–10 h of accumulated operational time.

To the best of our understanding, poor performance was not an intrinsic feature of the reduced scale but rather the indirect consequence of the imperfections due to the operational, assembly, and manufacturing difficulties just listed. As noted in the earlier sections, distortion of the magnetic field caused by overheating of the center pole piece is the most likely explanation. Unfortunately, iteration and improvement would require both diagnostics, which are not available at this scale, and a stronger motivation that cannot be justified in view of the intrinsic lifetime limitations of this microplasma device.

Acknowledgments

The authors wish to acknowledge the help and support of the academic and student staff at the Princeton Electric Propulsion and Plasma Dynamics Lab, specifically, Prof. Edgar Choueiri, Robert Sorenson, and John Ziemer, for allowing the use of their laboratory facilities and for their technical assistance. The authors also acknowledge the support of the staff at C. S. Draper Laboratory during the design and construction phases of this project, as well as funding from the Air Force Office of Scientific Research (Mitat Birkan, monitor).

References

[1]London, A. P., *A System Study of Propulsion Technologies for Orbit and Attitude Control of Microspacecraft*, M.S. Thesis, Dept. of Aeronautics and Astronautics, Massachusetts Inst. of Technology, Cambridge, MA, 1996.

[2]Khayms, V., and Martinez-Sanchez, M., "Design of a Miniaturized Hall Thruster for Microsatellites," AIAA Paper 96-3291, 32nd AIAA Joint Propulsion Conf., Lake Buena Vista, FL, 1996.

[3]Khayms, V., and Martinez-Sanchez, M., "Preliminary Experimental Evaluation of a Miniaturized Hall Thruster," IEPC Paper 97-077, 25th International Electric Propulsion Conf., Cleveland, OH, 1997.

Chapter 10

Development and Testing of a Low-Power Hall Thruster System

Jeff Monheiser,* Vlad Hruby,† Charles Freeman,‡
William Connolly,§ and Bruce Pote¶
Busek Co. Inc., Natick, Massachusetts

I. Introduction

WITHIN the past 10 years, there has been a significant increase in both the use and the study of mini- and microsatellites having initial launch masses of less than a few hundred kilograms. Initially the use of these satellites was almost exclusively the domain of university scientific and amateur missions, whereas commercial and military satellites got larger and more expensive. However, with the recent decrease in budgets and the increasing capabilities of electronics, commercial and military interests have now begun looking at replacing or augmenting some functions of large single satellites with constellations of mini- and/or microsatellites. Several constellations of small satellites, having initial launch masses of ~500 kg, are being developed for real-time applications such as worldwide mobile communications and non-real-time applications such as paging services, asset tracking and identification, and meter reading. In addition to these commercial uses of mini- and microsatellites, there exists a strong interest in the scientific community in using these satellites to accomplish very ambitious science missions.

To stimulate the development of mini- and microsatellites, the U.S. Air Force and NASA have initiated several component development programs, and NASA has established the Small Spacecraft Technology Initiative (SSTI). The goal of SSTI is to promote the development of new technologies that reduce the cost and time of getting civil and commercial missions from the drawing board to orbit. In addition, several governmental agencies sponsor both high school and university missions such as TERRIERS, CATSAT, and SNOE. Also under consideration is

Copyright © 2000 by Busek Co., Inc. Published by the American Institute of Aeronautics and Astronautics, Inc., with permission.

*Senior Scientist, Plasma Group.

†Chief Scientist.

‡Engineering Aid, Plasma Group.

§Chief Electrical Engineer, Busek Co. Inc., also President of Electronic Design Associates Inc.

¶Senior Engineer, Director of Hall Thruster Programs.

the use of mini- and microspacecraft to accomplish ambitious planetary missions such as the exploration of Saturn's rings and a Pluto flyby.

With this new interest in mini- and microsatellites, there is now a further need to develop low-power electric propulsion systems capable of satisfying both the primary and the secondary (station keeping and attitude control) propulsion requirements.[1] Considerable research is currently being conducted to advance the state of the art in pulsed plasma thrusters (PPTs),[2–5] colloid thrusters,[6,7] field ion emission sources,[8,9] and microelectromechanical systems.[10–12] Of these systems only PPTs can operate at a power level greater than a few tens of watts, making them the only option for primary propulsion on minisatellites having initial power levels of the order of a few hundred watts. As an alternative to the use of PPTs for primary propulsion of minisatellites, Busek Co. Inc. developed a 200-W Hall thruster system. Presented within are descriptions of the individual components of the propulsion system and performance data showing how the system operated over the input power range from 100 to 300 W.

II. Thruster System Description

A. 200-W Hall Thruster

Hall thrusters typically consist of a coaxial annular cavity in which a plasma is created by passing a current between an annular anode located on the upstream end of an otherwise dielectric discharge chamber and the externally located cathode. The neutral, gaseous propellant enters the discharge chamber via an annular manifold typically located within the anode. A radial magnetic field is applied, either by permanent magnets or through electromagnetic coils and ferromagnetic yokes. This radial magnetic field is sufficient in magnitude to capture the electrons and, together with the applied axial electric field E, force them to execute the so called $\boldsymbol{E} \times \boldsymbol{B}$ drift. Because of collisions between electrons and neutral propellant atoms, the electrons also drift axially toward the anode, while being radially confined by plasma sheaths present at the insulating walls. The azimuthal electron flux (also called Hall current) dwarfs the axial flux. Their ratio, which approximates the Hall parameter, exceeds 100 in typical Hall thrusters. This leads to high collisionality between electrons and neutrals and high ionization rate, yielding a high propellant utilization fraction. As a result, as little as 10% of the total propellant mass flow exits the thruster in the form of neutral atoms. Because the mass of the ions is so much greater than that of the electrons, they are unaffected by the magnetic field and are electrostatically accelerated from the thruster creating the desired reaction force. Since the plasma density is relatively low, collisions between ions and other heavy species are rare, hence the ions are accelerated to an exit velocity $c_i = \sqrt{2eV_a/m_i}$, where V_a is the ion accelerating voltage and m_i is the ion mass. Since there is no momentum loss for the ions, the ion accelerating voltage is given by the difference between the voltage where the ion was created and the voltage at the thruster exit. Because of the plasma neutrality facilitated by the presence of electrons throughout the entire ion acceleration, there is no space-charge limitation of the extracted ion current. This allows Hall thrusters to have a significantly higher thrust density compared with conventional gridded ion thrusters.

The major component of the 200-W propulsion system is the patent-pending Tandem Hall thruster[13] shown in Fig. 1. The nominal operational characteristics for this thruster are listed in Table 1, which shows that this thruster has a specific

Table 1 BHT-200-X2B operating specifications

Acceleration annulus mid-diameter	21 mm
Input power	207 W nominal 100–300 W
Discharge voltage	300 V nominal 200–400 V
Propellant mass flow rate	0.74 mg/s (Xe) nominal 0.30 to 1.01 mg/s (Xe)
Thrust	11.4 mN nominal 4 to 17 mN
Anode efficiency	42% nominal 20 to 45%
Anode specific impulse	1570 s nominal 1200–1600 s
Thruster mass	<1 kg
Thruster dimensions	10.5-cm diam, 12-cm length

impulse of 1570 ± 80 s and an efficiency of $42 \pm 3\%$. Both of these values are presented excluding the expellant flow through the cathode and the magnet power.

Because of the unique problems associated with scaling of existing Hall thruster geometries to low power levels,[14] one must employ some unconventional geometric scaling to achieve efficient operation at low power levels. As an example of conventional scaling techniques, Khayms and Martinez-Sanchez[14] describe a methodology that involves ensuring that the ratio of the mean free path for each particle to the thruster characteristic length, L, remains constant. This implies that each particle density should scale inversely with the characteristic length, i.e., $n_e, n_i, n_n \sim 1/L$. Similarly, they suggest that, to prevent high-energy electrons from being lost to

Fig. 1 BHT-200-X2B thruster.

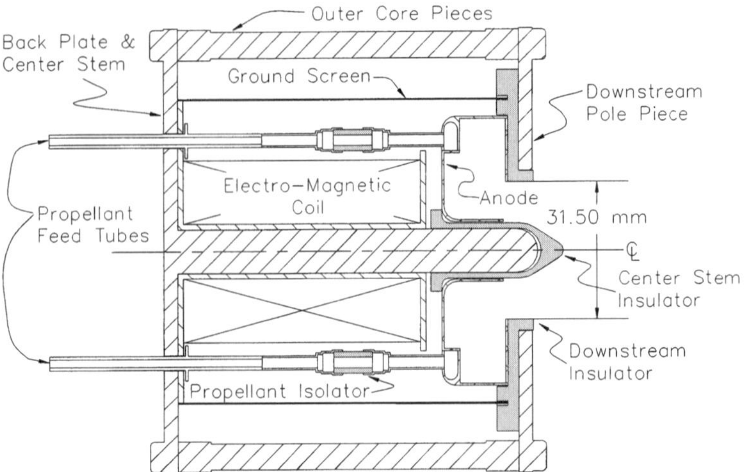

Fig. 2 Mechanical schematic of the 200-W thruster.

the discharge chamber walls, the ratio of the electron Larmor radius to the characteristic dimension must remain low and constant. This scaling relation implies that the magnetic field must increase linearly as the thruster power is decreased, i.e., $B \sim 1/L$. This condition is the most difficult to meet. As the thruster dimensions decrease, there is less ferromagnetic material to conduct the required magnetic flux. For example, a 1.35-kW SPT-100 thruster typically consists of an annular discharge chamber surrounded by inner and outer magnetic coils. To scale this thruster to 200 W requires an almost-sevenfold increase in the magnitude of the magnetic field and the same sevenfold decrease in the overall dimensions of the thruster. This implies that a 200-W SPT thruster would require a magnetic field of about 0.14 T and have a midcavity diameter of 14.8 mm. This very small diameter implies that the thruster lacks the volume necessary to include enough magnetic material and inner magnetic coils to create and conduct the high magnetic field. This scaling results in high particle fluxes within the discharge chamber that can significantly reduce the lifetime of the thruster. To mitigate these problems one must deviate somewhat from the conventional scaling laws.

Presented in Fig. 2 is a mechanical schematic of the BHT-200-X2B thruster that was designed by departing from the aforementioned conventional scaling where necessary. The result is a geometry that departs significantly from classical SPT designs. The magnetic structure, constructed from a material having a high permeability and saturation flux density, consists of the back plate, the center stem piece, the outer core pieces, and the downstream pole piece. This structure conducts the magnetic field created by a single electromagnetic coil and guides it across the radial gap between the center stem and the outer pole. In addition, this structure serves as the mechanical support to which the rest of the thruster components are mounted.

The electromagnetic coil is located in tandem with and upstream of the metallic discharge chamber, which also serves as the anode. The discharge chamber may be constructed from the same magnetic material as the magnetic structure, in which case it serves to shape the magnetic field at the thruster exit, improving the thruster performance and greatly extending its lifetime. To prevent electrical breakdown between the anode and the magnetic structure, two insulators are used. The first is

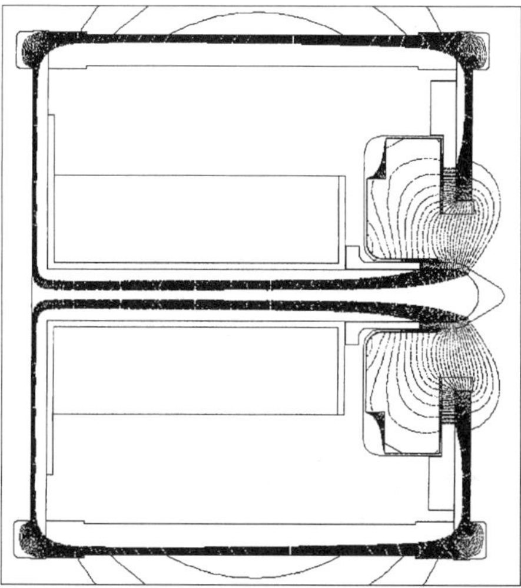

Fig. 3 Magnetic field line plot for the 200-W thruster.

the downstream insulator, which also serves to protect the downstream pole piece from impingement by high-energy beam ions, and the second is the center stem insulator. This insulator also serves to protect the center stem from high-energy ions and prevents shorting of the ion acceleration sheath to the magnetic structure.

The overall length and diameter of the thruster are 12 and 10.5 cm, respectively. This size creates the volume necessary for an efficient electromagnetic coil, and because of the large thruster diameter, the diameter of the discharge chamber/anode can be increased over that typically used for low-power Hall thrusters. This increase in the discharge chamber diameter creates a large discharge volume that serves to increase the propellant utilization by improving the propellant uniformity and increasing the ionization probability. Because of the magnetically permeable and electrically conductive discharge chamber walls, the interior volume is thought to be substantially free from fields that could drive ions into the chamber walls. This reduction in the ion flux to the discharge chamber walls reduces the heat load on interior surfaces of the discharge chamber. The ground screen is used to prevent ambient electrons from reaching the external surfaces of the discharge chamber, whereas the propellant isolators insulate the chamber/anode from the back plate potential.

Before the thruster geometry presented in Fig. 2 was finalized, several magnetic geometries were investigated using a commercial two- and three-dimensional, nonlinear, finite-element program called Maxwell from Ansoft Corp. This program was used to investigate if and where the magnetic material saturated, the maximum possible magnetic field achievable before the electromagnetic coil overheated, and the shape of the field within the thruster exit. Presented in Fig. 3 is a plot of the magnetic field lines computed for the nominal thruster operating conditions presented previously. If no potential sheaths are present at the interior surfaces of the metallic discharge chamber, the magnetic field lines should approximate lines of constant potential. This assumption is useful in the analysis of probable geometries for Hall thruster magnetic structures because the field lines can be used to estimate how ions are accelerated from the discharge chamber. Presented in Fig. 4 is a detailed view of the magnetic field/equipotential lines within the exit area of the thruster. Shown

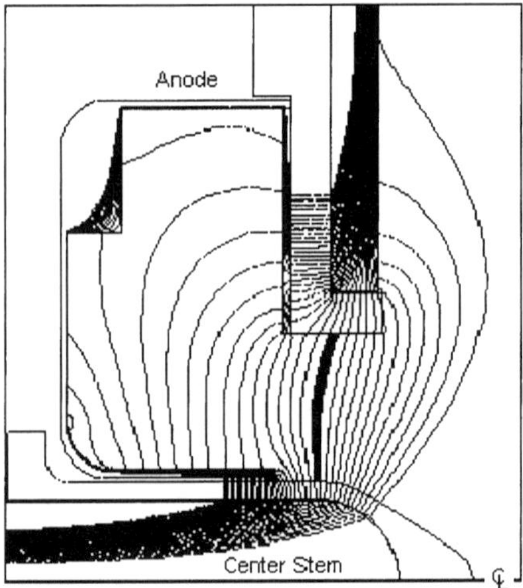

Fig. 4 Detailed view of magnetic field line within the thruster exit.

in the figure are the thruster centerline, the anode, and the center stem, along with a thick line that indicates a possible location of the ion acceleration sheath. If the ions are created with a minimal initial energy, then their trajectories should begin normal to this line. Changing the relative positions of the center stem, the anode, and the downstream pole piece allows a designer to vary this initial ion acceleration, resulting in a well-collimated ion beam. As an example, the field lines presented in the figure near the insulator surfaces either are normal to the surface or have a slight curvature that accelerates the ions either parallel to or away from the insulators. In addition, the overall curvature of the sheath acts as a plasma lens focusing the ions to a point downstream of the thruster. This focusing improves the divergence of the ion beam over that characteristic of more classical designs. In addition to the magnetic field lines affecting the ion beam divergence, the discharge plasma conditions may contribute to the beam divergence. At the most probable location for the ion-accelerating sheath, the boundary surfaces are covered by either the center stem or the downstream insulators. Since this surface is an isolator, a small potential sheath develops to attract electrons and repel ions, assuring that an equal number of ions and electrons reaches the insulator surface. This potential difference is typically of the order of a few times the electron temperature, which, for a Hall thruster, can be as high as 50 V. This sheath potential gives the ions some nonaxial acceleration, which, under the worst conditions, could be as high as the ratio of the sheath potential to the discharge voltage (i.e., $50/300$ for a 300-V discharge).

B. 1500-mA, Low-Power Hollow Cathode

One critical aspect of producing an efficient, low-power Hall thruster system is to develop a cathode that 1) is capable of delivering several hundred milliamps of current, 2) requires a minimal expellant flow, 3) develops a minimal voltage drop required to extract the discharge current, and 4) requires no heater or keeper power following its initial warmup. To this end, the BHC-1500-025 hollow cathode, shown in Fig. 5 and described in Table 2, was developed. This cathode is a conventional hollow cathode constructed from a 3.2-mm-diam refractory metal

Table 2 BHC-1500-025 specifications

Cathode tube diameter	3.2 mm
Emitter	Impregnated tungsten
Ignition time	<3 min
Standby mode	
Keeper current,voltage	700 mA, 20 V
Expellant flow rate	0.05 mg/s (Xe)
Operating mode	
Expellant flow rate	0.08 mg/s (Xe)
Emission current	700 mA
No keeper of heater power required	
Cathode mass	<200 g

tube with a refractory metal orifice. The size of the orifice and the keeper geometry were optimized to allow self-sustaining operation at very low flow rates [0.05 to 0.10 mg/s (Xe)] with no heater or keeper power. To establish operation at very low expellant flow rates and low discharge currents, significant effort was directed at thermal modeling of the cathode. From this modeling, it was determined that radiation and conductive heat transfer from the insert were the major contributors to its cooling below thermionic emission temperatures and extinguishing the discharge. To eliminate these effects, several thermal barriers were added to mitigate heat loss and prevent extinguishing of the discharge. The keeper current and voltage data are presented in Fig. 6 and they show the classic hollow cathode behavior of a negative current/voltage characteristic and low sensitivity to expellant flow rate when operating in the spot mode.[15] At a keeper current of 850 mA, the expellant flow variation from 0.068 mg/s (Xe) to 0.019 mg/s (Xe) resulted in a slight increase in the keeper voltage, from 20.1 to 20.7 V. At a keeper current of 1050 mA, as the expellant flow is decreased by the same amount, the keeper voltage increases from 15.4 to 18 V, which is most likely caused by the spot/plume mode transition. The typical operating conditions when used with the BHT-200-X2B thruster are an expellant flow rate of 0.08 mg/s (Xe), no keeper power, and no heater power. Rapid starting of the cathode can be accomplished by applying ~40 W of heater

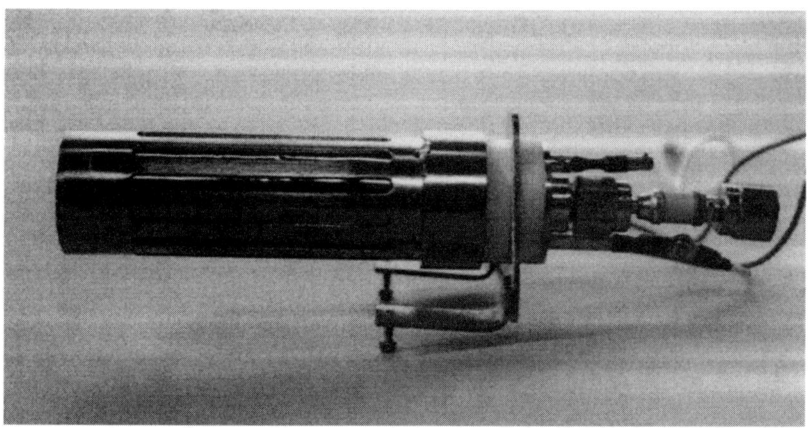

Fig. 5 BHC-1500-025 self-sustaining hollow cathode.

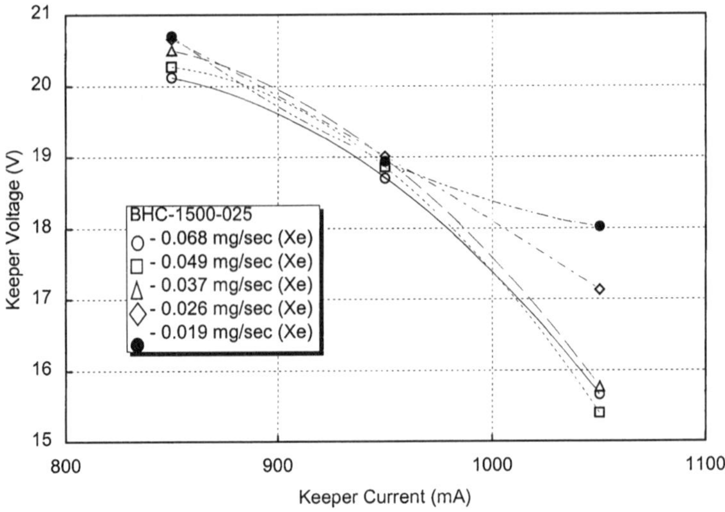

Fig. 6 Performance of the BHC-1500-025 miniature hollow cathode.

power while flowing 0.08 mg/s (Xe). Under these conditions, the cathode will start in less than 3 min at a voltage of less than 75 V. Once running, it will operate self-sustained in a standby mode at a flow rate of 0.05 mg/s (Xe), a keeper voltage of 20 V, and a current of 700 mA.

C. 400-W Power Processing Unit

An important part of the low-power propulsion system is the power processor unit (PPU). Its basic function, performed by the main DC–DC converter, is to transform the low-voltage spacecraft bus power to a form required to initiate and sustain the thruster discharge. A fully configured PPU typically incorporates a number of additional independent converters for other power handling functions including the electromagnet current supply, cathode heating and starting, propellant flow control and telemetry, etc. Our development focused on designing and constructing a breadboard DC-to-DC converter for the thruster plasma discharge.

Table 3 shows the major PPU/discharge converter specifications. The converter was designed to deliver up to 400 W of input power to the thruster. For test and

Table 3 Discharge converter major specifications

	Nominal	Range
Input voltage, V	28	24 to 35
Output voltage, V	300	20 to 350[a]
Output current, A	0.66	0 to 1.25[b]
Output power, W	200	0 to 400
Output current ripple	<5% resistive load	0.1 A p to p
Switching frequency, kHz	50	

[a]Operator selects V or I control.
[b]Short circuit protected to I_{max}.

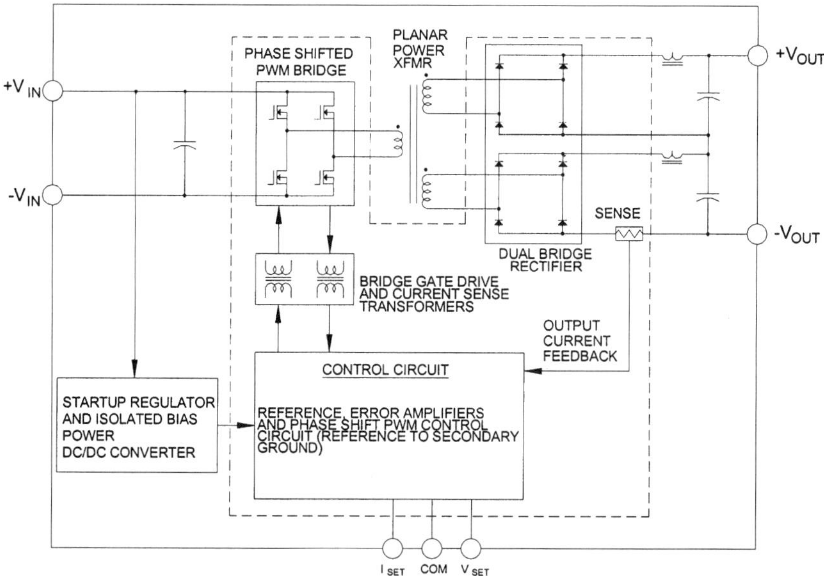

Fig. 7 Hall thruster discharge power supply/DC–DC converter schematic.

operating flexibility, the output voltage and current are real-time adjustable from 0 to 350 V and 0 to 1.25 A, respectively. The operator selects either voltage or current control mode with set points controlled by an external analog signal.

The power topology selected for the converter consists of a buck-derived pulse width modulated full Mosfet bridge followed by a dual bridge seriesed rectifier shown schematically in Fig. 7. Phase shift control is used for pulse width modulation of the Mosfet bridge to obtain resonant transitions for maximum efficiency. A switching frequency of 50 kHz was selected as a compromise between switching losses and size of the magnetics.

The primary bridge configuration is especially suited to high voltage and high power operation because it limits the voltage stresses on the Mosfets to the actual input line voltage. In addition to eliminating corona risk, this allows selection of lower resistance (R_{dson}) Mosfets than would be possible with a push–pull or forward converter topology. Further advantage of the bridge is its compatibility with resonant transition phase shift modulation. This modulation technique significantly reduces switching losses at high power levels. Additionally, since the gate drive waveforms are always symmetrical square waves, isolated gate drive signals are easily coupled to the bridge from an isolated control circuit using transformers. The rectifier consists of ultrafast, high-voltage diodes in a dual seriesed bridge to lower device stresses during reverse recovery.

The planar winding power transformer uses a combination of printed wiring board and flat sheet metal (copper) windings. Planar construction offers important features such as low skin effect losses and parameter repeatability. The flat conductors provide excellent current distribution at a high frequency yielding lower conductor losses at high currents than would be obtainable with a conventional toroidal device using magnet wire. Planar devices are also free of fabrication variables and yield the repeatable leakage inductance that is important to resonant transition switching.

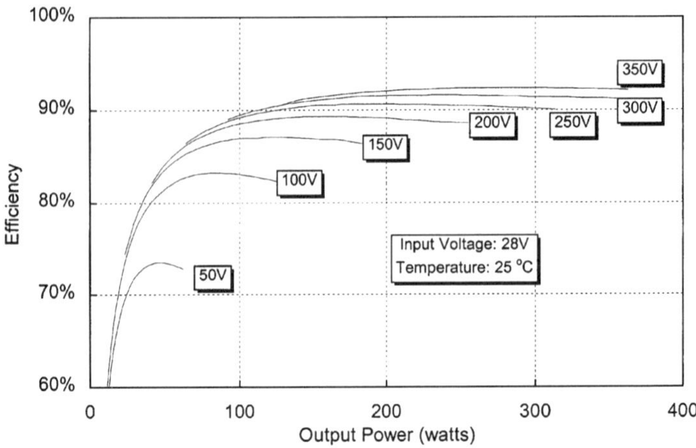

Fig. 8 Performance curve for the 400-W breadboard PPU.

Including a separate low-power DC/DC bias converter, powered from the 28-V DC main input bus, allows the complete PWM control circuit to be referenced to the secondary side while making its operation independent of the state of the main converter. The bridge gate drive transformers provide the required isolated feedback to the main converter's primary side. This results in inherently high noise immunity while avoiding the use of optical isolators and their associated radiation intolerance. The bias converter does not require precise regulation and therefore does not require secondary feedback. It does, however, provide imperviousness to the operational state of the main power converter and possible discharge oscillations.

Parts were selected based upon their potential for upgradability to space qualified or qualifiable components. The use of nonbipolar or CMOS components was avoided unless known radiation-tolerant full-temperature-range parts are available. Commercial plastic power Mosfets and other plastic-encased semiconductor components were used in the breadboard upon verification of the availability of hermetic radiation-tolerant parts that are essentially equivalent. General component selection was based upon the availability of similar or equivalent components with a military/space temperature range of -55 to $+135°$C.

Presented in Fig. 8 is a plot of the measured efficiency, including control power taken off the 28 V input, as a function of the output power for different output voltages. These performance characteristics were measured using a resistive load. Efficiency was measured at 92% or higher over the power range of interest from 150 to 350 W. The output voltage ripple was measured to be 1% at a voltage of 300 V and 670 mA current, i.e., for nominal 200 W thruster operation.

III. Thruster Performance

A. Facilities and Experimental Apparatus

Upon the completion of the development phase, significant effort was directed at determining the performance (i.e., thrust, specific impulse, and efficiency) of the thruster system over a wide range of operating conditions. To obtain the desired thrust data, the BHT-200-X2B thruster and BHC-1500-025 hollow cathode were

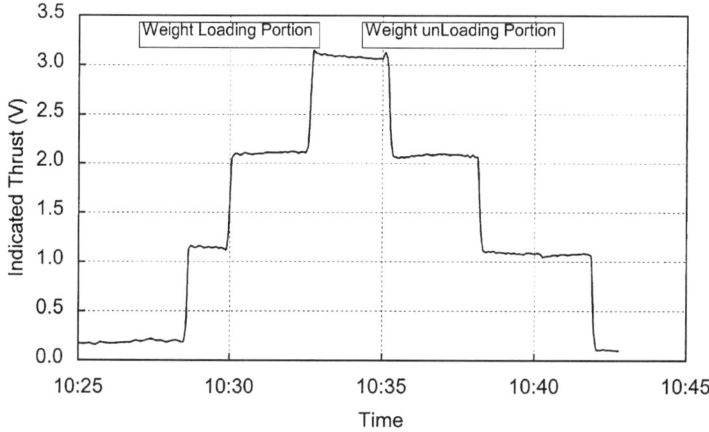

Fig. 9 Example of a calibration curve obtained both pre- and postexperiment.

mounted on a NASA Glenn inverted-pendulum-type thrust stand similar to that developed by Haag et al.[16] The output voltage from the linear variable differential transformer (LVDT), which is proportional to the thrust, and the inclination of the thrust stand (measured using a capacitive inclinometer) were both measured and recorded using a 12-bit computer data acquisition system. The experimental procedure used to calibrate the thrust stand prior to and post test involved applying a known load to the thrust stand and measuring the corresponding LVDT output voltage. A typical calibration curve obtained during an experiment is presented in Fig. 9. The figure shows that, as a known weight is applied to the thrust stand, a measurable deflection is obtained. To assure linearity of the calibration, the known weights are individually removed from the thrust stand, yielding a second measure of the deflection. In addition to this procedure for obtaining the conversion from indicated volts to millinewtons, the inclination of the thruster was continuously recorded to minimize the error associated with inclination changes induced by facility thermal effects. To quantify the effect of theses changes in inclination on the measured thrust, at the end of each experiment the output of the LVDT was recorded, as the inclination of the thrust stand was varied slightly from zero. A linear curve fit of this variation in the LVDT voltage was then used to correct the LVDT output for any experimental error induced by changes in inclination. Performing this error correction improved the accuracy of the thrust stand and reduced its error to ~1%, which is due almost exclusively to zero drift in the LVDT signal conditioning electronics.

All of the tests reported here were conducted in the T6 vacuum facility that is composed of two liquid nitrogen (LN_2)-cooled sections, one where the experimental apparatus is located and the second where the pumping is accomplished. The experimental section is 1.8 m in diameter and 1.8 m in length, whereas the pumping section is 2.4 m in diameter and 1.2 m in length. Pumping of the facility is accomplished using two pumps, the first is a 0.8-m-diam oil diffusion pump used to pump the low molecular weight gases and the second consist of four cryopanels cooled to 30 K used to pump the xenon. In full operation (i.e., diffusion pump plus cryopanels) the facility is capable of pumping 90,000 L/s of xenon, and with only the diffusion pump operating, the facility pumps ~8000 L/s of xenon. It should be noted that for all of the data to be presented, a LN_2 baffle was used to prevent any hot-oil vapor from reaching the experimental section of the facility.

The data obtained and presented in this chapter were recorded using a 12-bit, optically isolated, computer-driven data-acquisition system. The primary responsibilities of this system were measuring and recording the discharge power, setting and recording the MKS mass flow controllers, measuring the LVDT output and inclination, and recording the vacuum tank pressure along with the cathode and anode internal pressures. To assure validity of the performance data, the mass flow controllers were in-house calibrated on xenon using two independent methods, a constant-volume process and a constant-pressure process. These methods yielded calibration curves that were within 5% of each other for the mass flow rates of interest. All the data presented were obtained using a helium leak checked flow system and research-grade xenon purchased from Spectra Gasses having a purity of 99.999%.

B. Constant-Discharge Voltage Performance Data

Using the aforementioned experimental apparatus, the thruster performance was measured as the discharge power was varied from 100 to 300 W at a constant discharge voltage of 300 V. This variation in the discharge power was accomplished by linearly varying the thruster mass flow rate from 0.35 mg/s (Xe) to 1.01 mg/s (Xe) as shown in Fig. 10. To maximize the performance of the thruster at each operating point, the magnetic field was varied to yield the minimum discharge current. Presented in Fig. 11 are the measured thrust data obtained as the discharge power was increased over the aforementioned range. At a power of ~100 W the thruster delivers about 4 mN of thrust, which increases to 16 mN as the power is increased to almost 300 W. Presented in

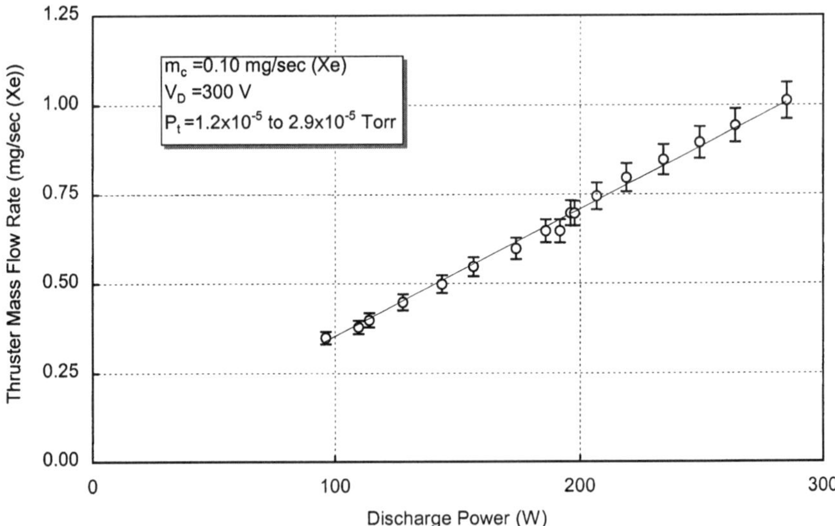

Fig. 10 Variation of the thruster mass flow rate as the discharge power is increased to almost 300 W.

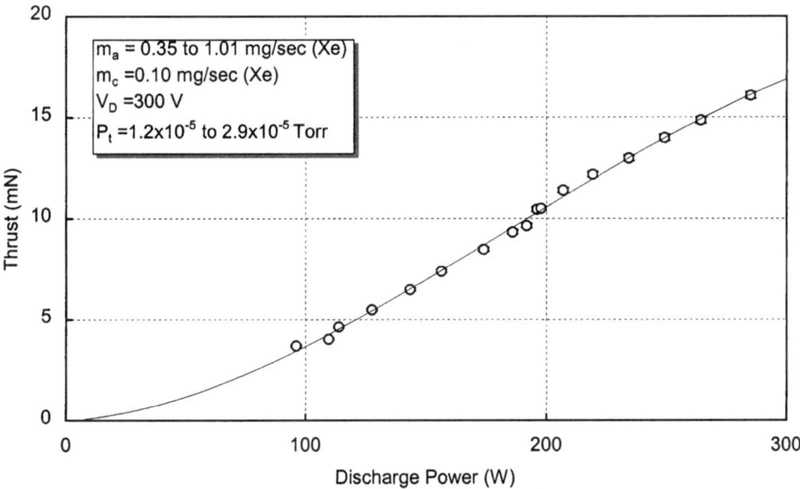

Fig. 11 Measured thrust for the BHT-200-X2B as the discharge power is varied from 100 to 300 W.

Fig. 12 are the total and anode specific impulse data obtained as a function of the discharge power. The total specific impulse includes the cathode and anode mass flow rate, while the anode specific impulse excludes the cathode flow. As Fig. 12 shows, the anode specific impulse data increase from 1050 s at a power of 100 W to ~1500 s at 200 W and the total specific impulse increases from ~800 to 1300 s over the same power range. The data also show that, as the discharge power is increased above 200 W, both specific impulses increase and appear to

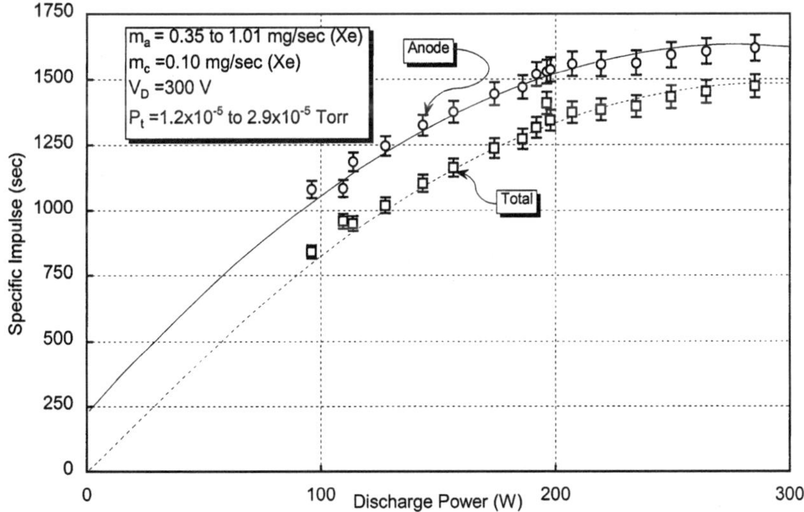

Fig. 12 Specific impulse vs discharge power.

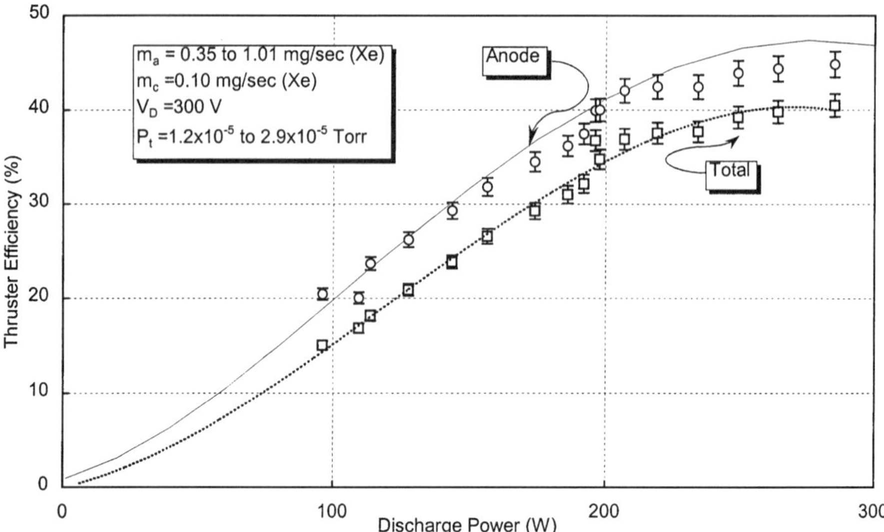

Fig. 13 BHT-200-X2B efficiency data for various discharge powers.

reach asymptotes of 1600 and 1500 s, respectively. Recalling that the specific impulse is proportional to the product of the propellant utilization and the mean ion speed, the data in Fig. 12 suggest that either or both values are saturating at their maximum values for discharge powers greater than 200 W.

Presented in Fig. 13 are the efficiency data computed using the experimental and curve fit data in Figs. 10 and 11. The open circles show how the anode efficiency, $\eta_a = T^2/2\dot{m}_a P_D$, where T is the thrust, $\dot{m}_a$ is the anode mass flow rate, and P_D is the discharge power, varies over the power range. The open squares represent the total efficiency, $\eta_t = T^2/2(\dot{m}_a + \dot{m}_c)(P_D + P_m + P_K)$, where $\dot{m}_c$, P_m, and P_K are the cathode mass flow rate, the magnet power, and the keeper power, respectively, measured over the same power range. The experimental data show that, for discharge powers from 100 to 200 W, the anode efficiency increases from 20 to 40%, whereas for powers above 200 W this efficiency further increases, from 41 to 45%. At a power of 200 W the thruster has a total efficiency of 35%, which is very good for a device of this power level. It should be noted that the good total efficiency is due in part to the large electromagnetic coil, which, for all the data presented, consumed less than 2 W. In addition, for the majority of the data presented, the cathode was operated with no keeper or heater.

C. Thruster Comparison to Current Sate of the Art

Currently there appear to be three thrusters that operate over the same or a comparable power range, the SPT-50 and the SPT-35, both manufactured by Research Institute of Applied Mechanics and Electrodynamics of Moscow Aviation Institute (RIAME), and a laboratory thruster developed by Soreq. Jacobson and Jankovsky[17] presented performance data for the SPT-35 showing that thruster operation at a discharge voltage and current of 200 V and 1 A produces 11.3 mN of thrust, an anode specific impulse of 1170 s, and an anode efficiency of 32%. The data just presented show that at the same power level the BHT-200-X2B produces a thrust of 11.4 mN, an anode specific impulse of 1570 s, and an anode efficiency of 42%. Manzella and Oleson[18] presented data showing that, as the SPT-50 discharge power is varied from 400 to 130 W at a constant voltage of 300 V, the thrust varies from 20.3 to 7.0 mN. In addition, they show that the anode efficiency varies from a maximum of 42% to a minimum of 35%, and the anode specific impulse varies from 1557 to 1392 s. These data are very close to the performance numbers presented for the BHT-200-X2B. A detailed comparison of the total specific impulse and total efficiency at 270 W reveals that the BHT-200-X2B thruster achieved a 24% higher total specific impulse and a 48% higher total efficiency than the SPT-50. These improvements in the total specific impulse and total efficiency over the SPT-50 result from both the efficient operation of the thruster and to the low-flow, self-sustaining hollow cathode. Finally, Ashkenazy et al.[19] presented data on their developmental thruster showing that, for discharge powers ranging from 215 to 520 W, they achieved anode specific impulses between 1320 and 1580 s, thrust levels between 13.5 and 39 mN, and anode efficiencies between 40 and 51%.

IV. Summary

Busek Co. Inc. has completed the engineering development of a low-power, nominal 200-W Hall thruster system, including the thruster, a self-sustaining hollow cathode, and a breadboard power-processing unit. Data were presented showing that, at a constant voltage of 300 V, as the discharge power was varied from 100 to 300 W, the thruster produced thrust levels ranging from 4 to 16 mN, specific impulses (excluding cathode flow) ranging from 1100 to 1600 s, and efficiencies (excluding cathode flow and magnet power) ranging from 20 to 45%.

Acknowledgments

This work was conducted under an Air Force phase II SBIR program monitored by the Air Force Research Laboratory. The authors thank Ron Spores and Keith McFall of the Air Force Research Laboratory for their support and technical monitoring of the contract.

References

[1]Mueller, J., "Thruster Options for Microspacecraft: A Review and Evaluation of Existing Hardware and Emerging Technologies," AIAA 97-3058, 33rd Joint Propulsion Conf., Seattle, WA, July 1997.

[2]Cassady, R. J., Willey, M. J., Meckel, N. J., and Blandino, J. J., "Pulsed Plasma Thruster for the New Millennium Space Interferometer Experiment DS-3," AIAA 98-3326, 34th Joint Propulsion Conf., Cleveland, OH, July 1998.

[3]Hoskins, W. A., Wilson, M. J., Willey, M. J., Meckel, N. J., Campbell, M., and Chung, S., "PPT Development Efforts at Primex Aerospace Company," 35th Joint Propulsion Conf., Los Angles, CA, June 1999.

[4]Burton, R. L., and Bushman, S. S., "Probe Measurements in a Coaxial Gasdynamic PPT," AIAA-99-2288, 35th Joint Propulsion Conf., Los Angles, CA, June 1999.

[5]Ziemer, J. K., Choueiri, E. Y., and Birx, D., "Is the Gas-Fed PPT an Electromagnetic Accelerator? An Investigation Using Measured Performance," AIAA-99-2289, 35th Joint Propulsion Conf., Los Angles, CA, June 1999.

[6]Martinez-Sanchez, M., et al., "Research on Colloid Thrusters," IEPC 99-014, 26th International Electric Propulsion Conf., Kitakyushu, Japan, Oct. 1999.

[7]Shtyrlin, A. F., "State of the Art and Future Prospects of Colloidal Electric Thrusters," IEPC 95-103, 24th International Electric Propulsion Conf., Moscow, 1995.

[8]Fehringer, M., Rüdenauer, F., and Steiger, W., "Space-Proven Indium Liquid Metal Field Ion Emitters for Microthruster Applications," AIAA-97-3057, 33rd Joint Propulsion Conf., Seattle, WA, July 1997.

[9]Marcuccio, S., Giannelli, S., and Andrenucci, M., "Attitude and Orbit Control of Small Satellites and Constellations with FEEP Thrusters," IEPC 97-188, 25th International Electric Propulsion Conf., Cleveland, OH, Aug. 1997.

[10]Janson, S. W., and Helvajian, H., "Batch-Fabricated Microthrusters: Initial Results," AIAA-96-2988, 32nd Joint Propulsion Conf., Lake Buena Vista, FL, July 1996.

[11]Mueller, J., et al., "Design, Analysis and Fabrication of a Vaporizing Liquid Micro-Thruster," AIAA-97-3054, 33rd Joint Propulsion Conf., Seattle, WA, July 1997.

[12]Janson, S. W., et al., "Batch-Fabricated Resistojets: Initial Results," IEPC 97-070, 25th International Electric Propulsion Conf., Cleveland, OH, Aug. 1997.

[13]Busek Co. Inc., Application for United States Letters Patent, "Tandem Hall Field Plasma Accelerator," 15 Dec. 1998.

[14]Khayms, V., and Martinez-Sanchez, M., "Design of a Miniaturized Hall Thruster for Microsatellites," AIAA 96-2992, 32nd Joint Propulsion Conf., Lake Buena Vista, FL, 1996.

[15]Mandell, M. J., and Katz, I., "Theory of Hollow Cathode Operation in Spot and Plume Modes," AIAA-94-3134, 30th Joint Propulsion Conf., Indianapolis, IN, June 1994.

[16]Hagg, T. M., and Osborn, M., "RHETT/EPDM Performance Characterization," IEPC-97-107, 25th International Electric Propulsion Conf., Cleveland, OH, Aug. 1997.

[17]Jacobson, D. T., and Jankovsky, R. S., "Test Results of a 200W Class Hall Thruster," AIAA 98-3792, 34th Joint Propulsion Conf., Cleveland, OH, 13–15, July 1998.

[18]Manzella, D., and Oleson, S., "Evaluation of Low Power Hall Thruster Propulsion," AIAA 96-2736, 32nd Joint Propulsion Conf., Lake Buena Vista, FL, 1996.

[19]Ashkenazy, J., Raitses, Y., and Appelbaum, G., "Low Power Hall Thrusters for Small Satellites," *Proceedings from Israel Annual Conf. on Aerospace Sciences*, Feb. 1997, pp. 424–434.

Chapter 11

Performance of Field Emission Cathodes in Xenon Electric Propulsion System Environments

Colleen M. Marrese* and James E. Polk†
Jet Propulsion Laboratory, California Institute of Technology, Pasadena, California
Kevin L. Jensen‡
Naval Research Laboratory, Washington, D.C.
Alec D. Gallimore§
University of Michigan, Ann Arbor, Michigan
Capp A. Spindt¶
SRI International, Palo Alto, California
Richard L. Fink**
Field Emission Picture Element Technology, Austin, Texas
and
W. Devereux Palmer††
Microelectronics Center of North Carolina, Research Triangle Park, North Carolina

Nomenclature

A_t = emitting area of microtip
A_{FN} = Fowler–Nordheim I term, A/V^2
a_{fn} = Fowler–Nordheim term, $e(V\text{Å})^2/fs$

Copyright © 2000 by the American Institute of Aeronautics and Astronautics, Inc. The U.S. Government has a royalty-free license to exercise all rights under the copyright claimed herein for Governmental purposes. All other rights are reserved by the copyright owner.

*Member of Technical Staff, Thermal and Propulsion Engineering Department.
†Group Supervisor, Thermal and Propulsion Engineering Department.
‡Research Physicist, ESTD.
§Professor, Aerospace Engineering Department.
¶Vacuum Microelectronics Program Director, Applied Physical Sciences Laboratory of the Physical Sciences Division.
**Director of Engineering.
††Principal Engineer, Materials and Electronic Technologies.

a_o = Bohr radius, Å
B_{FN} = Fowler–Nordheim I term, V
b = term in radius distribution factor for cathode
b_{area} = area factor, $Å^2$
b_{fn} = Fowler–Nordheim term, eV/Å
c = speed of light, Å/fs
c_o = term in radius distribution factor for cathode
d = atomic spacing, nm
E = ion energy, eV
E_{th} = energy threshold for sputtering, eV
e = electron charge, C
F_{tip} = electric field on axis of the microtip structure, V/cm
$\hbar$ = Plank's constant, eV-fs
I_a = current collected by the anode, A
I_e = field emission current, A
I_f = final cathode current after xenon exposure, A
I_g = current collected by the gate electrode, A
I_o = initial cathode current, A
I_{tip} = current emitted from a single microtip structure, A
J_{CEX} = total charge-exchange ion current density
J_{FN} = Fowler–Nordheim field emission electron current density, A/cm^2
J_{iCEX} = charge-exchange single ion current density
J_{iiCEX} = charge-exchange double ion current density
K = term in sputter yield model
k = electric field factor
M_I = incident ion atomic weight
M_{II} = target atom atomic weight
N = neutral xenon number density, atoms/cm^3
N_{tips} = number of tips in a field emission array cathode
n = number of ions striking the emitting area of a microtip per second
n_i = number of locally created singly charged ions striking the emitting area of a microtip per second
n_{iCEX} = number of singly charged charge-exchange ions striking the emitting area of a microtip per second
n_{ii} = number of locally created doubly charged ions striking the emitting area of a microtip per second
n_{iiCEX} = number of doubly charged charge-exchange ions striking the emitting area of a microtip per second
n_{iii} = number of locally created triply charged ions striking the emitting area of a microtip per second
n_s = number of atoms sputtered from the emitting area of a microtip per second
Q_i = cross-section for single ionization, πa_o^2
Q_{ii} = cross-section for double ionization, πa_o^2
Q_{iii} = cross-section for triple ionization, πa_o^2
Q = term in sputter yield model
r = radial distance from microtip structure, cm
r_i = radial position corresponding to ionization potential, cm
r_t = radius of curvature of a microtip structure, cm

r_g = radius of aperture in gate electrode, cm
r_m = maximum distance from a microtip within which an ion formed will hit the emitting area of the tip, cm
r_s = radial position corresponding to the sputter threshold voltage, cm
S_e = term used in sputter yield model
s = term used to define the deviation in tip radius from the minimum tip radius
s_n = term used in sputter yield model
T = temperature, eV
t_e = time to erode a single layer of atoms from a microtip emitting area, s
U_s = sublimation energy, eV
V_a = anode voltage with respect to ground, V
V_i = ionization potential, V
V_M = modulation voltage, V
$V_{\max}$ = maximum voltage in an I–V trace, V
$V_{\min}$ = minimum voltage in an I–V trace, V
V_{pk} = peak voltage, V
V_s = voltage corresponding to energy threshold for sputtering, V
V_t = ion temperature, eV
Xe^+ = singly charged xenon ion
Xe^{++} = doubly charged xenon ion
Xe^{+++} = triply charged xenon ion
Xe^+_{CEX} = singly charged xenon charge-exchange ion
Xe^{++}_{CEX} = doubly charged xenon charge-exchange ion
x_o = term in B_{FN}
Y = sputter yield, atoms/ion

Greek

α = term in sputter yield model
β_c = microtip cone angle, rad
β_g = field enhancement factor
Δs = tip radius spread factor
δ = term in B_{FN}
ε = term in sputter yield model
Σ = distribution factor associated with a spread in tip radii in a field emission cathode array
σ = half-cone emitting angle of a cathode tip, deg
ϕ_w = work function, eV
ϕ_τ = electron emission half-cone angle on a microtip, rad

I. Introduction

THE recent thrust toward developing sub-100 W electric propulsion (EP) systems warrants the development of a cathode to operate on a comparable scale.[1–3] The cathode in an electric propulsion system provides electrons to produce a plasma and neutralize the ion beam. Hollow cathodes are typically used with higher power Hall thrusters and ion thrusters, however, their propellant and heater requirements place lower limits on their size and power scalability.

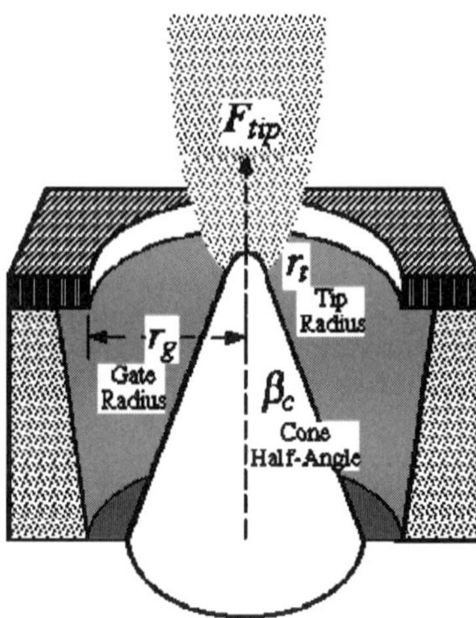

Fig. 1 Field emission tip and gate electrode structure.

State-of-the-art 1/8 in. hollow cathodes consume ~7 W and 0.1 mg/s of xenon for ~0.1 A.* Filament cathodes can also be used with Hall and ion thrusters, and field emission electric propulsion (FEEP) systems, however, they consume significant power, 0.2–10 W/mA. Thermionic cathodes have demonstrated 1 mA at 1.5–2 W.[4]

A field emission cathode is a promising alternative to thermionic cathodes. Thermionic cathodes require high operating temperatures for electrons to possess enough energy to overcome the potential barrier between the emitter and vacuum. Electrons tunnel out of FE cathodes when the potential barrier is lowered by electric fields in excess of 10^7 V/cm. This electric field strength is achieved with applied voltages less than 100 V because the cathodes are microfabricated with micron scale electrode and insulator dimensions and local field enhancement. Molybdenum field emitter array (FEA) cathodes have demonstrated performance of 100 mA with less than 1 mW consumed through the gate electrode.† FEA cathodes are very efficient and easily scalable in size and power. System complexity is also reduced by using field emission (FE) cathodes because they do not require heaters or propellant. Field emission cathodes are natural candidates for the electron sources in small and microscale electric propulsion systems.

Spindt-type FEA cathodes geometrically enhance the applied electric field. Electric fields at the emitting tips are typically greater than 4×10^7 V/cm for field emission. The field emission tip configuration that is typically employed is shown in Fig. 1; dimensions and materials are processor dependent. Microfabrication techniques are applied to deposit and etch insulating and conducting films and arrays of microcones on silicon wafers. Tip and gate aperture radii are on the order of 1 and 100 nm, respectively.[5] Packing densities greater than 10^8 tips/cm^2 have been achieved.[6,7] Materials most commonly used are silicon (Si) and molybdenum (Mo), however, thin carbide and carbon films on the microtips have been used to improve robustness of the tips, work functions, chemical inertness, and

---

*Private communication with Matt Domonkos, Nov. 1998.

†Private communication with Capp Spindt, Dec. 1998.

emission stability.[8–11] Mo cathodes have demonstrated 1 μA at 25 V from an array of 900 tips.[7] The same cathodes with cesiated tips demonstrated 1 μA at 10 V; however, cesium on the tips is typically not stable. Spindt has reported DC field emission up to 100 μA/tip from a 100 tip array.[12] A current of 2.5 μA has been measured from a single Si microtip structure at 25 V.[13] Emission current densities greater than 2000 A/cm^2 and lifetimes greater than 8000 h in ultra high vacuum (UHV) environments have been achieved.[14,15] The efficiency of these devices, defined by the ratio of anode current to anode and gate current, is typically high because the current collected by the gate electrode is commonly only 1/1000 of the emitted current. For optimum cathode performance operating pressures near 10^{-9} Torr are required after careful conditioning and seasoning.[16–18]

An alternative to the microfabricated tip arrays are diamond, diamond-like carbon, or carbon thin-film cathodes that are doped to improve their conductivity. The unique advantage to using doped carbon is its low electron affinity, or negative electron affinity (NEA).[19] These cathodes can emit electrons at macroscopic vacuum electric field strengths that are two orders of magnitude lower than those fields required by Si and Mo Spindt-type cathodes. Carbon-film cathodes are typically fabricated using plasma enhanced chemical vapor deposition (PECVD) or hot filament chemical vapor deposition (HFCVD). Carbon cathodes are more robust than the Spindt-type cathodes, however, to-date the Spindt-type cathodes have demonstrated higher current density and efficiency. A 40-mA current was measured from a 25-mm carbon-film cathode with an extraction field of 6.7×10^4 V/cm.[20] An emission current density of 100 mA/cm^2 was also achieved with an extraction field of less than 1.2×10^5 V/cm. Blyablin et al. demonstrated 500 mA/cm^2 from diamond film cathodes.[21] Cathodes with diamond grit in microfabricated gated structures have achieved 100 mA/cm^2 at 20 V; however, a significant portion of the current was collected by the gate electrode.[19]

The compatibility of EP and FE cathode systems is challenged by lifetime and space-charge current limitations. The successful integration of FE cathodes and EP systems requires cathodes capable of approximately 100 mA/cm^2 for 6000–10,000 h (with less than a 10% permanent decrease in current) in EP-system environments. Space-charge–current limited emission in electric thruster environments is addressed in Refs. 13 and 22. This chapter focuses on lifetime limitations induced by an electric thruster environment. In higher-pressure environments, similar to EP-system environments, cathode lifetime can be shortened by arcing between the gate electrode and tips and ion bombardment that sputters the cathode tip structure. Catastrophic arcing events can be prevented by using current limiting structures such as resistive layers,[23] field effect transistors (FET),[24] and vertical current limiting (VECTL)[25] architectures. Robust materials can be used to improve cathode sputter resistance and resistance to oxidation.[8,26,27]

Tip sputtering will be caused by two populations of ions. One population is created between the tips and gate electrode by the emitted electrons. The flux of this population of ions depends on the electron current density, electron energy, and facility pressure. The second ion population originates near the thruster where charge-exchange (CEX) collisions occur between beam ions and neutrals. The trajectories of beam ions will not intersect the cathode, however, they will create slow moving CEX ions that may be accelerated toward the cathode. The CEX ion current density to the cathode region of a 70-mm Hall thruster operating at 1.4 kW, 43 sc-cm of xenon (Xe), and 65 mN of thrust was determined to be 0.02 mA/cm^2

(Ref. 32). The CEX ion current density was measured by a Faraday probe flush with the physical exit plane of the thruster with a chamber pressure of 2×10^{-5} Torr. The neutral particle pressure measured by a neutral particle flux (NPF) probe[28] in the cathode region of the thruster was also approximately 2×10^{-5} Torr. These measurements are sensitive to facility pressure and thruster operation. Because the initial FE cathode and thruster experiments will occur in a similar facility and with a similar thruster, these estimates were applied to investigate the effect of thruster environment on FEA cathode performance and lifetime. A range of pressure and CEX ion current density values around these estimates will also be investigated to represent the cathode environments of small and mesoscale Hall and ion thrusters that have not yet been characterized.

FEA cathodes have demonstrated impressive performance in elevated pressure environments when start-up occurs at $\sim 10^{-9}$ Torr and the pressure is slowly increased.[29] Spindt et al.[29] showed that Mo FEA cathodes could operate in $\sim 10^{-5}$ Torr of O_2, Ar, He, H_2, and N_2 for hundreds of hours. During these exposure tests, the emission current dropped by about an order of magnitude, if at all. When the chamber was returned to the original UHV environment, the cathodes demonstrated full recoveries. A Mo FEA cathode was operated for over 300 h in air at 10^{-5} Torr. The current dropped by one order of magnitude during this exposure to air relative to the original value in UHV, but returned to its initial value when the pressure returned to UHV. The original current in this experiment was 20 μA, and the anode voltage was 6 kV. With 300 V on the anode, similar results were obtained. The results of another exposure test at 10^{-6} Torr of hydrogen showed a current increase by more than one order of magnitude with 300 V on the anode. When the hydrogen was removed from the vacuum chamber, the current returned to its original value. During exposure tests in He, Ar, and neon (Ne) environments with 300 V on the anode, no change in current was observed for pressure increases up to 10^{-5} Torr. During those experiments, air and hydrogen changed the work function of the cathode surfaces, whereas noble gas environments did not. During glow discharge cleaning of these cathodes at ~ 1 Torr of H + 10% Xe with 275–450 V on the anode, the emitter tips were destroyed and shorted to the gate electrodes. However, the same experiments performed with 10% Ne instead of Xe resulted in a cathode performance improvement.

Si FEA cathodes have demonstrated performance responses to increases in pressure that were similar to the experimental results obtained with Mo cathodes.[30,31] During exposures to O_2 and N_2 environments up to 10^{-6} Torr, with 50 μA of initial current, the current dropped by about an order of magnitude within a few seconds before stabilizing, and then returned to the original current when the chamber pressure returned to UHV. Exposure to He at 10^{-6} Torr resulted in a 25% drop in current in 1000 s. In all of these experiments, temporary work function changes resulted in temporary performance reductions. There was no permanent damage done to the cathodes in the forms of surface oxidation or tip sputtering.

Carbide cathodes have demonstrated surprising resistance to sputtering by ion bombardment. Experiments with single ZrC and HfC tips with operating voltages exceeding 1 kV showed that emission continued for several minutes at 10^{-4} Torr of Ar and O_2.[32] Although a glow discharge operating mode was attained, the tips were not destroyed during operation for a few minutes in this regime. The results of these experiments in elevated pressure environments are very promising, especially because some of the cathodes seemed to be undamaged during operation.

The performance of thin-film cathodes is not as sensitive to sputtering by ion bombardment because the critical FE surface is at the interface of the substrate and the film, and hence is protected by the film itself.[19,33,34] Diamond cathodes are very resistant to oxidation while some carbon films have demonstrated performance degradation in oxygen environments. Thin-film carbon cathodes have demonstrated half lives of 7000 h in sealed and gettered glass envelopes, where the half-life is defined as the time during which the current dropped by one half of its original value. Accelerated decay rates are observed when the cathodes are operated in oxygen and water environments. Elevated hydrogen environments seem to have no effect on the cathode performance. The diamond grit cathodes have operated at 400 V at pressures greater than 750 mTorr of nitrogen without being damaged.[19] Although performance degradation has been observed for the diamond and carbon-film cathodes while operating in oxygen-rich environments,[34] it was anticipated that the cathode performance would be much more stable in the inert gas environment of a xenon plasma thruster.

In the following sections of this chapter, the performance issues associated with a cathode operating with a Hall thruster are discussed. Models of tip sputter rates and cathode performance were used to predict the effect of xenon environments on cathode performance degradation. Experimental results are presented from cathode testing in xenon environments. Modeling results are compared with experimental results. The models were used to study the cathode performance degradation in the more complex thruster environments and determine the operating voltage limitations. This study concludes with a discussion on the compatibility limitations of FEA cathodes and electric propulsion systems, required cathode configurations, and recommendations for future research.

II. FEA Cathode Performance Modeling

To expedite FEA cathode development for EP systems and reduce development costs, a FEA cathode performance decay model was developed and used in conjunction with experimentation. The model discussed in this section incorporates three previously existing models with slight modifications that predict cathode emission currents, sputter yields of cathode materials by Xe ions, flux of ions to the surface and removal rate of material, and an additional model to predict changes in tip radius from ion bombardment. With these models and experimental results, the lifetime of Si and Mo FEA cathodes can be predicted in Xe environments of EP systems.

The inputs to this performance model are I–V data obtained in UHV environments, cathode work function ϕ_w, half-cone angle of the microtip structures β_c, the energy threshold for sputtering cathode material by ion bombardment E_{th}, gate aperture radius r_g, and number of tips N_{tips}. The I–V data are used in a model developed by Jensen to estimate the effective tip radius r_t and distribution of radii in the array.[35] The Jensen model is also used to predict the emission current from an array. To determine current changes in time caused by changes in tip radius, the change in tip radius in time must be determined. The tip radius increases in time are approximated by assuming that each layer removed from the emitting area of the tip increases the tip radius by the atomic spacing in the tip material:

$$\frac{dr_t}{dt} = \frac{d}{t_e} \tag{1}$$

where the atomic spacing in the material is d (Å), and a monolayer of material is removed from the tip emitting area with an effective tip radius r_t (Å) in t_e (s). A model developed by Brodie was used to determine t_e.[36] Material sputter yields were determined using the model developed by Yamamura et al.[37] Tip radii are represented statistically by an effective tip radius and spread in tip radius across an array Δs. The radius of the sharpest tips in the array will change at the fastest rate so that the uniformity in the array improves in time. The change in the spread in tip radii in time, shown in Eq. (2), is a function of the change in effective tip radius in time and the maximum tip radius in the array:

$$\Delta s(t) = \frac{r_{t\,\max}}{r_t(t)} - 1 \tag{2}$$

$I[r(t), \Delta s(t)]$ can then be determined using Jensen's emission model for FEA cathodes.

The following three sections present the details of the Jensen model to predict cathode parameters from I–V data and cathode emission current, the Brodie model to predict the removal rate of material from the tip emission area, and the model developed by Yamamura et al.[37] which predicts the sputter yields of Xe ions bombarding Si and Mo targets.

A. Field Electron Emission Model

The analytical and statistical model used to predict the emission current from a field emission tip and gate electrode structure as shown in Fig. 1 was developed by Jensen. The model is described in much more detail in another paper.[35] The emission model takes a statistical approach to estimating the current emitted by FEA cathodes because the emitting area of a tip, current distribution in that area, number of tips contributing to the measured current, and current per tip are impossible to determine exactly for any FEA cathode. Work function, tip radius, emission cone half-angle, and cone structure half-angle assume a range of values in a FEA cathode. Because each tip is processed in parallel and exposed to the same environments, the cone structure half-angles and work functions for all of the tips in the array should be very similar. The experimental results discussed previously showed that the emitting area of the tip depends on the cleanliness of the tip and its radius of curvature. Other studies have shown that field emission from a cathode can originate from protrusions of atomic dimensions, even on a cone structure with a radius of curvature of several nanometers.[9] Emission from a cathode microtip is not uniform within the emitting area.[32,38] The emitting area of tips in an array is then best represented statistically by an effective tip radius and distribution of tip radii in the array of emitters. These parameters can be determined from experimentally acquired I–V data and the Jensen FEA current model, which has been successfully used to model the performance of Si and Mo FEA cathodes.[39] Jensen models the tip of an array element as a hyperbolic shape; the field (eV/Å) at the apex of the tip can then be approximated by

$$F_{\text{tip}} \approx \beta_g V_g \tag{3}$$

where V_g is the voltage on the gate electrode with respect to the emitting cone. The electric field at the tip surface increases with decreasing tip radius r_t and half-cone angle β_c by the field enhancement factor β_g (1/Å),

$$\beta_g \approx \left(\frac{\pi}{\ell n(kr_g/r_t)} - \tan^2 \beta_c \right) \frac{1}{r_t} \tag{4}$$

where

$$k \approx \frac{1}{54}\left(86 + \frac{r_g}{r_t} \right) \cot(\beta_c) \tag{5}$$

With these relationships, F_{tip} (eV/Å) can be approximated with an analytical expression. The current density is approximated by the Fowler–Nordheim (F–N) equation for J_{FN} (e/fs-Å^2),[38]

$$J_{\text{FN}}(F) = a_{\text{fn}} F^2 \exp(-b_{\text{fn}}/F) \tag{6}$$

where the Fowler–Nordheim coefficients, b_{fn} (eV/Å) and a_{fn} (e/eV^2 fs), are given in atomic units by

$$b_{\text{fn}} = 0.642\phi^{3/2} \tag{7}$$

$$a_{\text{fn}} = (116\phi)^{-1} \exp\left(14.3994 b_{\text{fn}}/\phi^2\right) \tag{8}$$

When the current density is integrated over the surface of the emitter, the tip current in microamperes can then be determined by

$$I_{\text{tip}}(V_g) = b_{\text{area}} J_{\text{FN}}(F_{\text{tip}}) \tag{9}$$

The area factor b_{area}(Å^2) is the ratio between the current from the tip and the current density on-axis, and depends upon the magnitude of the gate voltage as

$$b_{\text{area}} = 2\pi r_t^2 \cos^2(\beta_c) \left(\frac{F_{\text{tip}}}{b_{\text{fn}} + \sin^2(\beta_c) F_{\text{tip}}} \right) \tag{10}$$

The current emitted from an array of tips can be defined by

$$I_{\text{array}}(V_g) = N_{\text{tips}} \Sigma(\Delta s, V_g) b_{\text{area}}(V_g) J_{\text{FN}}[F_{\text{tip}}(V_g)] \tag{11}$$

where Σ is the distribution factor associated with a spread in tip radii Δs. The tip radii are distributed according to $r(s) = r_t(1+s)$, where $0 \le s \le \Delta s$. For a single tip emitting, $\Delta s = 0$. The field for a tip of radius $r(s) = r_t(1+s)$ is then $F(s) \approx (1 + c_o s) F(0)$, where $F(0) = F_{\text{tip}}$, and

$$c_o = \frac{1}{\pi \beta_g r_t} \left(\tan^2(\beta_c) + \beta_g r_t \right)^2 - 1 \tag{12}$$

The natural log of $I_{\text{tip}}(s)/I_{\text{tip}}(0)$ is approximated as a linear function in s and results in

$$\Sigma(\Delta s, V_g) = \frac{\exp(\Delta s b) - 1}{\Delta s b} \tag{13}$$

$$b = \frac{c_o b_{\text{fn}} + (3c_o + 2)\beta_g V_g}{\beta_g V_g} - \frac{c_o \beta_g V_g}{b_{\text{fn}}} \tan^2(\beta_c) \tag{14}$$

Invoking a Legendre least square analysis on

$$I_{\text{array}} \cong A_{\text{FN}} V_g^2 \exp\left(\frac{-B_{\text{FN}}}{V_g}\right) \tag{15}$$

B_{FN} and A_{FN} are

$$B_{\text{FN}} = \frac{b_{\text{FN}}}{\beta_g} + \frac{\delta^2 + 2x_o}{x_o^3} \tag{16}$$

$$A_{\text{FN}} = N_{\text{tips}} 2\pi r_t^2 \cos^2(\beta_c) a_{\text{fn}} \frac{\beta_g^3}{x_o b_{\text{fn}}} \exp\left(2 + \frac{4}{3}\frac{\delta^2}{x_o^2}\right) \Sigma\left(\Delta s, x_o^{-1}\right) \tag{17}$$

where

$$\delta = \frac{V_{\text{rf}}}{V_{\text{pk}}(V_{\text{pk}} - 2V_{\text{rf}})} \tag{18}$$

$$x_o = \frac{V_{\text{pk}} - V_{\text{rf}}}{V_{\text{pk}}(V_{\text{pk}} - 2V_{\text{rf}})} \tag{19}$$

$$V_{\text{rf}} = \frac{V_{\max} - V_{\min}}{2} \tag{20}$$

Experimental data are used to estimate r_t and Δs. If I–V data are presented on a Fowler–Nordheim plot ($\ln(I/V^2)$ vs 1/V), $A_{\text{FN}}(A/V^2)$ is the y-intercept and $B_{\text{FN}}(V)$ is the slope of the curve. From these parameters and an estimated work function, effective tip radius r_t and distribution parameter Δs can be determined. Once the effective tip radius and spread in tip radius throughout the array of emitters are known, the current from the array can be calculated. In an array, $\Delta s = r_{t\,\max}/r_t - 1$, where $r_{t\,\max} = r_t(1 + \Delta s)$ is the maximum radius and r_t is the minimum radius. Sputtering of the cathode tips will decrease the spread in tip radii to improve the uniformity of the array. It is assumed that $r_{t\,\max}$ is constant during the exposure and that Δs changes in time with Xe exposure as r_t changes, as Eq. (2) shows. The change in effective tip radius in time depends upon the removal rate of material from the tip emission area. The time required to remove a monolayer of material from the tip emission area is determined using the tip sputtering model discussed in the next section.

B. Tip Sputtering Model

Brodie developed the field emission cathode tip sputtering model described in this section. It is described in more detail in another paper.[36] This model is capable of estimating sputter rates of microtips by ions generated by the emitted electrons. It is used here to estimate the change in tip radius in time when operating in Xe thruster environments. Brodie used average sputter yields and ionization cross-sections in his calculations for high operating voltages, >100 V. He also only considered the effect of singly charged ions. This study is primarily investigating low voltage operation where ion and electron energies are low and ionization cross-sections Q and sputter yields Y are exponentially dependent on ion and electron energies. Sputter yields and ionization cross sections are very sensitive to energy at the low operating voltages employed in this study; therefore, average values were not used in the calculations of the removal rate of material from the emission area of the tips. Also, Xe^+, Xe^{++}, and Xe^{+++} were considered in the tip sputtering model in this investigation. The time t_e required to erode a single layer of atoms from the emission area A_t is given by

$$t_e = \frac{A_t}{d^2 n_s} = \frac{2\pi r_t^2(1 - \cos\sigma)}{d^2 n_s} \tag{21}$$

where $d = 0.416$ nm for the atomic spacing of evaporated Mo and $d = 0.313$ nm for Si. The removal rate of material from the tip emitting area is n_s (atoms/s). An experimental investigation of single carbide tips showed that σ, the half-cone emission angle, ranged from 8 to 44 deg for eight tips with an average value of 28 deg.[32] A half-cone emission angle of 37 deg is used because it significantly simplifies the mathematics of the sputter rate calculations and is a reasonable estimate to represent tip half-cone emission angle in an array. The half-cone emission angle varies significantly for tips in an array. Gomer claims that the current density drops to approximately 10% of the value on axis by 45 deg off axis.[38]

The removal rate of material from the tip emitting area depends on the flux of ions, energy of the ions, and the dependence of material sputter yields on ion energy. Both the ion population created near the tips and the CEX population created near the thruster contribute to the flux of ions to the tips. Xe^+ and Xe^{++} were included in the CEX ion flux and Xe^+, Xe^{++}, and X^{+++} were considered in the population of ions created locally. The number of ions striking the emitting area per second n is given by

$$n = n_{i\mathrm{CEX}} + n_{ii\mathrm{CEX}} + n_i + n_{ii} + n_{iii} \tag{22}$$

$$n = \frac{A_t}{e}\left(J_{i\mathrm{CEX}} + \frac{1}{2}J_{ii\mathrm{CEX}}\right) + \int_{r_i}^{r_m} \frac{I_e}{e} N[Q_i(V_r) + Q_{ii}(V_r) + Q_{iii}(V_r)]\,\mathrm{d}r \tag{23}$$

and the number of atoms sputtered from the emitting area per second, n_s, is given by

$$n_s = \frac{A_t}{e}\left(J_{i\mathrm{CEX}} Y(V_g + V_c) + \frac{1}{2}J_{ii\mathrm{CEX}} Y[2(V_g + V_c)]\right)$$

$$+ \int_{r_s}^{r_m} \frac{I_e}{e} N[Q_i(V_r)Y(V_r) + Q_{ii}(V_r)Y(2V_r) + Q_{iii}(V_r)Y(3V_r)]\,\mathrm{d}r \tag{24}$$

The field-emitted electron current is I_e and the number of molecules per cubic centimeters in the chamber is $N = 3.55 \times 10^{16} P$ molecules/cm^3, where P is the pressure in Torr at room temperature. The cross-sections for single, double, and

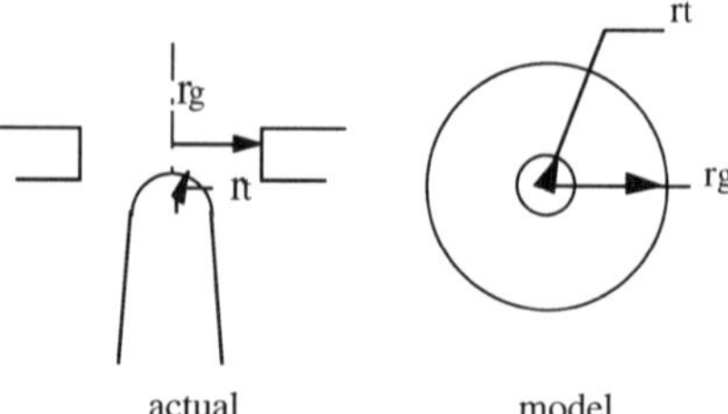

Fig. 2 Actual cathode configuration and the configuration used in the model to predict potential field near the tip and electric field at the tip

triple ionization of the ambient gas by electrons of energy eV_r is $Q_i(V_r)$, $Q_{ii}(V_r)$, and $Q_{iii}(V_r)$, respectively. The sputter yield in atoms per ion of energy eV_r is $Y(V_r)$. The CEX current density for single ions is $J_{i\text{CEX}}$ and for double ions is $J_{ii\text{CEX}}$.

In this cathode sputtering model, the cathode configuration is approximated as an isolated sphere of radius r_t at ground potential, as shown in Fig. 2, and the gate electrode as a concentric sphere with radius r_g. The radial potential distribution on the axis of the tip is V_r, and can be approximated by

$$\frac{V_r}{V_g} = \left(\frac{r_t}{r} - 1\right)\left(\frac{r_t}{r_g} - 1\right)^{-1} \tag{25}$$

This simple concentric spheres model gives remarkably good results compared with exact predictions obtained by computer solutions of Laplace's equation and geometric boundary conditions because, close to the emitting area of the tip, the field lines are radial and fall off inversely as the square of the distance.[40] The Jensen model more accurately predicts the electric field at the tip surface, which is more important when calculating the field emission current from a tip. However, the Brodie model is sufficient for predictions of the potential distribution near the tip to determine ion production near the cathode. For this analysis, the electric field and current over the emitting area of the tip are assumed to be uniform.

The ion flux to the tip depends on the local gas temperature and the positions at which ionization occurs. If an ion is formed with initial cross radial velocity v_t corresponding to a temperature of V_t (eV), then only the ions formed within a radial distance r_m will strike the emitting area of the tip, where

$$r_m = r_t\left(\frac{V_g}{V_t}\right)^{\frac{1}{3}} \tag{26}$$

The ion temperature in the farfield plume is estimated to be 0.03 eV. This assumes that the ions are in equilibrium with their surrounding neutral particles that are at the temperature of the vacuum chamber walls. At a fixed pressure, the flux of ions to the emitting area of the tips will decrease if a higher ion temperature is assumed. Therefore the assumption that $V_t = 0.03$ eV will provide an upper limit on dr_t/dt.

The significant radial positions used in this model are shown in Fig. 3. The radial position at which the energy of the emitted electrons reaches the threshold of ionization [$V_r(r_i) = V_i =$ ionization threshold] is r_i. The radial position from which an ion, if formed, gathers sufficient energy to reach the threshold for sputtering

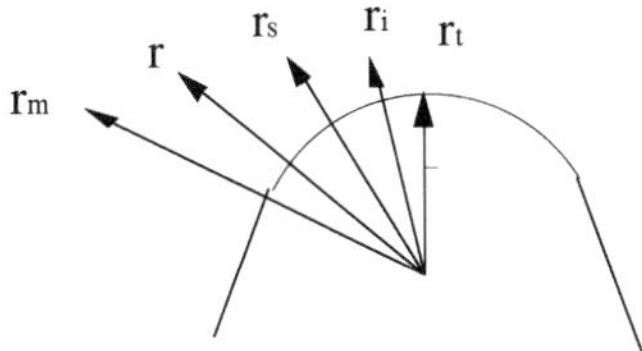

Fig. 3 Significant radial positions used in the tip sputter rate model.

when it strikes the tip surface [$V_r(r_s) = E_{\text{th}}$ = sputtering threshold] is r_s. The radial position beyond which the average ion formed will not strike the emitting area of the tip is r_m.

The calculations made by Brodie employed average values of V_r and then used corresponding Q and Y values. In this study, data curves and interpolation were used to estimate these parameters because these parameters are exponentially sensitive to ion and electron energies in the energy range relevant in this investigation. The integral for n_s was then evaluated numerically for hundreds of points. Ionization cross-sections used in the calculations were obtained from Krishnakumar and Srivastava.[41] Interpolation was used to determine values between data points. Sputter rates for xenon ions on molybdenum and silicon targets were used in the model because the thruster environment is primarily Xe. The relationship between sputter yield and ion energy is discussed in the next section.

C. Sputter Yield Model

Sputter yield and energy threshold values used in this study for Xe ions bombarding Mo and Si targets with a zero-deg angle of incidence were calculated using a model developed by Yamamura et al.[37] using fit parameters Q from Matsunami et al.[42] The electric field within r_m of the tip is primarily radial. Therefore the ion incidence angle can be approximated by zero. Because experimental data exist only for ion energies above 100 eV, the accuracy of the model is unknown at energies below this level. The energy of ions bombarding the emitting area of the cathodes in the configurations of interest will primarily be below 100 eV.

In the model developed by Yamamura et al.,[37] the sputter yield is defined as

$$Y(E) = 0.42 \frac{\alpha Q K_1 s_n(\varepsilon)}{U_s[1 + 0.35 U_s s_e(\varepsilon)]} \left[1 - \left(\frac{E_{\text{th}}}{E}\right)^{\frac{1}{2}}\right]^{2.8} \tag{27}$$

and the energy threshold for sputtering is defined by,

$$E_{\text{th}} = \left(1.9 + 3.8 \frac{M_{\text{I}}}{M_{\text{II}}} + 0.134 \left(\frac{M_{\text{II}}}{M_{\text{I}}}\right)^{1.24}\right) U_s \tag{28}$$

where the subscripts I and II refer to the incident and target particles, respectively. Q is an empirically derived parameter for a specific ion and target combination, U_s (eV) is the sublimation energy of the target material, and E (eV) is the ion energy. Q values are presented for different ion target combinations by Matsunami et al.[42]

Table 1 Ion and target material parameters used in the sputter yield model

	Xe	Mo/MoXe	Si/SiXe
M	131.3	95.94	28.10
Z	54	42	14
U_s, eV	—	6.82	4.63
Q	—	0.84	0.78
E_{th}, eV	—	49	91

Other parameters used in the model are

$$\alpha = 0.08 + 0.164\left(\frac{M_{\text{II}}}{M_{\text{I}}}\right)^{0.4} + 0.0145\left(\frac{M_{\text{II}}}{M_{\text{I}}}\right)^{1.29} \tag{29}$$

$$K_1 = 8.478\frac{Z_{\text{I}}Z_{\text{II}}}{\left(Z_{\text{I}}^{2/3} + Z_{\text{II}}^{2/3}\right)^{1/2}}\left(\frac{M_{\text{I}}}{M_{\text{I}} + M_{\text{II}}}\right) \tag{30}$$

$$\varepsilon = \frac{0.03255}{Z_{\text{I}}Z_{\text{II}}\left(Z_{\text{I}}^{2/3} + Z_{\text{II}}^{2/3}\right)^{1/2}}\left(\frac{M_{\text{II}}}{M_{\text{I}} + M_{\text{II}}}\right)E \tag{31}$$

$$s_n(\varepsilon) = \frac{3.44\varepsilon^{1/2}\,\ell n(\varepsilon + 2.718)}{1 + 6.355\varepsilon^{1/2} + \varepsilon(-1.708 + 6.882\varepsilon^{1/2})} \tag{32}$$

$$S_e = k_2\varepsilon^{\frac{1}{2}} \tag{33}$$

$$k_2 = 0.79\frac{(M_{\text{I}} + M_{\text{II}})^{3/2}}{M_{\text{I}}^{3/2}M_{\text{II}}^{1/2}}\frac{Z_{\text{I}}^{2/3}Z_{\text{II}}^{1/2}}{\left(Z_{\text{I}}^{2/3} + Z_{\text{II}}^{2/3}\right)^{3/4}} \tag{34}$$

The parameters used in this investigation for Mo and Si targets bombarded by xenon ions are shown in Table 1. It also shows the E_{th} values predicted by the Yamamura model. Other models predict even higher values for E_{th}, while some experimental measurements extrapolated to low energies yield E_{th} estimates for a Mo target and a Xe ion combination that is 27 eV.[43,44] The relationships between energy and sputter yield determined by these models are shown in Figs. 4 and 5 for Si and Mo, respectively. The predicted sputter yield has been experimentally validated at energies above 100 eV only. The E_{th} values have never been experimentally validated.

III. Cathode Experimental Performance Evaluations

The performance of a Mo cathode from SRI International, Si cathodes from MCNC, and a carbon thin-film cathode from FEPET was evaluated in xenon environments for this study. Experimental results discussed earlier in this chapter showed that the xenon pressure near the cathode of a Hall thruster at 1.4 kW is approximately 2×10^{-5} Torr. Because initial testing with FEA cathodes and EP systems will occur with this or a similar thruster, cathodes were tested in xenon environments with pressures up to 2×10^{-5} Torr. Initial experiments were cautiously

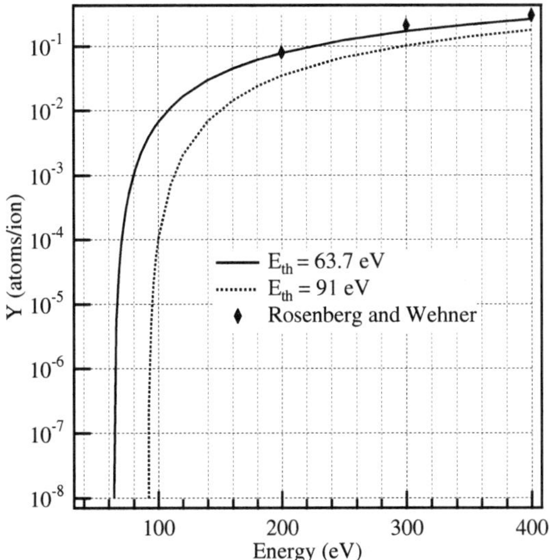

Fig. 4 Sputter yield for a Si target and Xe ions. Theoretical values for two values of E_{th} are determined by the Matsunami and Yamamura models[37] and experimental values are measured by Rosenberg and Wehner.[45]

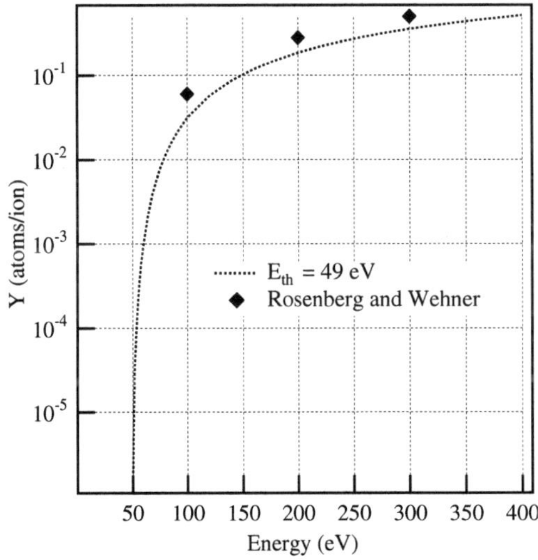

Fig. 5 Sputter yield for a Mo target by Xe ions. Theoretical values are determined by the Matsunami and Yamamura models[37] and experimental values are measured by Rosenberg and Wehner.[45]

conducted at lower pressures. The performance of these cathodes was not optimized in these experiments because of time constraints. Much better performance is expected if the cathodes are conditioned more carefully and operated at higher voltages. These experiments focused on operation near the voltage thresholds for sputtering and, in some experiments, at higher voltages to operate at current levels employed in similar experiments performed elsewhere ($\sim$20 μA).

The objectives of these experiments were to obtain an understanding of how these cathodes respond to Xe environments, under which operating condition performance degradation occurs, and what lifetime can be expected in the environments of Hall and ion thrusters. This section discusses the experiments and results and how the performance models were used to obtain a better understanding of the sputter yields of Mo and Si targets when bombarded by Xe ions. Theoretical performance of FEA cathodes is compared with experimental results.

A. Experimental Apparatus

The test facility consisted of a 0.0664-m^3 vacuum chamber that was pumped by turbomolecular, ion, and sublimation pumps to pressures as low as 7×10^{-11} Torr. Three Keithley 480 picoammeters were used to measure the current emitted through the base and collected by the gate electrode and anode. The picoammeters on the anode and gate electrodes floated electrically to minimize the measurement of stray currents. Pressure, current, and voltage signals were recorded using a National Instruments LabVIEW data acquisition system. Pressure was measured using a Varian UHV-24 nude ionization gauge. The cathodes and the ionization gauge were positioned at the same distance from the xenon inlet orifice and vacuum pumps to minimize pressure gradients between the cathodes and pressure gauge. The electrical configuration employed during testing of the three types of cathodes is shown in Fig. 6. A 10-MΩ resistor was used on the gate electrode for FEA cathodes, and 1 MΩ was used with the carbon film cathode.

B. Silicon FEA Cathodes

The Si cathodes used in these experiments consisted of 16,000 tips with 2 μm apertures on 10 μm centers in chromium gate electrodes. Four cathode arrays were fabricated on a single chip with four gate electrodes and a cathode base common

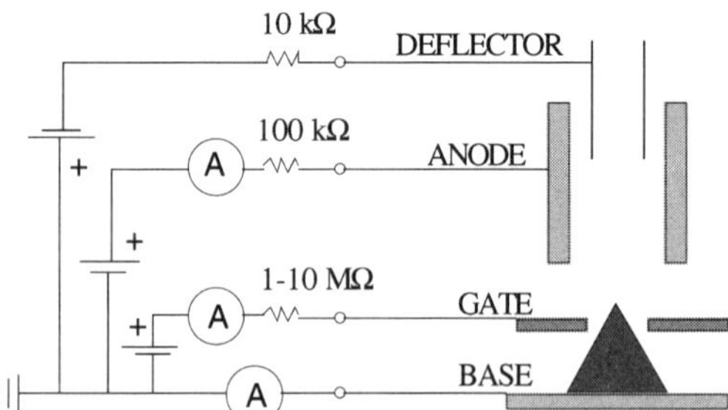

Fig. 6 Electrical configuration of the cathode experiments. The FEPET cathode does not employ the tip configuration.

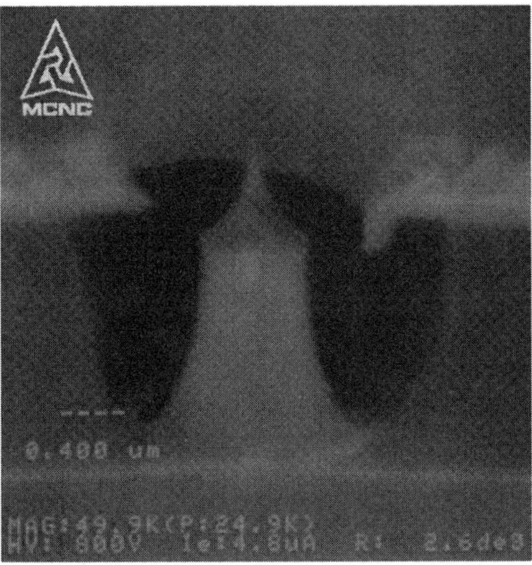

Fig. 7 Single tip in the Si field emission array cathodes fabricated by MCNC.

to each of the arrays. The chip was mounted to a T0-5 head. Each tip in the arrays used in the experiments is similar to the tip shown in Fig. 7.

The first experiment investigated the effect of Xe exposure on the work function of the cathode. The cathode was exposed to Xe at 2×10^{-4} Torr for 1 h to provide a dose of Xe to the surface of the cathode greater than it would experience during the exposure tests with the cathode operating. Data obtained during this and repeated experiments show that Xe coverage did not change the work function of the Si cathode surface.

An experiment similar to those conducted at Microelectronics Center of North Carolina (MCNC) was conducted at the Jet Propulsion Laboratory (JPL) where the cathode was exposed to 7×10^{-6} Torr of Xe with an initial current of 20 μA. The gate voltage in this experiment was 86 V and the anode voltage was 400 V. The Yamamura model predicted that E_{th} for sputtering Si with Xe ions is 91 eV, therefore, sputter damage at this voltage was not expected (it was originally thought that double ions would make a negligible contribution to the cathode performance degradation). The response of the cathode current to the increase in pressure is shown in Fig. 8. During the 2 h of operation in UHV after the exposure test, the cathode current only continued to decay. The experiment was repeated on another cathode with the same configuration. During this exposure the initial current was 22 μA with 75 V on the gate electrode and 500 V on the anode. The same results were obtained. During the 10 h at UHV that followed the second exposure, the current never recovered to the original level. The results suggest that the tips were damaged from ion bombardment. Experiments conducted with anode voltages below 90 V also showed performance degradation caused by sputtering, therefore high-energy ions generated between the gate and anode were not primarily responsible for the tip sputtering observed.

Experiments were then conducted at lower gate voltages to determine the gate electrode threshold for sputtering. Irreversible performance degradation was observed during exposure to xenon at 2×10^{-5} Torr with gate voltages of 75 (V_a at

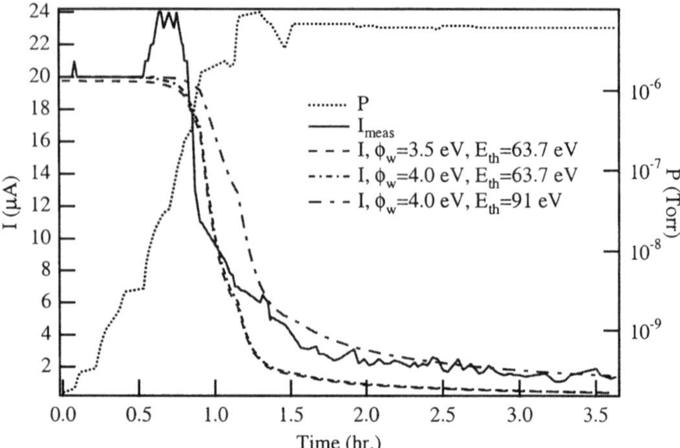

Fig. 8 Measured and theoretical Si cathode current response to an increase in Xe pressure up to 7 × 10^{-6} Torr with V_g at 86 V and V_a at 400 V.

100 V), 70 (V_a at 70 V) and 60 V (V_a at 80 V). The cathode performance with V_g at 70 V and V_a at 70 V is shown in Fig. 9. With 60 V applied to the gate electrode, no performance degradation was observed during a 1-h exposure to 2 × 10^{-5} Torr of Xe with 70 V applied to the anode; the current actually increased. The results of this experiment are presented in Fig. 10.

The performance and tip sputtering models were used to estimate the performance decay of the cathode in the xenon environment. Figures 8 and 9 show the measured and theoretical cathode performance degradation. The results are not sensitive to changes in work function in the range 3.5–4.5 eV, as shown in both figures. The sensitivity of the results to changes in work function is also shown in Fig. 9. This result is significant because there is some uncertainty in the value

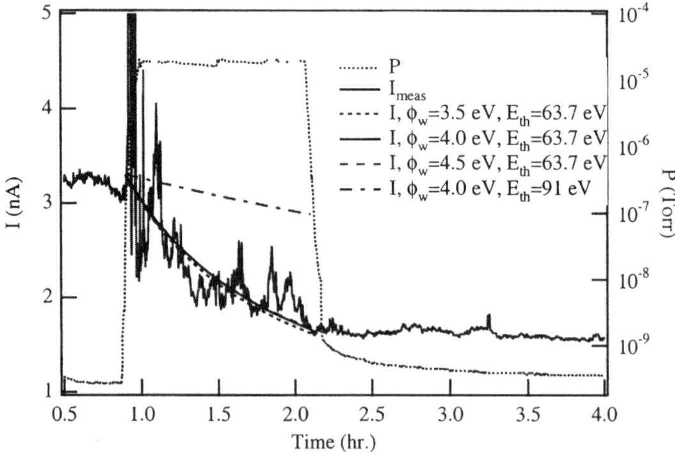

Fig. 9 Measured and theoretical Si cathode current response to an increase in Xe pressure up to 2 × 10^{-5} Torr with V_g and V_a at 70 V.

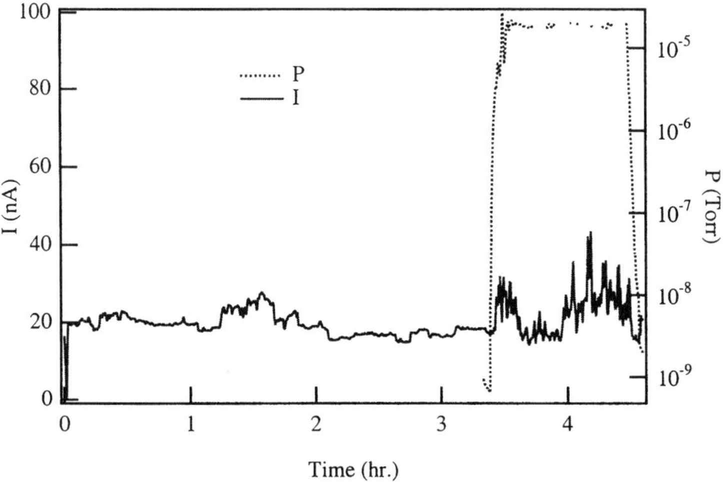

Fig. 10 Measured Si cathode current response to a 1-h Xe exposure with V_g at 50 V and V_a at 70 V.

of the work function, which depends on surface contamination, doping, and processing. The contribution to performance degradation by the Xe^{+++} is relatively insignificant. Xe^{++} are responsible for most of the performance degradation at gate electrode voltages below approximately 85 V. The relationship between the sputter yield and ion energy is required to determine the performance decay rate. To achieve good correlation between theory and experiment, it was concluded that E_{th} is approximately 63.7 eV, instead of 91 eV. Relationships between sputter yield and ion energy are shown in Fig. 4 for E_{th} values of 91 and 63.7 eV. The Yamamura and Matsunami models predict sputter yield and energy relationships for a wide range of ion target combinations that are fit through data that exist only above 100–200 eV. The accuracy of the sputter yield values below this energy is unknown. Sputter yield values calculated using the lower E_{th} value fit the experimental data taken at 200, 300, and 400 eV by Rosenberg and Wehner[45] better than the original estimate of 91 eV made by the Yamamura model.

C. Molybdenum FEA Cathodes

The Mo cathode consisted of an array of 50,000 tips with 0.9-μm gate aperture diameters and 4-μm centers fabricated on a 2000 Ohm-cm silicon wafer. A single element of the array is shown in Fig. 11. The high resistivity wafer protects the cathodes against arc damage by resistively regulating the difference between the tip and gate electrode potentials.

The first experiment conducted on this cathode investigated the effect of xenon exposure on the work function. The cathode was exposed to 2×10^{-4} Torr of xenon for 1 h to provide a dose of Xe to the surface of the cathode greater than it would experience during the exposure tests with the cathode operating. I–V traces were taken at UHV immediately before and after the exposure. The I–V trace taken after the exposure showed slightly better performance than before the exposure. The results of this experiment suggest that the work function did not

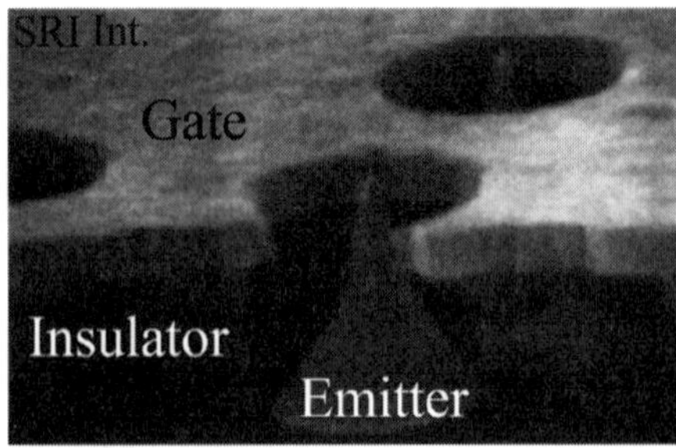

Fig. 11 Single tip in Mo field emission array cathodes fabricated by SRI International.

increase because of the adsorption of Xe or other contaminants. Any performance degradation observed during Xe exposure can therefore be attributed to changes in the geometrical configuration of the cathode.

For comparison, an exposure test similar to those conducted at SRI International was conducted at JPL at a gate voltage of 65 V and anode voltage of 80 V with ~18 μA for 1 h in 2×10^{-6} Torr of Xe. Figure 12 shows the current response to the xenon environment. The current dropped to one-third of its original value in 1 h. No current recovery was observed in 13 h at UHV that followed the Xe exposure, unlike the experimental results obtained at SRI International in Ar and air environments at even higher voltages.

An experiment was conducted to determine the gate voltage that corresponds to the onset of cathode sputter damage. The anode voltage in this experiment was 100 V, and the Xe pressure was 2×10^{-5} Torr. The gate voltage was increased in 1-V increments every 10 min, starting at 50 V. Performance decay caused by tip sputtering was observed during 10 min when the gate voltage reached 58 V, as shown in Fig. 13. This cathode had already been exposed to Xe environments for several hours and had suffered significant sputter damage. It is expected that

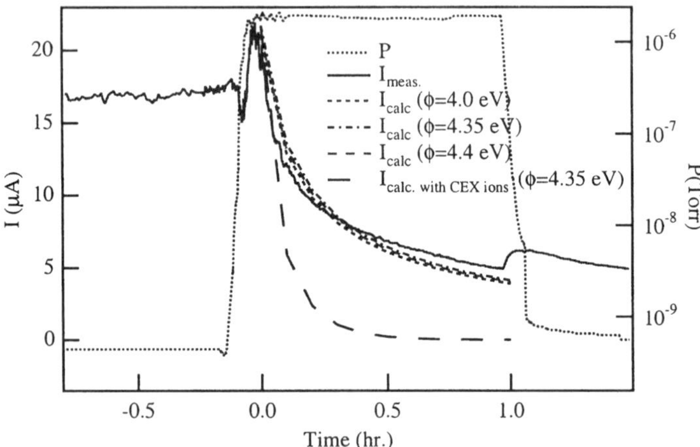

Fig. 12 Measured and theoretical Mo cathode current response to an increase in Xe pressure up to 2×10^{-6} Torr with V_g at 65.6 V and V_a at 80 V.

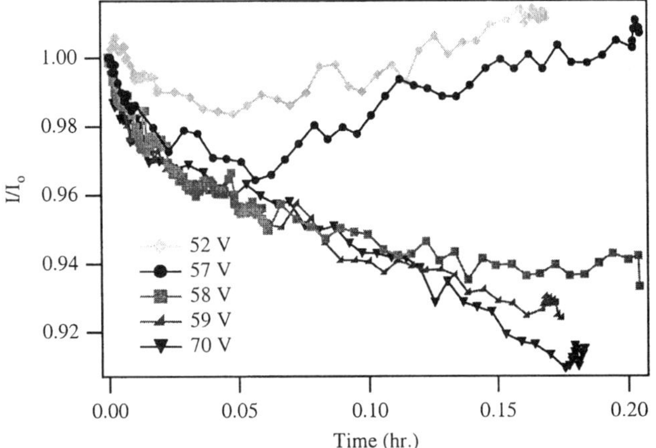

Fig. 13 Measured Mo cathode current response to increasing gate voltage with V_a at 100 V and 2 × 10^{-5} Torr Xe pressure, showing the gate voltage that corresponds to the onset of sputter damage.

other cathodes with the same configuration will respond somewhat differently. A cathode with sharper tips and the same number of tips emitting should show performance decay during 10 min at slightly lower voltages.

Several experiments were conducted with the gate voltage at 50 V. The maximum energy of Xe^+ and Xe^{++} in this configuration is 46 eV and 92 eV, respectively. In an exposure test that lasted for 5 h with the gate electrode voltage at 50 V and the anode voltage at 60 V, the current did not decrease below its original value until the fourth hour of the exposure. The results of this experiment are shown in Fig. 14. In an experiment at the same pressure with V_g at 50 V and V_a at 100 V, the current only increased during a 2-h exposure test. These results are shown in

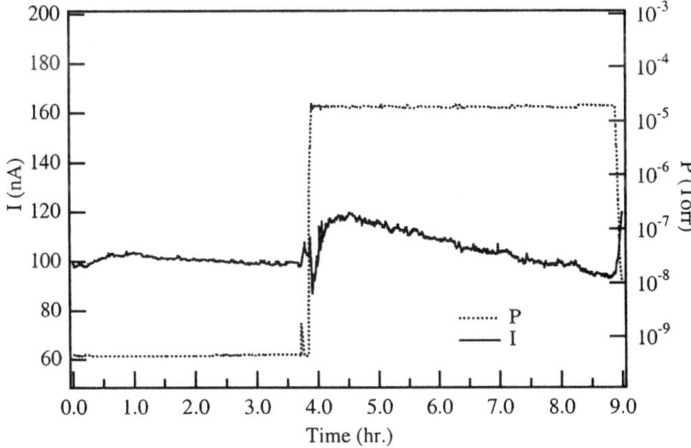

Fig. 14 Measured Mo cathode current response to 2 × 10^{-5} Torr of Xe for 5 h with V_g at 50 V and V_a at 60 V.

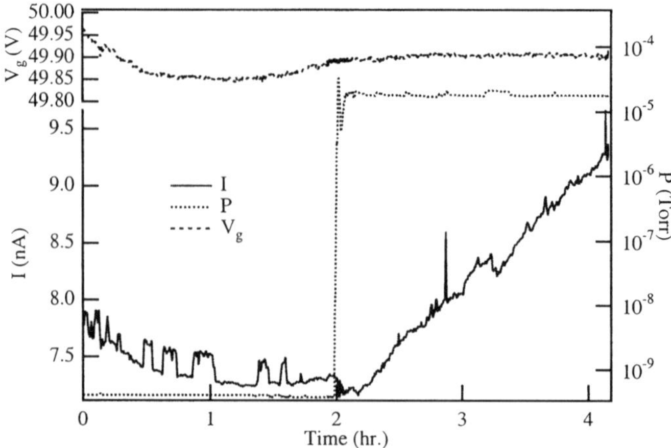

Fig. 15 Measured Mo cathode current response to 2 × 10^{-5} Torr of Xe for 2 h with V_g at 50 V and V_a at 100 V.

Fig. 15. Similar experiments yielded the same results. With the gate electrode at 50 V, the anode voltage was incrementally increased from 50 V to 100 V. At 70 V the current began to increase. This effect could be caused by defocusing of the ion beam impinging on the tips and tip sharpening.[32]

Experimental and theoretical data are compared in Fig. 12. Calculated performance degradation caused by Xe^+ and Xe^{++} created locally is presented as I. Xe^{+++} were not considered in the calculations because the ionization potential of Xe^{+++} is ~70 eV, greater than the operating voltage of the cathode. Xe^{++} were primarily responsible for the performance degradation observed. I–V data taken before the exposure were used with the Jensen model to estimate the effective tip radius r_t and a spread parameter Δs. The parameters used in the models are displayed in Table 2. Using the E_{th} value suggested by the Yamamura model, 49 eV, leads to good correlation between experimental and theoretical performance degradation results.

The work function of Mo is typically assumed to be 4.0–4.4 eV. The modeling results in Fig. 12 show that the cathode performance degradation is

Table 2 Cathode parameters before (I) and after (II) Xe exposure obtained from I–V data and predicted (III) by the performance degradation model

	I	II	III
ϕ_w, eV	4.35	4.35	4.35
βc, rad	0.26	0.26	0.26
r_g, Å	4500	4500	4500
B_{FN}	819	952	—
A_{FN}	1.3 × 10^{-3}	2.4 × 10^{-3}	—
r_t, Å	43.6	53.5	51.5
Δs	50	42.5	42.2

not very sensitive to the assumed work function within the range of possible values.

Table 2 shows the cathode parameters before (I) and after (II) the exposure, as extracted using the Jensen model and I–V trace. Table 2 also shows the cathode parameters determined by the performance degradation model (III), using the initial parameters (I). Data in Table 2 show that the Xe ion bombardment caused an increase in r_t and a decrease in Δs; during the ion bombardment, the tip radii increased and the uniformity in tip radii improved.

The contribution of the CEX ions was also considered in the cathode performance decay rate, and is shown in Fig. 12. For these calculations it was assumed that the gate electrode will be 20 V below the plasma potential so that it will not collect electrons from the thruster discharge. Because the CEX ions are originating at plasma potential, they are accelerated through 20 V more than the potential between the gate electrode and cathode base, 85.6 V. The total CEX ion current density J_{CEX} was assumed to be 0.02 mA/cm^2, as measured.[32] It has been shown that 11% of the Xe ions are doubly ionized in a SPT-100 Hall thruster operating at 1.35 kW and 1600 s specific impulse.[46] Making this assumption about similar Hall thrusters, and assuming a specific impulse of 1600 s to determine singly[47] and doubly[48] charged CEX ion cross-sections, it can be shown that 1.5% of the total CEX ion current density measured consists of $\mathrm{Xe}_{\mathrm{CEX}}^{++}$. The current decay rate from the local and CEX ion populations is shown in Fig. 12; the CEX population of ions significantly increased the performance degradation rate. At the operating voltage employed, the $\mathrm{Xe}_{\mathrm{CEX}}^{+}$ were responsible for the majority of the damage done to the cathode. At lower voltages, $\mathrm{Xe}_{\mathrm{CEX}}^{++}$ dominate in this process. The CEX ions will also sputter material off of the sides of the cathode cones and onto the insulator walls between the tips and gate electrode, eventually shorting them together. With a gate voltage of 65 V, this short could be generated within several minutes.

D. Carbon-Film Cathodes

The cathode used in these experiments, which was fabricated and provided by Field Emission Picture Element Technology (FEPET), consists of a thin carbon film deposited on a ceramic substrate with a stainless steel extraction grid spaced 100 μm from the cathode. The size of the emitting area of the cathode is 0.25 cm^2. The cathode is mounted on a glass stem as shown in Fig. 16. The test configuration is shown in Fig. 6.

The efficiency of this type of cathode is much lower than the Spindt-type cathodes. The majority of the emitted electrons are intercepted by the gate electrode. About 20% of the 28-μA current emitted from the carbon film was typically collected at the anode during these experiments, however, the efficiency was often much higher at currents below 200 nA.

The cathode response to the increase in Xe pressure is shown in Fig. 17. The pressure was only increased to 2×10^{-6} Torr of xenon because of the close proximity of the ionization gauge and anode at 900 V. This configuration in high-pressure environments created charged particles that were collected by the electrodes at a much higher rate than the electrons emitted by the cathode, making it impossible to measure that current. The cathode demonstrated impressively stable performance in the Xe environment. Figure 17 shows that fluctuations in the current can be attributed to gate voltage fluctuations. The gate electrode voltage fluctuated between 835–850 V to cause the 2-μA fluctuations in the emitted current. In the

Fig. 16 Carbon-film cathode on a glass stem from FEPET.

high-pressure environments the ionization gauge affected the current measurements on all of the electrodes. The ionization gauge was responsible for ~250 μA to the anode and 15 μA to the gate electrode. With a 1-MΩ resistor on the gate electrode, this increase in current changed the gate voltage by 15 V. The ionization gauge was turned off and on several times during the high pressure exposure experiments to check the pressure. The large jumps in the gate voltage, and therefore current, can be attributed to the ionization gauge. The results of these experiments show that even at such high electrode voltages, the cathodes are resistant to performance degradation from ion sputtering. Two of these exposure tests were conducted at the same operating conditions. The first exposure test lasted for 4 h, and the second exposure lasted for 3 h. No performance degradation was observed during either of the experiments. I–V data taken before and after the exposure were identical. These data also show that the cathode performance was not affected by the exposures.

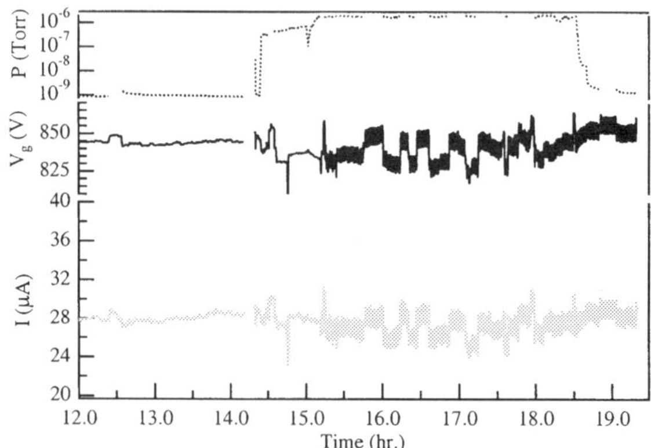

Fig. 17 Measured carbon-film cathode current response to a Xe pressure of 2×10^{-6} Torr with V_a at 900 V.

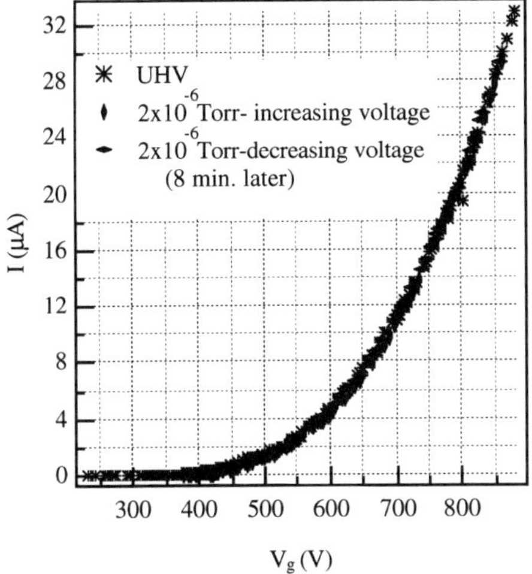

Fig. 18 Carbon-film cathode turn-on and turn-off data with V_a at 900 V in 2×10^{-6} Torr of Xe compared with turn-on data in UHV.

Figure 18 shows I–V data taken in UHV and Xe environments. The Xe pressure was increased to 2×10^{-6} Torr and stabilized while the cathodes were off. The cathodes were then slowly turned on in the Xe environment for ~8 min and then slowly reduced to zero current to obtain two of the data sets shown in Fig. 18. The cathodes turned on in 2×10^{-6} Torr of Xe no differently than in the UHV environment. This experiment was also repeated with the same results. This characteristic of the cathode is extremely valuable because, although the base pressure of facilities used in Hall thruster testing is typically between 10^{-7} and 10^{-6} Torr, the chamber pressure during thruster ignition is usually higher since propellant is flowing through the anode prior to start-up.

IV. Discussion

Two cathode technologies were tested to determine their performance response to Xe environments and limitations. The response of both the carbon film and Mo and Si FEA cathodes showed that their work functions were not increased by the Xe environment. These cathodes were operated in 10^{-5} to 10^{-6} Torr of Xe for several hours without catastrophic destruction. The carbon-film cathode was not sensitive to ion bombardment when operating at 850 V at 2×10^{-6} Torr of Xe. This cathode was incredibly robust; however its compatibility with EP systems is currently limited by high operating voltages and gate currents (1.5 μA/mW). Microfabrication of the base and gate structure to reduce the base-to-gate gap dimensions is recommended to improve efficiency through both operating voltage and gate current. The FEA cathodes were very sensitive to operating voltages and Xe pressure. Table 3 shows a summary of the data acquired during the experiments.

A FEA cathode performance model was developed incorporating the experimental results. It was determined that the E_{th} values for sputtering Mo and Si targets with Xe ions are approximately 49 eV and 63.7 eV, respectively. With this model, it was possible to determine the effect of the additional CEX ion flux on cathode

Table 3 Experimental highlights showing Si, Mo, and C cathode currents before (I_o) and after (I_f) xenon exposures at the pressures and voltages noted

Xenon pressure, T	V_a, V	V_g, V	I_o	I_f	Exposure time, h
			Silicon		
7×10^{-6}	400	86	20.0 μA	1.0 μA	1
2×10^{-5}	100	75	12.0 nA	8.3 nA	1
2×10^{-5}	100	75	11.4 nA	6.0 nA	2
2×10^{-5}	70	70	3.2 nA	1.7 nA	1
2×10^{-5}	80	60	700 nA	300 nA	1
2×10^{-5}	70	50	20.0 nA	20.0 nA	1
			Molybdenum		
2×10^{-6}	80	66	17.0 μA	5.0 μA	1
2×10^{-5}	60	50	100 nA	92.0 nA	5
2×10^{-5}	100	50	7.0 nA	9.2 nA	2
			Carbon		
2×10^{-6}	900	845	28.0 μA	28.0 μA	5

performance degradation rates and operating voltage limitations. The modeling results showed that the effect of the self generated Xe ion population on cathode performance was less significant than the effect of the Xe_{CEX} ion population. The erosion of the tips by the Xe ion population created locally was dominated by the Xe^{++} with gate electrode voltages below approximately 85 V.

The performance degradation model was also applied to determine the operating voltage limitations for cathode configurations which are more advanced than the cathodes tested. The cathode configurations tested were not optimized for the application. Cathodes with smaller gate apertures, sharper tips, higher packing densities, lower work functions, and better uniformity have demonstrated much better performance with higher current densities at lower operating voltages. Cathode performance that is much better than demonstrated in these experimental evaluations will be required for FEA cathodes to be compatible with EP systems in current, lifetime, and dimensions. Optimistic, but possible, next generation cathode configurations are described in Table 4. The operating voltage limits, lifetime, and dimensions of these cathodes are also shown in Table 4, as predicted by the performance model and determined by a 100 mA requirement and 5×10^7 tips/cm^2 packing density (tip packing density was set by VECTL architecture fabrication limitations). The initial currents, currents after 10,000 h or current at the lifetime limitation, and lifetime limitations are shown in the Table 4. These results are shown for a range of Xe pressures expected for small and mesoscale Xe EP systems and for Si and Mo cathodes. The Xe pressure does affect the operating voltage limit when the cathode is under bombardment from only the ions created locally, Xe^+ and Xe^{++}. Because the Xe^{++} are more damaging than the Xe^+ at these low voltages, the operating voltages are limited to values closer to the ionization potential of Xe^{++} than the energy thresholds for sputtering Mo and Si. The ions created locally will limit the gate electrode voltages of Mo and Si cathodes to approximately 37 V.

Table 4 Operating voltage limits, current, lifetime, and the length of a square cathode array L_c ($\beta_c = 0.26, r_g = 2000$ Å, $r_t = 40$ Å, $\Delta s = 2$, and $pd = 5 \times 10^7$ tips/cm^2 for thruster environments with and without a CEX ion population)

Case	ϕ_w, eV	V_g, V	N_{tips}	I_o, mA	I_f, mA	P, Torr	J_{CEX}, mA/cm^2	t, h	L_c, cm
1-Mo	4.35	38	9×10^8	103	97	2×10^{-6}	—	>10000	4.2
2-Mo	4.35	37	15×10^8	108	107	—[a]	—	>10000	5.4
3-Mo_{CEX}	4.35	5	3×10^{52}	103	91	—[a]	0.02	4000	$>10^{22}$
4-Mo_{CEX}	4.35	4	2.6×10^{64}	110	110	—[a]	0.02	>10000	—
5-Mo_{CEX}	4.35	5	3×10^{52}	103	100	—[a]	0.002	>10000	—
6-Si	4.05	39	9×10^7	101	96	2×10^{-6}	—	>10000	1.4
7-Si	4.05	38	1.4×10^8	103	100	2×10^{-5}	—	>10000	1.7
8-Si	4.05	37	2.3×10^8	108	108	—[a]	—	>10000	2.2
9-Si_{CEX}	4.05	13	3×10^{20}	104	94	—[a]	0.02	7000	$>10^6$
10-Si_{CEX}	4.05	12	9×10^{21}	107	107	—[a]	0.02	>10000	—
11-HfC	3.5	37	6.8×10^6	103					0.4
12-HfC_{CEX}	3.5	12	1.5×10^{18}	111					$>10^5$

[a]The cathode performance is insensitive to pressure at this operating voltage.

The effect of the CEX ion population on cathode performance and limiting voltages is also shown in Table 4 in cases 3, 4, 5, 9, and 10. At the low voltages considered, $\text{Xe}^{++}_{\text{CEX}}$ dominates in the erosion process and prohibitively limits the cathode operating voltages. These ions limit the operating voltages for Mo cathodes to 4–5 V. The energy of the $\text{Xe}^{++}_{\text{CEX}}$ at this operating point is 48–50 eV. They limit the operating voltages of Si cathodes to 12–13 V. The energy of the $\text{Xe}^{++}_{\text{CEX}}$ at this operating point is 62–64 eV.

Most of the dimensions of the cathodes shown in Table 4 are not compatible with EP systems because, at the limited voltages, the Mo and Si cathodes cannot provide 100 mA/cm^2. A mesoscale thruster with a ~25 cm^2 area will require 100–500 mA, and a small thruster with a ~60 cm^2 area will require 1–3 A. According to the results shown in Table 4, the cathode dimensions will be larger than the dimensions of the thrusters.

Lower work function and NEA materials could significantly improve the compatibility of these systems by providing higher current densities at the low voltage limits.[48–52] Solid carbide tips or carbide, carbon, and diamond films could be deposited on FEA cathodes. The sputtering and electron emission properties of these materials have not yet been characterized, therefore their performance in thruster environments cannot yet be predicted using the models. The performance of a HfC cathode could be significantly better than Mo and Si cathodes because of its low work function. At only 37 V, the HfC cathode could provide more than 100 mA/cm^2 as shown in Table 4. E_{th} for sputtering HfC has not yet been determined, however, it is believed to be higher than Mo and Si.[32] This cathode may be able to operate in an EP system environment under bombardment of both populations of ions with the required lifetime, performance, and dimensions if the CEX ion population does not hit the emitting tips. A cathode lens and ion repeller (CLAIR) has been designed to electrostatically shield the microscale tips from CEX ion bombardment.

CLAIR is a microfabricated electrode configuration that can be integrated into the FEA cathode fabrication process.[32] With CLAIR and low work function materials like HfC, much higher operating voltages can be tolerated, and required cathode current densities may be achievable.

V. Conclusions

Experimental and theoretical results showed that the performance of Mo and Si FEA cathodes operating in Xe environments is extremely sensitive to pressure and operating voltages. Specific environments were considered that represent some worst-case scenarios, however, some startling discoveries were made about the sensitivity of the cathode performance to a range of possible environments. FEA cathode erosion in a Xe environment is dominated by Xe^{++} at operating voltages below 85 V so that the operating voltage limitation is approximately the ionization potential of Xe^{++}, 37 V, at 2×10^{-5} Torr of xenon. At this voltage, Mo and Si FEA cathodes, with an optimistic configuration ($r_t = 40$ Å, $r_g = 2000$ Å, $\Delta s = 2$, and 5×10^7 tips/cm^2), cannot provide the required 100 mA/cm^2. Decreasing the local pressure by one order of magnitude only increased the tolerable operating voltage by 1 V, providing a fairly inconsequential decrease in cathode dimensions. If a CEX ion population contributes to the cathode erosion process at the low operating voltages of interest (<50 V), then the Xe^{++}_{CEX} dominates the Xe^{+}_{CEX} in the erosion process and limits the operating voltage to 12–13 V for Si and 4–5 V for Mo if J_{CEX} is 0.02 mA/cm^2. At these low voltages, the currents will be prohibitively low or cathode dimensions will be prohibitively large, even with optimistic cathode configurations. Decreasing J_{CEX} by one order of magnitude also only increased the tolerable operating voltages by 1 V. Even without the CEX ion population bombarding the cathode, Mo and Si cathodes may not be capable of providing 100 mA/cm^2 for 10,000 h in thruster environment.

The compatibility of EP systems and FE cathodes requires materials with work functions lower than the work functions of Mo and Si in addition to a configuration that can protect the cathodes from the CEX ion population to achieve the performance and lifetime requirements demanded by EP systems. NEA films on Mo and Si FEA cathodes have significantly improved the emission characteristics of Si and Mo FEA cathodes and demonstrated excellent stability in elevated pressure environments. The carbon-film cathode performance results discussed in this chapter provide some examples of the ruggedness of NEA FE cathodes. It is recommended that Si and Mo FEA cathodes with $r_t \approx 40$ Å, $r_g \approx 2000$ Å, $\Delta s \approx 2$, and 5×10^7 tips/cm^2, be coated with HfC, ZrC, or NEA carbon films to improve their performance at low voltage, CLAIR be used to protect the cathode from ions originating near the thruster, and a VECTL architecture be used to prevent arcing between the tips and gate electrode. Combining the FEA configuration with a carbide or carbon film, CLAIR, and a VECTL architecture, should improve the cathode performance to provide the required current density and lifetime efficiently.

Acknowledgments

The work described in this chapter was performed at the University of Michigan and Jet Propulsion Laboratory, and sponsored by the Ballistic Missile Defense Organization, the Air Force Office of Scientific Research (Mitat Birkan), and

NASA. Publication support was provided by the Jet Propulsion Laboratory and California Institute of Technology under a contract with NASA. The authors would also like to gratefully acknowledge Don Strayer for the vacuum facility and Bill Thogmartin, Bob Toomath, and Al Owens for their assistance in the laboratory. Reference herein to any specific commercial product, process, or service by trade name, trademark, manufacturer, or otherwise, does not imply its endorsement by the United States Government or the Jet Propulsion Laboratory, California Institute of Technology.

References

[1]Mueller, J., "Thruster Options for Microspacecraft: A Review and Evaluation of Existing Hardware and Emerging Technologies," AIAA Paper 97-3058, July 1997.

[2]Khayms, V., and Martinez Sanchez, M., "Design of a Miniaturized Hall Thruster for Microsatellites," AIAA Paper 96-3291, July 1996.

[3]Marcuccio, S., Giannelli, S., and Andrenucci, M., "Attitude and Orbit Control of Small Satellites and Constellations with FEEP Thrusters," International Electric Propulsion Conf. (IEPC), Paper 97-188, Aug. 1997.

[4]Tajmar, T., "3D Numerical Plasmasimulation and Backflow Contamination of a Cesium Field-Emission-Electric-Propulsion (FEEP) Emitter and Thermionic Neutralizer," Ph.D. Dissertation, Vienna Univ. of Technology Vienna, Austria, 1999.

[5]Brodie, I., and Schwoebel, P., "Vacuum Microelectronics Devices," *Proceedings of the IEEE*, Vol. 83, No. 7, 1994.

[6]Spindt, C. A., Holland, C. E., Schwoebel, P. R., and Brodie, I., "Field Emitter Array Development for Microwave Applications II," *Journal of Vacuum Science and Technology B*, Vol. 16, No. 2, 1998, pp. 758–761.

[7]Bozler, C. O., Harris, C. T., Rabe, S., Ratham, D. D., Hollis, M., and Smith, H. I., "Arrays of Gated Field-emitter Cones Having 0.32 μm Tip-to-Tip Spacing," *Journal of Vacuum Science and Technology B*, Vol. 12, No. 2, 1994, pp. 629–632.

[8]Mackie, W. A., Xie, T., and Davis, P. R., "Field Emission from Carbide Film Cathodes," *Journal of Vacuum Science and Technology B*, Vol. 13, No. 6, 1995, pp. 2459–2463.

[9]Charbonnier, F., "Arcing and Voltage Breakdown in Vacuum Microelectronics Microwave Devices Using Field Emitter Arrays: Causes, Possible Solutions, and Recent Progress," *Journal of Vacuum Science and Technology B*, Vol. 16, No. 2, 1998, pp. 880–887.

[10]Rakhshandehroo, M. R., "Design, Fabrication, and Characterization of Self-Aligned Gated Field Emission Devices," Ph.D. Dissertation, Univ. of Michigan, Tech. Rept. SSEL-284, Ann Arbor, MI, 1998.

[11]Li. Tolt, Z., Fink, R. L., and Yaniv, Z., "The Status and Future of Diamond Thin Film FED," Nikkei Microdevices' Flat Panel Display 1998 Yearbook, English translation published by InterLingua, 186, 1998.

[12]Spindt, C. A., Holland, C. E., Schwoebel, P. R., and Brodie, I., "Field Emitter Array Development for Microwave Applications," *Journal of Vacuum Science and Technology B*, Vol. 14, No. 3, 1996, pp. 1986–1989.

[13]Urayama, M., Ise, T., Maruo, Y., Kishi, A., Imamoto, R., and Takase, T., "Silicon Field Emitter Capable of Low Voltage Emission," *Japan Journal of Applied Physics*, Vol. 32, Pt. 1, No. 12B, 1993.

[14]Murphy, R. A., Harris, C. T., Matthews, R. H., Graves, C. A., Hollis, M. A., Kodis, M. A., Shaw, J., Garven, M., Ngo, M. T., and Jensen, K. L., IEEE International Conf. on Plasma Science, May 1997.

[15]Spindt, C. A., and Brodie, I., Technical Digest of the 1996 IEEE International Electron Devices Meeting (IEDM), 12.1.1 (1996); also C. A. Spindt, C. E. Holland, P. R. Schwoebel, and I. Brodie, IEEE International Conf. on Plasma Science, May 1997.

[16]Schwoebel, P. R., and Spindt, C. A., "Field-Emitter Array Performance Enhancement Using Hydrogen Glow Discharges," *Applied Physics Letters*, Vol. 63, No. 1, 1993, pp. 33–35.

[17]Schwoebel, P. R., and Spindt, C. A., "Glow Discharge Processing to Enhance Field-Emitter Array Performance," *Journal of Vacuum Science and Technology B*, Vol. 12, No. 4, 1994, pp. 2414–2421.

[18]Rakhshandehroo, M. R., and Pang, S. W., "Sharpening Si Field Emitter Tips by Dry Etching and Low Temperature Plasma Oxidation," *Journal of Vacuum Science and Technology B*, Vol. 14, No. 6, 1996, pp. 3697–3701.

[19]Geis, M. W., Twichell, J. C., and Lyszarz, T. M., "Diamond Emitters Fabrication and Theory," *Journal of Vacuum Science and Technology B*, Vol. 14, No. 3, 1996, pp. 595–598.

[20]Tolt, Z. L., Fink, R. L., and Yaniv, Z., "Electron Emission from Patterned Diamond Flat Cathodes," *Journal of Vacuum Science and Technology B*, Vol. 16, No. 3, 1998, pp. 1197–1198.

[21]Blyablin, A. A., Kandidov, A. V., Pilevskii, A. A., Rakhimov, A. T., Samorodov, V. A., Seleznev, B. V., Suetin, N. V., and Timofeev, M. A., 11th International Vacuum Microelectronics Conf., 1998.

[22]Marrese, C. M., Wang, J., Goodfellow, K. D., and Gallimore, A. D., "Space-Charge–Limited Emission from Field Emission Cathodes for Electric Propulsion and Tether Applications," *Micropropulsion for Small Spacecraft*, Progress in Astronautics and Aeronautics, Vol. 187, edited by M. Micci and A. Ketsdever, AIAA, Reston, VA, 2000, Chap. 18 (this volume).

[23]Itoh, S., Niiyama, T., Taniguchi, M., and Watanbe, T., "A New Structure of Field Emitter Arrays," *Journal of Vacuum Science and Technology B*, Vol. 14, No. 3, 1996, pp. 1977–1981.

[24]Grossman, K., and Peckerar, M., "Active Current Limitation for Cold-Cathode Field Emitters," Nanotechnology 5, 179–182, 1994, pp. 179–182.

[25]Takemura, H., Tomihari, Y., Furutake, N., Matsuno, F., Yoshiki, M., Takada, N., Okamoto, A., and Miyano, S., "A Novel Vertical Current Limiter Fabricated with a Deep Trench Forming Technology for Highly Reliable Field Emitter Arrays," *Tech. Digest of the IEEE-IEDM*, 1997, p. 709.

[26]Lindberg, P. A. P., and Johansson, L. I., "Work Function and Reactivity of Some Crystal Faces of Substoichimetric Transition-Metal Carbides," *Surface Science*, Vol. 194, 1988, pp. 199–204.

[27]Ishikawa, J., Tsuji, H., Yashuto, S., T., Kaneko, T., Nagao, M., and Inoue, K., "Influence of Cathode Material on Emission Characteristics of Field Emitters for Microelectronics Devices," *Journal of Vacuum Science and Technology B*, Vol. 11, No. 2, 1993, pp. 403–406.

[28]King, L. B., and Gallimore, A. D., "Gridded Retarding Pressure Sensor for Ion and Neutral Particle Analysis in Flowing Plasmas," *Rev. Sci. Instrum*, Vol. 68, No. 2, 1997, pp. 1183–1188.

[29]Spindt, C. A., Holland, C. E., Rosengreen, A., and Brodie, I., "Field-Emitter Arrays for Vacuum Microelectronics," *IEEE Transactions on Electron Devices*, Vol. 38, No. 10, Oct 1991.

[30]Palmer, W. D., Temple, D., Mancusi, J., Yadon, L., Vellenga, D., and McGuire, G. E., "Emission Current Measurements Under Flat Panel Display Conditions," 25th IEEE International Conf. on Plasma Science, June 1998.

[31]Temple, D., Palmer, W. D., Yadon, L. N., Mancusi, J. E., Vellenga, D., and McGuire, G. E., "Silicon Field Emitter Cathodes: Fabrication, Performance, and Applications," *Journal of Vacuum Science and Technology A*, Vol. 16, No. 3, 1998, pp. 1980–1990.

[32]Marrese, C. M., "Compatibility of Field Emission Cathode and Electric Propulsion Technologies," Ph.D. Dissertation, Univ. of Michigan, Ann Arbor, MI, 1999.

[33]Tolt, Z. L., Fink, R. L., and Yaniv, Z., "Electron Emission from Patterned Diamond Flat Cathodes," *Tech. Digest IVMC*, 1997.

[34]Fink, R. L., Thuesen, L. H., Li. Tolt, Z., and Yaniv, Z., "Lifetime and Stability of Diamond Field Emission Devices," *Diamond Films and Technology*, Vol. 8, No. 6, 1998, p. 429.

[35]Jensen, K. L., "An Analytical Model of an Emission-gated Twystrode Using a Field Emission Array," *Journal of Applied Physics*, Vol. 83, No. 12, 1998, pp. 7982–7992.

[36]Brodie, I., "Bombardment of Field-Emission Cathodes by Positive Ions Formed in the Interelectrode Region," *International Journal of Electronics*, Vol. 38, No. 4, 1975, pp. 541–550.

[37]Yamamura, Y., Matsunami, N., and Itoh, N., "Theoretical Studies on an Empirical Formula for Sputtering Yield at Normal Incidence," *Radiat. Effects and Defects in Solids*, Vol. 71, No. 1-2, 1983, pp. 65–86.

[38]Gomer, R., *Field Emission and Field Ionization*, Harvard Univ. Press, Cambridge, MA, 1961.

[39]Jensen, K. L., Mukhopadhyay, P., Zaidman, E. G., Nguyen, K., Kodis, M. A., Malsawma, L., and Hor, C., "Electron Emission from a Single Spindt-Type Field Emitter: Comparison of Theory with Experiment," *Applied Surfaces Science*, Vol. 111, 1997, pp. 204–212.

[40]Everhart, T. E., "Simplified Analysis of Point-Cathode Electron Sources," *Journal of Applied Physics*, Vol. 38, No. 113, 1967.

[41]Krishnakumar, E., and Srivastava, S. K., "Ionization Cross-Sections of Rare Gas Atoms by Electron Impact," *Journal of Physics B: Atomic Molecular and Optical Physics*, Vol. 21, 1988, pp. 1055–1082.

[42]Matsunami, N., Yamamura, Y., Itikawa, Y., Itoh, N., Kazmuta, Y., Miyagawa, S., Morita, K., Shimizu, R., and Tawara, H., "Energy Dependance of the Ion-Induced Sputtering Yields of Monatomic Solids," *Atomic Data and Nuclear Data Tables 31*, 1984.

[43]Bohdansky, J., Roth, J., and Bay, H. L., "An Analytical Formula and Important Parameters for Low-Energy Ion Sputtering," *Journal of Applied Physics*, Vol. 51, No. 5, 1980.

[44]Stuart, R. V., and Wehner, G. K., "Sputtering Yields at Low Bombarding Ion Energies," *Journal of Applied Physics*, Vol. 33, No. 7, 1962.

[45]Rosenberg, D., and Wehner, G. K., *Journal of Applied Physics*, Vol. 33, 1962.

[46]King, L. B., "Transport-Property and Mass Spectral Measurements in the Plasma Exhaust Plume of a Hall-Effect Space Propulsion System," Ph.D. Dissertation, Univ. of Michigan, Ann Arbor, MI, 1998.

[47]Rapp, D., and Francis, W. E., "Charge Exchange Between Gaseous Ions and Atoms," *Journal of Chemistry and Physics*, Vol. 37, No. 11, 1962, pp. 2631–2645.

[48]Fetisov, I. K., and Firsov, O. B., "Resonance Charge Exchange of Doubly Charged Ions in Slow Collisions," *Soviet Physics JEPT*, Vol. 37, No. 10, 1960, pp. 67–68.

[49]Xie, T., Mackie, W. A., and Davis, P. R., "Field Emission from ZrC Films on Si and Mo Single Emitters and Emitter Arrays," *Journal of Vacuum Science and Technology B*, Vol. 14, No. 3, 1996, pp. 2090–2092.

[50]Mackie, W. A., Xie, T., and Matthews, M. R., "Field Emission from ZrC and ZrC Films on Mo Field Emitters," *Journal of Vacuum Science and Technology B*, Vol. 16, No. 4, 1998, pp. 2057–2062.

[51]Rakhshandehroo, M. R., "Design, Fabrication, and Characterization of Self-Aligned Gated Field Emission Devices," Ph.D. Dissertation, Univ. of Michigan, Tech. Rept. SSEL-284, Ann Arbor, MI, 1998.

[52]Lee, S., Lee, S., Lee, S., Jeon, D., and Lee, K. R., "Self-Aligned Silicon Tips with Diamond-Like Carbon," *Journal of Vacuum Science and Technology B*, Vol. 15, No. 2, 1997, pp. 457–459.

[53]Jung, J. H., Ju, B. K., Lee, Y. H., Jang, J., and Oh., M. H., "Emission Stability of a Diamond-Like Carbon Coated Metal-Tip Field Emitter Array," *Journal of Vacuum Science and Technology B*, Vol. 17, No. 2, 1999, pp. 486–488.

Chapter 12

Electric Breakdown Characteristics of Silicon Dioxide Films for Use in Microfabricated Ion Engine Accelerator Grids

Juergen Mueller,* David Pyle,† Indrani Chakraborty,‡
Ronald Ruiz,§ William Tang,¶ Colleen Marrese,** and Russell Lawton§
Jet Propulsion Laboratory, California Institute of Technology
Pasadena, California

I. Introduction

A STRONG interest has arisen recently within the space community to develop micropropulsion devices capable of delivering very low thrust levels and impulse bit values while featuring engine sizes and masses orders of magnitude smaller than are available with current technologies.[1] Applications for such devices would span the propulsion needs of some of the smallest and largest spacecraft currently being envisioned by NASA and the U.S. Air Force, ranging from primary and attitude control of microspacecraft to precise positioning control of spacecraft constellations for interferometry or military Earth-observing radar missions and compensation of solar pressure-induced torques on large inflatable spacecraft. Microspacecraft,[2] typically defined as spacecraft having wet masses of a few tens of kilograms or less, are being considered to reduce overall mission cost. Since launch costs are determined to a large extent by spacecraft mass and may contribute as much as 30% to the cost of a mission, microspacecraft will be less expensive to launch, thus allowing for more frequent access to space. Microspacecraft may also be deployed in fleets or constellations, charting entire regions of space and reducing mission risk by not relying on a single craft.

However, microspacecraft require radically new approaches in design, both on the system and on the component level. While significant progress in that

Copyright © 2000 by the American Institute of Aeronautics and Astronautics, Inc. The U.S. Government has a royalty-free license to exercise all rights under the copyright claimed herein for Governmental purposes. All other rights are reserved by the copyright owner.

*Advanced Propulsion Technology Group.
†JPL Academic Part-Time, University of Texas.
‡MEMS Group, Microdevices Laboratory.
§Failure Analysis Group.
¶Formerly Supervisor, MEMS Group, Microdevices Laboratory; currently at DARPA.
**Advanced Propulsion Technology Group.

direction is being made in the instrument, attitude sensing, and data handling and storage areas, for example, propulsion still appears to be lagging in this regard, offering only limited hardware choices able to fit the design constraints imposed by microspacecraft with respect to mass, size, and power.[1] Virtually all propulsion subsystem areas, such as attitude control, primary propulsion, and feed system components, are still needed for suitable design solutions. Within the primary propulsion area, high specific impulse options appear to be of particular interest, since their ability to conserve considerable amounts of propellant mass may have a significant bearing on microspacecraft design.[1]

Other space applications, such as the aforementioned interferometry-class or space inflatable spacecraft, also may have a need for small, lightweight engine technology able to provide a continuous, low-level thrust to offset solar pressure-induced disturbance torques, for example. Depending on the mission, these thrust levels may range from as little as a few micronewtons[3] to a few millinewtons. Because of the fairly long durations of thrust required over the course of the missions, high specific impulse devices may again be desirable.

Currently among the most mature high specific impulse propulsion technologies is ion propulsion, as evidenced by the recent flight of this technology on NASA's Deep Space 1 (DS-1) mission. Presently available engine technology, however, is relatively large, extending from beam diameters of 30 cm in the case of the DS-1 engine down to about 10 cm, and requiring power levels on the order of several thousands to several hundreds of watts.[1] Thus, there exists a need to miniaturize this technology further to make it more amenable to the aforementioned mission categories. Other competing thruster options, such as field emission electric propulsion (FEEP) and pulsed plasma thruster (PPT) technologies, exist and also have reached very high levels of maturity up to the point that one or both are expected to play a role in the types of mission applications discussed.[3] However, micro-ion engine technology, if it can be successfully developed, will be characterized by a unique combination of high specific impulse capability, the use of inert, noncontaminating propellants, higher thrust-to-power ratios than obtainable with FEEP or PPT systems, and, unlike in the case of the inherently pulsed PPT devices, a continuous mode of operation.

At the Jet Propulsion Laboratory (JPL) a feasibility study is currently under way to investigate the potential of reducing ion engine sizes dramatically below current state-of-the-art levels. Engine diameters in the 1- to 3-cm range and thrust levels in the submillinewton to few-millinewton range are being targeted. To arrive at a functional ion engine of such a small size, however, several feasibility issues will need to be investigated and overcome. Among these are the sustainability and efficient operation of high surface-to-volume ratio plasma discharges, the replacement of hollow-cathode technologies with lower power-consuming and easier-to-miniaturize cathode systems to function as both engine cathodes and neutralizers (such as field emitter array technology[4]), miniature accelerator grid system fabrication and operation, and the fabrication of miniaturized power conditioning units and feed system components.

In this chapter, the feasibility of microfabricated grid designs is investigated from the perspective of obtainable grid breakdown voltages. To this end, the breakdown characteristics of typical insulator materials used in the microfabrication field, such as silicon dioxide, are studied. Although the motivation for this investigation was the evaluation of this material as a grid insulator in ion engine accelerator systems, data obtained in this study may also be applicable to other micropropulsion systems. For example, microcolloid or micro-FEEP systems might have a need for such insulator materials as well if microfabricated versions of these thrusters types were to be pursued.

II. Microfabricated Grid Design Issues

Although the targeted micro-ion engine diameters of 1–3 cm as such may not require microfabrication techniques to machine the engine body, a case can be made to investigate the feasibility of micromachined grid designs. Smaller-diameter engines allow grids to be spaced much more closely with respect to each other since the amount of electrostatic stress-induced grid deformation will be less. Placing grids closer with respect to each other will increase the grid perveance, proportional to $1/d^2$, with d being the grid spacing. Thus, higher beam currents could be extracted from the engine for a given voltage, extending its performance range provided that sufficiently large ion densities can be provided in the discharge chamber. However, ion optical considerations generally require grid aperture diameters to be scaled down in size with the grid spacing to avoid ion impingement on grids causing potentially engine life-threatening grid erosion. Smaller aperture diameters, and the requirement to place apertures of the various grids (screen, accelerator, and, potentially, decelerator) of a grid system concentrically with respect to each other, in turn, require tight machining tolerances. Current, "macromachined" grids are fabricated within 0.05-mm or 50-μm tolerances, representing a limit in most cases for many conventional machining techniques, such as electric discharge machining (EDM) and laser drilling. Using microfabrication techniques, however, much smaller tolerances can easily be obtained. In addition, the ability to produce entire batch-fabricated grid systems, not requiring any additional assembly and grid alignment procedures, weighs in favor of microfabrication approaches as well.

In this chapter, the feasibility of grids based on silicon-based MEMS (microelectromechanical systems) machining techniques is explored. While other microfabrication possibilities exist, silicon-based MEMS techniques were investigated first because of the considerable heritage and experience available with this technique, as well as its demonstrated ability to produce extremely small feature sizes within very tight tolerances of 1 μm or less. However, MEMS fabrication of accelerator grids opens up a host of fabrication- and operations-related issues. Foremost among them is the selection of appropriate grid materials, suiting both microfabrication and grid operation needs, in particular, with respect to sputter erosion and voltage stand-off characteristics. The grid insulator material, for example, isolating the screen and accelerator voltages from each other, will have to be able to stand off voltages of the order of 1.3 kV or more over distances of the order of a few microns. This assumption is based on current ion engine designs. Propulsion requirements for future microspacecraft missions remain very unclear at this point. While the possibility exists that ion engines with lower grid voltages (and consequently reduced specific impulse performance and engine efficiencies) may still be a possible thruster option for microspacecraft, a conservative approach would aim at least to maintain current, state-of-the-art grid performances at this stage of the development.

It is the scope of this study to investigate the feasibility of silicon dioxide as a grid insulator material. Silicon dioxide was chosen since it exhibits good electric insulating characteristics compared to other materials used in silicon-based MEMS fabrication and is already widely used in the microfabrication field. To study

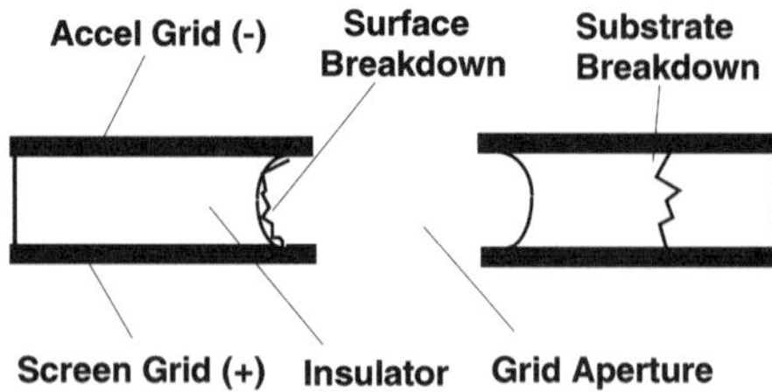

Fig. 1 Anticipated grid breakdown modes.

the suitability of silicon oxide for this application, both bulk electric breakdown characteristics and electric breakdown characteristics along its surface needed to be studied. This is evident on inspecting Fig. 1. As can be seen, both modes of electric breakdown, substrate (or bulk) and surface, are possible in a typical grid design. The latter may occur along the walls of grid apertures. Two sets of experiments were conducted using specially designed silicon oxide breakdown test chips to study systematically both modes of electric breakdown and are described in detail below.

The measurements performed here can be regarded only as a first step in evaluating MEMS-based grid designs. For example, other insulators, deposited in modified and previously untried processes, may result in different breakdown characteristics. Regardless of the type of insulator used, a key feasibility issue with respect to microgrid designs such as the one shown in Fig. 1 is the possibility of coating the insulator material along the exposed grid aperture wall surfaces with conducting, sputter-deposited material, thus shorting the grid. Shadow-shielding around grid spacers is commonly used in grid systems today, and similar concepts will need to be explored for microfabricated grids and integrated into the batch fabrication process. Another important feasibility issue is sputter erosion of the various MEMS grid materials under consideration. However, insulator grid breakdown was regarded as a logical starting point for a MEMS grid feasibility investigation, to be followed up, if successful, in later studies by topics of considerably higher degrees of complexity in fabrication, such as insulator shielding.

III. Previous Related Research

It may seem surprising at first that a detailed study of breakdown behavior of oxide films is necessary since a substantial amount of research has already been performed in this area over the past several decades. However, a closer examination of the available literature reveals that results obtainable from past research may not be directly applicable to the problem studied here. Most previous research work on breakdown characteristics has focused on studying the electric breakdown of gate oxides in MOSFET (metal-on-silicon field effect transistor) applications. These gate oxides are typically very thin, less than 0.1 μm thick, and the required minimum breakdown voltages range into the tens of volts and, thus, are significantly lower than the kilovolt voltage range considered for grid applications. One particular type of oxide most frequently considered for gate oxide applications is thermal oxide. This oxide layer is created by directly oxidizing the silicon surface in an oxygen furnace (dry oxide), sometimes aided by the addition of steam (wet oxide) to increase film growth rates.[5]

Studies on breakdown strengths of thermal oxides have been performed by Osburn and Ormond,[6,7] Osburn and Weitzmann,[8] Klein,[9] Chou and Eldridge,[10] Soden,[11] Fritzsche,[12] Worthing,[13] and Yang et al.[14] Typically two types of breakdowns were observed by all researchers: the so-called primary and the intrinsic, or final, breakdown. Primary breakdown field strengths range from approximately 200 V/μm to as high as 1000 V/μm, whereas final breakdowns follow a more sharply peaked distribution ranging between approximately 800 and 1000 V/μm.[6] In some cases, final breakdown strengths as high as 1400–1500 V/μm have been observed for extremely thin oxides.[7] Primary breakdowns are thought to be triggered along defects in the oxide.[9] As currents and heat dissipation increase locally, melting and evaporation of local material result. Using very thin electrodes (less than 0.3 μm in the case of Osburn's and Ormond's experiment[7]), the electrode will be destroyed through evaporation of electrode material near the breakdown location, thus representing a "self-healing" breakdown since no electrical contact can be maintained between the two electrodes due to the loss of conductive material. This allows all defect-related breakdown sites to be eliminated until the intrinsic, or final, breakdown is reached. This breakdown strength thus corresponds to the dielectric strength of ideal, defect-free oxide material. Different theories evolve around this final breakdown and both thermal breakdown[9] and electronic breakdown mechanisms[7] due to electron avalanches have been proposed. Chou and Eldridge[10] have succeeded in fabricating virtually defect-free thermal oxides and eliminated primary breakdowns, resulting in final breakdown strengths of 600–700 V/μm and up to 1000 V/μm for thermal oxide coated with phosphorsilicate glass, filling pits in the oxides that were believed to have triggered breakdowns.

While it thus appears possible to achieve rather high electric breakdown strengths using carefully prepared thermal oxides, absolute voltages that can be stood off with these oxides may, however, be rather limited. This is due largely to the fact that thermal oxides are typically grown only up to thicknesses of about 1 μm, possibly somewhat larger, but almost always less than 2 μm. The reason for this limitation can be found in the thermal oxidation process. The surface is oxidized directly, i.e., no oxide layer is deposited onto the silicon surface, and the oxide layer instead grows partly into the silicon, using the substrate silicon to form the oxide.[5] Since new oxygen arriving at the surface now has to penetrate an increasingly thicker oxide layer to form an oxidation reaction with the underlying silicon, diffusion limitations will eventually result in increasingly longer process times until the process finally becomes impractical. Therefore, even using Chou's and Eldridge's[10] breakdown values for defect-free oxides, the obtainable voltages that can be stood off for oxides less than 2 μm thick may thus be somewhat marginal assuming that voltages of 1.3 kV will be required for grid applications and an adequate additional margin of safety will have to be maintained. If, as was the case in most of the experiments conducted, much lower-voltage primary breakdowns occur, stand-off voltages would be insufficient for ion engine grid applications. The process of "self-healing" breakdowns, while appropriate in experiments addressing fundamental research, would not be suitable for operational ion engine grids since the massive erosion of thin electrode material would lead to grid destruction. In addition, thermal oxides will need to be grown directly on silicon surfaces, thus limiting the choice of substrate materials to silicon only.

Other oxides that have been investigated in the past are RF sputter-deposited oxides. These oxides can be grown to much larger thicknesses (several microns) since the silicon surface is coated with externally supplied, sputter-eroded silicon oxide material. Limitations with respect to thickness arise eventually as thick oxides develop intrinsic compressive stresses that may lead to delamination of oxide from its substrate material. Pratt[15] performed dielectric strength measurements on RF sputter-deposited oxides, however, given targeted applications in the electronics industry, focused only on very thin oxides. Pratt[15] measured dielectric strengths ranging from 1000 V/μm at 0.07 μm to about 220 V/μm at 0.7 μm. This trend of decreasing electric breakdown field strength is noteworthy and has also been noted for thermal oxides. While breakdown voltages typically still increase with increasing oxide thickness, the trend toward lower electric breakdown field strengths for thicker oxides limits this increase. In the case of Pratt's experiment, the breakdown voltage at 0.7 μm can be calculated as about 150 V.

Klein and Gafni[16] reported electric breakdown field strengths for vapor-deposited oxide films on glass slides, fabricated by evaporation of silicon monoxide in an oxygen atmosphere. Silicon dioxide and silicon monoxide layers were created. The silicon dioxide layers were up to 0.49 μm thick and yielded breakdown strengths of 490 V/μm, or about a 250-V voltage stand-off capability. Silicon monoxide layers of up to 5 μm were deposited and resulted in electric breakdown field strengths of 192 V/μm, thus yielding a voltage stand-off capability of just under 1000 V. Silicon monoxide breakdown field strengths were found to be lower than those for silicon dioxide for comparable oxide thicknesses. Again, as in the case of thermal and sputter-deposited oxides, a trend toward lower breakdown field strengths with increasing oxide thickness was noted.

The survey of the literature thus established the need for a more targeted investigation of thick oxides capable of delivering stand-off voltages comparable to typical grid voltages with acceptable margins of safety. Chemical vapor-deposited (CVD) oxides are known to produce good electric insulation and can be deposited to thicknesses up to about 5 μm. While some breakdown data for these oxides can be found in the literature, the sources very often do not list the film thickness or list data for relatively thin films only given the focus of applications in the semiconductor electronics field. More detailed information was required on breakdown characteristics of thick oxide films, taking into account the previously observed dependence of breakdown field strength on oxide thickness, surface breakdown data, and temperature dependence of the breakdown strength of these oxides since grid operating temperatures may range between 300 and 400°C.

Therefore, a systematic study of breakdown strengths of CVD low-temperature oxides (LTO) was initiated. Preliminary results were reported in an earlier work.[17] Those tests were conducted with a limited amount of test chips and thus provided only a very preliminary database. Although tests in Ref. 17 were initially targeted only to provide substrate, or bulk, electric breakdown field strengths, and tests were therefore conducted in atmosphere for simplicity, unintended electric breakdowns along the surface were also noted during those experiments. Surface electric breakdown field strengths at the gap distances encountered (about 200 μm) were low, ranging only around 2 V/μm. This necessitated further development of this experiment. First, test chips intended for the measurement of substrate breakdowns had to be redesigned to eliminate the parasitic surface breakdowns, and a more systematic examination of surface breakdowns had to be initiated. The latter tests were conducted under vacuum conditions to eliminate any gas breakdown or surface contamination effects. The following sections describe this new set of experiments in detail.

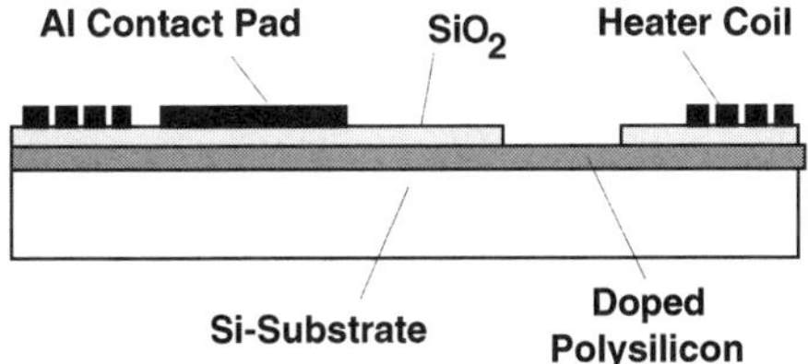

Fig. 2 Schematic of the substrate breakdown chip.

IV. Description of the Experiment

The experiments (substrate, or bulk, and surface breakdown) were conducted with two types of test chips. A total of about 200 chips was tested. Of these, 100 chips were used for substrate breakdown tests and 100 chips for surface breakdown tests. The chip type used for substrate breakdown is shown in Fig. 2. Each chip is 1×1 cm^2 in size and consists of a silicon substrate wafer (400 μm thick) onto which a thin layer (0.3 μm) of doped polysilicon is deposited (about 22 $\Omega/\square$ resistivity). Next, a layer of LTO oxide, using a low-pressure CVD (LPCVD) silane/oxygen process, is deposited up to a thickness of 3.9 μm at around 450°C. Poly and oxide deposition was performed at the University of California/Berkeley. Some chips tested were poly- and oxide-deposited at the University of California/Los Angeles (UCLA) earlier using a similar process, yielding oxide thicknesses of a maximum of 2.7 μm. Samples of the latter batch of chips were also used in previous tests reported in Ref. 17.

Depending on the desired oxide thickness, the oxide layer is etched back. Next, a via is etched into the oxide to provide access to the underlying polysilicon layer, which will form one of the two electrodes. Finally, a 0.25-μm-thick aluminum layer is deposited onto the chip, then patterned and etched to form the second electrode as well as a heater coil. This (square-shaped) heater coil is shown in Fig. 3 and is used to heat the chip for breakdown testing at elevated temperatures. Temperatures up to 400°C have been achieved with this design at power levels of

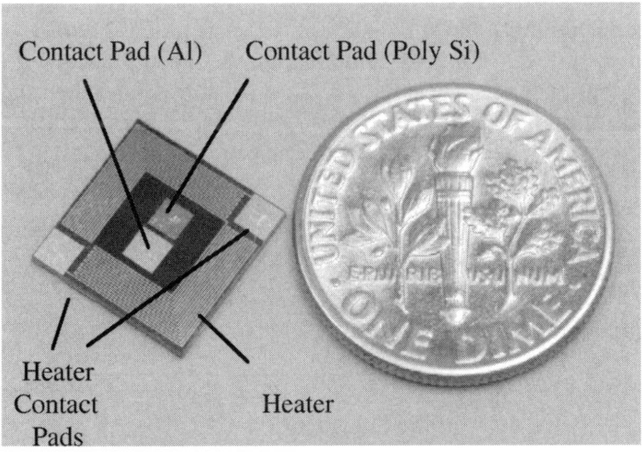

Fig. 3 View of the substrate breakdown test chip.

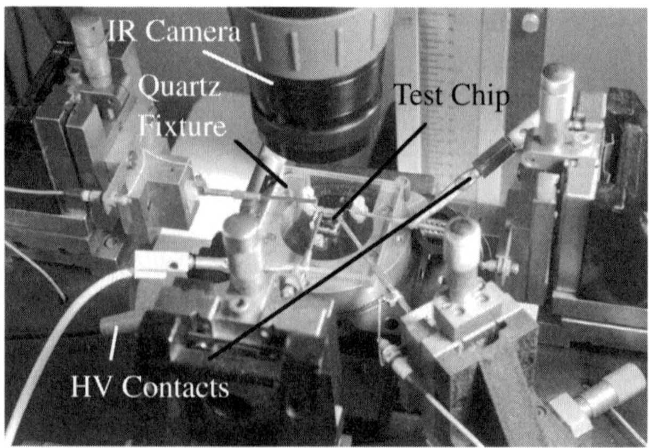

Fig. 4 Substrate breakdown test setup.

about 11 W (160 V, 0.07 mA). Small variations in heater coil performance were found from chip to chip.

The substrate breakdown tests were performed under atmospheric conditions by placing the test chips into a specially designed quartz fixture, which in turn was placed underneath an infrared (IR) camera (see Fig. 4). The IR camera was used for temperature measurements but was also able to record arcing on the chip at ambient temperature. The IR image was recorded on videotape for later test evaluation. The chip was contacted via a probe station featuring four adjustable probe tips. Two tips served as high-voltage leads, while the remaining two were used to contact the heater coil. Unfortunately the range of the probe tips was not large enough to test entire wafers. Therefore, wafers had to be diced into individual chips and the chips were tested one by one.

The design of the surface breakdown test chip varied slightly from the substrate breakdown chip design. The surface breakdown chip design is shown schematically in Fig. 5. The chip is of the same size as the substrate breakdown chip and very similar in appearance to the chip in Fig. 3, however, it features smaller contact pad areas. In the case of the surface breakdown test chip, no doped polysilicon layer was deposited onto the silicon substrate. Instead, LTO oxide (same process as described above) was deposited directly onto the substrate wafer. Following this was an aluminum deposition (same thickness as above) and then patterning and etching of the aluminum. Aluminum pads were placed between 100 and 600 μm apart, in 100-μm increments. Later in the course of the experiment it was found that testing of molybdenum contact pads was considered desirable, and accordingly

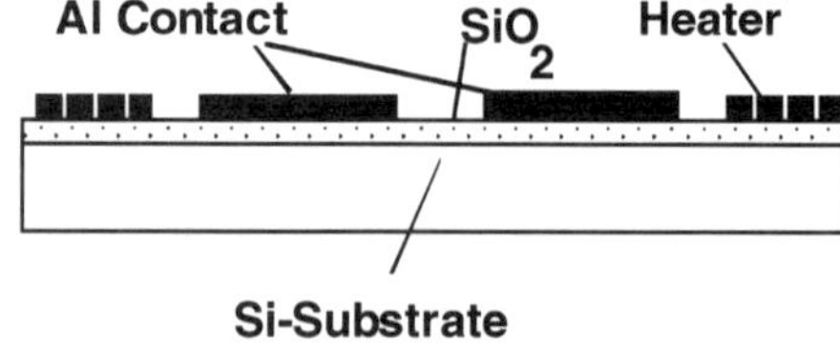

Fig. 5 Schematic of the surface breakdown chip.

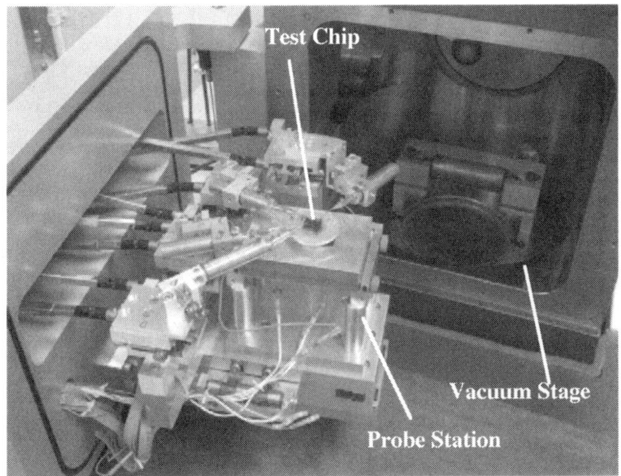

Fig. 6 Experimental setup for surface breakdown.

chips featuring contact pads made from this material were fabricated. Pads on that set of chips were separated by 5, 10, 20, 100, 200, and 300 μm, taking into account new data found with the previously described aluminum chips indicating much higher surface breakdown strengths than measured in earlier tests.

To simplify the fabrication process, the surface breakdown test chips also featured a 3.9-μm-thick oxide layer that allowed the wafers to be fabricated in the same oxidation run as the wafers bound for substrate breakdown chip fabrication. In the course of the tests it was noted that the thick oxide had suffered localized surface delaminations in the form of circular, droplet-shaped protrusions due to the high intrinsic stresses in the thick layer. Since it was uncertain how these delaminations would affect surface breakdown strengths, another set of surface breakdown chips featuring a 2-μm-thick oxide layer, and free of surface delaminations, was also fabricated. Tests were performed with both sets of chips to determine the effect of surface morphology on surface breakdown characteristics.

The surface breakdown chips were mounted in a different probe station, also featuring four probe tips, that could be attached to a scanning electron microscope (SEM) vacuum stage (see Fig. 6). Pressures as low as 1×10^{-6} Torr could be obtained (background gas was air), although the majority of tests was performed at about 3×10^{-5} Torr, measured using the SEM stage pressure gauge. This pressure level could be reached rather quickly using the existing pumping facilities. The vacuum stage of the SEM was turbomolecular pumped.

Both breakdown experiments, substrate and surface, were conducted using a portable DC Hypot device by Associated Research, Inc. (Model 5220A). This device is capable of delivering up to 15-kV voltage at currents of 2 mA or less. Voltages were recorded with a separate voltmeter (Simpson 260 Series 4). Currents were registered with an ampmeter provided with the Hypot device. This current scale was calibrated and known to be accurate within 3–5%. Prior to breakdown of the chips, however, it was noted that most of the current registered (in the microampere range) was flowing through the voltmeter, as current levels were severely influenced by voltmeter settings. During breakdown, however, currents typically ranged as high as 0.5 mA and voltmeter effects were negligible by comparison. Voltage to the chip was applied in increments of 100 V and held for several seconds. If no breakdown occurred, the voltage was increased by another increment. Voltage increments were reduced to 50 V in some cases when it was assumed that breakdown was imminent (based on experiences gained with previously tested chips).

All chips were cleaned after dicing inside the microfabrication cleanroom facilities in an acetone ultrasonic bath for 10 min to remove contaminants and remaining photoresist traces, followed by an isopropyl alcohol rinse to remove remaining acetone residues and a dry. The chips were finally subjected to an oxygen plasma etch at 200 W for 10 min to remove remaining organic residue. The chips were then sealed inside plastic trays. The chips were left sealed inside those trays until the moment of usage. At this time they were subjected to the laboratory environment either for the duration of the test (substrate breakdown) or, in the case of the surface breakdown tests, for the duration it took to install one chip onto the probe station and pump down the system, typically a few minutes.

V. Substrate Breakdown Tests

A. Oxide Thickness Dependence

Determining breakdown field strength with respect to oxide thickness is crucial in the evaluation of LTO oxides for use in ion accelerator grids. As discussed in Section III, electric breakdown field strengths typically vary with oxide thickness, and simple extrapolation of a breakdown field strength obtained for one oxide thickness to much different thicknesses may not be appropriate. Chips with oxide thicknesses of 1, 1.5, 2, 2.7, and 3.9 μm were tested. The breakdown field strengths vs thickness are plotted in Fig. 7. Typically four to six chips were tested for each

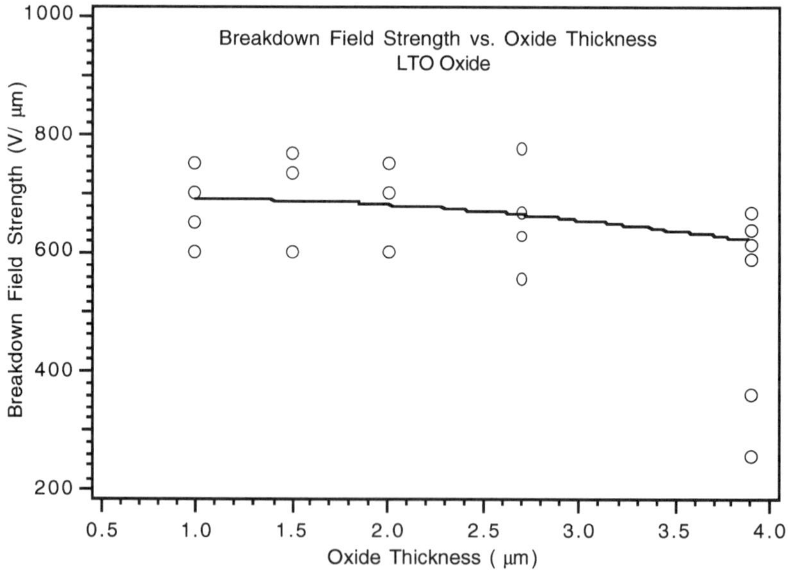

Fig. 7 Electric breakdown field strength vs LTO oxide thickness.

oxide thickness. Each data point shown in Fig. 7 corresponds to the test data obtained with one of these chips. As can be seen, for the thicknesses studied here a small downward trend in breakdown field strength can be noted with increasing thickness. Breakdown field strengths range from approximately 600–750 V/μm at a 1-μm oxide thickness to about 600–650 V/μm at 3.9 μm. Two data points significantly below those values can be found for two 3.9-μm chips. These values may likely be due to oxide defects. The curve fit shown in Fig. 7 excludes these two data points. Breakdown voltages can thus be found between 600 and 750 V at a 1-μm oxide thickness, approaching 2000 V at a 2.7-μm thickness, and reaching values as high as 2500 V at a 3.9-μm oxide thickness.

Oxide thicknesses were measured using an instrument by NanoSpec that derives the thickness from the reflected light intensity off the oxide film of the test specimen and a bare (no oxide) silicon reference wafer. The resulting interference pattern allows determination of the oxide film thickness. Calibration of this method using a known oxide thickness yielded an accuracy of about 1%. Using this technique, oxide thicknesses of the breakdown test chips were to be found accurate within 0.2 μm. This will result in an error for the determined breakdown field strength of about 20% for the 1-μm oxide thickness, decreasing to about 5% for the 3.9-μm oxide thickness.

In addition, an error was incurred in the voltage measurement, due mostly to the measurement technique used. Since voltages were increased in 50- or 100-V increments, the actual breakdown voltage was therefore determined only within 50 or 100 V. (Smaller increments would have resulted in higher accuracies. However, because of the large number of test chips and multitude of voltage recordings per chip, smaller increments would have resulted in very large data sets, extending beyond the level of effort and experimentation time that could be afforded in these tests.) For 1-μm films, where breakdown voltages were of the order of 600–750 V, this resulted in another error of just under 10 or 20%, respectively, depending on the voltage increment (50 or 100 V). For larger oxide thicknesses and correspondingly higher breakdown voltages, the voltage error decreased accordingly, too as low as about 2 or 4%, respectively, in the case of 3.9-μm oxide films.

Total errors in the breakdown field strength measurements may therefore range from about 40% for 1-μm films to just under 10% for 3.9-μm films and may be at least partially responsible for the data scatter. Other effects, such as oxide defects, or locally roughened surface morphology, resulting in uneven aluminum electrode surfaces and locally varying electric field strengths, may also have played a role in the scatter of breakdown field strength data.

It can be estimated from the data set obtained that LTO oxide thicknesses of about 3 μm or greater are fully sufficient to stand off typical grid operating voltages, as far as substrate breakdown is concerned. Later it is shown that, for the associated surface breakdown for this thickness, a different conclusion may have to be drawn.

The data obtained in this study for LTO CVD oxide were compared with data obtained for different oxides from the previously reviewed literature (see Section III). Breakdown field strengths for various oxides at different thicknesses are compared in Fig. 8. Breakdown field strengths much greater than the ones obtained in this study have been recorded in almost every case found in the literature, however, at much lower oxide thicknesses. This increase in breakdown field strength with decreasing oxide thickness appears to become more pronounced with thinner oxides in all cases, independent of the oxide considered, although numerical values

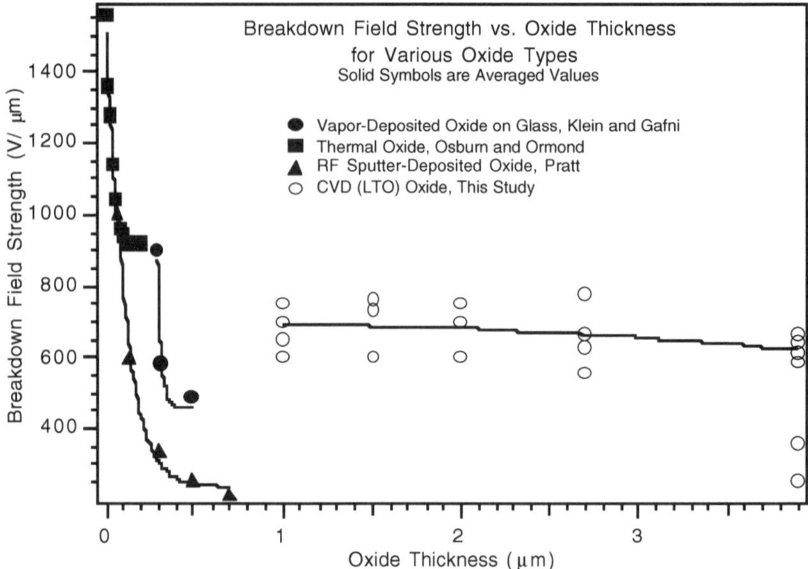

Fig. 8 Breakdown field strengths for various oxides vs oxide thickness.

vary from oxide to oxide. Thermal oxide breakdown strengths are particularly remarkable, which is precisely the reason for their extensive use as gate oxides in MOSFET applications. It should be noted, however, that the values listed in Fig. 8 for thermal oxide, taken from Ref. 10, are the aforementioned intrinsic, or final, breakdown values, and primary breakdown values due to oxide defects are typically significantly lower.

The value of this investigation becomes evident when plotting the obtained breakdown voltages vs oxide thickness, as shown in Fig. 9. Because of the availability of thicker LTO oxides, achievable breakdown voltages are much higher for LTO oxides than for any other oxide considered in this comparison. Even if breakdown voltages for thermal oxides were to be extrapolated into the 1- to 2-μm thickness range (roughly the maximum obtainable thermal oxide thickness), obtainable breakdown voltages would be marginal for ion engine grid applications, and LTO oxides, due to their larger achievable thicknesses, will still outperform thermal oxides, as well as all other oxides considered. These results displayed in Fig. 9 thus very clearly validate the necessity of this study.

B. Temperature Dependence

Attempts were made to perform measurements with respect to the temperature dependence of oxide breakdown strengths using chips with oxide thicknesses of 2.7 and 3.9 μm, respectively. These oxide thicknesses had previously (see above) been shown as sufficient to withstand typical grid voltages at room temperature. However, since these tests were performed under atmospheric conditions for reasons of simplicity and to have access to the IR camera, and higher voltages are required to cause breakdown in the thicker oxides, heavy arcing was noted on and

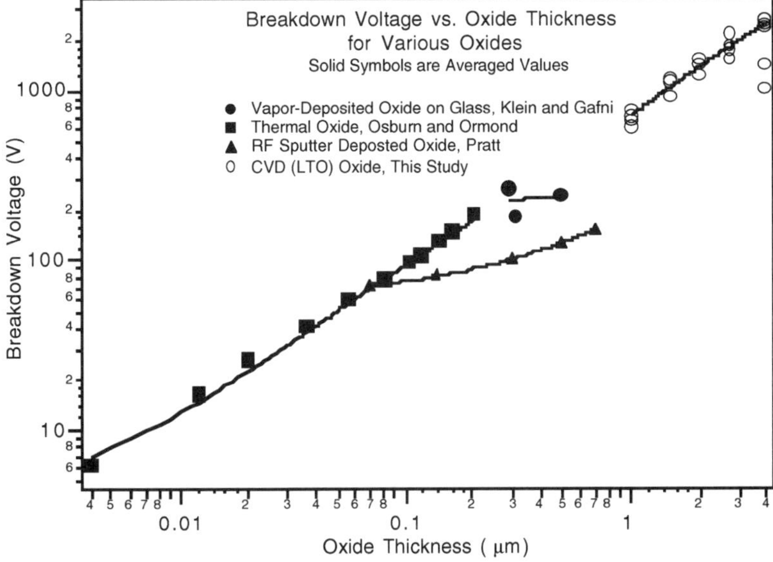

Fig. 9 Breakdown voltages vs oxide thickness for various oxides.

above the chip surface. Arcing was noted between different locations on the chip, between probe tips and the chip, and between probe tips. The arcing was found to be clearly more pronounced at higher temperatures and may have been due to a Paschen breakdown.

However, since, in the case of the 1-μm sample, the required substrate breakdown voltages are much lower, these problems were not encountered using that class of chips. Therefore, the temperature dependence of oxide breakdown field strengths was determined for, and limited to, chips using 1 μm. In Fig. 10 the determined breakdown field strengths for this oxide thickness at various temperatures are shown. Temperatures were varied from ambient (23°C) to as high as 400°C. Typical grid temperatures for conventional (macrosized) grids range between 300 and 400°C. As can be seen, breakdown field strengths decrease slightly with temperature. At ambient temperature, breakdown field strengths range around 600–750 V/μm (and breakdown voltages accordingly around 600–750 V for a 1-μm-thick oxide sample). At 400°C, the breakdown strength has fallen off to 500–650 V/μm, corresponding to a breakdown voltage range of 500–650 V. This corresponds to a drop in breakdown strength and voltage of about 15%. Note, however, the large scatter of data, making this quantitative conclusion a preliminary one. The same error considerations, as outlined in the previous section, apply here.

It cannot be excluded that breakdown field strengths for thicker oxides may have behaved differently with respect to temperature. However, this would require a geometry (thickness)-dependent temperature effect, an assumption that appears not to be obvious. Furthermore, the drop in breakdown field strength, at least for the smaller oxide thicknesses, is so low, and the margins for the larger thicknesses with respect to breakdown voltages so great, that substrate breakdown temperature effects are currently not considered a serious impediment to proper ion engine grid function.

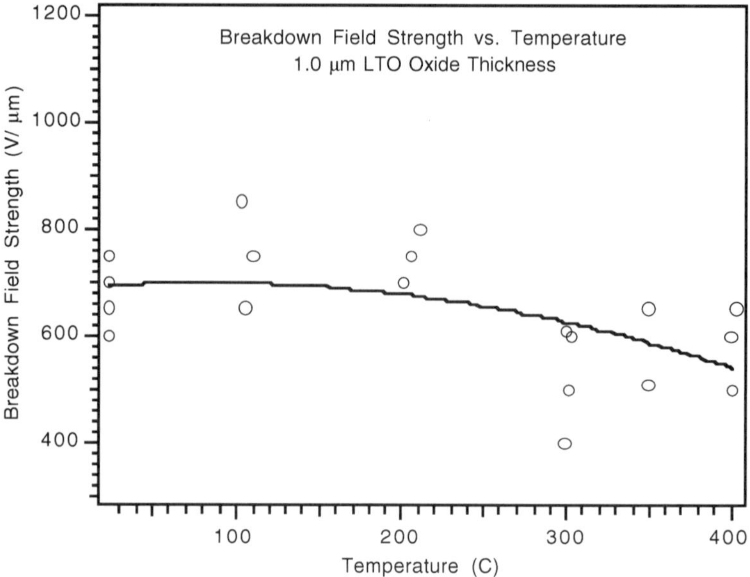

Fig. 10 Electric breakdown field strength vs temperature for 1-μm LTO oxide.

C. Visual Post-Test Inspection of Test Samples

Electron microscope scans were taken of various test samples after the breakdown tests to determine their failure mechanisms. Figure 11 shows a typical oxide breakdown. It is located at the edge of the aluminum contact pad area, which is still recognizable in the lower part of the photograph although heavily eroded in the immediate vicinity of the breakdown. Note the relatively large size of this breakdown, extending to approximately 30 μm in diameter. The oxide thickness

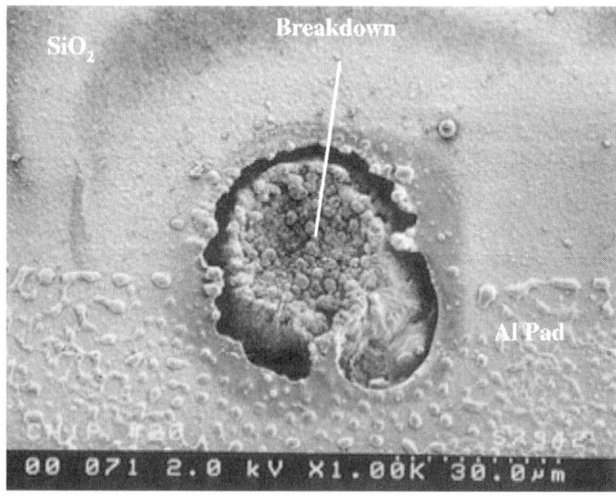

Fig. 11 Electric breakdown at aluminum contact pad edge (2.7-μm oxide thickness).

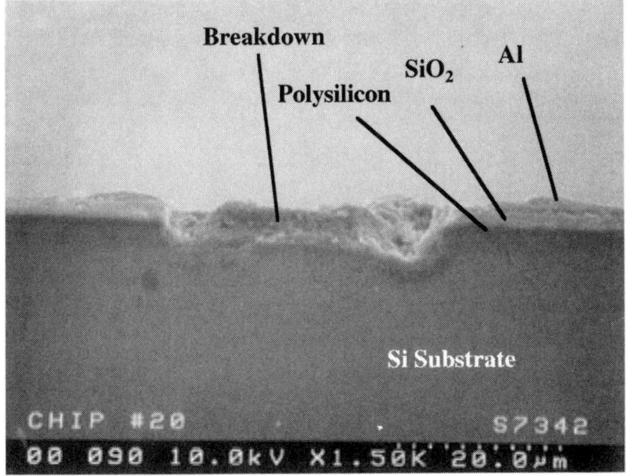

Fig. 12 Side view of the breakdown shown in Fig. 11.

in this case was 2.7 μm. Electric breakdown occurred at 1800 V. The oxide used in this case was of the batch provided by UCLA.

Breakdowns at the contact pad edges and contact pad corners by far outnumbered breakdowns at other pad locations. Similar observations were made by Soden[11] during his investigation of the dielectric strength of thermal oxides. Soden attributed this fact to the lack of defects in the oxides. If defects would have triggered a breakdown, one would expect the breakdown sites to be distributed more randomly. The fact that breakdowns instead occur predominantly on contact pad edges and corners is an indication that these may be intrinsic breakdowns, triggered by the higher electric field strength in these regions. Small inhomogeneities on the contact pad surface or slight variations in the oxide thickness may trigger breakdown at one particular location along the contact pad edge vs another. The high number of breakdowns observed on contact pad edges leads us to believe that the oxides used in our tests were mostly free of defects.

Figure 12 shows a side-on view of the breakdown shown in Fig. 11, clearly indicating that the oxide layer, visible as the lightly colored layer just above the darker-colored silicon substrate, has been penetrated. (The polysilicon layer, being only 0.3 μm thick, is hardly visible in the photograph and appears as a very thin black line just between the silicon substrate and the oxide in the original.) As can be seen, besides the destruction of the oxide layer, substantial damage has also been done to the silicon substrate located directly below the breakdown area. This damage is due to the substantial local Joule heating during breakdown. Although no temperature measurements on the arc were performed in this study, Klein,[9] in performing spectroscopic temperature measurements on the breakdown arc, determined arc temperatures of the order of 3900–4500 K for thermal oxide breakdowns. If similar temperatures were to occur in LTO breakdowns as well, these values would certainly be sufficient to melt the silicon substrate, having a melting temperature of about 1400°C.

Figure 13 shows a spectral (X-ray fluorescence) analysis of the distribution of elements surrounding the breakdown shown in Figs. 11 and 12. Three picture segments show the distribution of silicon (top right), aluminum (bottom left), and

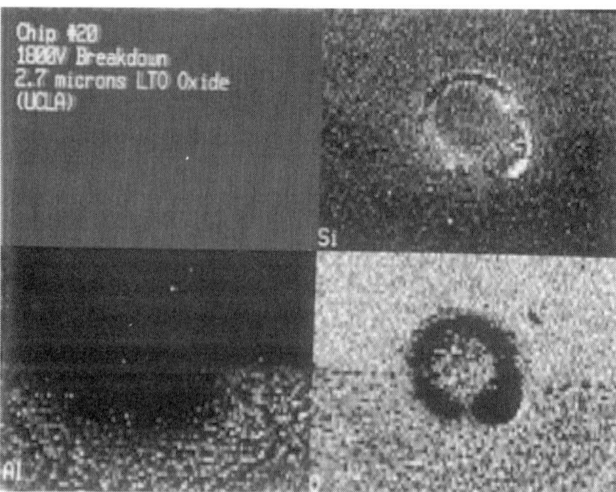

Fig. 13 Spectral (X-ray fluorescence) analysis of the breakdown shown in Fig. 11.

oxygen (indicative of silicon oxide, bottom right) as seen from a top-view position similar to the one shown in Fig. 11. As can be seen by inspecting the top-right segment, silicon is clearly visible through the gap in the oxide layer, which shows up as a dark ring-shaped structure in the oxygen scan in the bottom-right segment, indicating the lack of oxide here. This, together with the visual evidence presented in Fig. 12, also gives a clear indication that a breakthrough to the underlying silicon/polysilicon layers has indeed taken place. Also visible in these scans is the heavy erosion of the aluminum contact pad (located in the lower half of the picture segments). While some aluminum traces can still be found in this area (see lower-left picture segment), the silicon oxide, onto which the aluminum contact pad was deposited, is clearly visible in this area now as well (see lower-right picture segment).

In the case shown in Figs. 11–13, as in all breakdown cases recorded during this set of experiments, a permanent short was noted after breakdown. Voltages typically collapsed to values ranging around a few tens of volts or less (after having been as high as several hundred or even thousands of volts just prior to the breakdown) and currents in excess of 0.5 mA were measured. The short is likely caused by the severe disturbances noted in the breakdown area, as shown in Fig. 12, mixing elements of the various chip layers, thus providing electrical contact.

Figure 14 shows another breakdown mode of a chip featuring 2.7-μm-thick oxide, with the breakdown also occurring at 1800 V, as in the case of the chip depicted in Figs. 11–13. This chip was fabricated using the oxide provided by Berkeley. A peculiar meandering pattern can be noted on the chip surface in areas that have seen heavy aluminum pad erosion. The sequence of events, as documented by the IR camera and recorded on tape, was as follows. Breakdown occurred first at a contact pad edge location in the top left corner of the pad area. The probe tip contacted the pad area in the location shown. After breakdown at the contact pad edge, the aluminum pad eroded outward from the initial breakdown location, with the eroded aluminum pad edge recessing until it reached the probe tip location. At this point the erosion process stopped. The voltage dropped from 1800 V prior to breakdown (at low, microampere current values believed to be conducted largely through the voltmeter), to about 500–600 V at about 0.5 mA during the surface

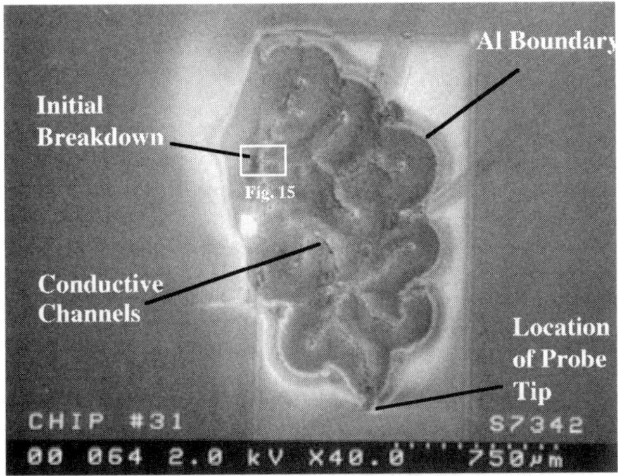

Fig. 14 Propagating breakdown pattern showing "tree-shaped" conductive channel formations (2.7-μm oxide thickness).

erosion/arcing process, and, finally, to the aforementioned few to few tens of volts, depending on the chip sample, at currents of about 0.5 mA, creating a permanent short. Current and voltage values, except for the initial breakdown voltage of course, were found to be typical for most chips breaking down in this fashion, except for the ones using the thinnest oxides (1 μm; see below).

At first glance, the meandering erosion pattern seems to point to a pure surface breakdown phenomenon as a result of arcing between the exposed grounded polysilicon layer (after the initial breakdown) and the eroded aluminum pad edge, which is held at a high positive voltage. This surface arcing between the breakdown area and the aluminum layer could then be thought to continue to generate heat, which causes the aluminum layer to ablate further until the probe location is reached, representing the minimum path of resistance to the high-voltage supply. A more detailed study, however, reveals a more intricate process.

Figures 15 through 17 show a detailed view of the initial breakdown area and the starting point of the meandering "tree-shaped" pattern that was observed on the chip surface. Figures 16 and 17 were obtained by dicing the chip along one of the "branches" of the "tree" pattern. In Fig. 16, the initial breakdown can be seen, revealing a structure similar to that of the breakdown shown in Fig. 12. Again, a penetration of the oxide layer combined with a significant disturbance of the various layers of the chip (aluminum, oxide, polysilicon, and silicon substrate) can be observed. Just to the right of the initial breakdown area shown in Fig. 16, however, along one of the surface breakdown "branches," additional penetrations of the oxide and cavities formed inside the silicon substrate can be noted. This pattern continues if one were to progress farther to the right of the location shown in Fig. 16, as shown in Fig. 17. Clearly, a large penetration of the oxide can be noted in the left half of Fig. 17. Additional cavities appear to be sealed by the oxide layer, however, it should be noted that dicing farther into the chip may have revealed these cavities to be "open" as well, thus quite possibly representing oxide penetrations as well. Thus, the process forming the meandering "tree" pattern on the surface of this chip is clearly not solely a surface phenomenon, but involves subsurface events as well.

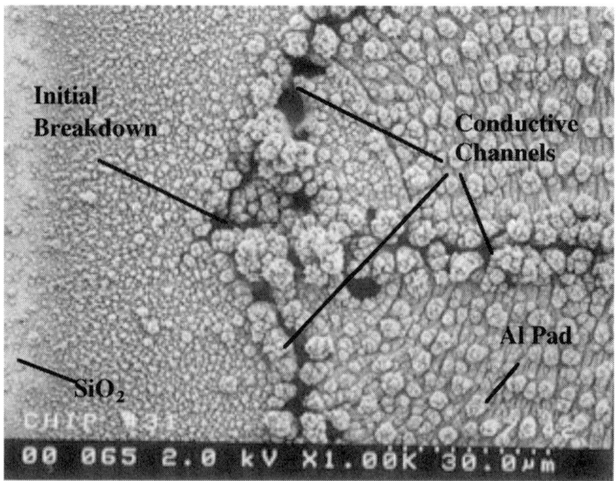

Fig. 15 Close-up of the initial breakdown location in Fig. 14.

Similar erosion patterns have been observed previously by Klein[9] during breakdown tests performed with thermal oxides. Klein termed these types of breakdowns "propagating breakdowns" and offered an explanation for their occurrence. According to Klein,[9] the breakdown starts at a single location, as observed in our experiments also. Due to the Joule heat produced by this initial breakdown, the conductivity of the insulator material may be slightly increased in the vicinity of the initial breakdown location, causing another breakdown to occur in an area immediately surrounding the initial breakdown location. The process now continues, causing the "branch" pattern to form. Since, as noted in this study, a current of approximately 0.5 mA is constantly flowing between the two electrodes during this erosion process, a voltage drop (however minute) is expected to occur along the uneroded aluminum pad area, extending from a high value at the location of

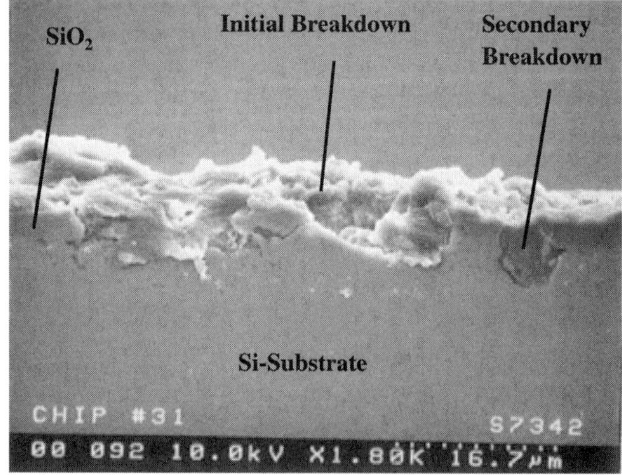

Fig. 16 Side view of the initial breakdown area shown in Fig. 15.

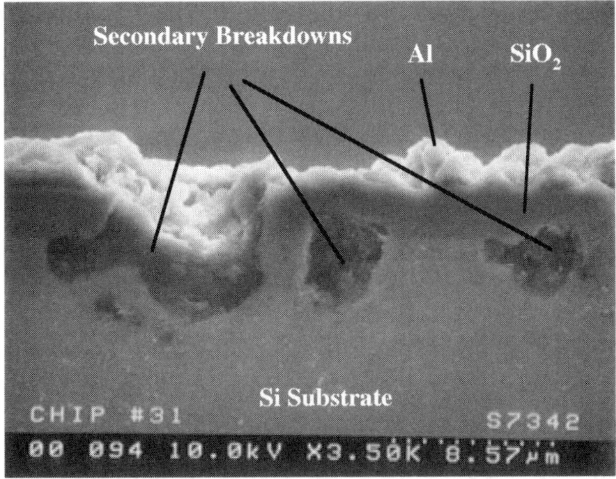

Fig. 17 Side view of a conductive channel segment, located to the right of the formation shown in Fig. 16.

the contacting probe tip to a low value in the proximity of the eroded pad edge. Thus, a preferential direction is given for subsequent breakdowns to occur (toward higher voltage values) until one of the "branches" finally connects with the probe tip location. It should be noted that the observed erosion process could be stopped anytime after the initial breakdown and the low-voltage short would still have been observed, as demonstrated in various test runs.

For chips featuring oxide thicknesses of 2.7 μm and greater, all breakdowns observed fell in either of the two categories described in Figs. 11 and 14. However, an entirely different breakdown pattern was found to be characteristic for oxides with a thickness of 1 μm, as can be observed in Fig. 18. This figure shows a chip featuring an oxide thickness of 1 μm after a 750-V breakdown. The same

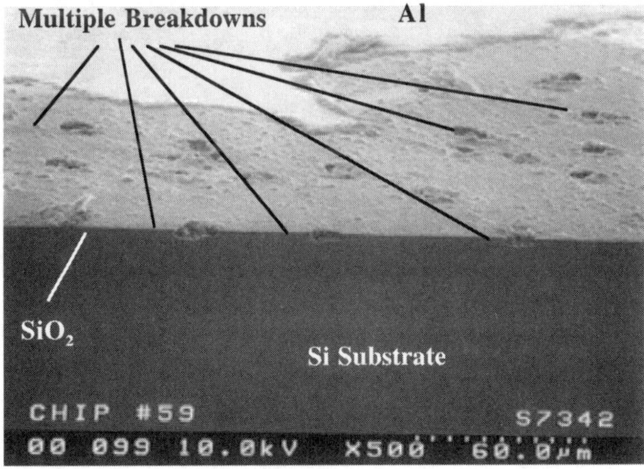

Fig. 18 Example of multiple breakdown locations for thin oxides (1 μm).

type of breakdown as shown in Fig. 18 was noted for all 1-μm chips. As can be seen, multiple breakdown locations can be recognized distributed over an area that was again located close to the contact pad edge. Several breakdowns had again occurred very near to this edge.

All breakdowns again penetrate the silicon oxide layer deep into the silicon substrate, causing the previously noted severe disturbance of the chip material in this area, again leading to a permanent short after breakdown. Current and voltage characteristics for the shorts in 1-μm chips were about 0.3–0.4 mA and 0.15 to about 12 V, respectively, with one value being as high as 150 V. The breakdown patterns shown in Fig. 18 did not occur instantaneously, but required several seconds to develop, with arcing starting near the edge or corner of the contact pad and then progressing toward the probe tip location. In the case of the chip shown in Fig. 18, however, this process stopped on its own after reaching the state depicted in the figure. Current and voltage characteristics for these chips during this arcing process were about 0.3 mA and 400 V and, thus, as for the case of the shorts, slightly lower than in the case of thicker oxides.

Again, this type of breakdown pattern was observed previously by Klein[9] in his study of dielectric strengths of thermal oxides and was attributed by Klein to the same thermally triggered breakdown process as described above. However, the different appearances of the two classes of propagating breakdown patterns shown in Figs. 14 and 18, respectively, warrant a closer examination. One obvious difference between the chips exhibiting these different propagating breakdown behaviors is the much smaller oxide thickness (1 vs 2.7 μm) in the case of the chips shown in Fig. 18 vs the chip shown in Fig. 14. Several tests were performed to examine how breakdown patterns for intermediate oxide thicknesses would appear. The results of one of these tests is shown in Fig. 19. The breakdown pattern exhibited on this chip appears to be somewhat of a cross between the two classes identified above: while multiple, separated breakdowns did occur near the edge, almost all of these breakdowns show rudimentary "branch" growth emanating from the breakdown locations.

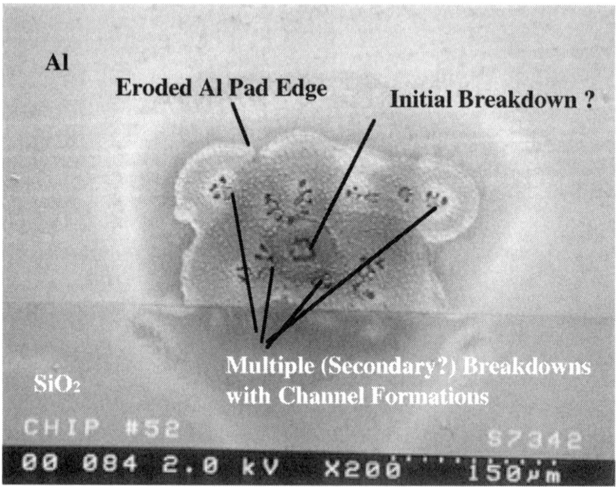

Fig. 19 Multiple breakdowns for a chip featuring 1.5-μm oxide. Note fewer breakdowns and start of channel formations.

We believe that an explanation for this behavior may be found in possibly different thermal conduction processes in chips of different oxide thicknesses. Silicon dioxide is a poor thermal conductor compared to silicon, the thermal conductivity being 1.4 W/mK in the case of oxide vs about 150 W/mK for silicon. Given that the destruction found underneath the initial breakdown locations involves the silicon substrate, heat conduction away from the initial breakdown site is likely to occur both through the oxide as well as through the silicon. In the case of thinner oxides heat conduction away from the initial breakdown site may therefore be enhanced, allowing higher temperatures to be reached farther away from the initial breakdown site, leading to breakdown there and resulting in a more widely scattered breakdown pattern. Since those locations closer to the high-voltage probe tip will carry the majority of the current as it seeks the path of lowest resistance, the current passing through the original breakdown site subsides and, with it, heat dissipation to immediately adjacent areas decreases as well. Therefore no additional breakdowns in its immediate neighborhood occur. The ultimate location of the individual breakdowns, apart from the temperature profile, may then be determined by small variations in oxide thickness or inhomogeneities on the contact metal surface. Since a current is observed to be constantly flowing between the two electrodes (polysilicon and aluminum), a voltage drop will extend from the high-voltage probe tip location on the aluminum pad to its eroded edge. Thus, a preferential direction for further breakdowns is provided (toward areas of higher voltage), until the position of the high-voltage probe tip has been reached. As in the case of the previously discussed class of breakdowns, the breakdown process can be stopped immediately after initial breakdown by turning off the voltage.

VI. Surface Breakdown Tests

A. Dependence on Gap Distance

As noted in Section I and indicated in Fig. 1, in an ion engine accelerator grid arcing may also occur along the insulator oxide surface. Previous tests performed by the authors under atmospheric conditions[17] had led to parasitic surface breakdowns when performing substrate breakdown tests. The resulting surface breakdown voltages were a troublesome 2 V/μm over gap distances of about 200 to 300 μm. Surface breakdown field strengths that low, if applied over a 5-μm-thick oxide layer (corresponding to roughly the maximum LTO oxide thickness that can be deposited), would be wholly insufficient for typical ion engine grid applications. Thus, a more thorough investigation of surface breakdowns along LTO oxide surfaces was conducted. These tests were performed in a vacuum system, as outlined in Section IV. Unless noted otherwise, breakdown tests were performed at a vacuum pressure of 3×10^{-5} Torr.

Given the low measured breakdown field strength in earlier experiments,[17] initial tests were performed with contact pads separated by gap distances of 100, 200, 300, 400, 500, and 600 μm. Results obtained from these tests are shown in Fig. 20. As can be seen, surface breakdown electric field strengths range from about 20 V/μm at a 100-μm gap distance to as low as 3–4 V/μm at a 600-μm gap distance between the aluminum pads. At values between 200 and 300 μm, electric breakdown field strengths are about 10 V/μm, thus clearly higher than for breakdown under atmospheric conditions.

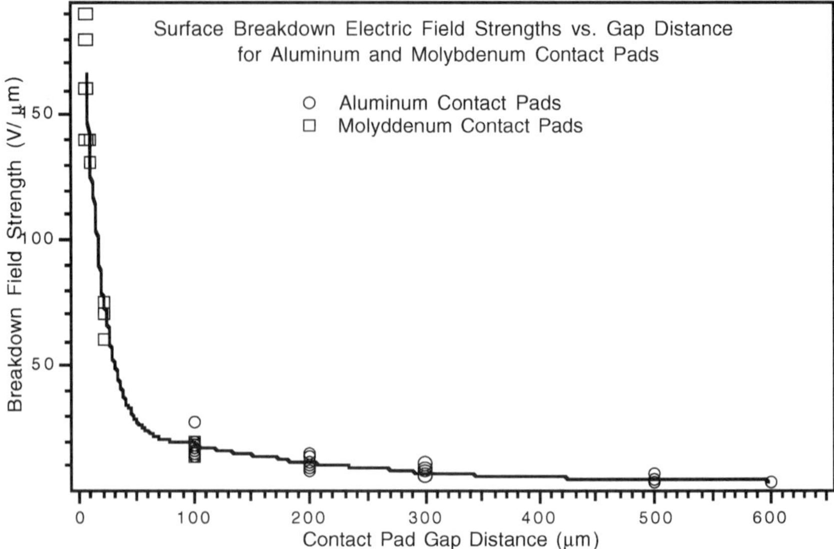

Fig. 20 Surface breakdown electric field strengths for LTO oxide using aluminum and molybdenum contact pads vs pad gap distance.

Even these increased breakdown field strengths, however, are still too low for ion engine grid applications. Suspicions were raised that the use of aluminum, which has a tendency to form hillocks on its surface, may have led to a decreased voltage stand-off capability as a result of these surface roughnesses.[18] Aluminum had been used in the fabrication of these test chips because of its ready availability in our cleanroom facilities, its ease of use in the microfabrication process, the past experience with its use as a MEMS material, and its good sticking abilities. In addition, Osburn and Ormond,[7] in performing experiments aimed at determining substrate breakdown field strengths for thermal oxides, had tested various electrode materials, including aluminum and molybdenum, and had found no difference in breakdown behavior.

To resolve the remaining doubts chips using molybdenum contact pads were fabricated. In addition, due to the noted increase in breakdown field strength with decreasing gap distance for the chips using aluminum pads, the mask design for the molybdenum chips was changed. Now, in addition to gap distances of 100, 200, and 300 μm, gap distances of 5, 10, and 20 μm were included to perform tests at these lower gap distances as well. The data obtained are also plotted in Fig. 20 (open squares) and represent the steeply inclined part of the curve. Two remarkable findings can be noted. First, in testing molybdenum chips at a 100-μm gap distance, it was noted that there is no apparent difference in surface breakdown field strength compared with that of chips featuring aluminum contact pads. Data for the 100-μm gap distance for both types of contact pads almost overlap identically at about 20 V/μm. These results obtained for surface breakdown experiments on LTO oxides thus mirror experiences gained by Osburn and Ormond[7] with substrate breakdowns of thermal oxides.

Second, on decreasing the gap distance, a remarkable increase in breakdown field strength can be noted. At least three measurements were taken for each gap,

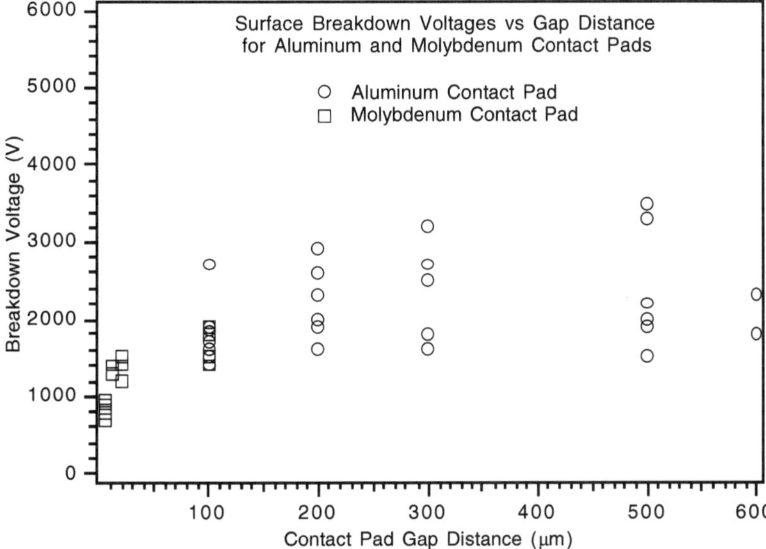

Fig. 21 Surface breakdown voltages for LTO oxides using aluminum and molybdenum contact pads vs pad gap distance.

with results repeating each other with comparably little scatter in the data. This increase in surface breakdown field strength toward lower gap distances thus mirrors a similar behavior found for substrate, or bulk, breakdown of many other oxides (compare with Fig. 8). Note that two curve fits were used for the breakdown field strength data—one for the molybdenum data and another for the aluminum data—yet both curves appear to match up very well at the 100-μm data points.

This increase in electric breakdown field strength is encouraging, however, still not quite sufficient for ion engine accelerator grid use, as can be seen by inspecting Fig. 21. For gap distances of 5 μm, representing the approximate maximum LTO oxide thickness that can be deposited, breakdown voltages remain just below 1000 V. Thus, a new electrode geometry was explored. A set of surface breakdown test chips was fabricated featuring an oxide undercut extending below the (molybdenum) contact pad (see Fig. 22). The undercut was achieved through a buffered oxide etch (BOE). It was hoped that this undercut would 1) increase the breakdown surface path, thus increasing the surface breakdown voltages, and 2) eliminate the sharp 90-deg edge of the pad in direct contact with the oxide, thus hopefully decreasing the local field strengths and therefore delaying the onset of breakdown to larger voltages. This electrode design was influenced by cold cathode designs. Using similar designs, Spindt* has reported breakdown voltages of about 250 V/μm.

Table 1 lists the results obtained with a set of chips featuring an oxide undercut as shown. A 5-μm gap was tested. For comparison, data obtained for chips

*Spindt, C., Personal communication with C. Marrese, SRI International and Jet Propulsion Lab., June 1998.

Table 1 Surface breakdown voltages and field strengths for a 5-μm gap with and without an oxide undercut

Breakdown voltage, V	Breakdown field strength, V/μm
Oxide undercut (5-μm *gap*)	
600	120
600	120
900	180
900	180
1300	260
No oxide undercut (5-μm *gap*)	
700	140
700	140
700	140
800	160
900	180
900	180
950	190

with the same gap distance, but without an oxide undercut, were tested as well. Both sets of chips were fabricated from the same wafer and were exposed to the same fabrication processes and conditions, except for the oxide etch in the case of the chips featuring the undercut. As can be seen by inspecting Table 1, although one single data point obtained for a chip featuring an undercut resulted in a record breakdown field strength of 260 V/μm (corresponding to a 1300-V breakdown voltage), the remainder of the results are not very convincing. As a matter of fact, in some cases the breakdown field strengths and voltages obtained with chips featuring undercuts are lower than for chips not featuring an undercut, although this may be an effect of the scatter of data. Thus, the undercut does not appear to be effective, at least for the current chip geometries. It is possible that the oxide etch may also affect the molybdenum and may have increased the surface roughnesses, counteracting any desired effects the change in geometry may have caused, if any. It is interesting to note that breakdown voltages obtainable with cold cathode

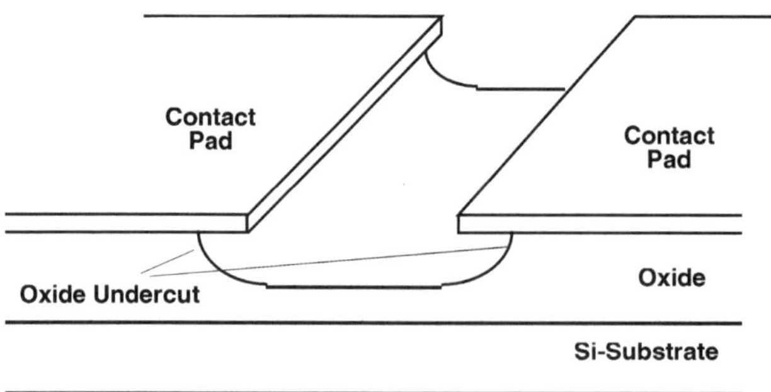

Fig. 22 Attempt to achieve an oxide undercut of a molybdenum pad.

arrays, featuring similar electrode/insulator geometries, yield breakdown values as high as 250 V/μm, however, at oxide thicknesses of about 1 μm (see footnote, previous page). It is unclear whether breakdown occurs through the substrate or along the surface in these cases. If due to surface breakdown, these data would compare rather well with data obtained in this study, taking into account that surface breakdown values for a 5-μm gap were about 200 V/μm and a further increase in breakdown strength would be expected for a 1-μm gap, following the observed trends in breakdown field strength with gap distance.

Spot checks were performed to verify the gap distances between the contact pads specified in the fabrication process. Gap distances varied between 0.2 and 2 μm from the nominal value. The highest deviations in the case of 5-μm gaps were measured to be 0.75 μm, corresponding to a 15% error in the breakdown field strength, and in the case of 100-μm gaps the difference from the nominal value was 2 μm, corresponding to a 2% error. Measurements on other chips resulted in percentage errors between those two values. As in the case of the substrate breakdown experiments, the voltage was increased in fixed increments in the cases of surface breakdown experiments by 100 V. Therefore, the actual breakdown voltage is known only within an error band of 100 V. For the lower breakdown voltages, such as in the case of the 5-μm chips, the associated error is about 10%, leading to a total error, including the gap distance error, of about 25% for the breakdown field strength in these cases. For larger gap distances, where breakdown voltages are higher, the voltage error reduces. In the case of a 100-μm chip, the error is about 5%, leading to a total error in the breakdown field strength of 7%, taking into account variances in the gap distance.

In the case of the 5-μm chips, more relevant for ion engine grid considerations, breakdown voltages remain marginal for grid applications, even taking into account these error calculations.

B. Paschen Breakdown Considerations

An experiment was conducted to determine the influence any remaining rest gases in the vacuum system might have had on the measurements, if any. In Fig. 23, breakdown voltages are plotted vs the product of the gas pressure inside the vacuum system and the gap distance. Using this representation, if arcing through the rest gas had been present, a Paschen-type curve should have resulted. All measurements were performed at a gap distance of 100 μm and taken at various stages during the pump-down process. This allowed for measurements at pressures ranging from 10^{-4} to as low as 10^{-6} Torr. Accordingly, pressure/gap products are extremely low, ranging from 10^{-8} to 2×10^{-6} Torr cm. Typically, these values would indicate a position far to the left of the minimum of the Paschen curve for commonly used rest gases that could have been present in the chamber (nitrogen, oxygen, water vapor traces). At these values, if a Paschen breakdown were present, breakdown voltages should have been much higher than observed and should have decreased dramatically toward larger pressure/gap product values. In inspecting Fig. 23, however, it is clear that this is not the case. No particular trend is visible among the data points, and only the usual scatter of the data, as observed for measurements taken at constant pressure and gap distance (see Fig. 21), is noted. Thus, it was concluded that the surface breakdowns observed were likely true surface effects and Paschen breakdown was not present.

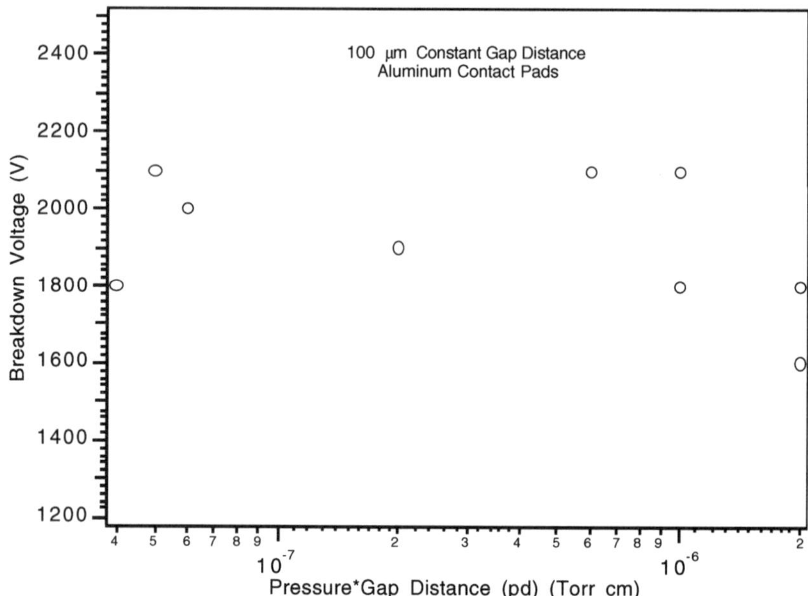

Fig. 23 Breakdown voltages for a constant gap distance at various background pressures.

C. Influence of Surface Morphology

During early surface breakdown measurements, chips fabricated from wafers featuring a 3.9-μm-thick oxide were used for reasons explained in Section IV. In the course of these experiments it was discovered that droplet-shaped surface features were present on the chip surface and, thus, in the gap area as well (see Figs. 24 and 25). Feature sizes ranged between 3 μm (Fig. 24) and less than 1 μm (Fig. 25) in diameter. Naturally, it was feared that these features could have had an influence on the data obtained and be at least partly responsible for the low surface breakdown strengths. It was quickly determined, through a combination of X-ray fluorescence spectral analysis and various standard cleaning techniques, that the surface features were not contaminations resulting from organic residue, photoresist, etc., but instead were stress delaminations caused by the large intrinsic stresses in the thick LTO layer. Consequently, wafers featuring thinner oxides (2 μm) were fabricated and used in subsequent tests.

However, using these chips, an unexpected opportunity presented itself to study the influence oxide surface morphology might have on surface breakdown characteristics. Chips of the original 3.9-μm LTO batch, chips fabricated by UCLA using a 2.7-μm oxide featuring fewer delaminations, and the latest Berkeley batch using a 2-μm oxide having no detectable delaminations were tested and the data obtained were compared. All surface delaminations inside the gap area were counted under an optical microscope and the average surface delamination densities in the gap area were calculated. These densities are believed to be accurate to within less than 10% or so, as counting such a multitude of features leads to miscounts, in particular,

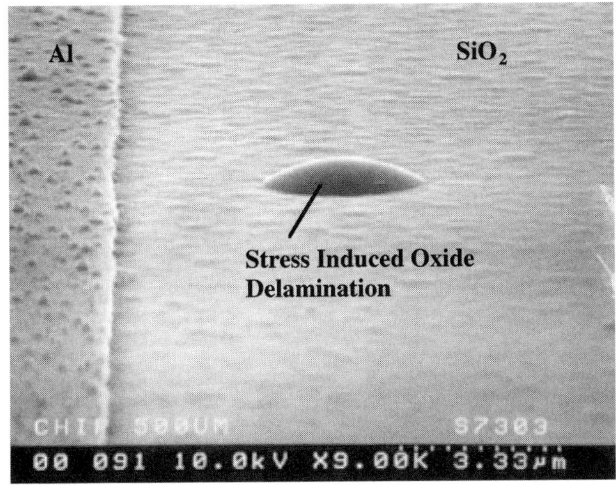

Fig. 24 Example of a stress-induced surface delamination (approx. 3-μm diam) on oxide.

since in some cases chips had already been tested and debris resulting from aluminum pad erosion had to be discerned from surface delaminations. However, this accuracy is believed to be sufficient, considering that a very wide range of surface delamination densities, ranging from 0 to as high as 4000/mm^2, was obtained.

Figure 26 shows the results for three gap distances: 100, 200, and 300 μm. No particular trend of breakdown field strength with respect to delamination density can be observed for either of the gap distances. The scatter in breakdown field data appears somewhat less pronounced for lower delamination densities, but the

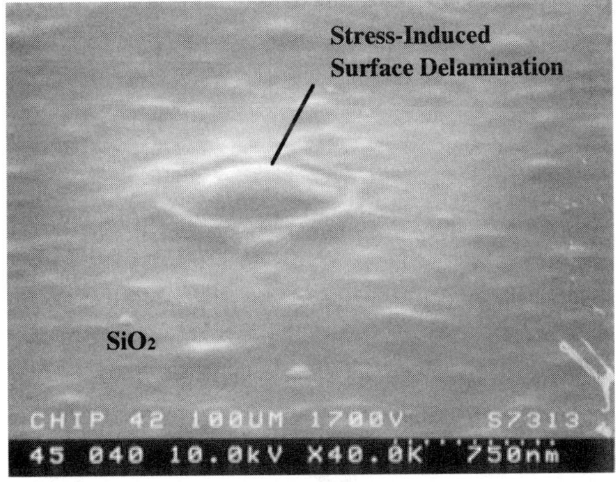

Fig. 25 Example of a stress-induced surface delamination (less than 1-μm diam) on oxide.

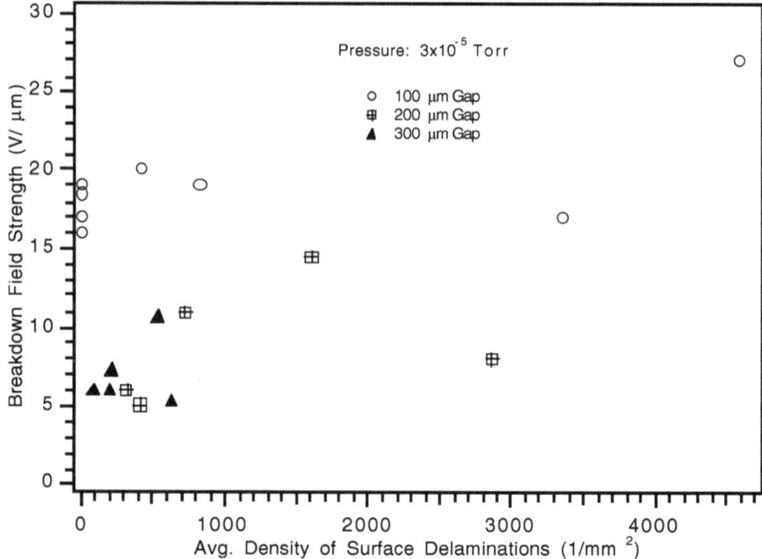

Fig. 26 Breakdown field strength vs surface delamination density.

differences remain small. There definitely appears to be no trend toward lower breakdown field strengths at higher delamination densities. Therefore, it was concluded that surface morphology of the type observed in Figs. 24 and 25 did not affect breakdown strengths. However, it should be pointed out that the particular surface features encountered here have relatively smooth shapes and comparably large radii of curvature.

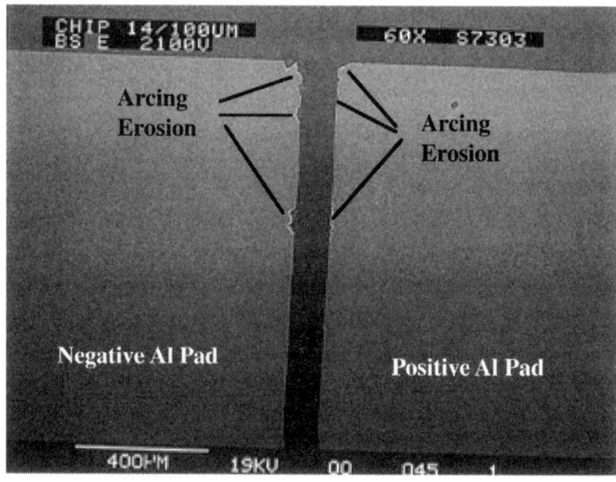

Fig. 27 Example of contact pad damage after surface arc breakdown (arcing voltage, 2100 V; gap, 100 μm).

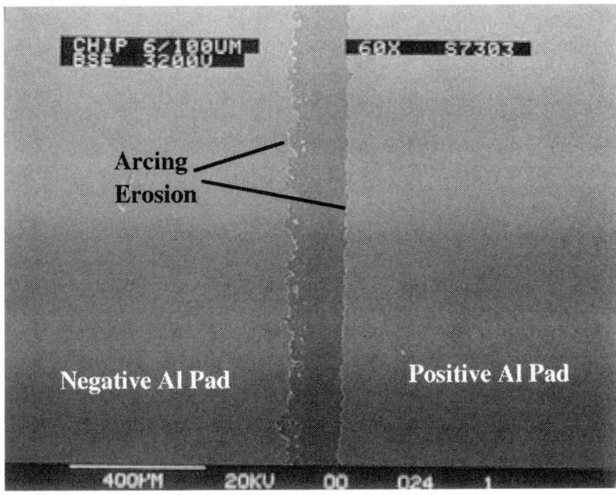

Fig. 28 Example of contact pad damage after surface arc breakdown (arcing voltage, 3200 V; gap, 100 μm).

D. Visual Post-Test Inspection of Test Samples

All surface breakdown test samples appeared very similar after breakdown. Examples of two chips imaged after breakdown are shown in Figs. 27 and 28. Both chips featured aluminum contact pads and a gap distance of 100 μm. Arcing occurred preferentially at the corners of the pad area but also at straight edge sections. Depending on the breakdown voltage, isolated burn marks, as in Fig. 27 (breakdown at 2100 V, or 21 V/μm), or extensive erosion along the entire pad edge, as in Fig. 28 (breakdown at 3200 V or 32 V/μm), can be observed. Damage is typically

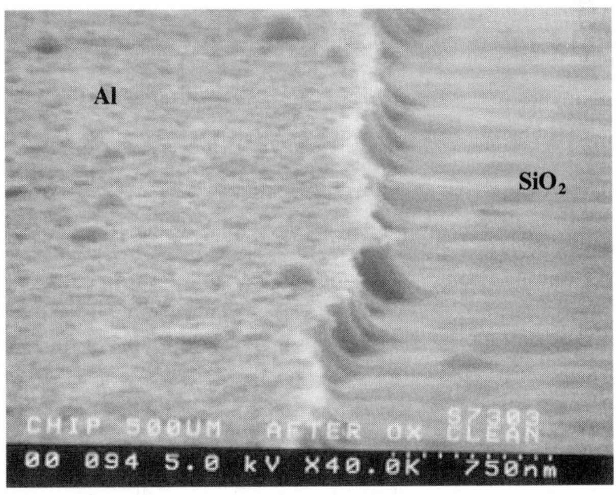

Fig. 29 Close-up of aluminum pad edge.

more intense on the negative pad (shown at the left in both figures) than on the positive pad. The cause for this behavior is not certain. One explanation offered was that electron field emission from microscopic tips along the negative pad edge may have led to local heating and thus increased erosion. An example of an (uneroded) aluminum contact pad edge is shown in Fig. 29. Tips protruding from the edge are small (approximately less than a few tenths of microns, representing state-of-the-art microfabrication/patterning technology), however, are sharply pointed. The amount of damage to the contact pads in all cases is of some concern in view of potential ion accelerator grid applications of these thin-film metallic deposits. Arcs could, if they did occur, cause substantial damage to a microfabricated grid.

VII. Conclusions

Microfabricated ion accelerator grids are being considered for use in micro-ion engines due to the unrivaled precision with which these components could be built. In particular, for grids, requiring a multitude of closely spaced apertures placed within tight tolerances to provide proper grid hole alignment and beam extraction, microfabrication techniques may be beneficial. However, fabrication of these grids will require the use of new materials, typically not used in the fabrication of conventional grids, requiring an investigation into the feasibility of such an approach.

Among the material properties to be studied is the dielectric breakdown strength of grid insulator materials. One of the most popular insulator materials used in the MEMS area is silicon dioxide. Most breakdown work in the past, however, focused on the evaluation of thin thermal oxides, such as for use as gate oxides in MOSFETs. While these oxides show excellent electric breakdown field strengths for thin layers, thermal oxide, due to its growth process, can typically be grown only up to a thickness not exceeding 2 μm. Over such thicknesses the total voltage that can be stood off is marginal with respect to ion engine grid applications. On the other hand, CVD LTO oxide can be deposited up to a thickness of possibly 5 μm. However, many details of the dielectric properties of LTO oxides, in particular, for very thick films and at elevated temperatures, were not known. Thus, a thorough investigation of these properties was initiated. Results of this evaluation remain mixed at this point of the investigation.

On the one hand, the substrate, or bulk, electric breakdown properties of LTO oxide were shown to be excellent. Voltages as high as 2500 V can be stood off over an oxide thickness of 3.9 μm, providing more than a sufficient margin of safety for grid applications. In addition, there are strong indications that the oxides used showed few to no defects that could lead to premature electric breakdowns. This was evidenced by the fact that breakdowns usually occurred near contact pad edges, rather than being randomly distributed, as would be expected if defects had caused these breakdowns. No particularly adverse temperature effects with respect to breakdown strengths were noted for LTO oxides either. Although a small drop in breakdown strength was determined for a 1-μm-thick oxide sample, decreases are small (approximately 15%) with temperatures increasing from ambient to 400°C. The breakdown voltages obtained compare very favorably to the corresponding literature data found for other oxides, such as thermal and sputter-deposited oxides. In the case of thermal oxides this is due mainly to the comparatively larger LTO oxide thicknesses that can be deposited.

On the other hand, the surface breakdown properties appear marginal at best. Although it was discovered that the surface breakdown electric field strengths increase significantly with smaller gap distances, reaching values of about 200 V/μm for 5 μm, the obtainable voltages over these distances remain relatively low (i.e., less than 1000 V). New grid/insulator geometries were therefore explored, based on cold cathode design features. Since breakdowns tended to occur predominantly along contact pad edges and near corners, it was reasoned that the field concentration at these locations plays a role in oxide breakdowns. Chips with oxide undercuts, extending underneath the contact pad edges, were therefore fabricated. The maximum obtainable field strength at breakdown obtained using these chips was 260 V/μm, but in most cases results comparable to those obtained with chips featuring no undercut were found, rendering the undercut quite ineffective. It is interesting to note that similar electrode/insulator configurations featuring oxide undercuts have been used in cold cathode arrays in the past and resulted in breakdown field strengths of up to 250 V/μm, similar to the data found here.

Thus, the surface breakdown results obtained so far do not look promising if viewed in the context of ion engine grids fabricated using silicon-based MEMS technologies. The possibility exists to operate an ion engine at lower grid voltages, thus sacrificing performance with respect to specific impulse and thruster efficiency, or to use these types of grids with lighter inert gas propellants, such as krypton or argon, rather than the typically used xenon. In these cases, the required grid voltages will be lower and MEMS-based grids may possibly be used even given the surface breakdown data obtained here. However, this approach would lead to performance reductions of the ion engine. At present, engine performance requirements for future microspacecraft missions remain very unclear. Therefore, a conservative approach would require grid performances equal to those of state-of-the-art "macroscopic" systems. Other grid design options should therefore be explored as well to maintain these performances. For example, the use of very thick PECVD (plasma-enhanced chemical vapor deposition) oxide films reaching thicknesses up to 15 μm as proposed by Alberta Microelectronic Corporation of Canada may be explored. Although PECVD oxide films typically do not have the same quality as LTO oxide films, the much larger thickness available with these films may well compensate for any defects. Finally, non-silicon-based microfabrication techniques to machine micro-ion engine grid systems are also under consideration and will be explored in the near-future.

Acknowledgments

The authors would like to thank Eunice Koo and James Bustillo of the Microfabrication Laboratories at the University of Berkeley, as well as Kevin Tsing of the University of California/Los Angeles, for performing the polysilicon and oxide growth processing steps for the wafers used in the experiment. The research described in this work was carried out by the Jet Propulsion Laboratory, California Institute of Technology, under a contract with NASA.

References

[1]Mueller, J., "Thruster Options for Microspacecraft: A Review and Evaluation of State-of-the-Art and Emerging Technologies," *Micropropulsion for Small Spacecraft*, Progress in Astronautics and Aeronautics, Vol. 187, edited by M. Micci and A. Ketsdever, AIAA, Reston, VA, 2000, Chap. 3 (this volume).

[2]Collins, D., Kukkonen, C., and Venneri, S., "Miniature, Low-Cost Highly Autonomous Spacecraft—A Focus for the New Millennium," IAF Paper 95-U.2.06, Oslo, Norway, Oct. 1995.

[3]Blandino, J., Cassady, R., and Sankovic, J., "Propulsion Requirements and Options for the New Millennium Interferometer (DS-3) Mission," AIAA 98-3331, 34th Joint Propulsion Conference, Cleveland, OH, July 1998.

[4]Marrese, C., Polk, J., Jensen, K., Gallimore, A., Spindt, C., Fink, R., Tolt, Z., and Palmer, W., "An Investigation into the Capability of Field Emission Cathode and Electric Thruster Technologies: Theoretical and Experimental Performance Evaluations," *Micropropulsion for Small Spacecraft*, Progress in Astronautics and Aeronautics, Vol. 187, edited by M. Micci and A. Ketsdever, AIAA, Reston, VA, 2000, Chap. 11 (this volume).

[5]Wolf, S., and Tauber, R., *Silicon Processing for the VLSI Era*, Vol. 1, Lattice Press, Sunset Beach, CA, 1986.

[6]Osburn, C. M., and Ormond, D. W., "Dielectric Breakdown in Silicon Dioxide Films on Silicon, Part I," *Journal of the Electrochemical Society*, Vol. 119, No. 5, 1972, pp. 591–597.

[7]Osburn, C. M., and Ormond, D. W., "Dielectric Breakdown in Silicon Dioxide Films on Silicon, Part II," *Journal of the Electrochemical Society*, Vol. 119, No. 5, 1972, pp. 597–603.

[8]Osburn, C. M., and Weitzman, E. J., "Electrical Conduction and Dielectric Breakdown in Silicon Dioxide Films on Silicon," *Journal of the Electrochemical Society*, Vol. 119, No. 5, 1972, pp. 603–609.

[9]Klein, N., "The Mechanism of Self-Healing Electrical Breakdown in MOS Structures," *IEEE Transactions on Electron Devices*, Vol. ED-13, No. 11, 1966, pp. 788–805.

[10]Chou, N. J., and Eldridge, J. M., "Effects of Material and Processing Parameters on the Dielectric Strength of Thermally Grown SiO_2 Films," *Journal of the Electrochemical Society*, Vol. 117, No. 10, 1970, pp. 1287–1293.

[11]Soden, J. M., "The Dielectric Strength of SiO_2 in a CMOS Transistor Structure," *Proc. 1979 Electrical Overstress/Electrostatic Discharge Symposium*, Sept. 1979, pp. 176–182.

[12]Fritzsche, C., "Der dielektrische Durchschlag in SiO_2-Schichten auf Silizium," *Zeitschrift für Angewandte Physik*, Vol. 24, No. 1, 1967, pp. 48–52.

[13]Worthing, F. L., "D-C Dielectric Breakdown of Amorphous Silicon Dioxide Films at Room Temperature," *Journal of the Electrochemical Society*, Vol. 115, No. 1, 1968, pp. 88–92.

[14]Yang, D. Y., Johnson, W. C., and Lampert, M. A., "Scanning Electron Micrographs of Self-Quenched Breakdown Regions in Al-SiO_2-(100) Si Structures," *Applied Physics Letters*, Vol. 25, No. 3, 1974, pp. 140–142.

[15]Pratt, I. H., "Thin-Film Dielectric Properties of RF Sputtered Oxides," *Solid State Technology*, Dec. 1969, pp. 49–57.

[16]Klein, N., and Gafni, H., "The Maximum Dielectric Strength of Thin Silicon Oxide Films," *IEEE Transactions on Electron Devices*, Vol. ED-13, No. 12, 1966, pp. 281–289.

[17]Mueller, J., Tang, W., Li, W., and Wallace, A., "Micro-Fabricated Accelerator Grid System Feasibility Assessment for Micro-Ion Engines," IEPC 97-071 Paper, 25th International Electric Propulsion Conf., Cleveland, OH, Aug. 1997.

IV. Electromagnetic Thrusters

Chapter 13

Pulsed Plasma Thruster Performance for Microspacecraft Propulsion

Rodney L. Burton,* Filip Rysanek,† Erik A. Antonsen,†
Michael J. Wilson,‡ and Stewart S. Bushman§
University of Illinois at Urbana–Champaign, Urbana, Illinois

Nomenclature

E_o = capacitor energy
ESR = capacitor equivalent series resistance
I_{bit} = thrust impulse/pulse
I_{sp} = specific impulse
L = inductance
L' = inductance gradient
M = mean molecular weight
m = ablated mass per pulse
T = thrust
$\bar{u}$ = mass-averaged velocity
$u_{f,s}$ = fast, slow particle velocity
V = voltage; volume of exhaust gas
Z_{ppt} = impedance of pulse plasma thruster
Z_{tot} = impedance of total circuit
α = species; fast particle mass ratio m_f/m
η_t = thruster efficiency based on thrust and I_{sp}
Ψ = current integral $\int I^2\,\mathrm{d}t$
Φ = late-time ablation factor
ε_I = ionization potential
ρ = mass density

Copyright © 2000 by the American Institute of Aeronautics and Astronautics, Inc. All rights reserved.

*Professor, Department of Aeronautical and Astronautical Engineering. Associate Fellow AIAA.
†M.S. Candidate, Department of Aeronautical and Astronautical Engineering. Student Member AIAA.
‡M.S. Candidate, Department of Aeronautical and Astronautical Engineering; currently at Primex Aerospace Company. Member AIAA.
§M.S. Candidate, Department of Aeronautical and Astronautical Engineering; currently at W.E. Research. Member AIAA.

Subscripts and Superscripts

dist = velocity distribution
div = flow divergence
EM = electromagnetic
f = fast
ff = frozen flow
h = heating
ℓ = loss
o = overall
ppu = power processing unit
s = slow
t = thruster
tr = energy transfer
z = ionization state, direction of thrust

I. Introduction

THE pulsed plasma thruster[1] (PPT) is a leading candidate to perform a number of propulsion tasks for microspacecraft, including orbit transfer, station-keeping, drag makeup, precision-formation flying, and attitude control. Microspacecraft (wet mass, <100 kg; power, <100 W) are being considered for single spacecraft missions, large-coverage constellations, and precision-formation constellations. Examples of the latter are space-based interferometry[2] and space-based radar.[3] For attitude control and station-keeping, impulse bits of ~1 μN-s are needed for milliradian and millimeter positioning accuracies. For drag makeup and orbit transfer, thrust levels in the millinewton range are needed. PPTs can be throttled over a wide range to provide this variation of thrust levels.

The PPT is a simple and rugged device and uses a few-microsecond-duration megawatt-level high-current pulse to evaporate and accelerate a solid propellant, usually Teflon, to produce a single impulse bit. Repetition of this process occurs at a frequency up to a few hertz to create an average thrust. The thruster (Fig. 1) is capacitor-driven and has a specific impulse in the 1000-s range, considerably

Fig. 1 PPT-7 coaxial pulsed plasma thruster with a boron nitride nozzle.

above that of chemical thrusters. It can be single-pulsed for precise positioning or can operate at up to ~100 W, a low-power region not accessible by most other types of electric propulsion devices. The thrust-to-power ratio (specific thrust) ranges from a maximum of 50 μN/W for coaxial PPTs to as low as 5 μN/W. The PPT can operate from 5 Hz down to a single pulse.

The PPT has a long history on spacecraft, because it is relatively easy to integrate into the system, uses a solid, nontoxic fuel, and is robust and reliable. Palumbo and Guman[4] have made the systems case for the pulsed plasma thruster, based on nearly 9000 h of flight operations with a PPT called the LES-6 thruster, from 1968 to 1978.

1) Zero warmup time, zero standby power
2) Inert and fail-safe—no unpowered torques or forces
3) Scalable to performance requirements
4) Usable on spinning or three-axis stabilized satellites
5) Solid propellant advantages:
   No tankage, feedlines, seals, mechanical valves
   Easily measured propellant consumption
   Zero-gravity, cryogenic, vacuum compatible
   Noncorrosive, nontoxic, long shelf life
   Not affected by rapid temperature changes
   Not affected by variable high-"g" loads
6) Discreet impulse bits compatible with digital logic
7) Variable thrust level
8) Performance compatible with attitude control and station-keeping requirements
9) Operation at large variation in environmental temperature
10) Thrust vector control capability

To these may be added a few disadvantages:

1) Solid Teflon located at thruster, making propellant feed geometry relatively inflexible
2) High internal losses giving thermal management problems
3) Electromagnetic interference (EMI) from pulsed discharge
4) Optical radiation interference with sensors and optics
5) Particles from PPT can backflow, impinge on other satellites
6) Pulse life requirements can exceed igniter plug and capacitor life
7) Low system specific impulse (impulse/system mass)

From the time of the early development of the PPT, two types have existed, the rectangular or parallel-plate PPT and the coaxial PPT. As shown in Fig. 2, the thrust and I_{sp} characteristics are considerably different for these two devices. Plasma acceleration in the rectangular version is dominated by electromagnetic ($\vec{j} \times \vec{B}$) forces, while coaxial PPTs generally are dominated by electrothermal (gasdynamic) forces. In some devices the two forces are roughly equal. In this chapter the characteristics of the PPT are discussed as applied to microspacecraft.

II. PPT Performance for Micropropulsion

PPTs traditionally operate in a power range below 100–200 W and provide impulse bits in the 50–1000-μN-s range. PPT thrust is generated by two mechanisms: gasdynamic and/or electromagnetic. Gasdynamic thrust is typified by the pulsed

Table 1 Gasdynamic and electromagnetic PPT thruster performance

	Gasdynamic	Electromagnetic
Geometry	Coaxial	Coaxial
Propellant	Teflon	Teflon
Energy, J	50	50
Specific thrust, μN-s/J	40	20
Specific impulse, s	500–1000	1000–1500
Max. thrust at 100 W, mN	4.0	2.0
Thrust efficiency	0.10–0.20	0.05–0.12

creation of a high-pressure plasma that provides thrust by pressure forces, and electromagnetic thrust is characterized by high currents that produce high $\vec{j} \times \vec{B}$ forces. The performance of these two classes of thrusters is compared in Table 1.

The principal design features of coaxial gasdynamic PPTs are a cylindrical cavity having an ablating wall, a closed end formed by a central electrode, and an electrically insulating nozzle. The discharge energy is deposited in the cavity by a nonreversing current pulse on a few-microsecond time scale and leaves the cavity by plasma outflow and wall heat loss. For a fast discharge pulse, most of the energy is transferred to the cavity walls, which, if lined with propellant, will evaporate into the cavity to raise the pressure.[5,6]

A. Electromagnetic Impulse Bit

The electromagnetic PPT, in contrast to the gasdynamic PPT, operates at a greatly reduced plasma density. The thruster is designed with an order-of-magnitude larger plasma volume and lower ablated mass, so that the pressure in an electromagnetic PPT is a few atmospheres, of the order of the magnetic pressure $B^2/2\mu_o$. The lower mass and density results in an $I_{\rm sp}$ higher than that of the gasdynamic PPT.

The impulse bit of an electromagnetic PPT is given by

$$\int T\,\mathrm{d}t = \frac{1}{2}L' \int I^2\,\mathrm{d}t \equiv \frac{1}{2}L'\Psi \tag{1}$$

where L' for a uniform current distribution is given below. The current integral

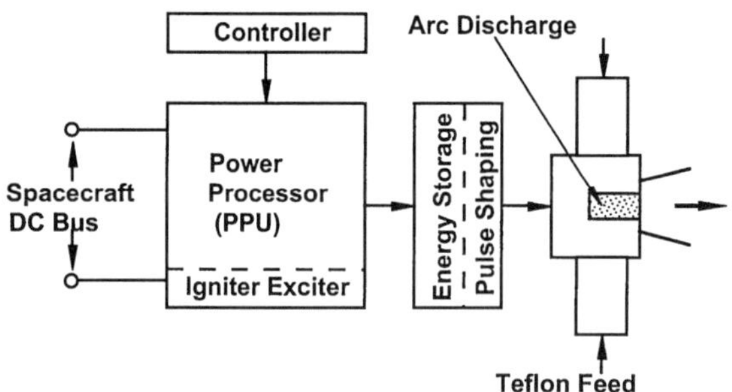

Fig. 2 Schematic of the pulsed plasma thruster system.

$\Psi = \int I^2 \, dt$ is related to the stored energy and the total impedance Z by $\Psi = E_o/Z$, so that a low Z is required for a high impulse bit. The specific thrust (μN-s/J) is then

$$\int T \, dt / E_o = L'/2Z \tag{2}$$

L' is typically 0.2 μH/m, so for $Z_{m\Omega} = 20$ mΩ, the specific impulse bit is 10 μN-s/J.

B. Gasdynamic Impulse Bit

The thrust generated gasdynamically by pressure forces cannot be predicted with a simple model. The following effects occur simultaneously.

1) The discharge energy evaporates and heats Teflon propellant, creating a pressure field that acts on the thruster surfaces, creating gasdynamic thrust.

2) A fraction of the discharge energy is lost as radiation, in electrode sheath drops, and to the walls by convection, cooling the plasma and decreasing the pressure.

3) An unsteady rarefaction wave moves into the heated plasma, accelerating it and decreasing the pressure. The plasma velocity is reduced by wall friction.

4) The gasdynamic thrust is increased by using a nozzle, the performance of which is reduced by heat loss, friction, frozen flow, and unsteady flow effects.

5) Following the current pulse Teflon ablation can continue (late-time ablation), resulting in a significant fraction (up to 40%) of the mass exiting the thruster well after the pulse, at a pressure too low to contribute to the gasdynamic impulse bit.

Faced with the flow features listed above, a prediction of the gasdynamic impulse bit cannot be easily made. As an upper limit, the thruster pressure can be estimated from $p = (\gamma - 1)E/V$, where E is the energy delivered to the plasma, V is the plasma volume, and $\gamma \sim 1.3$ for Teflon plasma. This pressure can be multiplied by a suitable area and by a characteristic time (acoustic travel time or current pulse length) to estimate roughly the gasdynamic impulse bit. Experimentally, the gasdynamic impulse bit can be derived from a thrust stand measurement of the total impulse bit, subtracting off the electromagnetic impulse bit $\frac{1}{2}L'\Psi$. An experimentally validated numerical model based on the above physics can also be used.

C. Defining Thruster Efficiency

A simplified schematic of a PPT propulsion system, consisting of a power processing unit (PPU), a capacitor, a transmission line, and an accelerator is shown in Fig. 2. The accelerator consists of the electrodes and insulators, provision for feeding the propellant (Teflon or other) into the device, an igniter plug, and, in some cases, a nozzle.

Rather than defining efficiency based on the current and voltage at the input terminals, PPTs define thruster efficiency η_t in terms of the energy stored in the capacitor:

$$\eta_t = \frac{\text{thrust energy}}{\text{capacitor stored energy}} = \frac{\frac{1}{2} u_e \int T \, dt}{E_o} \tag{3}$$

This definition includes velocity distribution losses, discussed below, and divergence losses due to exhaust beam spreading.

If it is assumed that the exhaust mass is monoenergetic and flows parallel to the thrust axis at a single ejection velocity $\overline{u}$, the exhaust kinetic energy can be written in terms of the thrust and specific impulse,

$$\frac{1}{2}m\overline{u}^2 = \frac{1}{2}u_e \int T\,\mathrm{d}t \tag{4}$$

which leads to the definition of PPT thruster efficiency:

$$\eta_t = \frac{1}{2}u_e\left(\int T\,\mathrm{d}t/E_o\right) = \frac{\left[\int T\,\mathrm{d}t\right]^2}{2mE_o} \tag{5}$$

where the specific thrust $\int T\,\mathrm{d}t/E_o$ is equivalent to the thrust-to-power ratio.

The usefulness of Eq. (5) is that the quantities $\int T\,\mathrm{d}t$ (impulse bit), m (mass per pulse), and E_o are directly measurable indicators of performance. Hence, Eq. (5) allows the comparison of different pulsed thrusters in a consistent fashion.

For example, Fig. 3 plots the specific thrust vs specific impulse for various PPTs. The constant curves of thruster efficiency, as defined in Eq. (3), are also shown. The efficiency of a given thruster design depends in a complex way on I_{sp} as discussed below.

The overall efficiency equation for the PPT is $\eta_o = \eta_{\mathrm{ppu}} \times \eta_t$, where η_t is the product of five subefficiencies, determined by the pulse energy transfer, heat loss, frozen flow, exhaust beam divergence, and exhaust velocity distribution:

$$\eta_t = \eta_{\mathrm{tr}} \times (1 - \eta_\ell) \times \eta_f \times \eta_{\mathrm{div}} \times \eta_{\mathrm{dist}} \tag{6}$$

D. Efficiency Definitions

The power processing unit efficiency η_{ppu} is the fraction of energy delivered to the power processing unit that is delivered to the capacitor. Energy not delivered to the capacitor includes PPU heat loss and the energy required to run the spark igniter system and thruster-related housekeeping functions. The PPU is a solid-state

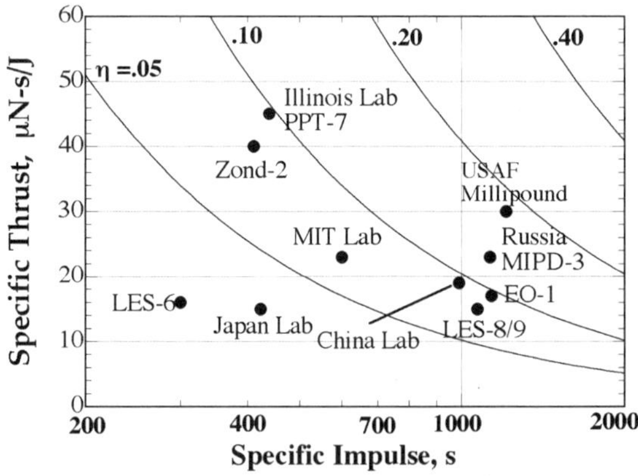

Fig. 3 Specific thrust vs specific impulse for PPTs.

low-to-high voltage DC–DC converter and, as defined here, includes the igniter circuitry. The PPU efficiency is 0.80–0.95, depending on the design, representing a 5–20% heat loss. The PPU triggers and powers the igniter plug, which typically has a stored energy of <500 mJ and a spark energy of tens of millijoules.

The energy transfer efficiency η_{tr} is the fraction of the energy in the capacitor delivered to the arc discharge and is determined by the equivalent series resistance (ESR) of the capacitor and the impedance Z_{ppt} of the PPT. Z_{ppt} is associated with ohmic heating ηj^2 and electromagnetic flow work $\vec{u} \cdot \vec{j} \times \vec{B}$. If the PPU is modeled as a constant-parameter simple LRC circuit, the total impedance around the circuit is

$$Z_{\text{tot}} = Z_{\text{ppt}} + R_{\text{trans}} + \text{ESR} \tag{7}$$

where R_{trans} is the resistance of the transmission line, typically $\sim$1 mΩ, and $Z_{\text{tot}} = E_o/\Psi$. The series circuit transfer efficiency for negligible R_{tr} is

$$\eta_{\text{tr}} = Z_{\text{ppt}}/Z_{\text{tot}} = \frac{1}{1 + \text{ESR}/Z_{\text{ppt}}} \tag{8}$$

Clearly $Z_{\text{ppt}} \gg \text{ESR}$ is required for a high transfer efficiency. However, Z_{ppt} cannot be made arbitrarily large since, from Eq. (2), doing so reduces the specific thrust.

The accelerator efficiency η_{acc} is the fraction of the energy delivered to the arc discharge converted to thrust energy. The losses in this process are associated with losses from voltage sheaths, wall heating, frozen flow, wall drag, beam divergence, and velocity distribution:

$$\eta_{\text{acc}} = (1 - \eta_{\text{loss}}) \times \eta_f \times \eta_{\text{div}} \times \eta_{\text{dist}} \tag{9}$$

The PPT plasma discharge loses heat to the electrodes, insulators, nozzle exit, and propellant by sheath losses, conduction, and radiation. The energy deposition in the sheaths is lost to the electrodes as heat. Most of the heat and radiation transfer to the Teflon is captured at the surface and used to evaporate mass, which is then heated to the discharge temperature. The energy e_{gas} required to ablate and sublimate the Teflon into the gas phase is $e_{\text{gas}} = h_d + h_f$, comprising the depolymerization or "unzipping" of the polymer chain h_d and the phase transition h_f to a gas.[7] For Teflon, $e_{\text{gas}} \approx 1.5 \times 10^6$ J/kg. The remaining heat loss appears as nozzle radiation or raises the temperature of the propellant and thruster, which then reradiate to space or conduct heat to the spacecraft.

Despite its importance, few data are available for heat loss in PPTs. Spanjers et al. measured steady-state propellant temperatures.[8] Kamhawi and Turchi[9] measured a total radiation heat loss of 18–23% on a PPT[10] with an 8% thruster efficiency, but this loss included that from capacitor heating. Heat loss has also been measured in coaxial PPTs and has been found to depend on the insulator thickness. Tests with a coaxial thruster have shown that a large reduction in heat loss can be achieved with a radially thick insulator around the cavity discharge.

The remaining energy not lost in the sheaths or to heat transfer is used to accelerate the propellant, associated with either the ion frozen flow efficiency η_f^+ or the neutral frozen flow efficiency η_{fn}. These frozen flow efficiencies are calculated in terms of a total enthalpy from the known ionization potentials of the ions of the Teflon gas (C + 2F). For a mean charge state Z,

$$h_o = \frac{5}{2}T_{\text{ex}} + (\varepsilon_i)_z + \frac{1}{2}u_e^2 \tag{10}$$

Table 2 Estimated particle frozen flow efficiency for T = 2 eV, LES-6

Particle	u (km/s)	$5/2T$, eV	ε_{ion}, eV	$\langle u^2 \rangle/2$, eV	h_o, eV	η_f
Slow neutrals						
C	3	5	0	0.9	5.9	0.15
F	3	5	0	0.9	5.9	0.15
Fast particles						
C	10 ± 5	5	0	6.2	11	0.56
F	10 ± 5	5	0	10	15	0.67
C^+	25 ± 5	5	11	39	55	0.71
F^+	20 ± 5	5	17	40	62	0.65
C^{++}	35 ± 5	5	24	77	106	0.73
F^{++}	30 ± 5	5	35	89	129	0.69

with all quantities in electron volts. The frozen flow efficiencies are then

$$\eta_f^+, \eta_{\text{fn}} = \left(\frac{u_{\text{ex}}^2}{2h_o} \right)_{+,n} \tag{11}$$

The frozen flow efficiency must be integrated over the exhaust species and velocity distributions. As an example, η_f is estimated (Table 3) for LES-6 velocities in Table 2. The temperature is estimated by assuming that the particles are in thermal equilibrium. The mass-averaged neutral velocity is 2.6 km/s, and since the LES-6 thruster has no nozzle, the neutral particles will have a mean velocity of the order of the sound speed. Allowing for divergence or "cosine" loss due to thermal expansion perpendicular to the thrust axis, the sound speed cannot exceed 5 km/s, and the temperature will be a maximum of about 2 eV.[12]

The mean frozen flow efficiency of the slow neutral particles is 0.15, and that of the fast particles is 0.68. Fast C and F neutrals may be accelerated electromagnetically and then undergo a loss of charge by recombination and/or charge exchange during acceleration. The higher velocity achieved by multiple fast particles keeps their efficiency roughly constant despite their higher ionization energy. The difference in these efficiencies for various particle charge states suggests that the ions are accelerated by a different mechanism (i.e., electromagnetic) than for the slow neutrals (i.e., gasdynamic).

Exhaust particles are ejected over a time of $<100\ \mu$s. Taking the exhaust from a single pulse once it has left the PPT, a control volume V can be created around the cloud of particles. The total mass for one pulse is

$$m = \int_V \sum_\alpha \rho_\alpha \, \mathrm{d}V \tag{12}$$

where each α is a different neutral or ionized species in the cloud. I_{bit} is expressed as

$$I_{\text{bit}} = \int_V \sum_\alpha \rho_\alpha (u_x)_\alpha \, \mathrm{d}V \tag{13}$$

where u_x is the velocity component along the thrust axis. Kinetic energy is

$$\text{KE} = \int_V \sum_\alpha \rho_\alpha \left(\frac{u_\alpha^2}{2} \right) \mathrm{d}V = \frac{1}{2} m \overline{\langle u^2 \rangle} \tag{14}$$

Table 3 Thruster subefficiencies from experiment and the two-stream model (TSM)

Thruster	η_{tr}	$(1-\eta_\ell)$	η_f	η_{div}	η_{dist}	Φ	C_N
Coaxial	0.93	0.76	TSM	0.93	TSM	1.0	1.25
Rectangular	0.90	0.70	TSM	0.93	TSM	0.6	1.0

Still using the control volume, the exit velocity u has the axial component u_x and a perpendicular component u_p (swirl is assumed to be zero). Total kinetic energy can be expressed as

$$\frac{1}{2}m\overline{(u^2)} = \frac{1}{2}m\overline{\left(u_x^2\right)} + \frac{1}{2}m\overline{\left(u_p^2\right)} \tag{15}$$

where the directed kinetic energy is

$$\frac{1}{2}m\overline{\left(u_x^2\right)} = \int_V \sum_\alpha \rho_\alpha \left(\frac{u_x^2}{2}\right)_\alpha \mathrm{d}V \tag{16}$$

The divergence efficiency is expressed as the ratio of the directed kinetic energy to the total kinetic energy:

$$\eta_{div} = \frac{\frac{1}{2}m\overline{\left(u_x^2\right)}}{\frac{1}{2}m\overline{(u^2)}} = \overline{\left(u_x^2\right)}/\overline{(u^2)} \tag{17}$$

Applying the species summation scheme to the frozen flow efficiency,

$$\eta_f = \frac{\overline{(u^2)}}{2\overline{h}_o} = \frac{\frac{1}{2}m\overline{(u^2)}}{\sum_\alpha m_\alpha \overline{h}_{o_\alpha}} \tag{18}$$

Combining expressions and referring to Eq. (1),

$$\eta_t = \eta_{tr} \times (1-\eta_{loss}) \times \eta_f \times \eta_{div} \times \eta_{dist} = \frac{\frac{1}{2}m\overline{(u^2)}}{E_o} \times \eta_{div} \times \eta_{dist} \tag{19}$$

Because the kinetic energy $\frac{1}{2}m\overline{(u^2)}$ is not readily measurable, the distribution efficiency introduces the measurable quantity $\overline{u}$ into the definition of thruster efficiency, allowing it to be determined from thrust stand measurements.

Using the mass-averaged velocity $\overline{u} = I_{bit}/m$ yields a thrust energy:

$$E = \frac{1}{2}m(\overline{u})^2 \tag{20}$$

The distribution efficiency, which quantifies profile losses in the thruster, is then

$$\eta_{dist} = \frac{(1/2m)(m\overline{u})^2}{\frac{1}{2}m\overline{(u^2)}} = \frac{(1/2m)\left(\int_V \sum_\alpha \rho_\alpha (u_x)_\alpha \,\mathrm{d}V\right)^2}{\frac{1}{2}\left(\int_V \sum_\alpha \rho_\alpha u_\alpha^2 \,\mathrm{d}V\right)} \tag{21}$$

where m is given in Eq. (12). Because of the integration over the exhaust mass control volume, Eq. (21) automatically includes effects such as velocity loss caused by viscous drag at the walls.

E. Two-Stream Model

While Eq. (3) is a reasonable statement of thruster efficiency for many devices such as the ion accelerator, for which beam spreading is minimal and all exhaust particles have about the same energy, it is a poor approximation for the PPT. An earlier approach that recognized that both electromagnetic and pressure forces were present was made in a single fluid model by Vondra and Thomassen,[11] who assumed that

$$\int T\,\mathrm{d}t = \frac{1}{2}\mu_o \frac{h}{w}\Psi + m\overline{c} \tag{22}$$

where m is the total ablated mass and $\overline{c}$ is the mass-averaged thermal velocity. Their model can be interpreted as an electromagnetic impulse bit plus a gasdynamic impulse bit, assuming that all the mass is ejected at velocity $\overline{c}$.

The two-stream model is proposed here in lieu of the nonavailability of complete exhaust velocity distribution data. It assumes two separately monoenergetic streams of fast and slow particles. The impulse bit is separated into fast and slow components:

$$\int T\,\mathrm{d}t = \int_f T\,\mathrm{d}t + \int_s T\,\mathrm{d}t \tag{23}$$

where

$$\int_f T\,\mathrm{d}t = m_f u_f \tag{24}$$

$$\int_s T\,\mathrm{d}t = m_s u_s \tag{25}$$

The ablated masses yield the relation

$$m_f + m_s = \Phi m \tag{26}$$

where m is the measured mass loss per pulse and Φ is a late-time ablation factor that takes into account low-velocity mass that exits the thruster at a very low velocity well after the pulse and does not contribute significantly to the impulse bit. The factor Φ is identical to the mass utilization efficiency η_m as defined by Stuhlinger,[12] here $\eta_m = (m_f + m_s)/m$. The amount of late-time ablated mass m_Φ is $m_\Phi = (1 - \Phi)m$. The two-stream expression for thruster efficiency is then

$$\eta_t = \frac{\left[\int T\,\mathrm{d}t\right]^2}{2mE_o} = \frac{[m_f u_f + m_s u_s]^2}{2mE_o} \tag{27}$$

The five equations (22–26) contain 11 unknown variables, requiring 6 more relations to close the system. Three variables, $\int T\,\mathrm{d}t$, m, and E_o, are obtained routinely during thrust stand performance measurements. Three remaining variables or conditions must then be determined.

The first remaining variable is determined by measuring the late-time ablation parameter Φ. If all the mass is accelerated and used to generate the impulse bit, $\Phi = 1$. Measurements on a rectangular Teflon PPT similar to the LES-8/9, called the XPPT-1,[13] indicate that about 40% of the mass is emitted well after the current pulse, implying that $\Phi = 0.6$. This parameter is poorly known for other PPT types but, in principle, can be found through experiment.

The second remaining variable is determined by making the assumption that the fast stream and its associated impulse bit $\int_f T\,\mathrm{d}t = m_f u_f$ is generated solely by the $\vec{j} \times \vec{B}$ force:

$$\int_f T\,\mathrm{d}t = \frac{1}{2}L' \int I^2\,\mathrm{d}t \equiv \frac{1}{2}L'\Psi \tag{28}$$

The current integral Ψ is determined from the pulse current, and L' is calculated from the thruster geometry and the current distribution in the thruster. For evenly distributed current sheets, for a rectangular PPT, $L'[\mu\mathrm{H/m}] = 0.6 + 0.4\,\ell\mathrm{n}[d/(b + c)]$, and for a coaxial PPT, $L'[\mu\mathrm{H/m}] = (\mu_o/2\pi)(\ell\mathrm{n}(r_a/r_c) + 3/4)$.

The last remaining variable is found from the observation that an ionized plasma colliding with a cloud of neutrals at a relative velocity u_c has a velocity limited by the neutral ionization potential V_i. This principle was first hypothesized by Alfven, who proposed that the neutrals become ionized when

$$u_c = \sqrt{2eV_i/m}$$

The relative velocity u_c has come to be called the Alfven critical velocity.

The Alfven hypothesis has been demonstrated to be correct at a high ion Hall parameter ($\Omega_i \sim 1000$) by Danielsson,[14] at an intermediate Hall parameter ($0.1 < \Omega_i < 30$) for hydrogen and nitrogen by Fahleson,[15] and for deuterium, oxygen, helium, neon, and argon by Angerth et al.,[16] and at a low Hall parameter ($\Omega_i < 1$) by Eninger.[17] The Hall parameter in a PPT is $\Omega_i \approx 1$, and the conditions are closest to the results of Fahleson and Angerth et al. The Larmor radius for carbon at the critical velocity is $\approx$1 mm. Calculating the critical velocity behavior for carbon and fluorine:

$$\begin{aligned} &\text{carbon } (V_i = 11.2\ \text{eV}): && u_c = 13.4\ \text{km/s} \\ &\text{fluorine } (V_i = 17.3\ \text{eV}): && u_c = 13.2\ \text{km/s} \end{aligned}$$

We therefore adopt $u_c = 13.3$ km/s for Teflon plasma.

Applying the critical velocity principle to the two-stream model,

$$u_f - u_s = u_c = 13.3\ \text{km/s} \tag{29}$$

The mass ratio–velocity schematic of the two-stream model is shown schematically in Fig. 4. The 11 variables of the two-stream model can now be determined. Introducing two nondimensional variables, the fast particle mass ratio $\alpha = m_f/m$ and the electromagnetic impulse bit fraction $\beta = (1/2)L'\Psi/\int T\,\mathrm{d}t$, the mass-averaged velocity is

$$\overline{u} = \alpha u_f + (\Phi - \alpha)u_s = gI_{\mathrm{sp}} \tag{30}$$

and since

$$\beta = \frac{m_f u_f}{m_f u_f + m_s u_s} \tag{31}$$

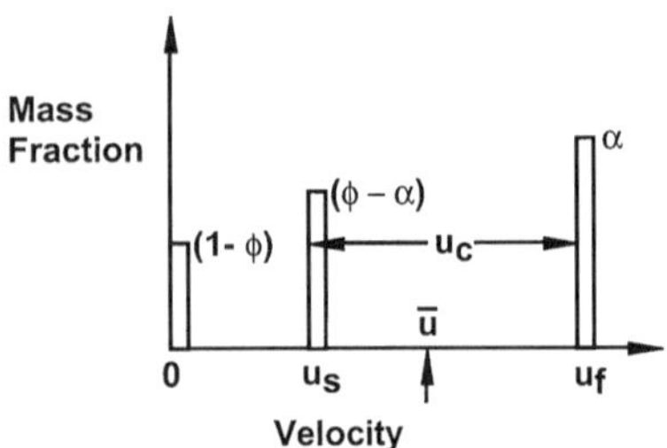

Fig. 4 Mass fraction–velocity schematic for the two-stream model.

the slow particle velocity is

$$u_s = \left(\frac{1-\beta}{\Phi-\alpha}\right)\overline{u} \tag{32}$$

For a purely gasdynamic PPT, $\beta = 0$ and $\alpha = 0$, so $u_s = \overline{u}/\Phi$. For a purely electromagnetic PPT, $\beta = 1$ and $u_s = 0$. Introducing the Alfven critical velocity condition from Eq. (29) gives a quadratic in the fast particle mass fraction α, the solution of which is

$$\alpha = \frac{1}{2}\left[(\Phi + \overline{u}/u_c) - [(\Phi + \overline{u}/u_c)^2 - 4\beta\Phi\overline{u}/u_c]^{\frac{1}{2}}\right] \tag{33}$$

This function allows α to be calculated from I_{sp}, thrust, and current data and gives $\alpha = 0$ for $\beta = 0$ and $\alpha = \overline{u}/2u_c$ for $\beta = 1$. The parameter α is plotted in Fig. 5 for $\Phi = 1$. The remaining variables m_s and u_f are then easily found.

We now apply the model to the calculation of thruster efficiency. The velocity distribution efficiency is given by

$$\eta_{\text{dist}} = \left[\frac{\alpha(\Phi-\alpha)}{\Phi\beta^2 + \alpha(1-2\beta)}\right] \tag{34}$$

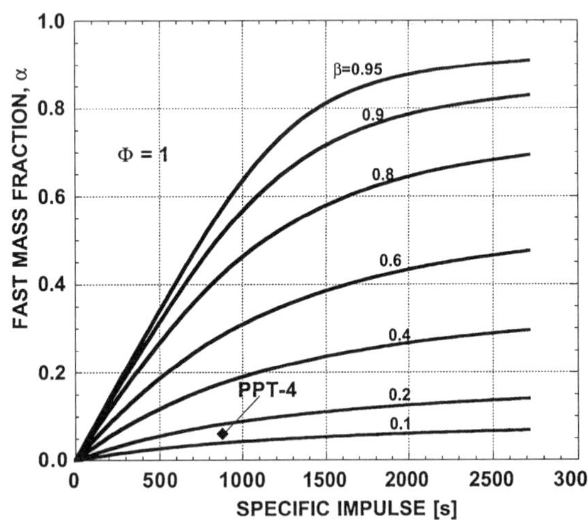

Fig. 5 Fast particle mass ratio α for $\Phi = 1$ vs specific impulse for values of the electromagnetic thrust fraction β. The model predicts $\alpha = 0.05$ for a gasdynamic PPT (PPT-4).

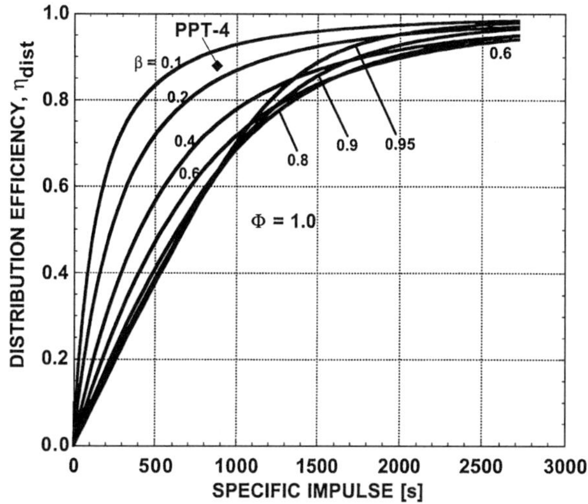

Fig. 6 Velocity distribution efficiency for Φ = 1 vs specific impulse for values of the electromagnetic thrust fraction *β*. The model predicts η_{dist} = 0.88 for a gasdynamic PPT (PPT-4).

and is plotted in Fig. 6 for $\Phi = 1$. This function gives $\eta_{dist} = \Phi$ for $\beta = 0$ and $\eta_{dist} = \overline{u}/2u_c$ for $\beta = 1$.

The frozen flow efficiency is given by mass-weighting the frozen flow efficiencies for fast and slow particles:

$$\eta_f = \alpha\eta_{ff} + (1 - \alpha)\eta_{fs}C_N \tag{35}$$

where, from Table 2, $\eta_{ff} = 0.68$ and $\eta_{fs} = 0.15$. The effect of a nozzle is included by applying a coefficient C_N to the gasdynamic (slow) particles. C_N is 1 for PPTs without nozzles and is >1 for PPTs with nozzles.

The thruster efficiency is estimated for PPTs by applying the two-stream model to Eq. (6). Values are adopted for rectangular and coaxial PPTs as shown in Table 3. The measured values of transfer efficiency η_{tr} and heat loss efficiency $(1 - \eta_\ell)$ for the coaxial thruster come from a heavily insulated version of the PPT-4 thruster. The reduced values of η_{tr} and estimated $(1 - \eta_\ell)$ for the rectangular thruster are due to the lower impedance and the large exposed electrode surface in this thruster type. The frozen flow efficiency is from Eq. (35). The divergence efficiency η_{div} is estimated for a 20-deg half-angle spread in a uniform exhaust plume. The distribution efficiency η_{dist} is from Eq. (34). The value of Φ for rectangular PPTs is based on experiment. For coaxial PPTs the Φ value has not been measured, and the value $\Phi = 1$ is taken from a plasma model of the PPT-4.[18] The nozzle coefficient C_N for coaxial thrusters is taken as 1.25 based on an area ratio of 4:1.

III. Discussion

Calculated efficiencies based on Table 3 are shown in Figs. 7 and 8. Data points and model predictions are also shown for existing PPTs based on their measured specific thrust and I_{sp} (LES-6, LES-8/9, EO-1, PPT-4).[1] The efficiencies predict generally higher values for coaxial PPTs, and this has been borne out in the

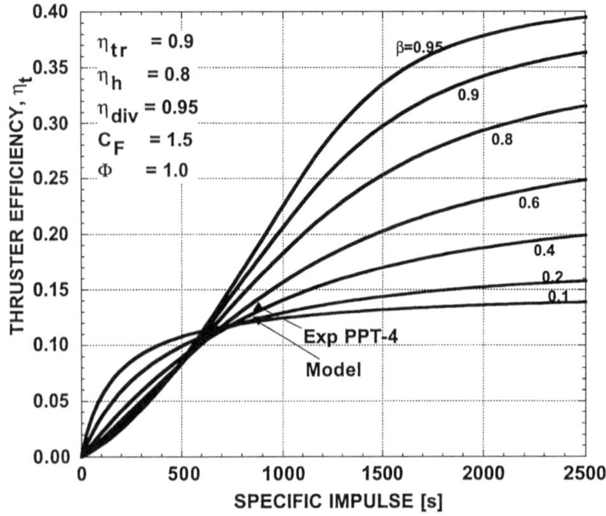

Fig. 7 Thruster efficiency for losses typical of a coaxial PPT and values of the electromagnetic thrust fraction β. The model shows agreement with the experimentally measured thruster efficiency.

laboratory. The performance of the EO-1 thruster (Fig. 8) is considerably higher than that predicted by the model, suggesting that the late-time ablation factor is not 0.6 but a higher value.

Also of interest is the specific thrust, expressed as impulse bit per joule [μN-s/J]. This quantity is derived from the relation $\int T \, \mathrm{d}t / E_o = 2\eta_t / g I_{\text{sp}}$ and is shown in Figs. 9 and 10, together with measured values and model predictions for several

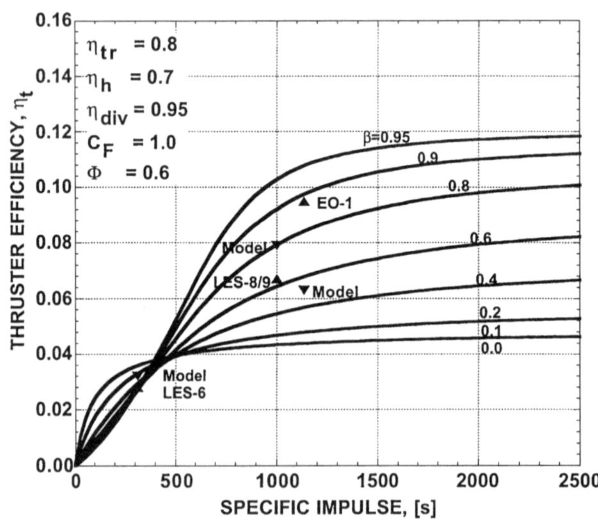

Fig. 8 Thruster efficiency for losses typical of a rectangular PPT and values of the electromagnetic thrust fraction β.

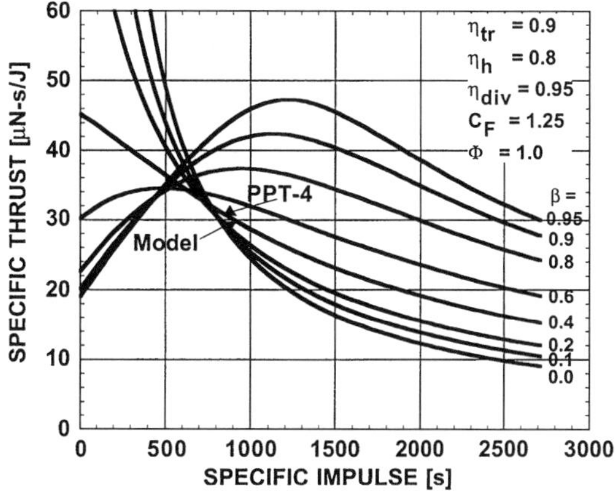

Fig. 9 Specific thrust for values typical of a coaxial PPT and values of the electromagnetic thrust fraction β. The model shows agreement with the experimentally measured value for PPT-4.

PPTs. The plot shows a maximum in specific thrust at a given β, moving to higher I_{sp} values as β increases. For PPT applications for which thrust is critical, Figs. 9 and 10 indicate the desired range of I_{sp}. Comparison with measured performance of several PPTs gives reasonable agreement with the model.

Acknowledgments

We acknowledge valuable discussions with G. Spanjers of the Air Force Research Laboratory at Edwards Air Force Base, R. Myers, A. Hoskins, and J. Cassady of Primex Aerospace Company, and E. Pencil of the NASA Lewis Research Center. This work was funded by the Air Force Office of Scientific Research, under Grant F49620-97-1-0138. M. Birkan is the Program Monitor.

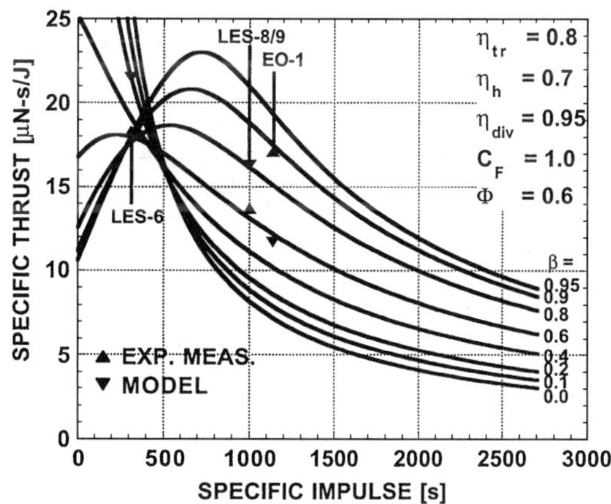

Fig. 10 Specific thrust for values typical of a rectangular PPT and values of the electromagnetic thrust fraction β. The model shows agreement with three rectangular PPTs.

References

[1]Burton, R. L., and Turchi, P. J., "Pulsed Plasma Thruster," *Journal of Propulsion and Power*, Vol. 14, No. 5, 1998, pp. 716–735.

[2]Blandino, J. J., Cassady, R. J., and Peterson, T. T., "Pulsed Plasma Thrusters for the New Millennium Interferometer (DS-3) Mission," IEPC 97-192, *Proceedings of the 25th International Electric Propulsion Conference*, Vol. 2, Electric Propulsion Rocket Society, Worthington, OH, 1998, pp. 1183–1190.

[3]Sedwick, R. J., Kong, E. M. C., and Miller, D. W., "Exploiting Orbital Dynamics and Micropropulsion for Aperture Synthesis Using Distributed Satellite Systems: Applications to TechSat21," AIAA Paper No. 98-5289, 1998.

[4]Palumbo, D. J., and Guman, W. J., "Continuing Development of the Short-Pulsed Ablative Space Propulsion System," AIAA Paper 72-1154, Nov. 1972.

[5]Bushman, S. S., "Investigations of a Coaxial Pulsed Plasma Thruster," M.S. Thesis, Rept. No. UILU 99-0509, Dept. of Aeronautical and Astronautical Engineering, Univ. of Illinois, Urbana, IL, 1999.

[6]Burton, R. L., and Bushman, S. S., "Probe Measurements in a Coaxial Gasdynamic PPT," AIAA Paper 99-2288, 1999.

[7]Guman, W. J., "Pulsed Plasma Technology in Microthrusters," Fairchild Hiller Corp., AFAPL-TR-68-132, Farmingdale, NY, Nov. 1968.

[8]Spanjers, G. G., Malak, J. B., Leiweke, R. J., and Spores, R. A., "The Effect of Propellant Temperature on Efficiency in a Pulsed Plasma Thruster," AIAA Paper 97-2920, July 1997.

[9]Kamhawi, H., and Turchi, P. J., "PPT Thermal Management," 25th International Electric Propulsion Conf., IEPC Paper 97-125, Cleveland, OH, Aug. 1997.

[10]Kamhawi, H., Turchi, P. J., Leiweke, R. J., and Myers, R. M., "Design and Operation of a Laboratory Bench-Mark PPT," AIAA Paper 96-2732, July 1996.

[11]Thomassen, K. I., and Vondra, R. J., "Exhaust Velocity Studies of a Solid Teflon Pulsed Plasma Thruster," *Journal of Spacecraft and Rockets*, Vol. 9, No. 1, 1972, pp. 61–64.

[12]Stuhlinger, E., *Ion Propulsion for Space Flight*, McGraw–Hill, New York, 1964.

[13]Spanjers, G. G., Lotspeich, J. S., McFall, K. A., and Spores, R. A., "Propellant Losses Because of Particulate Emission in a Pulsed Plasma Thruster, *Journal of Propulsion and Power*, Vol. 14, No. 4, 1998, pp. 554–559.

[14]Danielsson, L. "Experiment on the Interaction Between a Plasma and a Neutral Gas," *Physics of Fluids*, Vol. 13, 1970, pp. 2288–2294.

[15]Fahleson, U. V., "Experiments with Plasma Moving Through Neutral Gas," *Physics of Fluids*, Vol. 4, No. 1, 1961, pp. 123–127.

[16]Angerth, B., Block, L., Fahleson, U., and Soop, K., "Experiments with Partly Ionized Rotating Plasmas," *Nuclear Fusion Supplement, Part 1*, 1962, pp. 39–46.

[17]Eninger, J., "Experimental Investigations of an Ionizing Wave in Crossed Electric and Magnetic Fields," *Proceedings of the Seventh International Conference on Phenomena in Ionized Gases, Beograd*, Vol. 1, 1966, pp. 520–527.

[18]Keidar, M., Boyd, I. D., and Beilis, I. I., "Model of an Electrical Discharge in a Co-axial Pulsed Plasma Thruster," 26th International Electric Propulsion Conf., IEPC Paper 99-214, Japan, 1999.

Chapter 14

Pulsed Plasma Thrusters for Microsatellite Propulsion: Techniques for Optimization

Peter J. Turchi,* Ioannis G. Mikellides,† Pavlos G. Mikellides,‡
and Hani Kamhawi§
Ohio State University, Columbus, Ohio

Nomenclature

B	= magnetic (induction) field, T
d	= channel width, m
dm_D	= decomposed propellant mass, kg
dm_{EM}	= $j \times B$-accelerated mass, kg
E	= electric field, V/m
F	= thrust, N
h	= specific enthalpy, J/kg
J	= current, A
P	= power, W
Q	= specific enthalpy change, J/kg
T_s	= propellant surface temperature, K
u	= flow speed, m/s
V_{crit}	= Alfven critical speed, m/s
w	= mass flow rate per unit area, kg/s-m^2
W_f	= specific energy in internal states, J/kg
W_i	= specific ionization energy, J/kg
η	= thrust efficiency
η_s	= efficiency of delivering energy to thruster
μ_0	= permeability of free space, H/m
ρ	= mass density, kg/m^3

Copyright © 2000 by the American Institute of Aeronautics and Astronautics, Inc. All rights reserved.

*Adjunct Professor, Department of Aerospace Engineering and Aviation/Team Leader, Hydrodynamics and Pulsed Power Physics, P-22, Los Alamos National Laboratory. Associate Fellow AIAA.
†Post-Doctoral Researcher, Department of Aerospace Engineering and Aviation. Member AIAA.
‡Adjunct Assistant Professor, Department of Aerospace Engineering and Aviation. Member AIAA.
§Graduate Research Assistant, Department of Aerospace Engineering and Aviation. Member AIAA.

Superscript

$*$ = magnetosonic location

Subscripts

0 = propellant surface location
E = exhaust location

I. Introduction

PULSED plasma thrusters (PPTs)[1] have a long history of use for station-keeping on small satellites with limited electrical power. Application of PPTs to microsatellite missions represents a natural evolution in spacecraft propulsion systems. Simplicity, robustness, and ability to operate in pulses at a low average power establish PPTs as leading candidates for microsatellite missions, even though their thrust efficiency has been very poor.

The PPT, while simple in its embodiment, incorporates considerable complexity, as suggested by the sketch in Fig. 1. An arc discharge heats a solid propellant and provides a gradient of magnetic pressure that accelerates the ablated material. The arc current can rise in a fraction of a microsecond and continue for several microseconds in the oscillatory waveform of the traditional LRC circuit or for longer durations using inductive energy storage.[2] Heat diffuses into the propellant slab, causing decomposition of material, only a portion of which accelerates electromagnetically to speeds exceeding 10 km/s; the rest of the decomposed mass leaves at a much lower speed (<1 km/s), a large fraction of which consists of large polymer chains. These *macro*particles.[3] (vs molecular particles of various sizes) are surrounded by vapor in which a pressure gradient exists as the vapor expands away into the discharge. This gradient serves to impart a velocity to the macroparticles. In the immediate vicinity of the plasma discharge the vapor becomes electrically conducting and accelerates due to the Lorentz force. The nonconducting macroparticles are left with whatever speed they obtained before the vapor density decreased to the point that the drag force on them became negligible. This speed is much lower than that of the vapor/plasma, so the macroparticles leave the vicinity of the propellant surface after the current pulse is over.

Recent modeling of PPT operation using the MACH2 magnetohydrodynamics code[4] has identified and quantified the mechanisms that cause inefficient propellant utilization. Postpulse evaporation and macroparticle production account for the

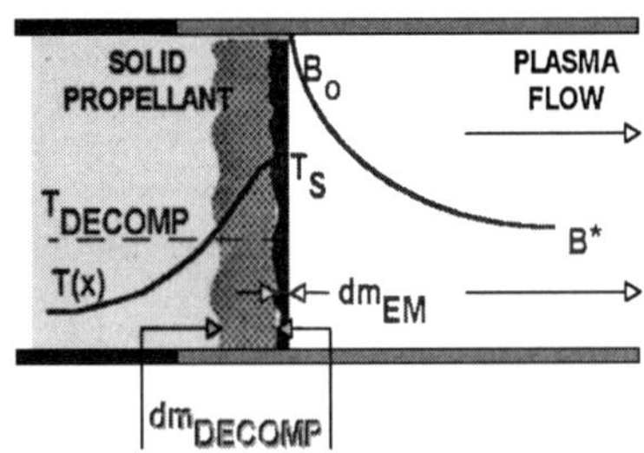

Fig. 1 Schematic depiction of PPT operation based on the idealized, quasi-steady, one-dimensional model.

major portion of the total mass loss. These processes contribute minimally to the thrust because the speed of the expelled mass is based on the temperature of the solid (<1000 K). An idealized analytic model prescribes the foundations for matching the current pulse and the PPT geometry to accelerate all the decomposed propellant electromagnetically, producing useful thrust at a high specific impulse. Numerical simulations with the MACH2 code indicate that waveforms from circuits based on inductive-energy storage will optimize thrust efficiency.

II. Numerical Modeling

The time-dependent, $2\frac{1}{2}$-dimensional, magnetohydrodynamic code MACH2 has been successfully employed to model PPTs.[5,6] The code has been augmented with a new ablation model that includes the interaction of the magnetohydrodynamic flow and the solid propellant. Specifically, the fundamental notion utilized for the development of this model is that vapor is created at the equilibrium vapor pressure based on the temperature of the solid surface. This temperature is calculated by accounting for the net heat flux to the surface due to energy transfer from the local plasma (conduction, convection) versus evaporation of the surface. The net heat flux serves as a boundary condition for a two-dimensional diffusion equation within a semi-infinite solid. This diffusion solver utilizes a second-order accurate numerical scheme in both space and time with the option of an adaptive grid for better gradient resolution at the ablating surface. The solid surface temperature, and thus the vapor temperature, is used to calculate the vapor pressure based on an appropriate Teflon vapor pressure curve.[7] This in turn implies a vapor density under the ideal gas assumption. The vapor temperature and density are then sufficient boundary conditions to calculate the velocity at the boundary and thus the mass flow rate.

Simulation of the LES-6[1] thruster (see Fig. 2) captures the experimental magnitude and trends for the impulse bit. The computed mass of ablated propellant (all of which participates in electromagnetic acceleration in these calculations) is much less than the experimental value for mass loss per shot (~10 mg at a stored energy of 1.85 J), as determined upon measuring the total mass lost (over many discharges) and then dividing by the total number of shots. The discrepancy in mass loss may result from two processes: late-time evaporation and macroparticle production, both of which degrade thrust efficiency. The late-time ablation (between discharge pulses) can also be calculated by MACH2 but depends on the base temperature of the Teflon propellant. The average heat flow in the slab at a given repetition rate and capacitor energy determines this base temperature. At a base temperature of 520 K, for example, propellant evaporation after the current pulse could account for as much as 55% of the total mass loss per shot.

In experiments[8] at 40 J/pulse and a repetition rate of 1 Hz, thermocouple measurements indicate a base temperature of 370 K. At this lower temperature value, evaporation between pulses contributes a negligible fraction to the total mass loss. Interrogation of the computed temperature profile within the solid Teflon, however, indicates that a significant portion of the propellant has been heated above the temperature for decomposition of the Teflon polymer. Based on a value for the temperature at which Teflon polymer chains break (~600 K),[7] the depth of solid propellant above this value implies 116 μg of decomposed Teflon. This amount is more than an order of magnitude larger than the experimentally measured mass

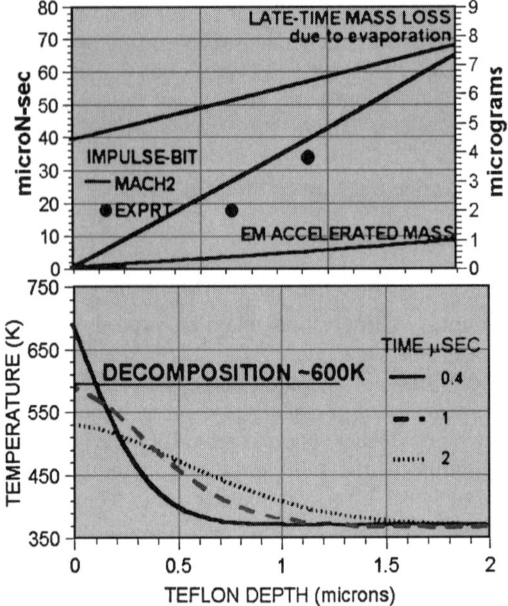

Fig. 2 MACH2 simulations of the LES-6 PPT identify and quantify propellant loss mechanisms.

loss at this energy level. (This discrepancy would increase if late-time evaporation at elevated base temperatures occurs.) The fraction of decomposed mass that may actually be liberated into the discharge in steady vs pulsed experiments is a function of a variety of processes on the molecular and fluid level. While some macroparticles may in fact penetrate soon enough into the plasma discharge to evaporate and reduce their size, cooling near the solid surface after the end of the current pulse may permit some reattachment of material back to the surface for particles that have not moved far enough away. The spectrum of macroparticle sizes, speeds, and rates of resolidification in this pulsed environment is too difficult to ascertain with the present theoretical tools and knowledge of transport properties. The value of the decomposition temperature of Teflon as found in the literature (~600 K) is therefore refined in the present pulsed environment to a higher value for purposes of optimization studies. Comparison of the experimentally measured mass loss for LES-6 with the MACH2 simulation suggests an effective decomposition temperature for Teflon (in PPTs) of 673 K. With the temperature profile in the solid predicted by MACH2, the decomposition depth can be computed separately. Based on the steady, sublimation value of 600 K, 0.18 μm has the opportunity to decompose, while the depth based on 673 K for pulsed operation (e.g., LES-6) is 0.0155 μm.

III. Idealized Model

A quasi-steady, one-dimensional, idealized model provides key insights into PPT operation. The magnetic field at the propellant surface B_0 is defined by the current J and the width of the rectangular channel d: $B_0 = \mu_0 J/d$ (see Fig. 1). In the limits of a high magnetic Reynolds number and a magnetic pressure greatly exceeding the plasma pressure, magnetohydrodynamic flow from the channel into a field-free vacuum implies that a magnetosonic condition exists in the channel. This condition corresponds to a flow speed equal to the local value of Alfven sound speed: $u^* = B^*/\sqrt{(\mu_0 \rho^*)}$, where ρ is the mass density and asterisked quantities refer to (magneto-)sonic values. From the derivation in the Appendix, the value of the magnetic field at the magnetosonic point is proportional to the magnetic field at the propellant surface: $B^* = B_0/\sqrt{3}$. The Appendix also calculates the plasma speed at the magnetosonic point:

$$u^* = 1.468\sqrt{2Q^*} \tag{1}$$

where Q^* is the change of flow enthalpy per unit mass due to electrical dissipation between the stagnation and the magnetosonic points. (If ionization of the plasma dominates this change, u^* is about 50% higher than Alfven critical velocity.) At fixed u^*, the magnetic pressure difference, $(B_0^2 - B^{*2})/2\mu_0$, from the propellant surface to the location of the magnetosonic point defines the mass flow rate per unit area $\dot{m}/A$ required to maintain the electrical discharge and flow in steady state. The current and channel width thus prescribe the mass flow per unit area at the propellant surface that must be provided by ablation.

Adjacent to the propellant surface, the mass flow rate per unit area is limited by a choking condition, based on the usual (thermal) sound speed. For a calorically perfect flow, with a stagnation pressure equal to the equilibrium vapor pressure of the propellant, this condition depends only on the surface temperature T_s. The surface temperature, in turn, scales the temperature profile in the solid propellant. The fraction of the energy deposited in the solid that prescribes the surface temperature and the subsequent profile is quite small compared with the total energy deposited in the plasma; the ionization energy per unit mass, for example, greatly exceeds the vaporization energy. Comparison of the temperature profile with the temperature value for decomposition of the solid propellant indicates the opportunity for mass loss from the propellant slab. Only a portion of this mass, however, may be needed to sustain the steady-state position of the discharge. Indeed, for a constant current, the depth of material required by the discharge increases linearly with time, while diffusion theory for the heat pulse suggests that the depth of decomposed mass will increase as the square root of time. The upper sketch in Fig. 3 depicts this difference in depths in the case of a constant discharge current, for which the idealized model predicts constant values of surface temperature and ablation rate. If the discharge current ends before the curves intersect, a portion of the decomposed propellant will not be accelerated electromagnetically. Figure 4 displays the depth for decomposition and depth of material required by the discharge for the LES-6 waveform. (A decomposition temperature of 600 K is used here. The higher value of 673 K provides similar behavior but corresponds to the experimental mass loss in LES-6.) Figure 3 also displays a sketch for a current pulse that is not constant, indicating the opportunity for the depth of a material needed by the discharge to merge with that for thermal diffusion. Such a merger corresponds to the evolution of an electromagnetically powered deflagration wave[9] that propagates into the propellant. The model therefore suggests directions for improving poor propellant utilization by implying that there exist current pulse times that will utilize the major portion of the decomposed propellant for electromagnetic acceleration.

Further interrogation of the MACH2 computations provides the relative magnitudes of the different heat transfer mechanisms that cause ablation and supports the assumption of excluding radiative heat transfer. In all cases examined, with operating energies of several joules, heat conduction has proven to be the

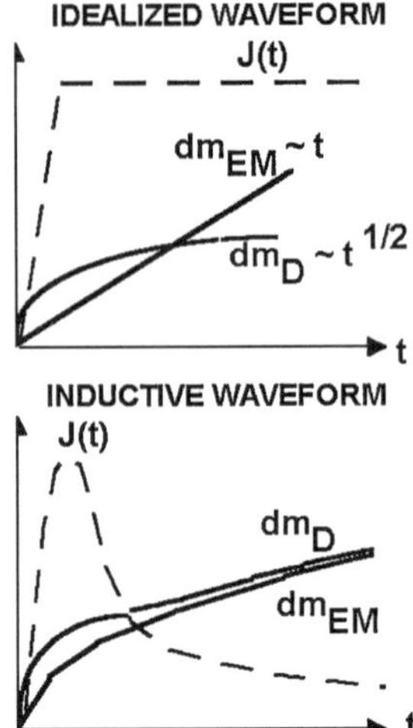

Fig. 3 Qualitative evolution of electromagnetically accelerated mass and decomposed mass for idealized and inductive waveforms.

dominant heat process. In particular, the heat flux to the surface due to conduction $[\mathcal{O}(10^8\ \mathrm{W/m^2})]$ is two orders of magnitude larger than convection $[\mathcal{O}(10^6\ \mathrm{W/m^2})]$ and upper estimates of radiation heat flux. Radiative heat transfer is not included in the calculations due to computational expenditures; specifically, the nature of the Teflon opacity coefficients involves steep gradients leading to a hostile numerical environment. However, independent calculations utilizing the MACH2-SESAME opacity tables prescribe that radiative heat flux does not exceed $1 \times 10^6\ \mathrm{W/m^2}$ and total radiation energy losses are of the order of millijoules.

IV. Confirmation of the Idealized Model

By providing in MACH2 a current waveform that rises quickly to a steady value, the principal elements of the idealized model can be checked. Computed evolution of the pertinent variables (see Fig. 5) shows that indeed steady-state operation

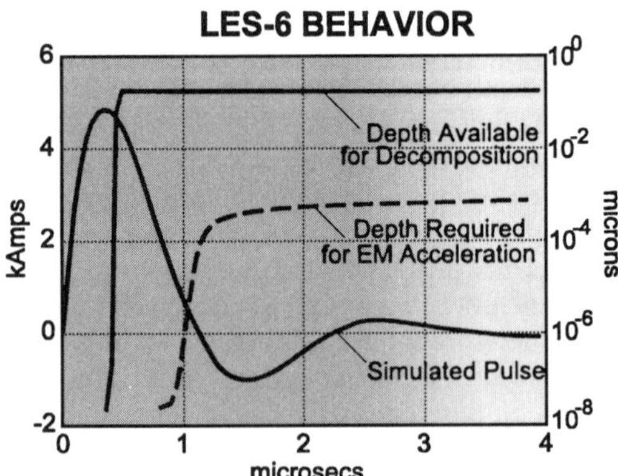

Fig. 4 MACH2 calculation of electromagnetic and decomposition depths for two effective decomposition temperatures.

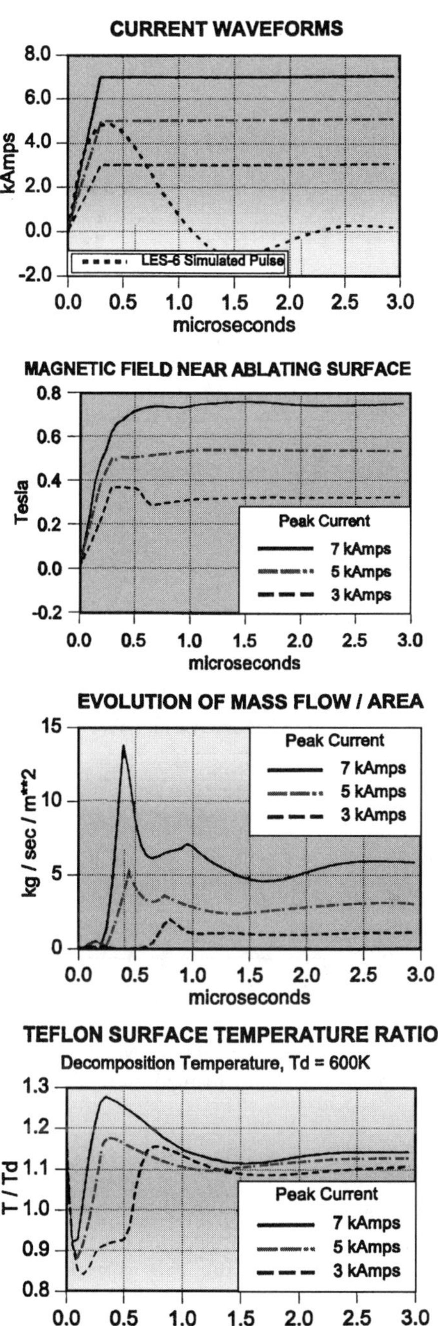

Fig. 5 MACH2 numerical simulations with idealized current waveforms (top) confirm steady-state operation.

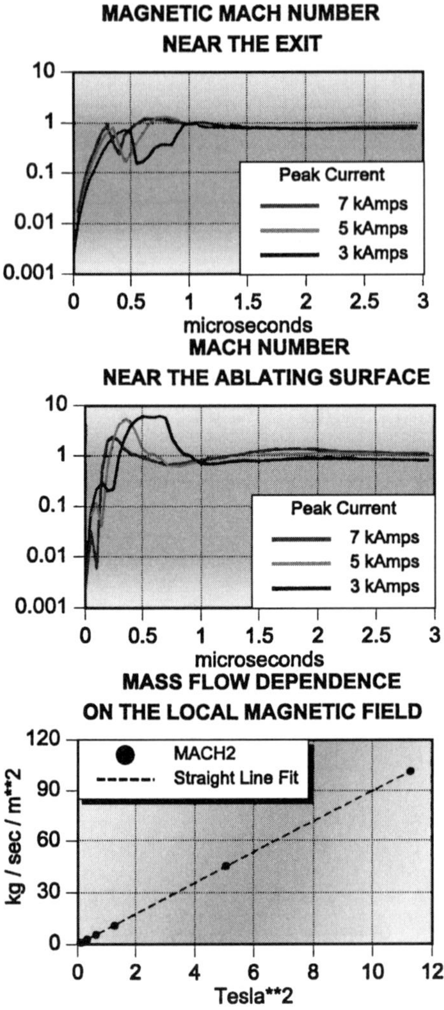

Fig. 6 MACH2 evolution of magnetic and thermal Mach numbers to sonic conditions. Scaling of $\dot{m}/A$ with magnetic field at the propellant's surface.

is achieved within 3 μs (a few times the current rise time used here). Figure 6 indicates that the magnetosonic condition is attained in all three cases of current amplitude, along with the expected thermal sonic condition adjacent to the propellant's surface. The location of the magnetosonic point, just downstream of the first portion of the discharge current, does not vary significantly with interelectrode space or electrode length. The mass flow rate per unit area scales as the square of the magnetic field in agreement with the notion of sustaining a magnetic pressure difference established by the discharge, $\dot{m}/A \sim B_0^2$. This scaling is maintained for much higher peak currents than the three test cases (3, 5, and 7 kA) displayed in detail here. The computed magnitude of the mass flow rate is within 15% of the value calculated by the idealized model. The agreement between MACH2 and the idealized model lends credence to both the utility of the analytical model and the accuracy of MACH2.

V. Optimized Current Waveforms

The qualitative profiles depicted in Fig. 3 suggest that sufficiently long pulse times will ensure that all of the available decomposed mass participates in the electromagnetic acceleration process. Numerical estimates, however, indicate that, if current levels are held constant, pulse times must exceed a few hundred microseconds to achieve such behavior (at least for surface magnetic fields of less than 5 T). For microspacecraft applications, the available energy per pulse will not permit these times with conventional circuit elements. Use of inductive energy storage techniques can provide discharge durations of 10–50 μs at initial capacitor energies in the range of 50 J or less. For the PPT, such inductive-energy storage circuitry can be accomplished merely by using the PPT plasma to short the energy storage capacitor soon after the maximum energy has been delivered to an inductor in series with the PPT (i.e., just after voltage zero occurs on the capacitor). Figure 7 displays the arrangement we have developed along with the desired behavior of the current waveform.

For initial optimization surveys, the inductive-circuit waveform has been simulated by a typical LC current rise, followed by an L/R decay (see Fig. 8, top). MACH2 is used to handle the complex interactions of the discharge, plasma flow,

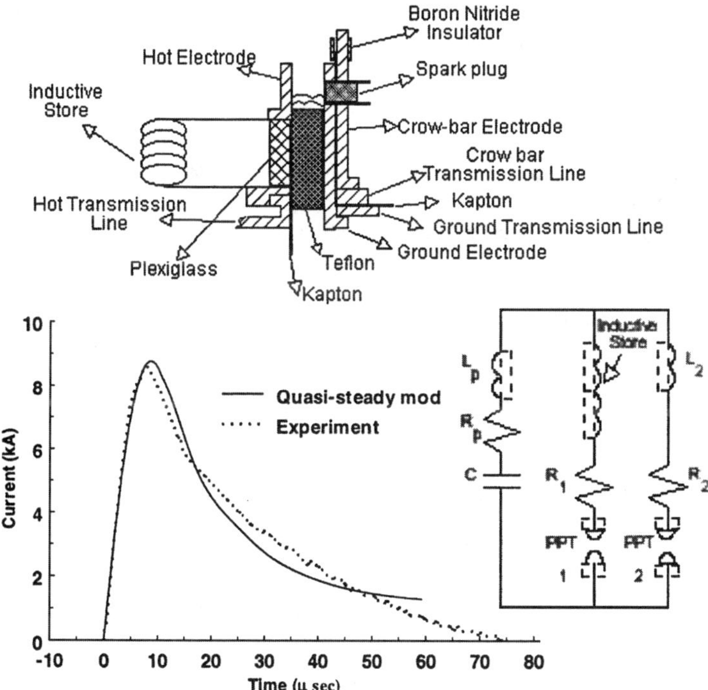

Fig. 7 Inductive-energy storage circuit arrangement and resulting current waveform.

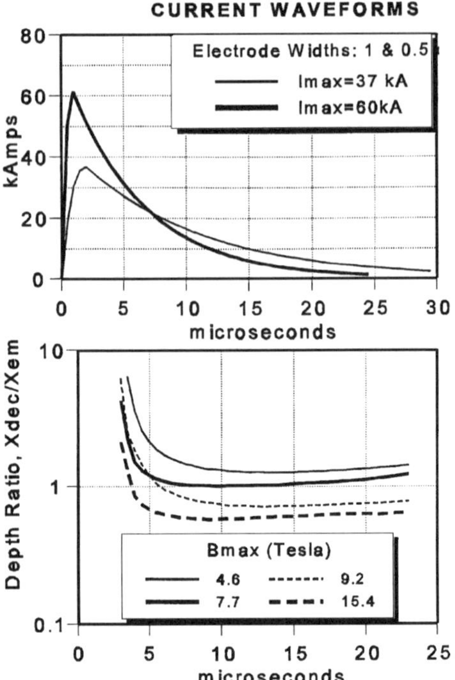

Fig. 8 MACH2 input current waveforms (top) and evolution of X_D/X_{EM} (bottom) for the rectangular PPT geometry. Width dimensions in centimeters.

and ablation process, as we seek to maximize the fraction of propellant accelerated electromagnetically. A combination of two current waveforms and two widths of thruster channel provides four values of the magnetic field at the surface B_0. The optimization is displayed in the lower part of Fig. 8, in terms of a ratio of depths for the two important processes: electromagnetic acceleration and solid decomposition. This depth ratio indicates the percentage of decomposed mass compared to that used by the discharge for electromagnetic acceleration. Consequently, a decrease in this ratio toward unity implies efficient mass utilization. Higher magnetic fields (peak current per width) at the propellant surface, and hence higher surface temperatures, provide substantial improvements in propellant utilization. It is noted that, at these elevated temperatures and prolonged pulse durations, the calculated decomposition depth is much less sensitive to uncertainties in propellant properties (than under the LES-6 operating conditions). More specifically, under typical optimizing conditions, error analysis based on a $\pm 5\%$ uncertainty in the equilibrium vapor constants and effective decomposition temperature reveals less than $\pm 15\%$ error in the determination of the decomposition depth.

Consequently, a basic scaling for efficient PPT performance requires matching the circuitry to the width (or circumference) of the thruster to exceed minimum values of magnetic field. This matching and the design of the electrical circuit parameters must include the temporal variation of the current, so that there is sufficient time for the ablation wave to merge with the position of the temperature value in the solid corresponding to propellant decomposition. Simply delivering a high current pulse to narrow electrodes for a short time may not be sufficient to achieve this two-part operating condition. Empirical exploration of PPT performance can readily miss the conditions for efficient behavior. Furthermore, the selection of propellants influences the matching of current and geometry by providing different values for decomposition temperature, thermal diffusivity, and equilibrium vapor pressure (vs surface temperature).

VI. Simulations in Coaxial Geometry

Close comparison of experimental data and theoretical modeling of the PPT in its traditional rectangular configuration is severely hampered by three-dimensional effects. Optical measurements cannot use Abel inversion techniques, and theoretical calculations, even with state-of-the-art computer codes, are simply inadequate. This situation has prompted us to focus on PPTs in coaxial geometry, in hopes of avoiding the complexities of three-dimensional effects.

MACH2 is once again utilized to improve a cylindrical PPT based on insights gained from optimization of the rectangular configuration. In particular, a maximum magnetic field of the order of 10 T was implemented near a 1-mm-radius cathode, using fast-rising, slowly decaying current pulses that peaked at 50 kA. Figure 9 depicts the geometric arrangement of the coaxial PPT. The temporal behavior of the current and the ratio of decomposed mass to mass that is electromagnetically accelerated are shown in Fig. 10 for a typical case. For a given current waveform, the mass ratio has a minimum for an exposed propellant area that is approximately a fifth of the interelectrode gap area. Larger exposed areas result in a greater amount of decomposed mass, due to the increased propellant areas at the larger radii. Higher mass ratios are also associated with smaller exposed areas, due to the deeper decomposition depths in the solid as a consequence of an elevated heat flux near the ablating surface. The latter is driven by the mass flow rate requirement at the magnetosonic point.

For a fixed propellant area, the minimum ratio of decomposed mass to mass that has been electromagnetically accelerated improves with increased pulse duration due to the prolonged electromagnetic acceleration as the propellant cools. The effects of both the exposed propellant area and the pulse duration on the optimization

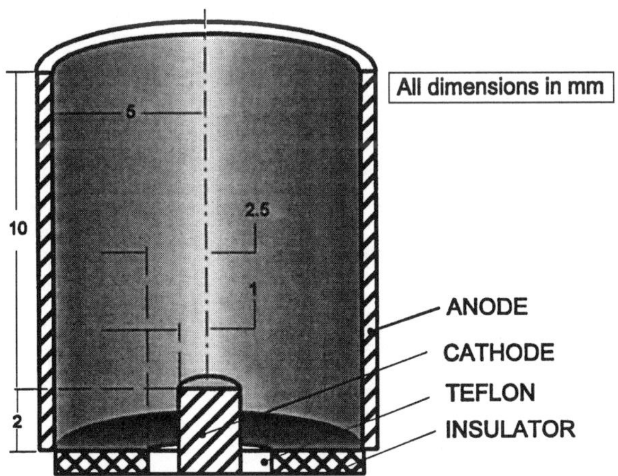

Fig. 9 Coaxial PPT geometry simulated by MACH2.

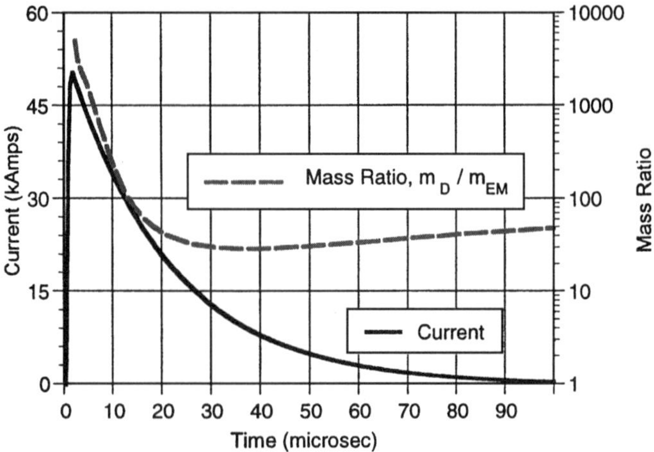

Fig. 10 Typical current waveform and corresponding mass ratio from the coaxial PPT simulations.

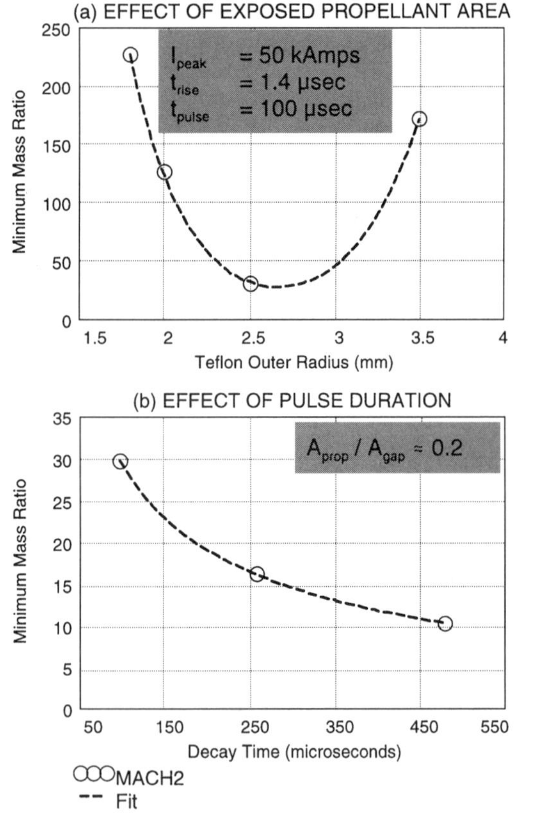

Fig. 11 Optimization of a coaxial PPT.

of this coaxial geometry are illustrated in Figs. 11a and b, respectively. It is evident that, although improved, the mass ratio is still much higher than desired. At the optimum propellant area and for a pulse time of about 100 μs, the amount of decomposed mass is still approximately 30 times higher than the amount accelerated. Further improvement is achieved with longer pulses, but in a manner that is inversely proportional to the square root of the pulse decay time (as suggested by Fig. 3 for times earlier than the idealized intersection point). This implies that, for the particular arrangement calculated, to eliminate the potential of any wasteful release of mass due to propellant decomposition, current waveforms of much longer durations than those attainable by typical circuit elements for PPT application to microsatellites should be implemented.

In contrast to the parallel-plate thruster, the principal difficulty with a coaxial PPT in this simple arrangement (resembling an ablation-fed magnetoplasmadynamic arcjet) is the decrease in magnetic field with radius. If the necessary magnetic field for efficient propellant utilization is achieved near the center conductor, the propellant at larger radii is not used properly. There will also be a tendency for the propellant surface to recede faster at smaller radii, which means a change in shape over the course of the mission (such variation in surface geometry, of course, is not new for solid-propellant rockets). Further explorations of coaxial arrangements,[10] using the MACH2 code guided by the principles derived from the idealized model, have permitted efficient propellant utilization in an inverse-pinch axisymmetric system.[11]

VII. Optimizing the Specific Impulse

Determination of the optimum specific impulse or exhaust speed value depends on the mission requirements. If not limited by drag or the so-called power supply penalty, higher exhaust speeds maximize the payload mass delivered to a desired trajectory or orbit. The Appendix suggests that the plasma speed at the magnetosonic location and speeds in the downstream exhaust scale with Alfven critical speed. The choice of propellant for the PPT can therefore significantly affect the available specific impulse value (in addition to playing a major role in the efficiency of propellant utilization for a given current pulse and thruster size).

For some missions (e.g., orbit-raising against atmospheric drag), the thrust-to-power ratio determines the optimum specific impulse. The thrust-to-power ratio varies inversely with the exhaust speed u_E:

$$\frac{F}{P} = \frac{u_E \Delta m}{u_E^2 \Delta m / 2\eta} = \frac{2\eta}{u_E} \tag{2}$$

We may write the thrust efficiency as the product of the efficiency of delivering energy to the thruster, η_s (i.e., due to external circuit elements), and the fraction of the exhaust energy in the form of directed kinetic energy vs energy lost to internal states (e.g., ionization), W_f:

$$\eta = \eta_s \left[\frac{u_E^2}{u_E^2 + 2W_f} \right] \tag{3}$$

Thus, the thrust-to-power ratio becomes

$$\frac{F}{P} = \frac{\eta_s u_E}{u_E^2 + 2W_f} \tag{4}$$

This maximizes for $u_E = \sqrt{(2W_f)}$. The lowest values of energy lost in the exhaust flow provide the highest thrust-to-power ratios. For high thrust-to-power ratio, plasma thrusters will always suffer in comparison to devices that do not require ionization of the working fluid, e.g., water resistojets. If ionization dominates W_f, the optimum exhaust speed would equal the Alfven critical speed. Thrusters such as the PPT and MPD arcjet, therefore, will tend to operate with exhaust speeds that exceed the value for the maximum thrust-to-power ratio.

VIII. Conclusions

The potential of the PPT to satisfy different mission constraints, ranging from a high thrust-to-power ratio for action against low-altitude drag to a high specific impulse for station-keeping or maneuvering at higher altitudes, could be achieved by matching the propellant to the mission need. Optimization begins with specification of the propellant in terms of its Alfven critical speed and continues with design of the thruster and propellant size to match the current waveform. The latter match demands consideration of the propellant properties and may obtain guidance from idealized models. Detailed design, including circuit behavior, requires computational tools, such as MACH2. Specifically, the use of MACH2 has identified and quantified the factors for poor mass utilization, confirmed the basic operation depicted by the idealized formulation, and provided specific operating conditions that optimize rectangular PPTs. In addition, it has provided insights on future avenues for optimization of coaxial configurations. The history of the PPT and electric propulsion indicates that simplicity is often more important than maximum performance in the selection of a spacecraft propulsion system. The insights and methods discussed in the present work should, nevertheless, expand the opportunities for new PPT systems.

Appendix: Plasma Speed at the Magnetosonic Point in the Limit of a Low β and a High Magnetic Reynolds Number

For a one-dimensional, steady flow, in the limit of a plasma pressure much lower than the magnetic pressure (low β) and a high magnetic Reynolds number (near the magnetosonic point), the equations of magnetohydrodynamics simplify considerably.

Continuity:

$$\rho u = w = \text{const} \tag{A1}$$

Momentum:

$$\rho u \frac{du}{dx} = -\frac{d}{dx}\left(\frac{B^2}{2\mu}\right) \Rightarrow wu + \frac{B^2}{2\mu} = \text{const} \tag{A2}$$

Energy:

$$w\frac{\mathrm{d}}{\mathrm{d}x}\left(h+\frac{u^2}{2}\right)=-\frac{\mathrm{d}}{\mathrm{d}x}\left(\frac{EB}{\mu}\right) \tag{A3}$$

At the magnetosonic point, $u = u^*$ equals the local Alfven sound speed:

$$u^*=\frac{B^*}{\sqrt{\rho^*\mu}} \tag{A4}$$

Substitution of this into the momentum equation provides the relationship of conditions at the magnetosonic point and the stagnation point ($u = 0$):

$$\frac{B_0^2}{2\mu}=\rho^*u^{*2}+\frac{B^{*2}}{2\mu}\Rightarrow\frac{B_0^2}{2\mu}=\frac{3B^{*2}}{2\mu} \tag{A5}$$

So,

$$B^*=\frac{B_0}{\sqrt{3}}$$

At the magnetosonic point, the mass density is

$$\rho^*=\frac{B^{*2}}{\mu u^{*2}} \tag{A6}$$

The mass flow per unit area is then

$$w=\rho^*u^*=\frac{B^{*2}}{\mu u^*} \tag{A7}$$

The electric field in the steady state, one-dimensional flow is uniform:

$$E=\text{const}=E^* \tag{A8}$$

In the limit of a high magnetic Reynolds number at the magnetosonic point,

$$E^*=u^*B^* \tag{A9}$$

The energy equation then becomes

$$\frac{B^{*2}}{\mu u^*}\left[\frac{u^{*2}}{2}+(h^*-h_0)\right]=\frac{u^*B^*(B_0-B^*)}{\mu} \tag{A10}$$

This equation provides the plasma speed in terms of the stagnation value of the magnetic field and the change in specific enthalpy, $Q^* = h^* - h_0$:

$$u^*=\sqrt{\frac{Q^*}{\sqrt{3}-1.5}}=1.468\sqrt{2Q^*} \tag{A11}$$

The Alfven critical speed, which has a long association with plasma accelerators, is

$$V_{\text{crit}}=\sqrt{2W_i} \tag{A12}$$

Thus, the plasma speed at the magnetosonic point is about 50% higher than the Alfven critical speed, if the energy needed for ionization dominates the change in flow enthalpy associated with electrical dissipation. The extent to which this is true depends on many factors within the plasma discharge, including heat transfer and nonthermal distributions.

Acknowledgments

The authors acknowledge the support of the NASA Glenn Research Center, the Air Force Office of Scientific Research, and the Ohio Supercomputer Center.

References

[1]Vondra, R. J., Thomassen, K., and Solbes, A., "Analysis of Solid Teflon Pulsed Plasma Thruster," *Journal of Spacecraft and Rockets*, Vol. 7, No. 12, 1970, pp. 1402–1406.

[2]Turchi, P. J., "Directions for Improving PPT Performance," *Proceedings of the 25th International Electric Propulsion Conference*, Vol. 1, Electric Rocket Propulsion Society, Worthington, OH, 1998, pp. 251–258.

[3]Spanjers, G. G., Lotspeich, J. S., McFall, K. A., and Spores, R. A., "Propellant Losses Because of Particulate Emission in a Pulsed Plasma Thruster," *Journal of Propulsion and Power*, Vol. 14, No. 4, 1998, pp. 554–559.

[4]Peterkin, R. E., Jr., and Frese, M. H., *MACH: A Reference Manual*, 1st ed., Air Force Research Lab., Phillips Research Site, Kirtland AFB, NM, 10 July 1998.

[5]Mikellides, P. G., and Turchi, P. J., "Modeling of Late-Time Ablation in Pulsed-Plasma Thrusters," AIAA Paper 96-2733, July 1996.

[6]Mikellides, P. G., Turchi, P. J., Leiweke, R. J., Schmahl, C. S., and Mikellides, I. G., "Theoretical Studies of a Pulsed-Plasma Microthruster," IEPC Paper 97-037, Aug. 1997.

[7]Wentink, T., Jr., "High Temperature Behavior of Teflon," AVCO-EVERETT Research Lab., Contract No. AF 04(647)-278, July 1959.

[8]Turchi, P. J., and Kamhawi, H., "PPT Thermal Management," AIAA Paper, 25th International Electric Propulsion Conf., Cleveland, OH, 1997.

[9]Courant, R., and Friedrichs, K. O., *Supersonic Flow and Shock Waves*, Interscience, New York, 1948, pp. 204–234.

[10]Mikellides, I. G., and Turchi, P. J., "Optimization of Pulsed Plasma Thrusters in Rectangular and Coaxial Geometries," IEPC 99-211 26th International Electric Propulsion Conf., Japan, 1999.

[11]Mikellides, I. G., "Theoretical Modeling and Optimization of Ablation-Fed Pulsed Plasma Thrusters," Ph.D. Dissertation, Ohio State Univ., Columbus, OH 1999.

Chapter 15

Laboratory Investigation of Pulsed Plasma Thrusters with Gas Valves

N. Antropov,* G. Diakonov,* O. Lapayev,* and G. Popov*
Research Institute of Applied Mechanics and Electrodynamics, Moscow, Russia

I. Introduction

THE advantages of pulsed plasma thrusters (PPTs), constant readiness for operation and capability for precise thrust and impulse control, are well known. That is why modern PPTs are designed for spacecraft attitude control and station-keeping systems. All these thrusters are devices of the erosion type, with an electromagnetic or electrothermal mechanism of plasma acceleration. PPT operation with liquid or gaseous propellants is in the stage of laboratory prototype development.

But in some cases, it is more advisable to use gas as the propellant. One obvious example is PPTs for an attitude control system of a spacecraft where the main propulsion system is fed by gaseous propellant [comprising stationary plasma thrusters (SPT), for example].[1] Application of low-power PPTs, using the same gaseous propellant as the main propulsion system and powered by a common power source combined with a power processing unit (PPU), would be the best solution from the mass characteristic point of view.

II. Electromagnetic Pulsed Gas Valves

The actuating mechanism of a pulsed gas valve comprises a fast-action converter, intended for transformation of capacitor energy into mechanical displacement. Induction-dynamic (IDM), electrodynamic (EDM), and electromagnetic (EMM) mechanisms are typically used.

With some assumptions it is possible to determine the relationship of the mechanism efficiency with its initial and final states. Assuming that Joule losses are absent and the period of natural oscillations in the discharge circuit is essentially less than the time of mechanism motion, efficiency is determined by[2]

$$\eta = 1 - \sqrt{L_0/(L_0 + \Delta L)} \tag{1}$$

where L_0 is the total initial inductance of the power supply circuit and ΔL is

Copyright © 2000 by the authors. Published by the American Institute of Aeronautics and Astronautics, Inc., with permission.

*Department of Applied Mechanics.

the change of inductance, caused by displacement of the movable parts of the mechanism.

The principle of IDM operation is based on the interaction of eddy currents, induced in the valve disk, with the pulsed magnetic field of the coil. The main advantage of an IDM is the high fast-action, limited only by the mechanical strength of the magnetic coil and other parts of the mechanism. As follows from Eq. (1), the efficiency of the IDM at small displacements of the disk, characteristic for the valve, can be estimated by

$$\eta \cup 1 - [1/\sqrt{1 + (x/h)}] \tag{2}$$

Thus the efficiency is determined by the ratio of the total mechanism travel x to the initial gap h between the conducting disk and the magnetic coil. To obtain maximum efficiency the coil should be made in the form of a plane disk and placed as close to the conducting disk as possible, aiming to increase the x/h ratio. As a rule, the travel of pulsed gas valves is 1–2 mm maximum; in this case, x/h is about 0.2 and efficiency is 10–15% maximum.

The principle of EDM operation is also based on the use of Ampere's forces, affecting a conductor with current in a magnetic field. However, unlike IDMs, EDMs use currents in coils instead of eddy currents.

EMM operations, based on the interaction of a magnetic field created by an electromagnet with a current in a core, are widely used in engineering.[3] However, in superfast-action mechanisms such as pulsed gas valves, they are rarely used. This is related to the fact that the force, created by an electromagnet, is proportional to the second power, and the mass of movable parts to the third power, of its linear dimensions. Therefore high acceleration of the armature can be obtained in a small-dimensional electromagnet with a low force. The authors developed a fast-action disk electromagnetic valve, a schematic of which is shown in Fig. 1.

The features of these valves are a plane armature and an inductor with a great number of concentric slots and poles, having alternating directions of current in the slots. In this case the armature thickness is determined by the width of the inductor poles and is independent of the force created by the electromagnet. That force as well as the mass of the armature is proportional to the square of the magnet operational surface, which allows the high fast-action to be obtained. If the

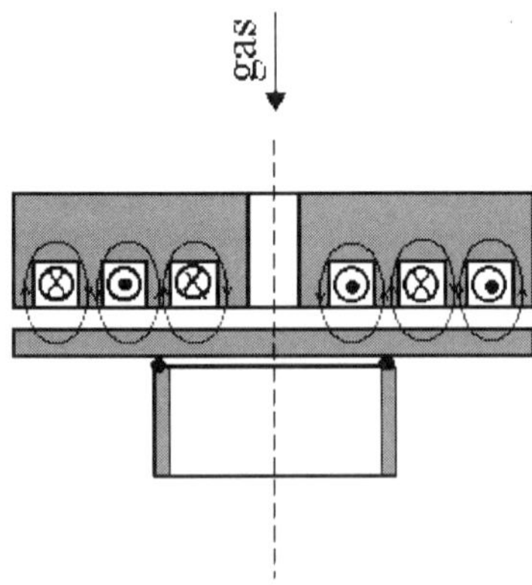

Fig. 1 Schematic of an electromagnetic pulsed gas valve.

discharge in the winding of the EMM is aperiodic, the efficiency of the mechanism can be approximately evaluated by[4]

$$\eta \sim 1 - \{1/[1 + (x/h)]\} \tag{3}$$

where x is the travel of the armature and h is the final air gap. In the given construction of the EMM the travel of the armature x is practically equal to the initial gap h_0 between the armature and the inductor. At $x = 0.5$ mm and $h = 0.1$ mm, the theoretical efficiency reaches 80%. Because of the high efficiency a simple, low-voltage circuit with electrolytic storage capacitors and semiconductor switches can be used.

Developed gas valves should meet the following three requirements: 1) The gas impulse duration should correspond to the duration of the current impulse in the PPT ($\leq 100\ \mu$s). 2) The power consumption per a gas mass unit should not exceed 10^8 J/kg. 3) The valve lifetime should not be less than 10^6 or even 10^7 cycles.

Known pulsed gas valves are based mainly on EMMs or IDMs. A valve disk of ferromagnetic material is a moving element in the case of an EMM. Electromagnets differ by their high efficiency and relatively low discharge currents, so they meet the second and third of the preceding requirements very easily. For example, the lifetime of electromagnetic injectors of diesel motors reaches 10^8–10^9 cycles. But as for fast-action, the available electromagnetic valves do not meet the requirements for PPT gas valves: the action period of the best of them is 5–10 ms. This is explained by the limitation of magnetic pressure on the moving armature by saturation of ferromagnetic materials. A gas impulse duration of more than 1000 μs is unacceptable for PPTs, because it causes too great a propellant mass loss.

On the contrary, electrodynamic valves, in which ferromagnetic materials are not used, allow one to obtain the required fast-action very easily (valves having a total action period of about 10 μs are known), but they do not absolutely meet the requirement of power consumption per a gas mass unit and resource. This is explained by the fact that their efficiency does not exceed 10% because of a number of features of the magnetic flow distribution. In the case of real electrodynamic gas valves, designers have not managed to make the power consumption per a gas mass unit less than 10^9 J/kg. This is absolutely unacceptable for PPTs, with a propellant consumption of about 10^{-7} kg per pulse and a discharge energy of less than 100 J. Besides, the low efficiency leads to high discharge currents in the valve circuit (of the order of kA) and, correspondingly, requires a complicated and insufficiently reliable high-voltage power supply circuit for the valve.

Electromagnetic pulsed gas valves of the disk type were developed by the authors during the program of coaxial (quasi-)stationary plasma accelerator (CSPA) development. Valves of this type were successfully used during many years of experimental work. The magnetic flow distribution used in them differs substantially from that in the classical electromagnet. The movable valve disk of the disk electromagnet is made in the form of a thin (0.1- to 0.5-mm) and light ferromagnetic disk. This allowed high fast-action comparable with the fast-action of EDMs. In addition, disk electromagnetic valves have, as do all electromagnets, a high enough efficiency, thus the power consumption per a gas mass unit does not exceed 10^8 J/kg. Figure 2 shows the flow rate characteristics for the valve at a pressure difference of 1 bar. The gas flow rate is precisely controlled by the gap h between the valve seat and the valve disk. Each curve corresponds to a different

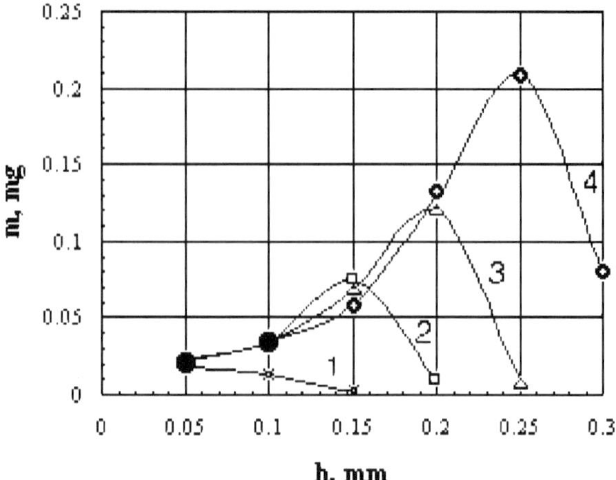

Fig. 2 Flow rate characteristics of the disk-type pulsed gas valve.

voltage on the capacitor: curve 1, 150 V; curve 2, 200 V; curve 3, 250 V; and curve 4, 300 V.

Figure 3 shows the time characteristic of the valve. The time period t is the period of the complete action of the valve moving part measured by a contact sensor. Because of elastic deformation of the rubber seal, the duration of the gas impulse is less than t by approximately 50 μs and is from 100 to 200 μs. Such a gas impulse duration may be matched by the discharge duration in the PPT, that being an insuperable problem until now.

III. Gas Propellant PPTs

It is easier to obtain a relatively long discharge, 100–200 μs in duration, in an electrothermal plasma thruster with a high ohmic resistance of the discharge gap. Besides, in this specific case the electrothermal acceleration mechanism has a number of substantial advantages, which will be described here. The external appearance of the electrothermal gas PPT laboratory model is shown in Fig. 4.

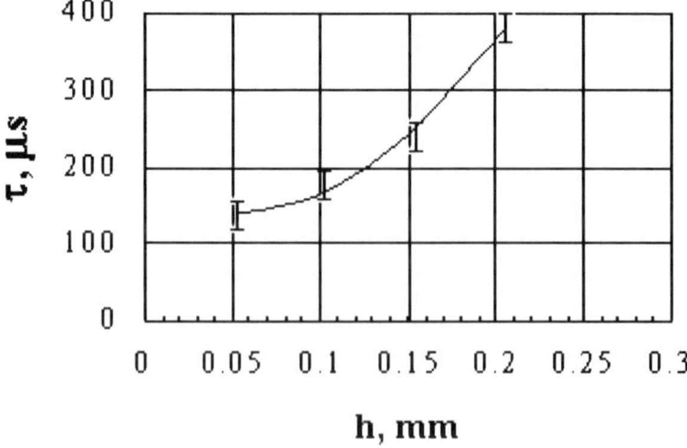

Fig. 3 Time characteristic of the valve.

Fig. 4 PPT with a gas valve.

A gas PPT requires a long experimental development, during which the corresponding passage cross sections, defining the channel gas dynamics, are selected. That is why, in the case of this model, there was the task of realizing the advantages of a pulsed gas valve in the simplest design.

The following devices are mounted inside the dielectric casing of the thruster: the gas valve, a long cylindrical ceramic discharge channel with electrodes, and an output nozzle. The thruster has rubber seals allowing it to operate both under vacuum conditions and connected to a flange of a vacuum chamber. The discharge channel length is 30 mm and the diameter is 5 mm. The outlet diameter of the nozzle is 10 mm.

The gas valve is the only controlled element in the thruster. Voltage from a capacitor is applied to the thruster electrodes and the discharge gap breakdown corresponds to the Paschen curve. In this case the breakdown should take place only at the moment when the gas impulse front reaches the outlet of the channel. Then at further discharge development, gas from under the valve will flow into the discharge gap, and the duration of the gas and current pulses will be matched. Breakdown at the gas impulse front is provided by the igniting electrode, to which the increased voltage from an additional capacitor of small capacitance is applied. In the case of correct selection of the passage cross sections for the channel and nozzle, it is possible to have a breakdown of the discharge gap after complete filling of the channel and at a voltage of about 400 V, which is a little higher than the voltage required according to the Paschen curve.

In the case of electrothermal acceleration the thrust efficiency of the thruster is defined mainly by the energy loss caused by the heat conductivity, radiation, and enthalpy of the outflowing plasma, and in contrast to electromagnetic acceleration, it does not depend on the discharge circuit inductance. The active resistance of the discharge gap, $R \sim 0.1\text{–}1.0\ \Omega$, is substantially higher than the total resistance of other parts of the circuit R_0. This is why, in a propulsion system comprising a low-voltage electrothermal PPT, it is advisable to divide the propulsion unit and the power unit spatially, keeping the condition $R > R_0$. The capacitive unit may be mounted separately from the thrusters and connected to the propulsion units by coaxial cables. One capacitive unit, combined with a power unit and able to provide simultaneous operation of two thrusters, may be used for supplying power to all thrusters of an attitude control system, which will result in substantial mass savings compared with current PPTs.

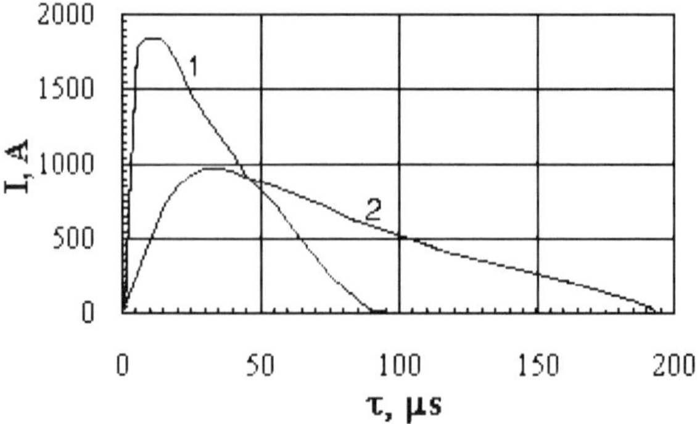

Fig. 5 Discharge current oscillograms.

Experimental development of the thruster was done using hydrogen and nitrogen. The gas flow rate per pulse and discharge currents in the valve circuit, ignition circuit, and main discharge circuit were measured during the test. Characteristic oscillograms for the current in the main discharge circuit for the cases of hydrogen (1) and nitrogen (2) are shown in Fig. 5.

The discharge is aperiodic and is characterized by a substantial excess of the discharge gap ohmic resistance over the wave resistance of the circuit. The form of the oscillogram for the current is determined by the circuit capacitance, C, inductance, L, and ohmic resistance of the discharge circuit, R. In this case $C =$ const and $L =$ const. Therefore, given the experimental current oscillogram, it should be possible to calculate the resistance of the discharge gap at any moment in time. As the ohmic resistance is determined by the temperature of the gas, it is possible using the experimental data to calculate the temperature and pressure of the plasma in a channel.

The acceleration of the gas in Laval's nozzle was considered, assuming that the flow is nonequilibrium (frozen flow). The gas dynamics of the thruster and, correspondingly, the processes of discharge expansion are determined by the flow cross sections of the inlet jet and outlet nozzle. The main purpose of the experimental work was to select such cross sections that provide a stationary breakdown of the discharge gap at the leading front of the gas impulse and the gas flow, matched by the discharge expansion under minimum erosion of the electrodes. In this case the diameter of the inlet jet cross section d_1 changed from 1.0 to 4.5 mm, and the diameter of the nozzle critical cross section d_2 from 2.0 to 3.0 mm. The best results were obtained at $d_1 = d_2 = 2.0$ mm. The characteristics of the thruster for thermal acceleration in Laval's nozzle are determined fully by the temperature and pressure at the nozzle inlet. The plasma temperature in the discharge channel can be determined adequately by the discharge current oscillogram. The resistance of the discharge gap alone is variable among the three parameters of the circuit, R, L, and C. It is determined by the plasma temperature in the channel and changes slightly during the discharge. When it is assumed, as a first approximation, that $R =$ const, the discharge current is determined by the second-order differential equation,

$$L\frac{d^2 I}{d_t^2} + R\frac{dI}{d_t} + \frac{1}{C}I = 0 \tag{4}$$

at initial conditions $I_{t=0} = 0$ and $U_{t=0} = U_0$, where I is the discharge current, U is the capacitor voltage, and L and C are the inductance and capacitance of the discharge circuit.

If $R > 2(L/C)^{\frac{1}{2}}$, the discharge is aperiodic, and the solution of Eq. (4) takes the form[2]

$$I = I_0 \exp(-\chi\tau)\frac{sh\sqrt{\chi^2 - 1}\tau}{\sqrt{\chi^2 - 1}} \tag{5}$$

where $\tau = t/\sqrt{LC}$, $\chi = (R/2)(C/L)^{\frac{1}{2}}$, and $I_0 = U_0(C/L)^{\frac{1}{2}}$.

The mean design resistance of the circuit R is taken so that the calculated current oscillogram conforms to the experimentally obtained one. Given the mean resistance of the discharge gap and its geometric dimensions, the plasma conductivity σ and then the temperature T in the thruster channel can be calculated. The relationships between the conductivity and the temperature, obtained experimentally for hydrogen and nitrogen, are given in the literature.[6,7] The temperature in the channel is 12,000–18,000 K for hydrogen and 10,000–14,000 K for nitrogen. Using the temperature T in the channel, the velocity of monatomic gas outflow from Laval's nozzle can be calculated by the equation

$$V = (5RT)^{\frac{1}{2}} \tag{6}$$

where $R = R_0/\mu$, $R_0 = 8.31 \cdot 10^3$ J/kmol · K is the universal gas constant, and μ is the mean molecular mass.

Relationships between hydrogen and nitrogen plasma masses and temperatures are well known. Equation (6) is applicable for nonequilibrium gas outflow (frozen flow). For calculation of the thrust impulse, apart from the mean outflow velocity, the mass m of gas, injected into the discharge, should be known. Generally speaking, this mass is less than the measured mass of gas passed through the valve per impulse m_v.

It follows from the time responses of the valve and discharge that $m \sim m_v$. More precise m valves can be estimated by the energy $W = CU^2/2$ input into

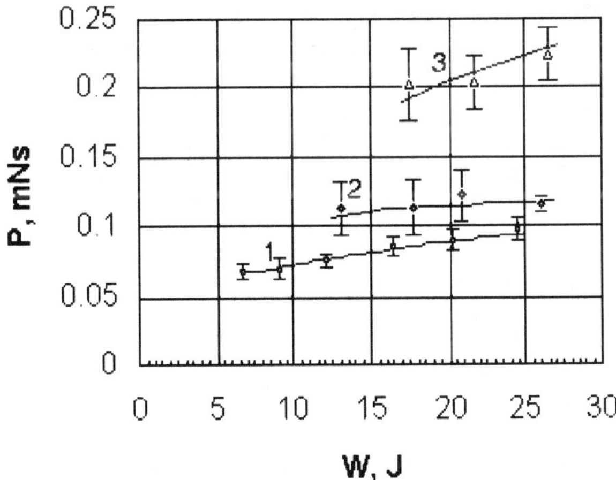

Fig. 6 Gas PPT impulse bit vs energy level.

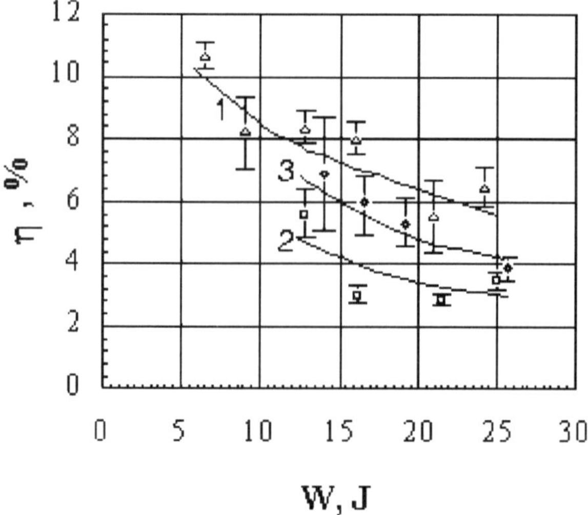

Fig. 7 Thrust efficiency vs energy level.

the discharge and by the enthalpy of the plasma at a given temperature. But this method of m determination can be used for hydrogen only, because of its rather low radiation losses.

Figure 6 shows the gas PPT calculated thrust impulse bit P as a function of the discharge energy W in the case of using hydrogen as the propellant. Calculations were made using oscillograms obtained experimentally for flow rates of $3.3 \cdot 10^{-9}$ kg (1), $6.3 \cdot 10^{-9}$ kg (2), and $27 \cdot 10^{-8}$ kg (3) of hydrogen per pulse. The errors of the submitted data are determined by the combination of the errors of the measurements and the errors of the published material properties.

Figure 7 shows the corresponding thrust efficiency η_t as a function of W while using hydrogen as the propellant. Mass flow rates are $3.3 \cdot 10^{-9}$ kg (1), $6.3 \cdot 10^{-9}$ kg (2), and $2.7 \cdot 10^{-8}$ kg (3) of hydrogen per impulse. The impulse becomes an order of magnitude higher while using nitrogen as the propellant, with the thrust efficiency being nearly the same. The impulse increases nonsubstantially and the efficiency decreases with an increase in the energy supplied to the discharge. This is explained by the substantial increase in plasma ionization losses with the temperature increase in the channel. Thus the discharge energy should be within 10–20 J at the characteristic gas flow rates for this thruster.

IV. Conclusions

Test results confirmed the principal possibility to develop a low-power propulsion system using gas-fed electrothermal PPTs. It is advisable to use such propulsion systems for the attitude control and angular stabilization system of a spacecraft equipped by a main propulsion system using gas as the propellant. Besides, it is possible to use liquid propellants with evaporators. An efficient model of a gas PPT was developed, which may be modified to obtain the required characteristics.

References

[1]Ziemer, J. K., Cubbin, E. A., Choueri, E., and Birx, D., "Performance Characterization of a High Efficiency Gas-Fed Pulsed Plasma Thruster," AIAA Paper 97-2925, 1997.

[2]Shneerson, G. A., *Fields and Transitional Processes in High-Current Apparatus*, Energoizdat, 1981 (in Russian).

[3]Seilly, A. H., "HELENOID Actuators—A New Concept in Extremely Fast Acting Solenoids," SAE Technical Paper Series, No. 790119, 1979.

[4]Bondaletov, V. N., "Determination of Electromagnetic Forces, Their Works and Electro-Dynamic Efficiency in Current Contours," *Elektritchestvo*, No. 1, 1966, pp. 57–60 (in Russian).

[5]Dyakonov, G. A., and Tikhonov, V. B., "Coaxial Quasi-Stationary Plasma Accelerator (QSPA) P-50A Experimental Results: Accelerating Channel Geometry and External Magnetic Field Effect on Plasma Flow Modes," *Plasma Physics*, Vol. 20, No. 6, 1994, pp. 533–540 (in Russian).

[6]Gross, P. A., and Eisen, C. L., "Some Properties of a Hydrogen Plasma," Fairchild Engine and Airplane Corp., Deer Park, NY, 1959.

[7]Finkelnburg, W., and Maecker, H., "Elektrische Bogen und thermisches Plasma," *Handbuch der Physik*, Bd. XXII, 1956, S. 254–444 (in German).

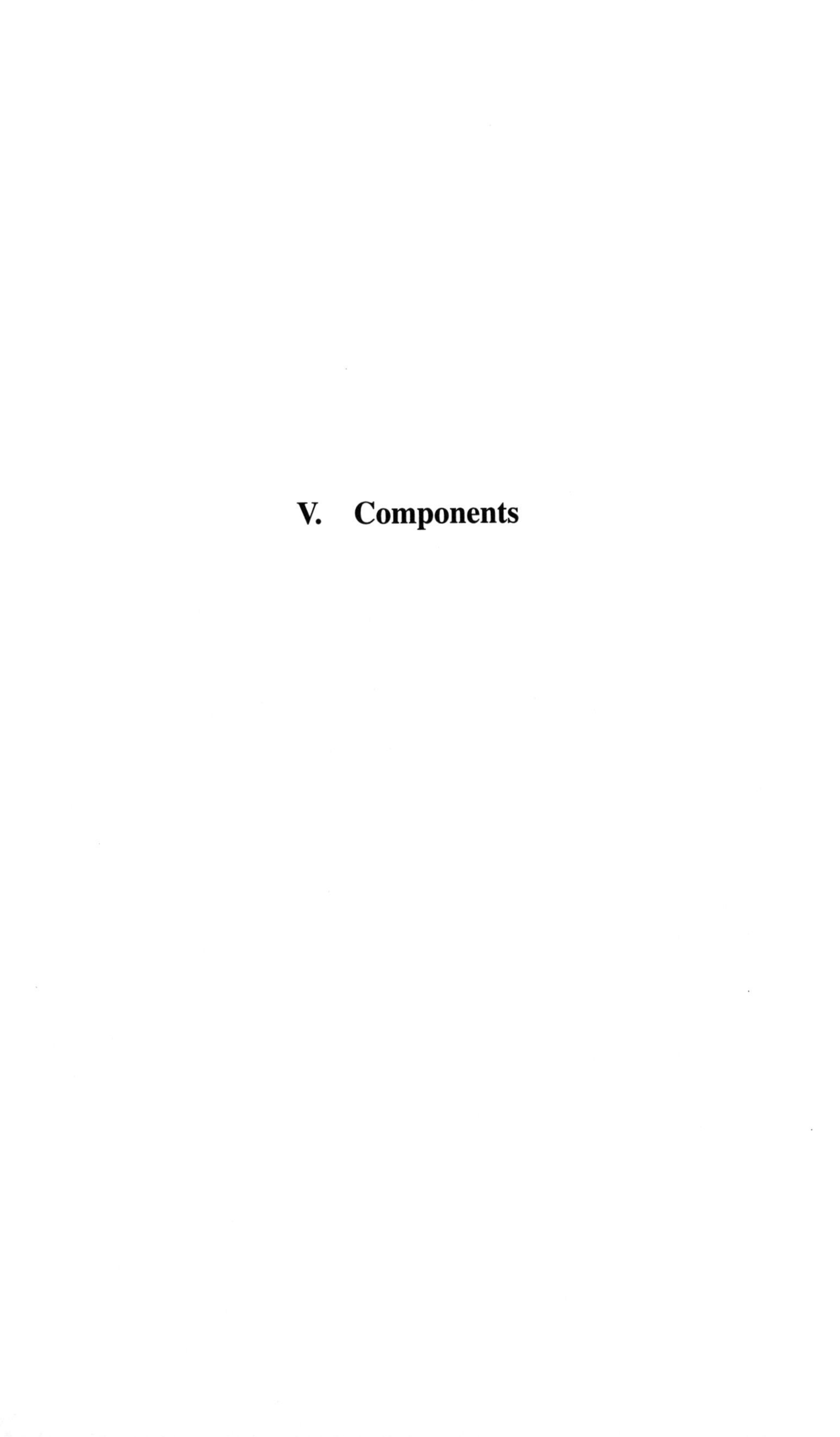

V. Components

Chapter 16

Fabrication and Testing of Micron-Sized Cold-Gas Thrusters

Robert L. Bayt*
United Technologies Research Center, East Hartford, Connecticut
and
Kenneth S. Breuer†
Brown University, Providence, Rhode Island

Nomenclature

a = speed of sound
C_d = coefficient of discharge
D = throat width
h_0 = nozzle height
$\dot{m}$ = mass flow rate
Re_t = Reynolds number based on throat conditions
T = thrust
t = throat property
U_e = exit velocity
X^* = X distance normalized by throat width
Y^* = Y distance normalized by throat width
δ^* = displacement thickness
$\delta_{99\%}$ = boundary layer thickness where velocity is 99% of core flow
ε = nozzle expansion ratio
η_{thrust} = thrust efficiency
μ = kinematic viscosity
ρ = density
θ^* = momentum thickness

Copyright© 2000 by R. L. Bayt and K. S. Breuer. Published by the American Institute of Aeronautics and Astronautics, Inc., with permission.

*Associate Research Engineer, Flow Systems Group. Member AIAA.
†Associate Professor, Department of Engineering. Senior Member AIAA.

I. Introduction

VARIOUS trends in the spacecraft industry are driving the development of low-thrust propulsion systems. These may be needed for fine attitude control or to reduce the mass of the propulsion system through the use of small lightweight components. As the number of satellites in a constellation increases, costs can be reduced by batch-fabricating individual spacecraft systems. MEMS (microelectromechanical systems) offer the capability to fabricate small fully integrated systems in large numbers.

Prior to micromachining, low thrust was achieved by fabricating the smallest nozzle possible through conventional machining and then running the device at a low chamber pressure to reduce thrust. However, the device Reynolds number scales with both the nozzle throat width and the chamber pressure (or chamber density):

$$Re_t = \frac{\rho_t a_t D_t}{\mu_t} = \frac{\dot{m}}{\mu_t h_0} \tag{1}$$

where the fluid properties are all defined at the throat condition. Thus, low thrust is accompanied by low Reynolds numbers. Mass flow rate can be substituted into this expression to yield the term on the right-hand side. The thrust of a nozzle is proportional to the momentum flux from the nozzle (the exit pressure component is small and may be neglected) and can be written

$$T \propto \dot{m} u_e \propto h_0 Re u_e \tag{2}$$

where u_e is the average exit velocity of the gas. The exit velocity is set by the geometry and remains roughly constant as the scale is reduced at moderate Reynolds numbers. Thus, if thrust is held constant, the Reynolds number increases as the feature size decreases for a constant exit velocity. The increasing Reynolds number is a direct result of the increasing chamber pressure needed to maintain a constant thrust while the scale is being reduced. Therefore, micromachining can be used to improve nozzle performance over its low thrust conventionally machined counterpart by allowing nozzles to be operated at higher Reynolds numbers. This implies higher thrust and mass flow efficiencies due to the reduction in viscous losses.

Nozzle performance at small scale has been studied on a number of occasions. Most notably, Rothe reported E-beam measurements of temperature and velocity profiles in a nozzle with a 5-mm throat.[1] A shockless transition to subsonic flow was observed, at a Reynolds number of 32, due to the viscous thermalization of flow energy. Grisnik et al.[2] investigated nozzles with throat diameters of the order of 650 μm. Each of these test cases was machined through conventional methods and was orders of magnitude larger than what is now available through MEMS. To reduce the thrust and hence minimize the impulse bit, these nozzles were run at low chamber pressures ($<$1 atm), resulting in low Reynolds numbers (500–9000) for a given thrust and hence a lower I_{sp} (58 s for 4.5 mN of thrust).

More recently, Janson et al.[3] presented results for batch-fabricated micronozzles that can achieve the same low thrust as their conventionally machined counterparts, with similar I_{sp} values. Janson's nozzle geometries were limited to converging–diverging orifices manufactured by anisotropic etching of silicon along crystalline planes using KOH and laser milling. This resulted in square nozzles with a 35.3-deg expansion, the angle of the flow exiting the nozzle relative to the centerline. The smallest geometry fabricated was 210 $\times$ 210 μm for a 10:1 expansion ratio. This large angle, combined with the sharp edge at the throat, resulted in probable flow separation and subsequent lower thrust efficiencies than have been achieved in the smoothly varying converging–diverging nozzles presented in this work. Typical nozzle designs maintain a 15–20-deg expansion and a throat radius of curvature twice that of the throat width. In addition, there is a divergence loss due to the component of the fluid momentum that is not along the thruster axis and is lost as useful thrust. This loss in performance increases with exit angle.

More advanced micromachining technologies, such as deep reactive ion etching (DRIE), allow arbitrary extruded geometries to be etched in the plane of the wafer. This affords flexibility in the nozzle geometry that can be used to design against flow separation and minimize divergence losses. By fabricating the nozzle in the wafer plane, the expansion ratio can be made arbitrarily large. The expansion ratio of nozzles etched along crystalline planes are limited by the thickness of the wafer since the throat area is set by the depth of the etch. Finally, there is flexibility in the nozzle thrust by adjusting the depth of the etch over the range of 50–500 μm. This changes the exit area without changing the expansion ratio and the exit Mach number. However, the nozzle aspect ratio (the ratio of nozzle height to local width) should ideally be large enough to minimize the influence of the endwall boundary layers.

This chapter presents a performance analysis of extruded two-dimensional micronozzles. The nozzles are fabricated by DRIE and tested for thrust and mass flow as a function of chamber pressure. Numerical simulations model the flow from the gas injection into the plenum through the exhaust of the nozzle. The numerical analysis is used to make performance predictions, and these predictions are compared with the experimental results. Ultimately, the impact of viscous losses on thrust performance is evaluated. This is quantified as the thrust efficiency, which is established as a function of the throat Reynolds number.

II. Fabrication

At the heart of this research program is a Surface Technology Systems (STS) deep reactive ion etcher, using the Bosch process.[4] The Multiplex ICP is an etcher that maintains tight control on ion directionality by using an inductively coupled plasma, which allows anisotropy to be maintained to great depths. Etched feature aspect ratios (depth to width) as high as 30:1 have been achieved. The micronozzle process flow is straightforward in principle. The challenge lies in attaining a highly variable geometry that maintains anisotropy over the full height of the structure. Complications include the fact that large features etch faster than smaller features, which causes the smaller features to distort when performing a deep etch. In addition, when small features of a constant geometry are etched, wall roughness can become accentuated.

The fabrication process is described in detail in Bayt et al.,[5] however a brief description is offered and is illustrated in Fig. 1. A clean wafer is protected with photoresist that can be applied with a thickness of up to 12 μm. The resist is patterned and developed and the resulting features are etched through the wafer (308 μm thick for the nozzles presented here). Typical etch rates are measured between 2 and 3 μm/min, depending on the feature size and etch recipe. The silicon wafers are mounted to quartz wafers with photoresist to prevent backside coolant from leaking when the features etch through the wafer and allow the cleared features to be detected by inspection.

The optimal geometry was eventually achieved by etching a feature of constant width to maintain a consistent loading. This is accomplished with a halo mask. A

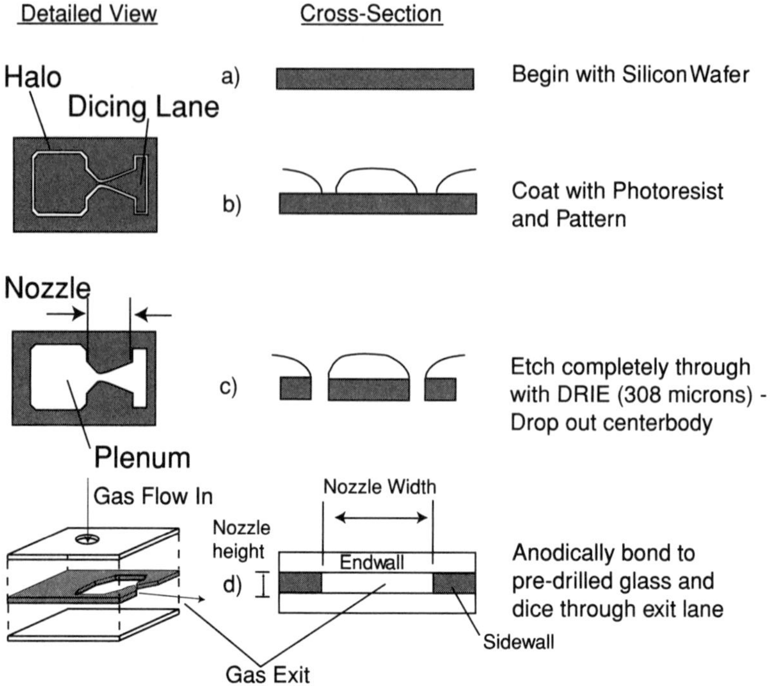

Fig. 1 Process followed for nozzle fabrication: a) A clean silicon wafer is b) coated with photoresist and patterned with a halo mask in the outline of the nozzle–plenum system. c) The wafer is etched by DRIE until the trench is through wafer and the centerbody falls out. d) The silicon is anodically bonded to glass.

10-μm trench outlining the feature of interest, called the halo, defines the nozzle. This allows the centerbody to fall out after the etch is completely through the wafer. Because of small feature distortion, best results are obtained for etches less than 300 μm in depth. Figure 2 is a scanning electron micrograph of a nozzle fabricated through this process. The anisotropy of this nozzle is striking. There is only a 1-μm variation in feature size from the top to the bottom of the wafer. The wall roughness is larger for this type of mask due to the small halo trench, which is of the order of 800 nm.

The etch process results in feature enlargement from the photoresist mask, which breaks down over time. For this reason, features that are 10 μm on the mask enlarge to 18 μm by the completion of the etch, however, this dimension varies by only 1 μm from the upper to the lower surface. In the future, it may be possible to mitigate this enlargement with use of an oxide mask. In addition to feature enlargement, there is a variation in etch rate across the wafer due to asymmetric etchant delivery to the chamber. Further characteristics of DRIE are given by Ayon et al.[6]

To achieve a high device performance, the gas must be injected into the nozzle with as low an entrance velocity as possible. To accomplish this, a settling chamber (or plenum chamber) is fabricated in parallel with and upstream of the nozzle. This chamber is a region of large volume (2×2 mm) that is pressurized and remains at constant pressure for the duration of the nozzle firing. Twelve nozzles and chambers can be fabricated on a 1-cm^2 die.

Fig. 2 Micrograph of a nozzle etched using a halo mask. The throat width is 19 μm with an expansion ratio of 5.4:1. The inlet area ratio is 7:1 and intersects a 2 × 2-mm plenum.

The flow channel is encapsulated by anodically bonding Pyrex (as described by Wallis and Pomerantz[7]) to the upper and lower surfaces. The upper surface has been ultrasonically drilled with a 1-mm hole, prior to bonding, to allow gas injection into the chamber. The anodic bonding is accomplished by contacting the clean silicon and Pyrex at 500°C with 1000 V and 2400 mbar of pressure applied across the stack. This task is accomplished with an Electronic Visions Aligner/Bonder, which allows the gas injection hole to be aligned with the chamber. Finally, the wafer is diced along lanes that intersect the nozzle but prevent the nozzle geometry from being distorted by inaccurate blade alignment. The dicing exposes the flow channel permitting the gas to discharge from the nozzle through the edge of the die.

Table 1 lists the geometries that were fabricated using this method. They are referred to throughout this chapter by their respective expansion ratios. The etches are 308 μm deep for all nozzles tested, except the 8.2:1 nozzle, which is 491 μm deep. All of the nozzles were fabricated with similar converging sections: a 7:1

Table 1 Fabricated nozzle geometry descriptions

Design throat width, μm	Design expansion ratio	Actual throat width, μm	Actual expansion ratio	Exit height-to-width aspect ratio
10	10:1	19	5.4:1	3.00
10	25:1	18	15.3:1	1.11
25	10:1	34	7.1:1	1.27
25	25:1	37.5	16.9:1	0.49
50	10:1	65	8.26:1	0.91

inlet area ratio, converging along a 45-deg half-angle. The diverging sections for all nozzles expand at a 20-deg half-angle.

Once the die has been fabricated, it must be packaged to interface with the macroscopic fluid delivery system. This is accomplished using a manifold, clamped to the silicon chip with a Parker-001 O-ring to prevent leakage. The manifold delivers the working gas through 0.25-in. stainless-steel tubing. The last frame in Fig. 1 shows the flow path of the nozzle. A valve and regulator are present upstream to control flow to the nozzle. Pressure in the chamber is read from a transducer that is integrated into the manifold. The pressure drop in the flow channels leading to the die is less than 0.1 psi at the tested flow rates.

III. Numerical Simulation

To assess the performance of the nozzles prior to fabrication and testing, numerical simulations establish a benchmark with which the experimental work is compared. Because of the nature of the problem, flow through an extruded nozzle, a two-dimensional simulation is used to evaluate the core viscous nozzle flow. The numerical calculation is for steady-state conditions through a finite-volume simulation of the Navier–Stokes equations. The geometry is nondimensionalized by the throat width, and the Reynolds number from Eq. (1) is used to scale the influence of viscosity.

The numerical analysis implements Van leer's flux-splitting scheme[8] for spatial discretization combined with a Jameson fourth-order Runge–Kutta timestepping algorithm.[9] The inlet conditions are set by the characteristic boundary conditions derived from the chamber properties of temperature and pressure with a fully axial velocity. The entire exit plane (subsonic and supersonic region) is extrapolated from the interior due to the parabolic nature of the boundary layer. A similar analysis was performed by Kim[10] for a low-Reynolds number resistojet, which also employed extrapolated outflow conditions. The walls are held at the chamber temperature due to the large volume of silicon present. Also, a zero wall-normal pressure gradient is enforced since no slip velocity is enforced at the walls, as described by Bayt et al.[11] A grid resolution study determined that grids of 125×160 and 230×240 are necessary for the 5.4:1 and 16.9:1 area ratio nozzles, respectively.

IV. Experimental Testing

The test setup utilizes grade 5.0 nitrogen regulated to chamber pressures ranging from 5 to 100 psia. A 0.5-μm filter is in line to prevent contamination from entering the nozzle. The flow rate is measured using a Teledyne–Hastings HFM-200 0- to 1000-sccm flowmeter, which is accurate to 1% of full scale, and the system is verified to be free of leaks before tests are run. Pressure is measured with an Omega PX-303 pressure transducer, which is accurate to 0.3% of full scale over the range of 0–300 psia.

Thrust tests were performed at the Aerospace Corporation using similar mass flow and pressure sensors. The thrust stand was accurate to 0.5 mN; however, it was undamped and oscillations limited the reading to ± 1 mN of accuracy. The thrust tests exhausted to a large volume chamber that was held to 100 mTorr by mechanical blowers over the duration of the firing. The test of the 8.2:1 nozzle was performed at MIT on a thrust stand with damping, accurate to ± 0.5 mN. Exhaust pressures were held to less than 50 mTorr for the duration of the firing.

V. Results and Discussion

A. Inlet Flow

To assess the validity of the boundary conditions imposed in the numerical simulation, a separate model was used to determine whether the pressure drop in the plenum is negligible and if the gas is axially injected into the nozzle. Since the gas enters through a hole perpendicular to the nozzle plane, a two-dimensional numerical model of the flow around a corner was developed using FLUENT/UNS, a commercial CFD code. The mass flow set by the nozzle throat is the inlet boundary condition. For a nozzle with a 19-μm throat at 10 atm of pressure, the throat Reynolds number is 2764. The streamlines for this case are shown in Fig. 3. A separation bubble forms on the upper and lower surface due to the discontinuity in wall curvature at the corner. The reattachment point is defined as the axial location at which the transverse velocity is less than 1% of the axial velocity and is marked in Fig. 3 for this operating condition.

The Reynolds number based on the channel height varies from 18 to 370, which corresponds to throat Reynolds numbers of 175 to 3721. A calculation of total

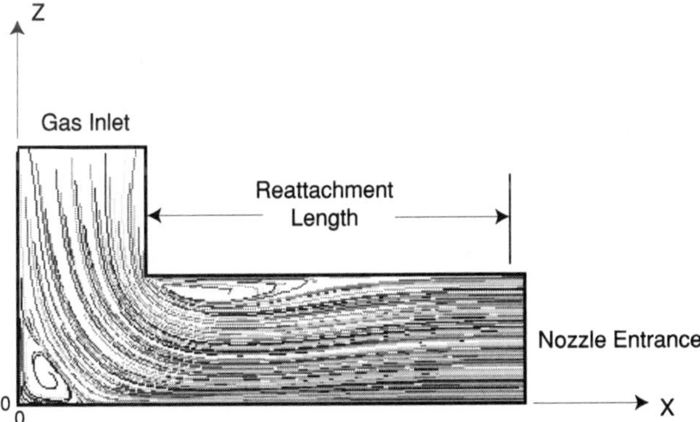

Fig. 3 Streamlines for the flow traversing the settling chamber for a Reynolds number of 1450 based on the duct height. The separation bubble mixes out before the nozzle entrance is reached.

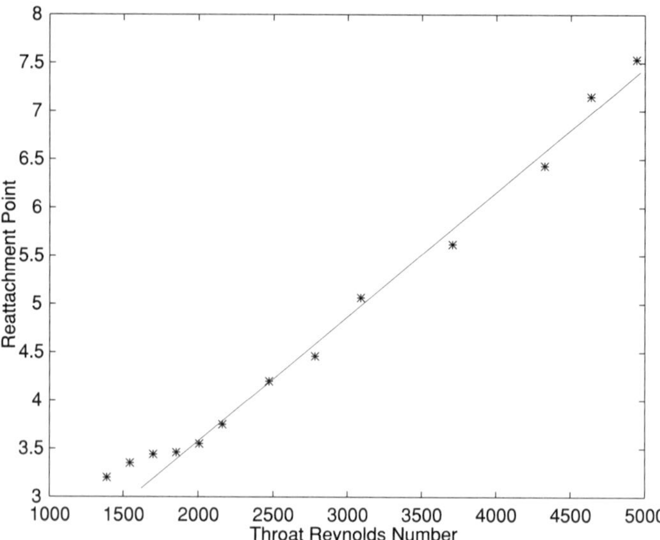

Fig. 4 Variation of reattachment point with throat Reynolds number. The reattachment point is defined as the location at which the *Z* velocity is 1% of free stream.

pressure as a function of distance downstream shows a negligible drop (<0.01%) over the length of the plenum for all cases. Also of interest is the point at which flow becomes fully axial. Figure 4 depicts the variation of the normalized reattachment length with the throat Reynolds number. The normalized plenum length for the fabricated nozzles is 8.1. Thus, the assumption of axially injected flow and total injected pressure relative to the measured pressure is valid over the range of Reynolds numbers tested.

B. Flowfield Analysis and Boundary Layer Calculation

Figure 5 depicts the calculated Mach number distribution through the 16.9:1 nozzle. This run was performed at a Reynolds number of 1940 and corresponds to a nozzle with a 37.5-μm throat width, a chamber pressure of 50 psia, and a chamber temperature of 299 K. The exit Mach number in the inviscid core is 4.24, which deviates slightly from the inviscid quasi–one-dimensional value of 4.5. The coefficient of discharge, or mass flow efficiency, is the ratio of the actual mass flow to the theoretical mass flow for a given chamber condition. This quantifies the blockage associated with the boundary layers. For the case in Fig. 5, the numerical calculation predicts a C_d of 95.9% and a corresponding thrust of 5.98 mN. Separation does not occur over the range of Reynolds numbers tested because the nozzles exhaust to vacuum, and the presence of a supersonic inviscid core flow is sufficient to maintain a favorable pressure gradient.

Once the state variables are computed throughout the domain, they can be used to compute the boundary layer characteristics as well as to assess the validity of the laminar flow assumption. The numerical data are used to compute displacement δ^* and momentum thickness θ^* at each axial station. Since this is a compressible

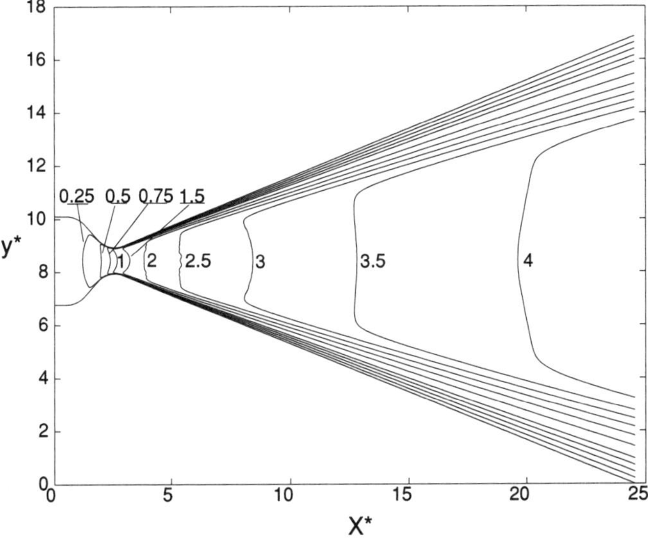

Fig. 5 Mach contours for a Reynolds number of 1940 in a nozzle with a 16.9:1 expansion ratio. This is equivalent to a nozzle with a 37.5-μm throat operating at a 50-psia chamber pressure.

flow, the definition of δ^*[12] is

$$\delta^* = \int_0^{\delta_{99\%}} \left(1 - \frac{\rho u}{\rho_e u_e}\right) \mathrm{d}y \tag{3}$$

The variables are integrated along a normal from the wall ($y = 0$) to the edge of the shear layer, which is defined at $\delta_{99\%}$. Within the shear layer, the flow is nearly parallel to the wall and therefore only the component of velocity parallel to the wall is used in the computation of the boundary layer thickness. Figure 6 depicts the variation of the displacement thickness, normalized by one-half the local width, from the throat to the exit for various Reynolds numbers.

One concern is whether the boundary layer flow along the nozzle expansion is laminar or turbulent. This is difficult to assess accurately without a full stability calculation. However, we can make some assessments by comparison with classical supersonic flows over flat plates. According to Mack[13] the critical Reynolds number (based on momentum thickness) for the stability of a flat plate boundary layer at high Mach-number flows is approximately 250. For the range of nozzles analyzed, the highest Reynolds number found was 125. This indicates that it is not likely that the flow undergoes transition to turbulence. If the flow were to transition to turbulence, the blockage due to the boundary layers will increase, which will lower the effective area ratio and hence the thrust efficiency. As the experimental results indicate, there is a good correlation with the laminar model at high Reynolds numbers, which implies that turbulence is not present.

The displacement thickness represents the displacement of the wall streamline due to the mass flow deficit in the boundary layer. This blockage directly influences the performance by altering the effective geometry of the nozzle. Figure 7 depicts the effective area ratio as a function of the geometric area ratio for the

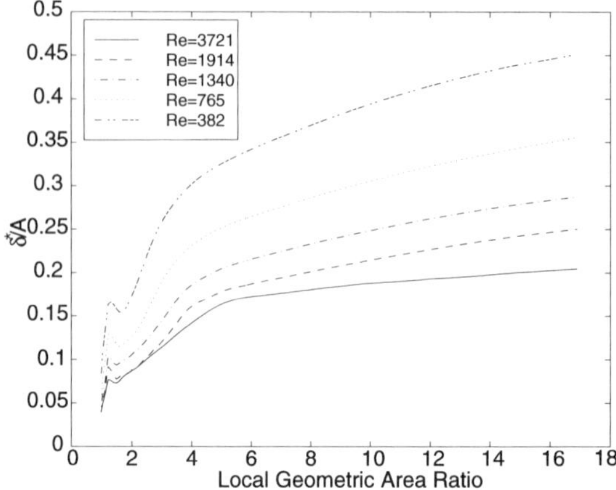

Fig. 6 Displacement thickness is computed from numerical results for the 16.9:1 nozzle. Displacement thickness is a fraction of one-half the local area (width) and plotted relative to its *x* location as defined by the geometric area ratio.

two-dimensional simulation of the 16.9:1 nozzle. This is computed by subtracting the displacement thickness from the local width between the contoured walls.

In addition to the coefficient of discharge, the performance parameters of interest are thrust and I_{sp} efficiency. The efficiencies are ratios of the predicted (or measured) performance to the ideal performance. The thrust and mass flow parameters are computed by numerically integrating the state variables from the computational analysis at the exit plane. The ideal is computed from the isentropic assumptions

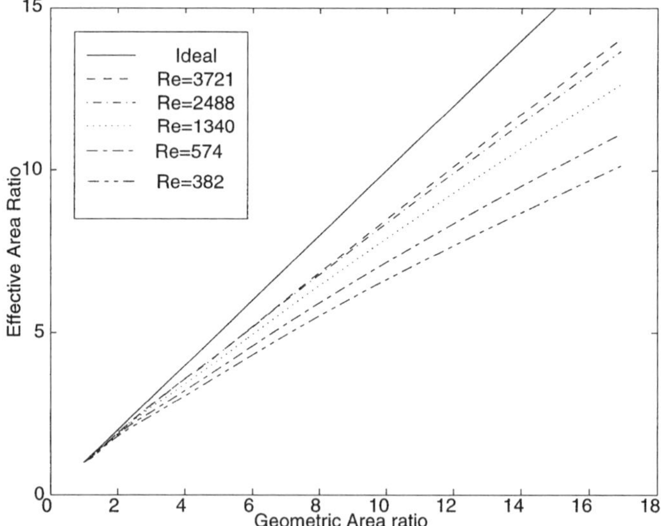

Fig. 7 The local area ratio is adjusted to compensate for boundary layer growth. This two-dimensional effective area ratio is plotted relative to the geometric area ratio using the displacement thickness computed in the two-dimensional simulation.

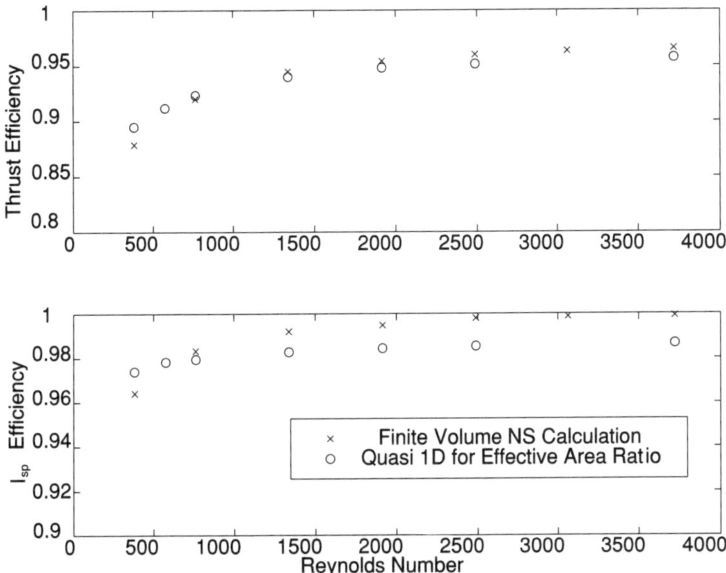

Fig. 8 Thrust and I_{sp} efficiencies are computed from numerical simulations. These are compared with inviscid calculations for similar nozzles operating at the same effective area ratio as depicted in Fig. 7. Numerical calculations are accurate to 2% as shown in this comparison.

of a quasi–one-dimensional nozzle. However, to gauge the viscous effects, this ideal is corrected for the nozzle divergence angle. The ideal exit momentum is distributed over the nozzle divergence angle, and then only the axial component is used to compute the ideal thrust. Thus, the efficiency represents only the viscous losses and not the divergence losses.

The thrust and I_{sp} efficiencies are plotted in Fig. 8 as a function of Reynolds number. The circles represent the integration of the CFD data to determine performance. The crosses represent data based on an inviscid quasi–one-dimensional analysis using the effective area ratio, which is derived from the displacement thickness calculated earlier. Figure 8 illustrates the utility of the effective area ratio. Once δ^* is known, the efficiencies of a nozzle can be predicted within 2% by using the one-dimensional calculation and effective area ratio.

C. Experimental Results

Figure 9 compares the mass flow measurements with the numerical calculation for 16.9:1 nozzles. In addition to the uncertainty in the mass flow meter, which is 0.5% of full scale, the feature geometry (i.e., the throat width) can be measured in plane only to within 0.5 μm. The uncertainty in the geometry results in a 2.5% uncertainty in the theoretical mass flow, which is set by the throat width. At the highest Reynolds numbers, the test data agree well with the model predictions. As the pressure is reduced, and hence the Reynolds number is reduced, the experimental C_d decreases faster than the CFD results predict. This discrepancy is probably due to the influence of the endwall boundary layers not modeled in the

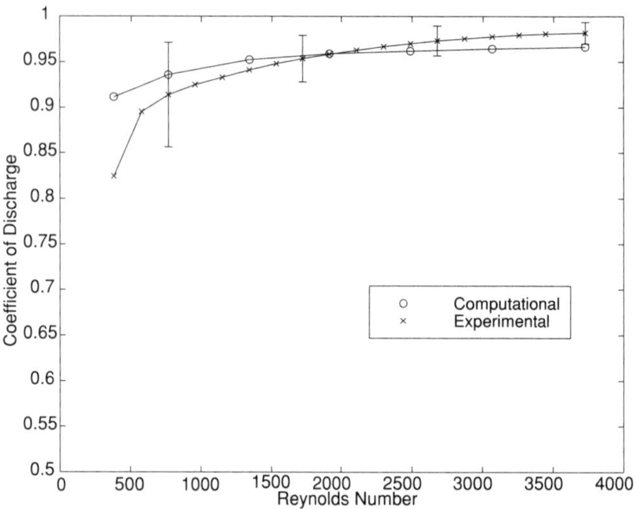

Fig. 9 Mass flow results for both the numerical simulation and the experimental testing for a nozzle with a 37.5-μm throat and a 16.9:1 expansion ratio.

two-dimensional simulation. Since the boundary layer thickness at the throat is 1.5 μm, this would represent an additional 1% loss in mass flow efficiency due to the boundary layers developing on the flat endwalls. Since the displacement that occurs at the choke point (where the mass flow is determined) is relatively small, the influence of the endwall boundary layers on the mass flow is also small and the coefficient of discharge remains consistent with the numerical data at all but the lowest Reynolds numbers studied.

Finally, the nozzle thrust was measured, exhausting to vacuum conditions. For the 16.9:1 nozzle, 11.3 mN of thrust was generated at a chamber pressure of 97.2 psia. The mass flow of this device was 834 sccm, which results in an average exit velocity of 650 m/s, or an I_{sp} of 66.1 s. This is approximately Mach 4.1, compared with the isentropic value of 4.5. The Mach number is calculated based on a speed of sound that the numerical calculation predicts would be present at the exit. Thus, the thrust tests verify that supersonic flow has been achieved.

The thrust test results are summarized in Fig. 10. The I_{sp} is a direct indication of the momentum exchange for this device, and it is a measure of the nozzle performance. The I_{sp} should remain insensitive to chamber pressure and thrust in an inviscid device. The rapid decrease in I_{sp} at the lowest thrust levels is due to the low-Reynolds number viscous effects that are present at the low chamber pressures. As expected, the highest-area ratio nozzles perform the best, due to their higher exit velocity. However, a crossover occurs at 2 mN, and the 18-μm nozzle outperforms the 37.5-μm nozzle of a larger area ratio. Since both nozzles are the same height, the smaller nozzle has a larger exit aspect ratio. Therefore, the influence of the endwall boundary layers is less in the 18-μm case and results in a larger effective area ratio and higher exit velocity than in the 37.5-μm case. This is reflected in the higher performance at low Reynolds numbers, where the boundary layers are thickest and have the most influence.

Having verified that the exit velocity is supersonic, the goal is to determine whether the nozzle performance is predicted by the two-dimensional numerical calculations. The performance parameter of interest is the thrust efficiency, or the

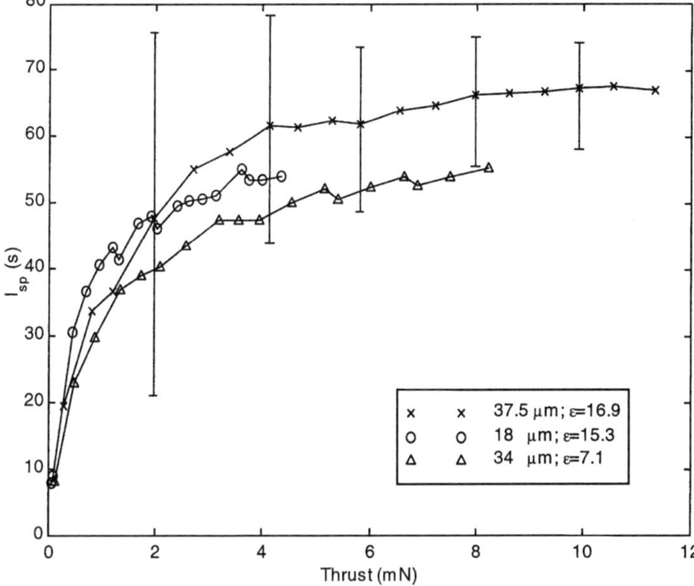

Fig. 10 Variation of I_{sp} with thrust for several nozzle geometries. The thrust measurements are accurate to ±1 mN.

ratio of the actual thrust to that of a quasi–one-dimensional nozzle corrected for divergence losses. Figure 11 depicts the variation of thrust efficiency with Reynolds number for the experimental and numerical data for the 8.2:1 and the 16.9:1 nozzle. The trends between the numerical and the experimental data are similar; however, there is a large variation in the magnitude of the efficiencies. The two-dimensional numerical calculations indicate that thrust efficiency decreases much quicker than the coefficient of discharge. This should be true since it is a function of the exit velocity squared, whereas mass flow is a function of velocity at the throat. However, the measured thrust efficiency decreases much faster than the model predicts. The results are within the uncertainty of the measurements at high Reynolds numbers but deviate 15% beyond the uncertainty of the model at low Reynolds numbers. A corrected model that accounts for endwall boundary layers, and is described in the following section, is also depicted in Fig. 11.

D. Endwall Boundary Layer and Plume Effects

As seen in all of the measurements to this point, there is a large deviation from the numerical calculations at low Reynolds numbers. Intuition would attribute this to the endwall boundary layer growth, which is not modeled in the two-dimensional simulation. These effects would be largest when the boundary layer is thickest at low Reynolds numbers. To first order the endwall boundary layers should develop at the same rate as those on the contoured nozzle sidewalls, since they are driven by the same edge condition represented by the inviscid core flow. By applying the displacement thickness computed by the two-dimensional simulation to the nozzle endwalls, an effective area ratio for the full three-dimensional geometry can be analyzed. However, by reducing the effective area, the core velocity should drop, which

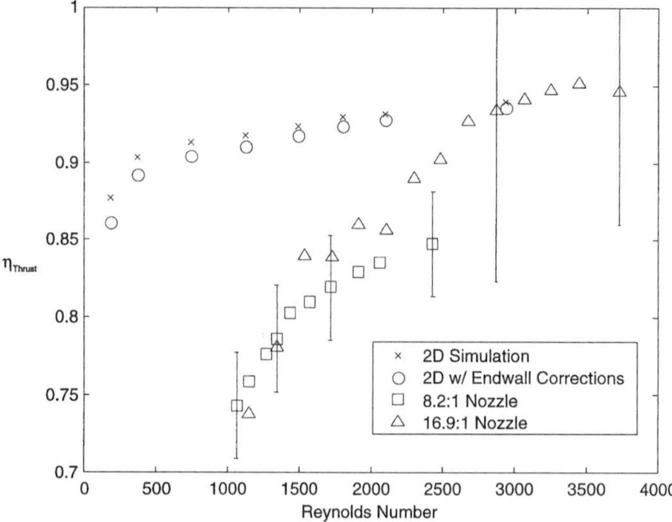

Fig. 11 Variation of thrust efficiency with Reynolds number for the two-dimensional simulation and corrected for the blockage that occurs on the endwalls. The endwall correction is a quasi–one-dimensional analysis applied to the effective area ratio from Fig. 12.

in turn would lower the displacement thickness. This would also cause the sidewall boundary layer thickness to adjust based on the deficit in the endwall boundary layer, which currently appears as the corner area being neglected on the endwalls. To get a true effective area ratio, this process must be converged upon by iteration. At this point, the process is simplified by considering only the effect of the boundary layers on the upper and lower surface, which is valid if the displacement thickness is much less than nozzle height. Figure 12 depicts the variation of the effective area ratio with Reynolds number for each axial location along the nozzle. Inset in Fig. 12, a scale diagram of the exit plane with the displacement thickness superimposed shows the effective area ratio for the nozzle operating at a Reynolds number of 371.

Because the thrust performance is a function of the conditions at the nozzle exit, the growth of the boundary layers results in a larger blockage in this region and has a greater impact on efficiency. For a Reynolds number of 3721, a 63-μm displacement thickness would result in a 56% reduction in thrust area, which would reduce the effective area ratio of the nozzle from 16.9:1 to 8.2:1. Such a reduction would result in an exit Mach number of 3.5 (664 m/s), compared with an inviscid exit Mach number of 4.5. This compares well with the exit velocity derived from thrust measurements of 650 m/s. For a Reynolds number of 371, the effective area ratio of the 8.2:1 nozzle is 5.7, which would result in only a 3% decrease in thrust efficiency and is not sufficient to explain the deviation from the model. The corrected model is also indicated in Fig. 11. Since the results correlate well at high Reynolds numbers and deviate at low Reynolds numbers, turbulence is ruled out as the root cause of this discrepancy.

There have been several attempts at modeling the flow from nozzles at this thrust level, though there is a lack of test data, especially for MEMS-fabricated geometries. Grisnik et al.[2] presented a TDK analysis with their work but could not

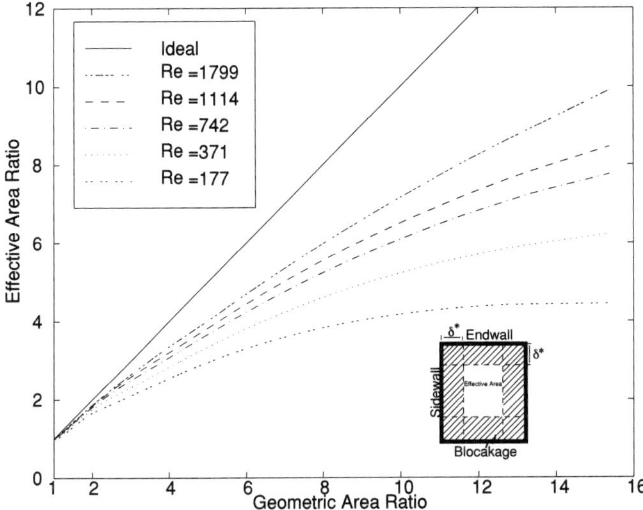

Fig. 12 Variation of effective area ratio with displacement thickness applied to both contoured sidewalls and flat endwalls. Inset: The displacement thickness applied to the exit plane and the effective area that results for the 8.2:1 nozzle operating at *Re* = 371.

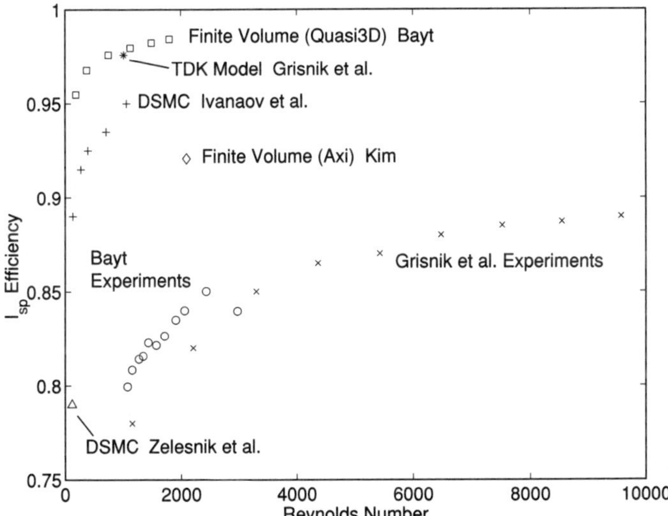

Fig. 13 This is a comparison of the various experimental[2] and modeling[10,14,15] efforts. The model with endwall corrections from this work is labeled Quasi3D. The experiments compare well with each other, but there are unmodeled effects causing the disparity with the simulations.

make a good correlation with their data. As mentioned before, Kim[10] performed an axisymmetric Navier–Stokes calculation for low-Reynolds number resistojets. Ivanov et al.[14] performed Navier–Stokes and DSMC calculations for geometries similar to Janson's experiments. Though their DSMC calculations produced results similar to the Navier–Stokes simulation presented here, they claim that there is a significant overprediction when extrapolated outflow boundary conditions are used. In their studies, the extrapolated outflow conditions resulted in less of an expansion than the DSMC solution with a portion of the plume modeled. With the plume modeled, there is a much lower pressure along the wall, and less thrust. In addition, streamlines in the plume diverge from the nozzle centerline to a far greater extent than those found in the simulations that end at the exit plane. This will incur further divergence losses. This difference is 3% at Reynolds numbers over 1000 and 8% at 120. Finally, Zelesnik et al.[15] performed a DSMC calculation for the Grisnik geometry, but at very low Reynolds numbers. They also modeled a portion of the plume and noted a similar loss in efficiency to Ivanov. Thus, the predominant causes of the deviation are thought to be the plume model and the three-dimensional nature of the flow, which is predominantly the endwall boundary layer effect. Figure 13 summarizes current and previous modeling and experimental efforts. Though the modeling results are consistent and the experimental results are consistent, there is a disparity between the two.

VI. Conclusions

The present results demonstrate that supersonic flow can be achieved in micron-scale contoured devices to be used in micropropulsion systems. The numerical calculations indicate that the performance of these nozzles can surpass that of both conventionally machined and KOH etched nozzles of similar thrust. This is because contoured throat features are smaller (which result in higher Reynolds numbers for a constant thrust) and the expansion ratios are larger than have been fabricated by KOH etched nozzles. The nozzles are numerically simulated as two-dimensional because of the anisotropy afforded during nozzle fabrication by deep reactive ion etching. This allows features to be etched that are of a constant geometry between the upper and the lower surfaces.

Thrust measurements indicate that Mach 4.1 flow is achieved for 16.9:1 expansion ratios, which is degraded from the 4.5 predicted by isentropic theory. The mass flow rates of these devices can be measured and produce similar repeatable results for nozzles of similar lengths and Reynolds numbers. The measurements indicate a reduction in mass flow efficiency with Reynolds numbers.

This thrust and I_{sp} efficiencies compare well with the numerical model at high Reynolds numbers and deviate at low Reynolds numbers due to the three-dimensional effects in the flow. The displacement thickness at the throat is 1 μm, which is a small fraction of the flow area and has a minimal impact on mass flow efficiency. The thrust efficiencies do not compare well with the theory, and the reasons for this discrepancy are thought to be the effects of the plume and three-dimensional geometry, neither of which was modeled in the present simulations.

Acknowledgments

The authors would like to thank Bill Tang and the Jet Propulsion Laboratory Microdevices Laboratory, as well as the Goddard Space Flight Center's Office of University Programs, through which this program was funded. In addition, we would like to thank Siegfried Janson and the Aerospace Corporation for the use of their thrust stand for measurements presented here. A special thanks is given to Vadim Khayms, who assisted with the MIT thrust stand measurements. Also, our gratitude is extended to Martin A. Schmidt and Arturo A. Ayon of the Microsystems Technology Laboratory for their assistance and insight in the development of the microfabrication process flow. Portions of this work were presented previously at conferences supported by the AIAA, the American Society of Mechanical Engineers, and the Institute of Electrical and Electronics Engineers.

References

[1]Rothe, D. E., "Electron-Beam Studies of Viscous Flow in Supersonic Nozzles," *AIAA Journal*, Vol. 9, No. 5, 1971, pp. 804–811.

[2]Grisnik, S. P., Smith, T. A., and Salz, L. E., "Experimental Study of Low Reynolds Number Nozzles," AIAA Paper 87-0092, May 1987.

[3]Janson, S. W., and Helvajian H., "Batch-Fabricated Microthrusters: Initial Results," *32nd AIAA Joint Propulsion Conference*, AIAA Paper 96-2988, July 1996.

[4]Bosch, R., "Method for Anisotropically Etching Silicon," U.S. Patents 4855017 and 4789720, German Patent 4241045C1.

[5]Bayt, R. L., Breuer, K. S., and Ayon, A. A., "DRIE-Fabricated Nozzles for Generating Supersonic Flows in Micropropulsion Systems," *Proceedings of the Sensors and Actuators Workshop*, Hilton Head, SC, pp. 312–315, June 1998.

[6]Ayon, A. A, Braff, R. A., Bayt, R., Sawin, H. H., and Schmidt, M. A., "Influence of Coil Power on the Etching Characteristics in a High Density Plasma Etcher," *Journal of the Electrochemical Society*, Vol. 146, No. 7, 1999, pp. 2730–2736.

[7]Wallis, G., and Pomerantz, D. I., "Field Assisted Glass-Metal Sealing," *Journal of Applied Physics*, Vol. 40, No. 10, 1969, pp. 3945–3949.

[8]Van Leer, B., "Flux-Vector Splitting for the Euler Equations," *Lecture Notes in Physics, 170*, Springer-Verlag, Berlin, 1982.

[9]Jameson, A., Schmidt, W., and Turkel, E., "Numerical Solution of the Euler Equations by Finite Volume Methods Using Runge-Kutta Time Stepping Schemes," AIAA Paper 81-1259, 1981.

[10]Kim, S. C., "Calculations of Low-Reynolds-Number Resistojet Nozzles," *Journal of Spacecraft and Rockets*, Vol. 31, No. 4, 1994, pp. 259–264.

[11]Bayt, R. L., Ayon, A. A., and Breuer, K. S., "A Performance Evaluation of MEMS-Based Micronozzles," *33rd AIAA Joint Propulsion Conference*, AIAA Paper 97-3169, July 1997.

[12]White, F. M., *Viscous Fluid Flow*, 2nd ed., McGraw–Hill, New York, 1991, p. 524.

[13]Mack, L. M., "Boundary-Layer Linear Stability Theory," *Special Course on Stability and Transition of Laminar Flow*, AGARD Rept. 709, 1984.

[14]Ivanov, M. S., Markelov, G. N., Ketsdver, A. D., and Wadsworth, D. C, "Numerical Study of Cold Gas Micronozzle Flows," *37th Aerospace Science Meeting and Exhibit*, AIAA Paper 99-0166, Jan. 1999.

[15]Zelesnik, D., Micci, M., and Long, L., "Direct Simulation Monte Carlo Model of Low Reynolds Number Nozzle Flows," *Journal of Propulsion and Power*, Vol. 10, No. 4, 1994, pp. 546–553.

Chapter 17

Micro-Isolation Valve Concept: Initial Results of a Feasibility Study

Juergen Mueller,* Stephen Vargo,† David Bame,‡
Indrani Chakraborty,† and William Tang§
Jet Propulsion Laboratory, California Institute of Technology
Pasadena, California

I. Introduction

THERE currently exists a strong interest within the aerospace community to build ever smaller spacecraft to reduce the cost of space missions and afford more frequent launches. Since launch costs may contribute as much as 30% to the total cost of a space mission, yet are determined to a large extent by spacecraft mass, the use of smaller spacecraft may have a dramatic impact on the overall mission cost. Most recently, this trend toward smaller and lighter spacecraft has accelerated as demonstrated by the introduction of the microspacecraft concept, typically understood as a spacecraft with a mass of a few tens of a kilogram or less.[1] Besides potentially offering reduced mission cost, microspacecraft may also allow new and unique mission profiles to be flown, including, for example, constellations of microspacecraft, charting entire regions of space simultaneously and cost-efficiently. The use of microspacecraft may also increase the reliability of a mission by off-loading experiments from a single large spacecraft to a fleet of microspacecraft. The loss of a single or a few microspacecraft may not jeopardize the entire mission.

However, such dramatic decreases in spacecraft weight and size will require the development of radically new approaches in the design of spacecraft components. An area in need of special attention in this pursuit is propulsion. Currently existing propulsion hardware, with the exception of a few new developments, will likely not meet the design constraints imposed by many microspacecraft with respect to mass, size, and power.[2] At present, several activities, in various stages of research

Copyright © 2000 by the American Institute of Aeronautics and Astronautics, Inc. The U.S. Government has a royalty-free license to exercise all rights under the copyright claimed herein for Governmental purposes. All other rights are reserved by the copyright owner.

*Advanced Propulsion Technology Group.
†MEMS Group, Microdevices Laboratory.
‡Propulsion Flight Systems Group.
§Formerly Supervisor, MEMS Group, Microdevices Laboratory; currently at DARPA.

and development, are under way in the propulsion field to address these issues.[2] Most of this work naturally focuses on thruster hardware. However, improvements in valve technologies, with the goal to meet the stringent mass, size, and power constraints expected to be found on a microspacecraft, are also crucial to the success of micropropulsion concepts since it is important to ensure that the entire propulsion system weight and volume is reduced.

In this chapter a newly proposed, normally close isolation valve concepts is discussed that is based in its fabrication on MEMS (microelectromechanical systems) technologies, resulting in a valve body approximately $1 \times 1 \times 0.1\,\mathrm{cm}^3$ in size and weighing but a few grams, excluding fittings and packaging. This isolation valve, which can be opened only once, will serve to seal a propulsion system and provide zero leakage prior to actuation. Thus, the micro-isolation valve will serve the same function as a conventional pyrovalve. However, as shown below, no pyrotechnic actuation will be required in the micro-isolation valve concept. Propulsion system isolation is of particular importance for many interplanetary missions, where propulsion systems may not be activated until many years into the mission, following a long interplanetary cruise. During this time, propellant leakage will have to be avoided. For microspacecraft, in particular, due to the limited onboard propellant supply, leakage rates will have to be minimized. Liquid propellants may be used in many applications, resulting in significantly reduced leak rates over gaseous propellants. However, in some cases the use of gaseous propellants may be unavoidable, such as for certain electric propulsion applications requiring xenon gas or for cold gas attitude control if spacecraft contamination concerns require the use of very benign propellants, such as nitrogen, for example. Currently, no alternative exists to the relatively heavy and large-scale pyrotechnically actuated valves typically used in these types of applications to seal the propulsion system. The micro-isolation valve is targeted to fill this gap.

MEMS technologies have recently gained increased attention in microspacecraft component designs due to their potential for achieving degrees of miniaturization otherwise unattainable. While more traditional metal-fabricating technologies have resulted in impressive reductions in component mass and size,[3] MEMS components offer the potential of a highly integrated, extremely small propulsion system through either the use of chip-to-chip bonding or the integration of various propulsion components onto the same chip, such as thrusters, filters, and valves.[2] This integration scheme may even include the necessary control and power conditioning electronics in silicon-based systems.[2] Extremely small size and weight, as well as minimal external interfaces, which should simplify integration into the spacecraft, would characterize the resulting packaged propulsion module and the costs associated with its integration considerably. Unfortunately, currently available MEMS valve technology, as provided by the nonaerospace industry, does not appear to meet many requirements of spaceflight application with respect to leakage rates, valve actuation times, required bus voltages, or robustness of design.[4] The proposed isolation valve concept is designed to address these shortcomings, in particular with respect to leakage rates.

In addition, applications of the micro-isolation valve may be found in more conventionally sized propulsion systems requiring only low flow rates, compatible with the to-be-expected small flow dimensions that could be provided on a chip. Certain electric propulsion systems, such as advanced ion and Hall thruster systems, requiring substantial dry weight reductions in its feed system designs, may benefit from such a valve. In the following sections the micro-isolation valve concept is introduced, key feasibility issues are identified, and initial tests and analysis aimed at addressing these issues are discussed.

II. Description of the Concept

A. Concept

The micro-isolation valve in its current form is a micromachined, silicon-based device that relies on the principle of melting a silicon plug, doped to enhance its electrical conductance, which in the valve's normally closed position blocks the valve flow passage. Melting of the plug will open the valve and will be achieved by passing an electric current through it and resistively heating it. The valve will thus serve a similar function as a normally closed pyrovalve, providing an essentially zero leak rate prior to actuation by completely sealing the flow passage. Unlike a pyrovalve, however, the valve proposed here will not rely on pyrotechnic actuation, thus avoiding the potential for pyroshocks as well as simplifying valve integration.

A schematic of the valve is shown in Fig. 1. It consists of two basic components: the silicon chip, featuring all of the flow passages and valve inlet and outlet, and a Pyrex cover to seal the flow passages while allowing a view of the internal design of the chip for experimental evaluation of the concept. Later versions may be entirely assembled from silicon. The silicon–Pyrex bond is achieved by means of anodic bonding, a standard bonding technique in the microfabrication field by which silicon and a special grade of Pyrex (Dow Corning 7740) are placed in immediate contact with each other. Applying pressure and an electrostatic potential across the bond surface at a temperature of approximately 450°C causes the two chips to fuse together. The bonding mechanism is believed to be due to the formation of a thin silicon oxide layer along the bond surface and is thus chemical in nature. Very strong bond strengths can be obtained using this technique as shown below.

The silicon side of the chip features the valve-internal flow channels, the plug, and a filter and will be batch-fabricated from larger silicon wafers. Channels in the

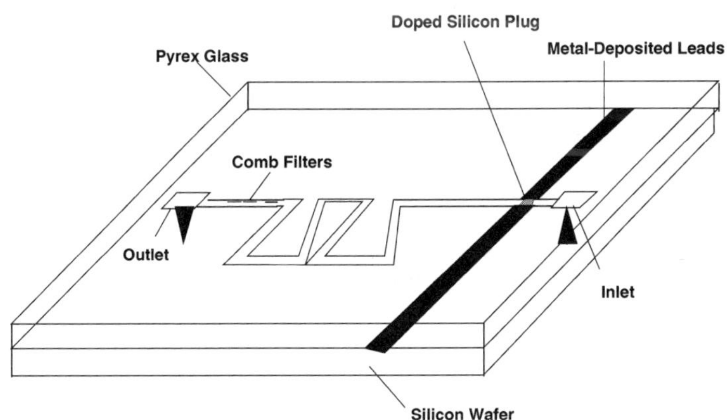

Fig. 1 Schematic of the micro-isolation valve concept.

chip are fabricated using deep trench reactive ion etching (DRIE) techniques. This etching process is highly anisotropic and allows deep features to be etched into the chip with very straight wall sections up to aspect ratios as high as 30:1. Metal (gold) leads deposited onto the silicon substrate, partially overlapping the doped-silicon plug region, will connect the plug to an external valve-opening circuitry. Propellant entering the valve chip will flow through a short channel section etched into the silicon side of the chip until it reaches the plug. Passing an electrical current through the plug will melt and/or vaporize it, and propellant located upstream of the valve inlet will push the plug debris downstream, thus opening the valve.

To prevent plug debris from contaminating flow components located downstream of the isolation valve, potentially clogging propellant lines, contaminating valve seats, or otherwise interfering with the proper function of those components, it is crucial to trap the debris within designated, noncritical regions of the valve without reclosing the flow path again. It is speculated that, due to melting rather than cold fracture of the plug, the debris count may be reduced, and fewer, larger debris particles may be produced, which will be easier to trap. Nonetheless, filtration and other debris trapping schemes will be required.

Figure 1 shows one potential flow path configuration designed to accomplish this task. Here the flow path goes through a series of S-shaped turns designed to trap molten plug debris in the corners of the etched channel. Oversizing the channel, in particular, near the corners, will avoid clogging. Other configurations may be explored as the experimental program progresses, such as parallel flow passages for redundancy, for example. A comb filter integrated into the flow path downstream of the plug will serve to trap debris that may not have been condensed at the flow path walls but, instead, has solidified in the propellant stream, or is due to fracture. Using MEMS-based techniques, it is expected that very small filter ratings may be produced, into the micron range.

B. Key Feasibility Issues

Several key feasibility issues for the micro-isolation valve can immediately be identified and will need to be addressed in the ensuing research program. Among these are the following.

1) Plug melting. Melting of the plug will need to be achieved within acceptable energy constraints. Energy storage devices, such as capacitors, may be used to boost power levels over the valve actuation period. Melting should be achieved quickly to limit heat conduction losses to the remainder of the chip, where high temperatures could lead to thermal stresses, in particular, between bonded components that feature a coefficient of thermal expansion (CTE) mismatch. In the case of the discussed laboratory devices this mismatch may occur between silicon and Pyrex, in particular, above temperatures of about 300°C, where the CTE values of Dow Corning 7740 Pyrex and silicon begin to diverge. Even in the case of future, all-silicon versions of this valve, mismatches will still occur between the valve and the packaging.

2) Pressure handling capabilities. The micro-isolation valve chip will be required to maintain high internal pressures, in particular, in the case of gaseous propellant applications, such as for some electric propulsion systems. The typical gas storage pressure in an electric propulsion xenon feed system is about 2000 psia (13.6 MPa). Since factors of safety of 1.5 are typically required, burst pressures may have to be as high as 3000 psia (20.4 MPa). This poses a major design challenge given that the chip consists of silicon and glass. Of particular interest in this context is also the plug. Thermal considerations, alluded to above, will drive the plug dimensions to smaller widths to minimize power requirements for melting. Pressure requirements, on the other hand, will drive the plug design into the opposite direction.

3) Contamination-related issues. Trapping of plug debris inside the micro-isolation valve chip is crucial to the success of this valve concept. No debris can be allowed to propagate downstream into other flow components, in particular not onto valve seats that may be located downstream of the isolation valve. For micropropulsion applications, in particular, these valve seats may themselves be very small in size, thus resulting in tight filter rating requirements.[4] In principle, micromachined comb filters may offer a solution in this regard and will need to be experimentally verified.

Both items 1 and 2 are addressed in this paper and show promising results. Given the interrelationship of thermal as well as structural (burst pressure) considerations in the plug design mentioned above, plug melting and valve burst pressure tests were addressed simultaneously in this study. Contamination and filtration issues will be considered in follow-on testing. In the following sections burst pressure tests are discussed, and initial plug melting tests, proving the feasibility of the concept from this viewpoint, are reported.

III. Burst Pressure Tests

A. Test Chip Design

As mentioned in Section II, valve plug design is governed by two predominant, yet conflicting, requirements. Thermal considerations, as discussed above, will favor a thinner plug to reduce the power requirements to melt the barrier. Pressure requirements, on the other hand, will drive the design to larger plug thicknesses. Therefore, a series of tests was first conducted to determine burst pressures for different plug thicknesses and valve body configurations. Successful designs would then be subjected to plug melting tests, reducing the number of required valve designs to be tested by eliminating designs with low pressure tolerances.

An experimental, rather than a numerical, approach was chosen to evaluate pressure handling capabilities of valve chips due to the expected statistical variation inherent in such tests, possibly depending on small material defects, which would have exceeded modeling capabilities. Strengths of the anodic bonds between the silicon and the Pyrex along the top of the barrier would also have been difficult to model accurately, yet may influence the results considerably.

To perform these tests, a series of dedicated test chips was fabricated. Two chip test chip iterations were tested, referred to as Batch 1 and Batch 2, shown in Figs. 2 and 3, respectively. An actual chip is shown in Fig. 4. In both cases, the chip design focuses solely on the plug region and the optimization of the plug design and, thus, does not yet contain any design features to trap plug debris. Besides being used for burst pressure tests in this set of experiments, the Batch 2 design was also used in plug vaporization tests. Common to both chip types (Batch 1 and Batch 2) is a straight, 4-mm-long channel section with a cross section of $300 \times 300\,\mu\text{m}^2$. In all cases the plug is located in the center of the chip, dividing the channel section into two sections of equal length. Several plug designs were tested, ranging in thickness from 10 to 100 μm for both types of chips. The channel is connected to an inlet and outlet through which gas can enter and exit the chip from the silicon surface. The chips are sealed with an anodically bonded Pyrex

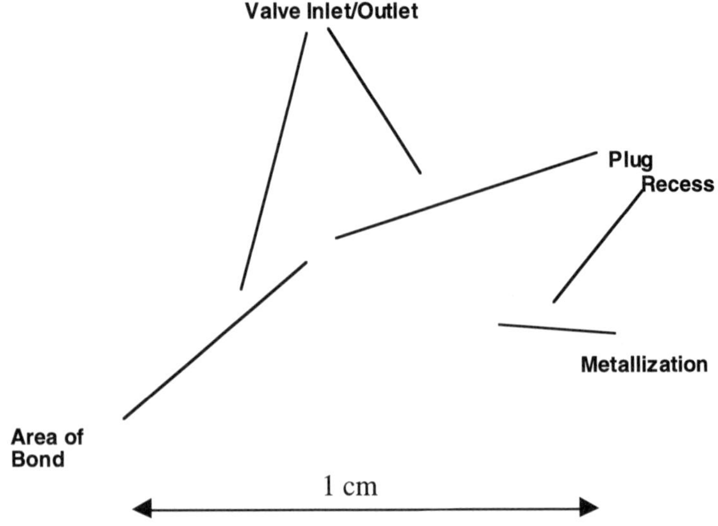

Fig. 2 Schematic of a Batch 1–type test chip.

cover. This seal necessitates the fabrication of two recesses into the chip, which feature the electric leads to the plug, visible as the lightly colored rectangular regions in Fig. 4. Since metal deposition may be as thick as several tenths of a micron, depositing the metal lead directly onto a nonrecessed silicon surface would have led to leakage paths immediately adjacent to the metal deposits, as the Pyrex would have been forced to bend over them. Although doping of silicon is possible to provide electric contacts, its resistivity is higher than that of gold, and since the desire was to create the majority of the voltage drop in the plug region where heating was supposed to occur, gold was chosen as the connecting material.

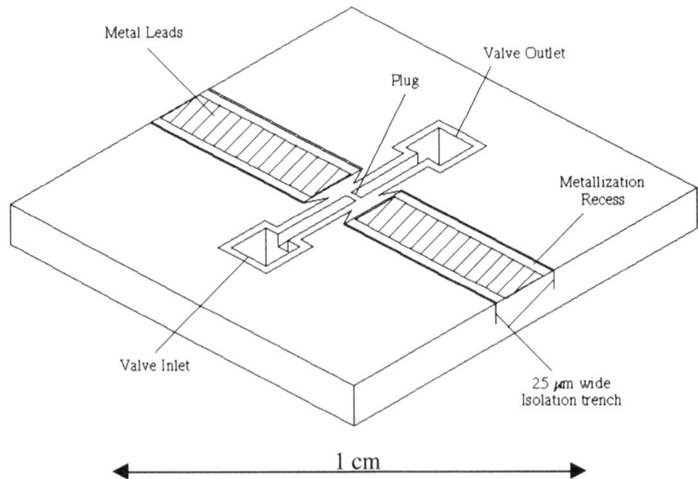

Fig. 3 Sketch of a Batch 2–type test chip.

Fig. 4 Micro-isolation valve test chip.

The chip itself can be electrically contacted near the edges, where notches in the Pyrex (see Fig. 4) are provided for this purpose. Contacting the chip near its edges is preferred, as it minimizes the wire length needed for wire bonding. These wires are typically very thin and fragile, and minimizing their lengths simplifies handling of the chip.

No current input was needed for the burst tests. However, since the recess may impact the pressure handling ability of the chip, it was also integrated into the chips used for burst testing. The reason for this may be seen in Fig. 5 (Batch 1

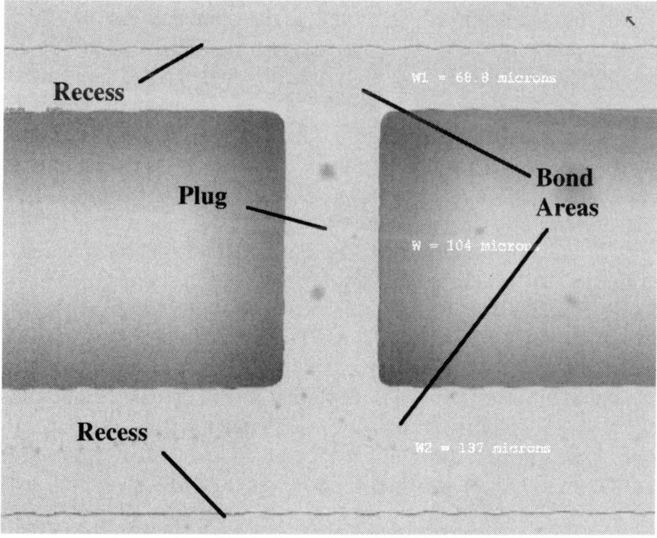

Fig. 5 Close-up of a plug (Batch 1, Chip 12).

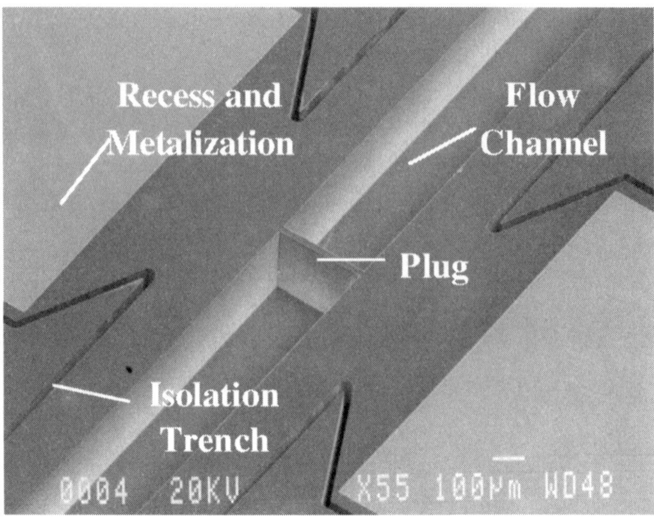

Fig. 6 Close-up of plug area (Batch 2 chip) with recess and isolation trench.

design) and also in Fig. 6 (Batch 2 design). Due to the recess, there exist regions of minimal bond width between the pressurized channel and the recess for the metal leads. Across these regions debonding and leakage may occur at very high internal pressures. As shown below, this did indeed occur at pressures exceeding 1,800 psig (12.2 MPa) for the Batch 1 design. For this reason, the Batch 2 design was developed.

One of the key differences between Batch 1 and Batch 2 chips can be found in the width of the metalization recess, being 4 mm for the Batch 1 design and only 1.5 mm for the Batch 2 design. This increased the bond area for the Batch 2 design and consequently, as shown below, the pressure handling capability of the chip. Another design feature added to the Batch 2 chip consisted of a so-called isolation trench. Wrapping around the recess area and the entire channel section is a narrow, 25-μm-wide trench, as shown in Fig. 6. This trench serves as an electrical insulation: to create a large enough opening in the barrier, the silicon surface has to be doped to a sufficient depth. More conventional doping techniques such as ion implantation, followed by a thermal drive-in, can provide sufficient doping to a depth of only about 10 μm. However, the advantage of this technique is that doping can be limited to certain regions of the chip, such as only in the plug region. However, using wafers featuring epitaxially grown doped layers covering their surface, doped silicon regions as thick as 40 μm may be obtained (so-called doped epiwafers). In this case, however, the entire chip surface is doped, not just the plug region. This could allow current to flow around the channel section to the opposite metal lead, potentially shorting the plug. To create a high-resistivity path for the current other than through the plug, the trench narrows the region along which current may flow around the channel section to a 200-μm-wide path, which is the separation between the isolation trench and the channel. (Obviously the trench may not intersect with the channel, as a leakage path would be created through the trench.) Thus, the Batch 2 chip design may be used for fabrication of MIV chips from doped epiwafers. Again, for burst testing, since no current flow was intended, this design feature would not have been needed. Indeed, chips used for burst pressure testing featured no doping at all, in either Batch 1 or Batch 2 cases. However, since the isolation trench may impact the pressure handling ability as it locally decreases bond widths, this design feature was also included in the burst tests.

The fabrication of the test chips consists of a combination of silicon etching and wafer bonding techniques. Desired features are first etched in silicon to create the device's structure and then the chip is pressure-sealed by anodically bonding a Pyrex wafer to the silicon. Etching of silicon is achieved using a Deep Trench RIE system available at JPL's Micro Devices Laboratory (MDL). This Deep Trench RIE system, which is manufactured by Surface Technology Systems, Inc. (STS), provides highly anisotropic etching parameters in silicon. The STS system can provide silicon etching rates of about 4.5 μm/min, aspect ratios of 30:1, and sidewall angles of $90 \pm <0.25$ deg. Figure 6 shows the results obtained with this etching technique. Note the straight channel walls and the thin, vertical barrier (i.e., "the plug") intersecting the channel.

B. Burst Test Setup and Procedure

The burst test setup is shown in Fig. 7. The chip was bonded to a stainless-steel fixture, connecting the inlet and outlet holes of the chip to two tube stubs featuring Swagelok fittings. One fitting was connected to the pressurant supply, while the other was connected to a leak detector, thus allowing for valve internal leak checks across the plug. External leaks were monitored by determining the pressure decay in the system; i.e., if no internal leak could be registered with the leak detector, yet the pressure decayed, the leak was determined to be external. The entire chip assembly is placed into a test barricade for protection during burst tests. The pressurant supply provides regulated helium pressure to the valve

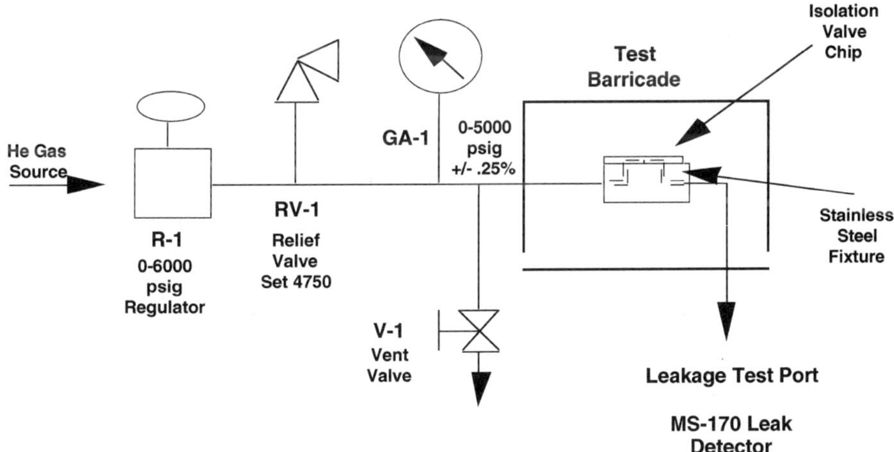

Fig. 7 Burst/leak test setup.

Table 1 Burst/leak test data for Batch 1–type chips

Chip I.D.	Plug width, μm	Burst pressure, psig	Burst/leak mode	Sidewall bond widths, μm
2	10 (nominal)	250	Internal leak	N/A
6	10 (nominal)	100	Internal leak	N/A
1	27.1	1050	Pyrex debond	111/68.8
13	38.8	900	Pyrex debond	22.5/68.3
3	58.9	1650	Pyrex debond	78/124
8	100	1750	Pyrex debond	122/64.1
12	104	1850	Pyrex debond	68.8/137

inlet. Pressure is measured with a 0- to 5000-psig pressure gauge, mounted directly upstream of the chip, to an accuracy of $\pm 0.25\%$. The test procedure begins with a slow purge and vent of the entire upstream pressurant system (up to the valve chip plug) to fill the system with helium. The pressure is then slowly increased in 100-psig steps and held at those increment levels to check for external (through pressure decay) and internal leaks across the chip barrier (with the leak detector).

C. Results

1. Batch 1–Type Chips

Results obtained for Batch 1–type chips are summarized in Table 1. This table lists a chip identification number, the burst pressure or the pressure at which leakage (external or internal) occurred, the mode of failure (external leak, internal leak, or burst), the plug width, and the bond widths to both sides of the channel, formed by the boundaries of the channel and recess, respectively (compare with Fig. 5). Although intended to be equal, a slight misalignment of masks caused the bond width on one side of the channel to be bigger than that on the opposite side. All the chip dimensions were measured under an electron microscope. This was possible, as shown below, because the Pyrex cover glass blew off the silicon side of the chip at the time of failure. In the cases were this did not occur (Chips 2 and 6), it was not possible to take the corresponding measurements since Pyrex is not transparent when observed under an electron microscope. Thus the nominal plug widths are listed in Table 1 for these chips instead, and sidewall bond widths could not be given.

As can be seen, chips featuring 10-μm-wide plugs suffered from internal leakage. Inspection of the chips after the test showed that the bond between the top of the barrier and the Pyrex had failed. The authors had noted before that good anodic bonds to very thin structures were difficult to achieve, and one of the goals of this test was to study the integrity of these thin bonds. Chips with somewhat larger bond widths (27 and 38.8 μm) performed better, however, burst failure of the chip occurred at the comparatively low pressures of 1050 psig (7.4 MPa) and 900 psig (6.1 MPa). In the case of Chip 13, failing at 900 psig (6.1 MPa), the bond widths next to the channel were rather small compared to those of the other chips, with a minimum bond width of only 22.5 μm, and may thus explain the lower burst

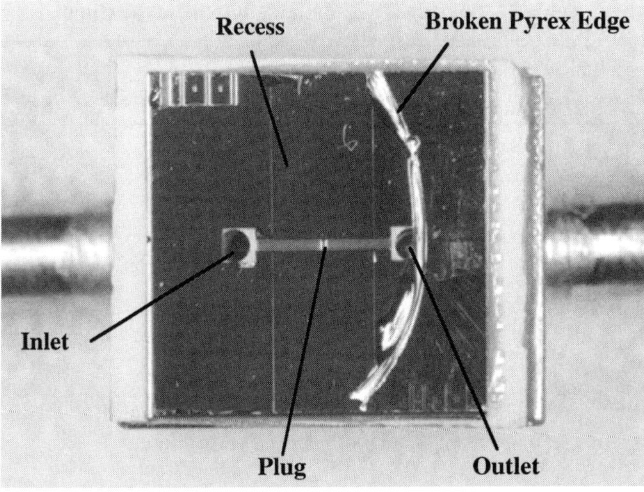

Fig. 8 Post-test image of a burst Batch 1–type chip. Note Pyrex blown off.

pressure. Chips with even thicker plug widths (59 and 100 μm) and larger bond widths along the channel walls performed better, reaching burst pressures as high as 1850 psig (12.6 MPa).

The burst failure mechanism for these chips was always the same and can be seen in Fig. 8. Chip 12 is shown in this figure. As can be noted, that portion of the Pyrex covering the inlet side of the chip was blown off. It thus was clear that the anodic bond between the silicon and the Pyrex had failed. Consequently, a redesign of the chip was undertaken, leading to the Batch 2 design described above, featuring narrower recesses. Results obtained with this set of chips are described below.

2. *Batch 2–Type Chips*

A total of 30 Batch 2–type chips was tested in three test series. Unfortunately, almost all chips in Test Series 1 and 2, with the exception of one chip each per test series featuring a 50-μm plug, failed at very low pressures, not exceeding a few hundred pounds per square inch gauge. Upon postinspection, it was noted that the plug barriers had thinned during fabrication toward the bottom of the channel and that the barriers had broken predominantly at these locations. Since the exact plug thickness could not be determined at these locations due to visual inaccessibility, all tests in Test Series 1 and 2 were dismissed, with the exception of the two tests performed with the 50-μm plug chips. Since in these cases the plugs did not fail, and, as shown below, the failure mode at burst was not related to the plugs, data obtained with these chips could be used.

Chips in Test Series 3 all featured very homogeneous plug thicknesses, and 15 chips were burst-tested in this series. Data obtained with these chips and the two remaining chips from Test Series 1 and 2 are listed in Table 2. A clear trend toward a higher burst pressure with increasing plug thickness can be noted. Plug thicknesses of 10 μm can maintain a pressure of a few hundred pounds per square inch gauge before the plug barrier breaks, 15-μm plugs can maintain up to about

Table 2 Burst/leak test data for Batch 2–type chips

Chip I.D.	Plug width, μm	Burst pressure, psig	Burst/leak mode	Test series no.
4	10	<150	Plug failure	3
6	10	325	Plug failure	3
9	10	500	Plug failure	3
C2	15	875	Plug failure, small internal leak	3
C4	15	1225	Plug failure	3
5	20	<150	Internal leak, plug intact	3
C3	20	2575	Plug failure	3
C5	20	2300	Plug failure and Pyrex failure	3
12	25	1950	Pyrex failure, plug intact	3
C5*	25	1970	Pyrex failure, plug intact	3
2*	35	2230	Pyrex failure, plug intact	3
3	35	2610	Pyrex failure, plug intact	3
2	50	2650	Pyrex failure, plug intact	3
C1	50	2725	Pyrex failure, plug intact	3
10	50	2825	Pyrex failure, plug intact	2
C4*	50	2850	Pyrex failure, plug intact	1
C4**	50	2900	Pyrex failure, plug intact	3

1000 psig, and 20-μm plugs can sustain pressures up to about 2300–2600 psig. Results of a typical plug barrier failure are shown in Fig. 9. It is interesting to note that, while portions of the barrier break away, some fragments of the barrier remain attached to the Pyrex cover, indicating the strength of the anodic bond.

However, beginning with 20-μm chips, and exclusively for the 25-, 35-, and 50-μm chips, the failure mode changes. For these chips, at burst pressures ranging as high as 2650–2900 psig, the plugs remained intact, however, a piece of Pyrex typically located directly above the inlet hole was blown out as shown in Fig. 10. Note that the Pyrex did not detach from the silicon substrate, as was observed for Batch 1–type chips. Thus, Pyrex failure, rather than anodic bond failure, was the cause for the burst of this type of chips, whereas the thicker plugs are able to withstand these pressure levels. However, it should also be noted that some of the valves (namely, Nos. 4 and 5 in Table 2) failed at very low pressures: in the case of valve 4 this may have been due to the small plug width; in the case of valve 5 a leak across the top of the plug barrier, where bonding is supposed to occur between the silicon material of the plug and the Pyrex cover, may have occurred.

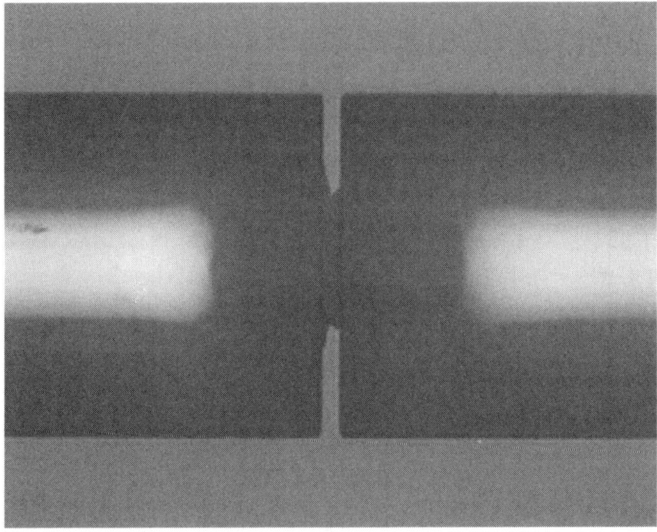

Fig. 9 Plug (barrier) failure for Batch 2–type Chip C3 (20-μm plug width; burst at 2575 psig).

Several conclusions may be drawn from these tests.

1) Burst pressures obtained with isolation valve chips, in particular, those featuring 50-μm-wide plugs, are impressive, considering that the valve chip is fabricated entirely from silicon and glass.

2) The increased bond area between the Pyrex and the silicon substrate in the Batch 2–type chips vs the Batch 1–type chips appears to have served its purpose of increasing the overall bond strengths, since Pyrex no longer debonds in the Batch 2 cases.

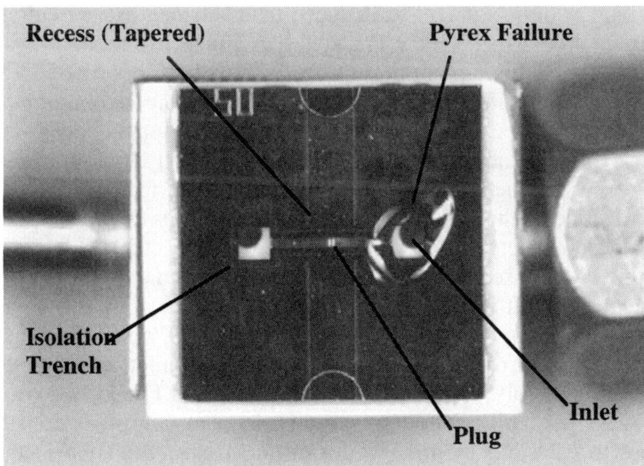

Fig. 10 Post-test image of a burst Batch 2–type chip. Note hole in Pyrex over inlet.

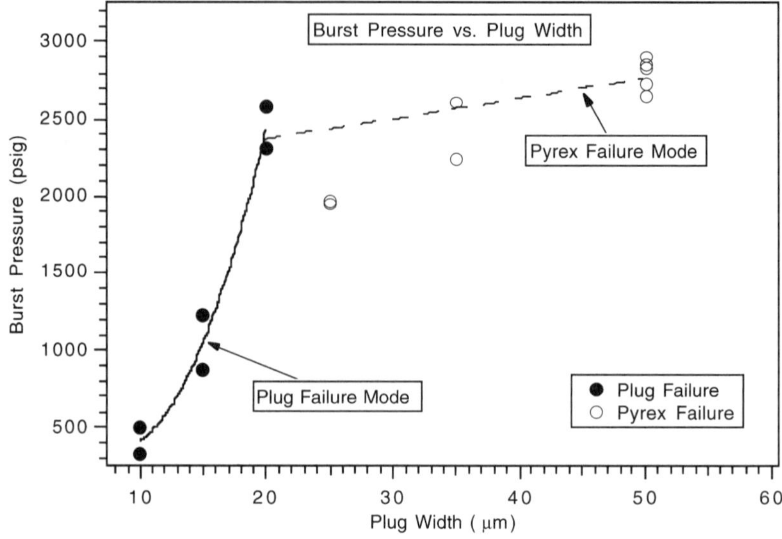

Fig. 11 Burst pressure vs plug width.

3) It seems possible that burst pressure values for Batch 2–type chips featuring plugs of at least 25-μm thickness could easily be extended to even higher values through the use of thicker Pyrex or a different cover material. Figure 11 displays obtained burst pressure values versus plug widths. Clearly, the two failure modes (plug failure for plugs $\leq$20 μm, Pyrex failure for plugs $\geq$20 μm) can be distinguished by the abrupt bend in the solid curve, carried on as a dashed line to the data points obtained for the 25-, 35-, and 50-μm plugs. Had the Pyrex cover not failed, an extrapolation of the solid curve would point to clearly higher burst pressure values for the plugs 25 μm and larger. Given the scatter of the data, however, it is not recommended that a quantitative burst pressure value for plug widths >20 μm be deduced from such an extrapolation. The current Pyrex thickness is 0.5 mm. Thicker Pyrex, possibly up to 1–1.5 mm, may be tested to verify this assumption. A Pyrex wafer thicker than ~1.5 mm may not be successfully bonded using currently available anodic bonding equipment. Beyond the current test program, which will require optical access to the valve interior for test evaluations, nontransparent, all-silicon chips may be explored, where Pyrex would be replaced with stronger silicon material. Note, however, the positive slope of the dashed curve indicating the Pyrex failure mode. A zero slope would be expected if Pyrex failure is the cause. It is unclear at this point whether this is a consequence of the scatter of the limited amount of data or a real physical effect. It was noted for Chips 12 and C5*, i.e., those chips featuring the lowest burst pressure values for any of the Pyrex failure mode cases, that a larger portion of the Pyrex located over the inlet side of the chip had been torn out, possibly pointing to a weakness in the Pyrex or a poorer anodic bond for these particular chips. Chips with identical plug sizes are typically located close to each other on the wafer during fabrication. If a material defect was present in the Pyrex material or bonding was not perfect at this location

during fabrication, it could explain the proneness to failure for these 20-μm-plug chips.

4) Depending on the application, plug sizes could be tailored to the actual pressure needs. A liquid propellant application, for example, typically featuring much lower feed pressures, of the order of 300 psia or so, may require only a 15- to 20-μm-thick plug, which would reduce the power requirements to melt the plug. Cold gas applications will require larger plug thicknesses, depending on the storage pressure. Ten-micron-thick plugs appear not to be very useful for either application.

IV. Plug Melting Tests

A. Test Chip Design

The Batch 2 design described in Section III was used in these tests. However, unlike in the burst pressure cases, the chips used here featured a 4-μm-deep doped epilayer across the silicon surface. This layer is adjacent to the silicon–Pyrex chip interface, lowering the resistance across the top 4 μm of the plug and, therefore, providing a preferential current path across the top of the plug.

B. Test Setup and Procedure

The test chips were mounted on the same stainless-steel fixture as used in the burst pressure tests and then installed into a probe station where they could be contacted electrically via two probe tips, as shown in Fig. 12. The chip was connected via this probe station to a capacitor ignition circuitry (valve driver). Capacitances available with this circuitry could be varied and ranged between 4 and 29 μF. The capacitors were typically charged to 100 V. One inlet of the fixture to which the chips had been bonded was connected to a 300-psig nitrogen pressurant supply.

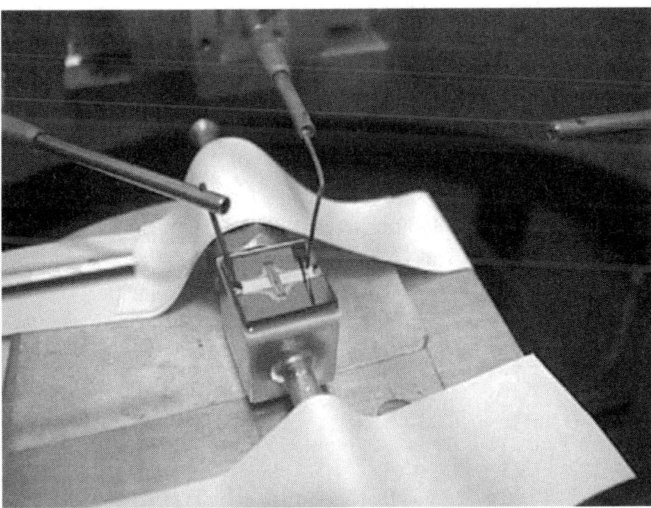

Fig. 12 Test setup for plug melting tests.

The valve exit was exposed to ambient. A dynamic pressure sensor, connected to the inlet port of the chip, was to record any pressure changes occurring as a result of the valve opening.

After the capacitor had been charged, it was connected to the valve by closing a switch. Current and voltage were registered on a storage scope (Tektronix TDS 744A) at sampling frequencies of typically 25 kHz. The pressure signal was also recorded with the scope, and any changes in pressure would allow for determination of the valve response time, i.e., the time elapsed between closing the switch in the capacitor–valve circuitry and the time at which the valve would open, indicated by the change in the pressure signal. Unfortunately, the pressure sensor ceased to operate during testing. It was noted previously, however, that at the time the valve opened, its resistance changed significantly, from its initial value, which could range anywhere between 15 and 25 Ω, depending on the plug width, to an almost-constant 38–40 Ω after valve actuation. It was assumed that, after actuation, once the doped plug had been removed, the current had to migrate through undoped remainders of the plug or around the channel section along the path defined by the isolation trench, leading to the increase in resistance. Subsequently, during data evaluation, this resistance change was calculated from the current and voltage data, and the point at which the new steady-state value of the resistance (38–40 Ω) was reached was assumed to be the point in time when the valve was fully open.

The time at which the valve was commanded open, on the other hand, could easily be detected by sharp and sudden voltage and current spikes. The power trace was calculated from recorded current and voltage data and the energy required to open the valve was calculated by integration of the power trace over the duration of the valve response by means of a simple trapezoidal integration. The average power over this time interval was obtained by dividing the energy required to open the valve by the response time.

1. Results

Ten valve chips were available for testing for this initial test run. Data obtained for seven of these chips are recorded in Table 3. Note, again, that chips are identified by part numbers that were stamped on the valve fixtures. Since these fixtures were recycled in burst and melting tests, some numbers in Table 3 may be identical to those in Tables 1 and 2, however, they obviously correspond to different chips since each chip can be used only once because of the nature of its operation.

Table 3 Results of plug melting tests

Chip I.D.	Plug width, μm	Resistance, Ω		Driver capacitance, μF	Response time, ms	Energy, J	Average power, W
		Pretest	Post-test				
12	50	16	—	29	0.6	0.124	206.7
2	35	17.8	39	29	0.68	0.115	160
13	35	21	—	19	0.4	0.069	157.3
3	25	25.5	40.5	19	0.32	0.076	211.9
7	25	20.5	39.0	16	0.28	0.057	177
C4	25	28	41.5	11	0.2	0.035	147.8
C3	25	24	41.5	6	0.12	0.016	99.5

Three of the ten available chips did not produce any data. One chip was successfully tested but data were not recorded by the oscilloscope. One chip, the first one in this test series, did not open due to inappropriate selection of operating parameters. These operating parameters were refined using a second chip that did finally open, however, only after several tries, so that the condition of the plug at the actual valve opening was not known. Learning experiences gained with these two chips then led to the successful and unproblematic testing of the remaining seven chips.

As reported in Table 3, chips with plug widths of 25, 35, and 50 μm were tested. The purpose of these tests was, apart from conducting a proof-of-principle demonstration, to determine a set of operating parameters required to open the valve successfully, such as the capacitance and voltage to which the capacitor was charged. Consequently, capacitances provided in the valve driver circuitry were varied. Voltages to which the capacitors were charged were held constant at 100 V in this first set of tests.

In the case of all of the chips tested, actuation of the valve was extremely fast. Response times varied between 0.1 ms for a 25-μm plug and about 0.6 ms for a 50-μm plug. These fast response times were considered critical. It was noted during a previous test run that too slow a heat addition to the chip would lead to heat diffusion across the chip aided by the excellent thermal conductivity of silicon, subsequently leading to thermal stresses along the silicon–Pyrex interface, which, in combination with the internal valve pressure, lead to delamination and destruction of the chip. No such behavior was observed during the experiments presented here.

Approximately 0.1 J of energy is required to actuate a 50-μm valve using a 29-μF capacitance, while a 25-μm valve using a 6-μF capacitor requires only 0.016 J, indicating the range of energy levels required to operate the valve. Note, however, that these data are very preliminary. In the case of the 25-μm plug, a few tests were conducted with ever-decreasing capacitor size, reducing the energy available for plug opening. Even in the lowest case tested (6 μF, 0.016 J), the valve still opened. In the case of the 50-μm plug, only one chip was successfully fired with a capacitance of 29 μF, and it is possible that lower capacitances and energy levels would still have resulted in chip openings.

Power traces for a 25- and a 50-μm chip are shown in Fig. 13. As can be seen, due to providing the energy to melt the plug in such a short time, power values peaked dramatically at the beginning of the actuation sequence. In the 50-μm case, up to about 1 kW is reached momentarily, and in the 25-μm case the peak value ranges around 200 W. As mentioned, providing high power values as quickly as possible to the valve is important, as it opens the plug sooner and provides less opportunity for heat to diffuse across the chip. Peak voltage levels range between 40 and 80 V and peak current levels between 4 and almost 13 A for the 25- and 50-μm chip, respectively, as shown in Figs. 14 and 15. These values may all seem high in view of microspacecraft applications, but note that the energy required to actuate the valve can be provided to the capacitor for storage by the microspacecraft bus over considerably longer time periods, reducing the required spacecraft bus power levels for this operation significantly.

The substantially different energy requirements for the valves represented in Figs. 13–15 (parts 12 and C3 in Table 2) can be explained not only by the different plug sizes (the larger plug obviously requiring more energy to melt) but also by the capacitances used. Figures 16–18 show the variation of response time, energy,

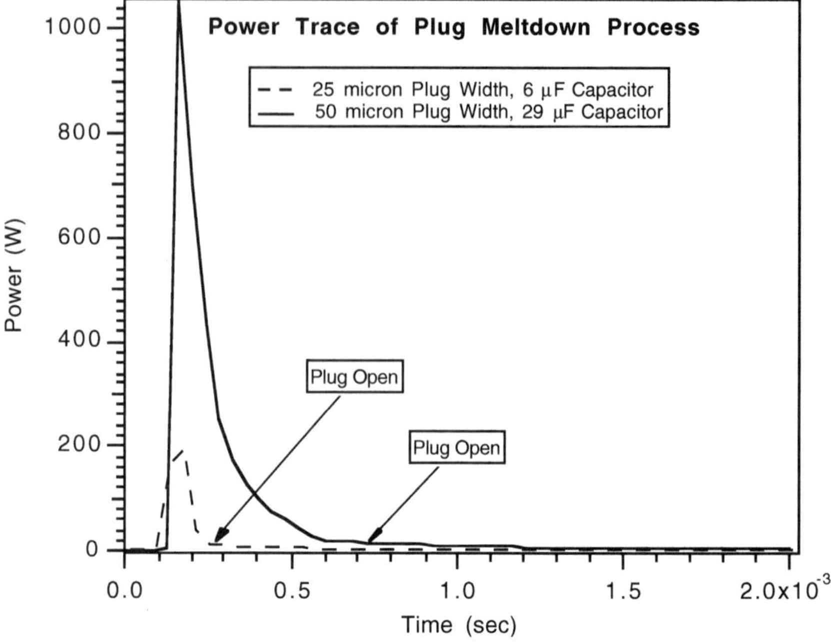

Fig. 13 Power trace to open valve.

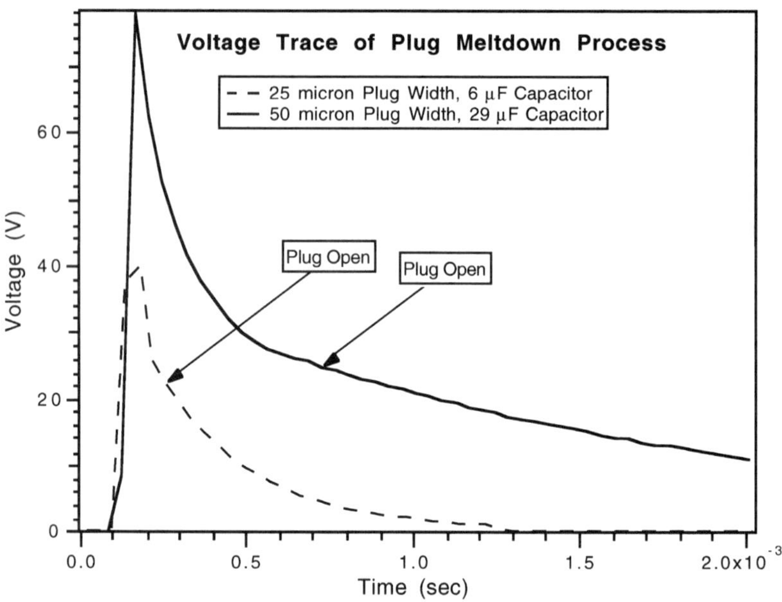

Fig. 14 Voltage trace corresponding to Fig. 13.

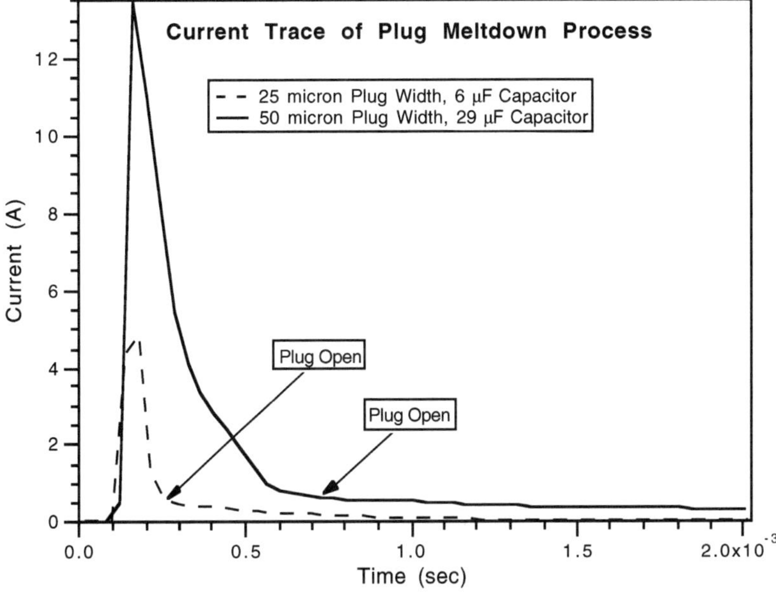

Fig. 15 Current trace corresponding to Fig. 13.

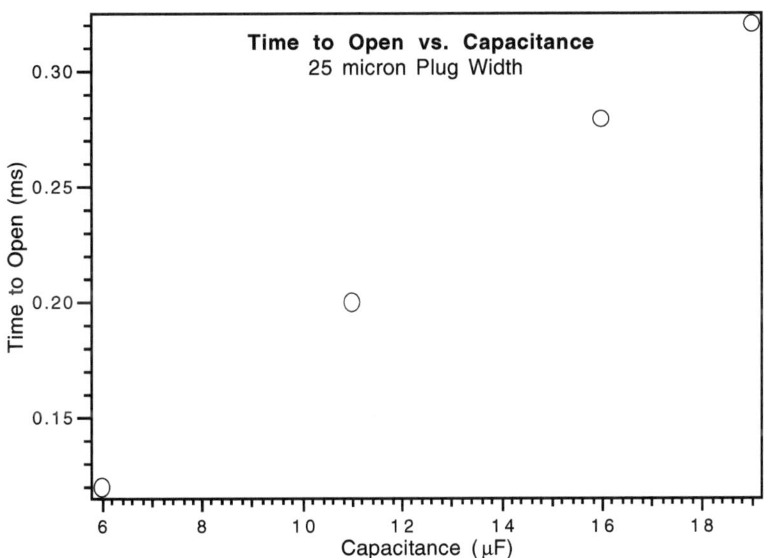

Fig. 16 Valve response time vs driver capacitance.

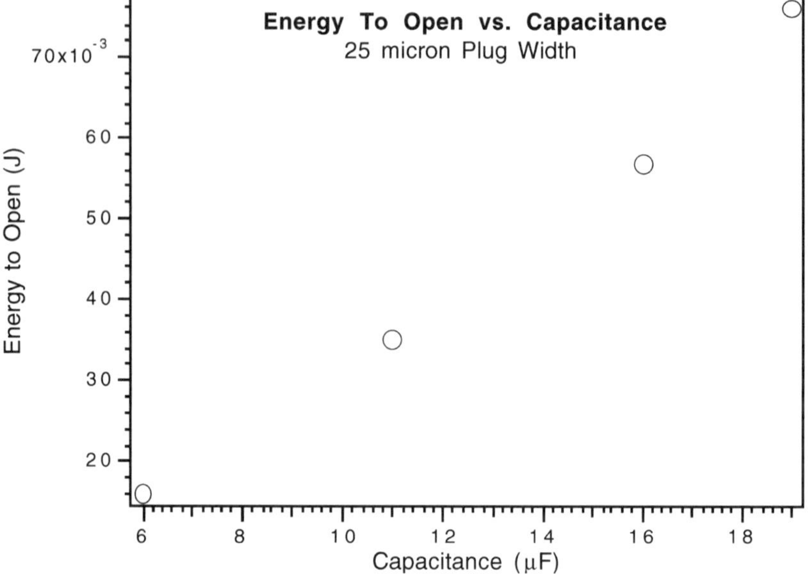

Fig. 17 Energy required to open valve vs driver capacitance.

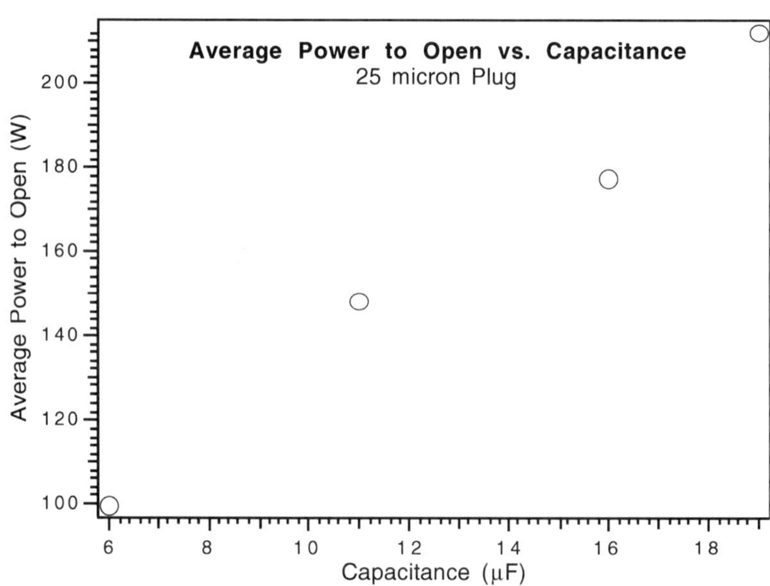

Fig. 18 Average power vs driver capacitance.

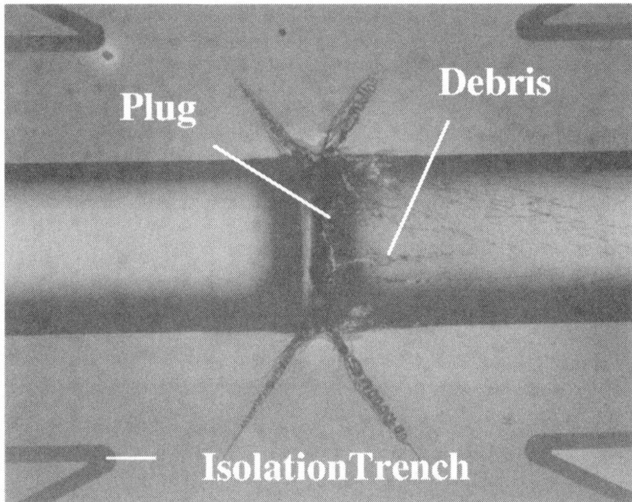

Fig. 19 Plug region after valve firing for Chip 12 (Table 3).

and average power level for 25-μm valves using different capacitances in the ignition circuitry. Inspection of Fig. 16 shows that the response time increases with higher capacitances. This is obvious, as the time constant of the capacitor circuitry increases, leading to a slower discharge. This causes more heat to diffuse into the remainder of the chip, away from the plug, leading to power losses and increasing both the energy and the average power required to actuate the valve (see Figs. 17 and 18). Thus, smaller capacitors, charged to higher voltages if need be to provide the necessary energy, appear to be beneficial.

A visual post-test inspection of the fired valves is shown in Figs. 19 and 20. As can be seen, although the plug has melted, the remainder of the chip appears to be in excellent condition. No melting has appeared anywhere on the chip except in the plug region. These results are impressive given the fact that silicon is a very good heat conductor (heat conductance of about 150 W/mK) and are a consequence of the approach taken, i.e., providing high power levels over very short time durations through a fast capacitor discharge. However, traces can be seen emanating from the corner regions of the plug. It is unclear at this point whether these are cracks in the Pyrex cover glass, caused by thermal stresses, or localized molten regions in the Pyrex or silicon caused by localized increased current flow. If these cracks/molten zones were longer, extending to the isolation trenches or into the recess area, they could result in leakage paths. However, at present these cracks/melting zones appear to be well contained.

Figure 20 shows an additional interesting result that will be of significance for future development work conducted with this valve type. This picture focuses on the bottom of the flow channel, and as a result, the top surface of the plug region is out of focus. As can be seen, plug debris has been pushed downstream as a result of the inlet pressure (300 psig) acting on the melting plug. The debris likely consists of molten silicon and molten Pyrex material. This debris was, of course, expected. Noteworthy, however, is the fact that the debris has solidified along the

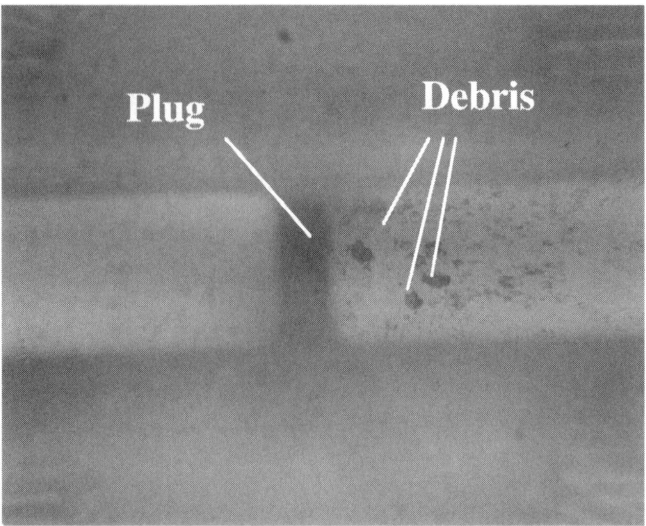

Fig. 20 Closeup of debris after valve firing (Chip 12, Table 3).

channel walls. This effect will obviously be exploited when incorporating debris filtration and trapping schemes into the valve design, through the use of either "splatter plates" or specially designed cavities that may provide ample surface area for debris condensation. Any remaining debris that may solidify in the gas stream may then be filtered out by integrated microfabricated comb filters, located farther downstream.

V. Conclusions and Future Work

Development of a newly proposed, microfabricated, normally closed isolation valve design was initiated. This valve relies on the melting of a potentially doped-silicon plug to open an otherwise sealed flow passage. Applications for this device may be found in micropropulsion feed systems, such as those envisioned for microspacecraft in the 10- to 20-kg class and below, or for larger propulsion feed systems where only low flow rates may have to be maintained, as in some electric propulsion systems (ion and Hall thruster). The micro-isolation valve would play the role of a normally closed pyrovalve in conventional systems, however, it does not rely on pyrotechnic actuation.

Several key feasibility issues need to be addressed in this valve development, such as the ability to melt a silicon plug at power levels acceptable in a microspacecraft environment, the ability to maintain high valve internal pressures to be of use for a variety of propulsion systems, and the ability to trap plug debris inside the chip through adequate filtration schemes. In the current study, both plug melting and pressure handling abilities of the valve were addressed. Both investigations are interrelated with respect to the plug design. Thermal considerations alone would lead to thinner plug widths, since these plugs will be easier to melt, whereas pressure handling considerations will drive the design to thicker plugs. Both issues were therefore addressed simultaneously. Filtration related investigations will follow as this program progresses.

Burst pressure tests of several valve test chips were performed to determine pressure handling abilities. Dedicated test chips were designed, aimed at investigating specifically the critical plug region of the valve. Different plug widths and chip designs were investigated. Building on earlier tests, the final design reached burst pressure values as high as 2850 psig. Pyrex breakage was determined as the failure mechanism for 50-μm plugs. Pyrex-to-silicon anodic bonds, however, were able to withstand even these high pressures. Given the fact that the entire valve is fabricated from silicon and Pyrex glass, the high burst pressure values obtained are quite remarkable. It is felt that using a thicker Pyrex or, in later experiments not requiring visual access to the chip, silicon material, an even higher burst pressure may be attained. Thinner plugs break at lower pressure levels. Ten-micron-thick plugs maintained a few hundred pounds per square inch gauge, whereas 15-μm plugs maintained up to 1000-psig pressures. Twenty-micron plugs were able to sustain up to 2600 psig. Thus, plug thicknesses may be tailored depending upon the application. In lower-pressure liquid feed systems, for example, thinner plugs may be used, which will require less power to melt, whereas higher-pressure cold gas applications will require thicker plugs.

Plug melting experiments showed that valve actuation is possible within acceptable energy levels. Fifty-micron plugs required about 0.1 J and 25-μm plugs about 0.01 J to open. Valve response times are very fast, ranging from 0.1 ms in the case of 25-μm plugs to about 0.6 ms in the case of 50-μm plugs. As a result of this fast actuation, peak power levels are very high, reaching almost 200 W in the 25-μm case and exceeding 1 kW in the 50-μm case. However, since these power levels are provided from a storage capacitor, they will not prevent these valves from being used on power-constrained microspacecraft. The energy levels required to open the valve can be provided to the storage capacitor by the spacecraft over much longer time periods, requiring substantially lower power levels.

Visual post-test inspection of the chips reveals that any melting occurring on the chip is localized in the plug region. The remainder of the chip remains intact. This is a direct consequence of providing the required energy to melt the plug over very short time durations, thus not providing enough time for significant thermal diffusion. This result is particularly noteworthy in view of the fact that the valve is machined from silicon, being an excellent heat conductor.

Future work will focus on additional plug melting characterization tests, expanding the preliminary database obtained so far. Plug debris filtration and trapping will also be investigated. The current test results show that plug debris, which was of course expected to be generated, solidifies on flow channel walls, a fact that will be exploited in future chip designs, featuring integrated debris trapping and filtration schemes.

System integration aspects will also need to be addressed. Current tests were performed with capacitances ranging between 6 and 29 μF charged to about 100 V. Valve operations clearly favor lower capacitances, as they lead to faster discharges and less diffusion heat losses, reducing the energy required to melt the valve. These capacitors may have to be charged to higher initial voltages, which will need to be provided by the driver circuitry of this valve, ideally on a chip scale. Initial design considerations, addressing these microfabricated circuitry issues are underway and look favorable.

Acknowledgments

The authors would like to thank Larry Rupel for performing the burst pressure tests, Sandee Chavez and John Rice for performing the packaging of the test chips, and Jim Kulleck and Ken Evans for performing many of the SEM scans. The authors would also like to thank Lilac Muller and Thomas George for contributing many ideas in the early design phase of this valve concept. Finally, we would like to thank Virendra Sarohia, Christopher Salvo, Barry Hebert, John Stocky, and Elizabeth Kolawa for their continued support. The research described in this chapter was carried out by the Jet Propulsion Laboratory (JPL), California Institute of Technology, under a contract with the NASA. The work was funded in part by the JPL Director's Research Discretionary Fund, and the support is gratefully acknowledged.

References

[1]Collins, D., Kukkonen, C., and Venneri, S., "Miniature, Low-Cost Highly Autonomous Spacecraft—A Focus for the New Millennium," IAF Paper 95-U.2.06, Oslo, Norway, Oct. 1995.

[2]Mueller, J., "Thruster Options for Microspacecraft: A Review and Evaluation of State-of-the-Art and Emerging Technologies," *Micropropulsion for Small Spacecraft*, Progress in Astronautics and Aeronautics, Vol. 187, edited by M. Micci and A. Ketsdever, AIAA, Reston, VA, 2000, Chap. 3 (this volume).

[3]Strand, L., Toews, H., Schwartz, K., and Milewski, R., "Extended Duty Cycle Testing of Spacecraft Propulsion Miniaturized Components," AIAA Paper 95-2810, San Diego, CA, July 1995.

[4]Mueller, J., "Review and Applicability Assessment of MEMS-Based Microvalve Technologies for Microspacecraft Propulsion," *Micropropulsion for Small Spacecraft*, Progress in Astronautics and Aeronautics, Vol. 187, edited by M. Micci and A. Ketsdever, AIAA, Reston, VA, 2000, Chap. 19 (this volume).

Chapter 18

Space-Charge–Limited Emission from Field Emission Cathodes for Electric Propulsion and Tether Applications

Colleen M. Marrese* and Joseph J. Wang†
Jet Propulsion Laboratory, California Institute of Technology, Pasadena, California
Alec D. Gallimore‡
University of Michigan, Ann Arbor, Michigan
and
Keith D. Goodfellow*
Jet Propulsion Laboratory, California Institute of Technology, Pasadena, California

Nomenclature

e	= charge on an electron, C
I_e	= field emission current, mA
I_{ee}	= current emitted by the cathode, mA
I_{trans}	= current transmitted through the sheath, mA
J_{ee}	= electron beam current density normalized to the ambient thermal electron current density
$J_{ee\,max}$	= upper limit on electron beam current density normalized to the ambient thermal electron current density
j_e	= plasma electron current density, mA/cm^2
j_{ee}	= electron beam current density, mA/cm^2
$j_{ee\,max}$	= upper limit on electron beam current density, mA/cm^2
j_i	= ion current density, mA/cm^2
k	= Boltzmann constant, J/K

Copyright © 2000 by the American Institute of Aeronautics and Astronautics, Inc. The U.S. Government has a royalty-free license to exercise all rights under the copyright claimed herein for Governmental purposes. All other rights are reserved by the copyright owner.

*Member of Technical Staff, Thermal and Propulsion Engineering Department.
†Senior Engineer, Thermal and Propulsion Engineering Department.
‡Professor, Aerospace Engineering Department.

m_e = electron mass
m_i = ion mass, kg
n_e = electron number density, electrons/cm^3
n_{ee} = electron beam electron number density, electrons/cm^3
n_{eo} = electron number density in the plasma, electrons/cm^3
n_i = ion number density, ions/cm^3
n_{io} = ion number density at the sheath boundary, ions/cm^3
r_c = cylindrical or spherical cathode radius, cm
r_{sh} = sheath radius, cm
T_e = electron temperature, eV
V_a = anode voltage with respect to ground, V
V_g = potential of the gate electrode with respect to the cathode emitting surface, V
V_t = ion temperature, eV
v_{ee} = velocity of electrons emitted from the cathode, m/s
v_i = ion velocity, m/s
x = position, cm

Greek

ε_o = permittivity of free space, F/m
η = voltage normalized with electron temperature
η_c = potential of gate electrode below the potential at the sheath boundary normalized with electron temperature
η_g = potential between the gate electrode and cathode normalized with electron temperature
η_o = potential across the presheath normalized with electron temperature
λ_D = electronic Debye length, cm
ρ = charge density, C/cm^3
$\bar{\rho}$ = charge density normalized to ambient plasma electron number density
ξ = position normalized with the Debye length
ξ_c = spherical or cylindrical cathode radius normalized with the Debye length
ξ_{sh} = position of sheath boundary normalized with the Debye length
ϕ = potential with respect to the potential at the sheath boundary, V
ϕ_c = potential of the gate electrode with respect to the potential at the sheath boundary, V
ϕ_{min} = potential minimum in the sheath with respect to the potential at the sheath boundary, V
ϕ_o = potential across the sheath, V

I. Introduction

SMALL, meso-, and microscale electric propulsion systems are currently being developed for class I (10–20 W, 10–20 kg), class II (1 W, 1 kg) and class III microspacecraft ($\ll$1 W, $\ll$1 kg)[1], also referred to as picoscale spacecraft, as well as small (1 kW, 100 kg) and large (>10 kW, >200 kg) spacecraft. For some applications these thrusters must operate at power levels below 100 W. These propulsion systems are required for continuous disturbance torque compensation, drag make-up, and primary propulsion.[2] Mesoscale Hall (<300 W, with a discharge chamber diameter less than 50 mm),[3,4] ion (<300 W, with a discharge chamber diameter less

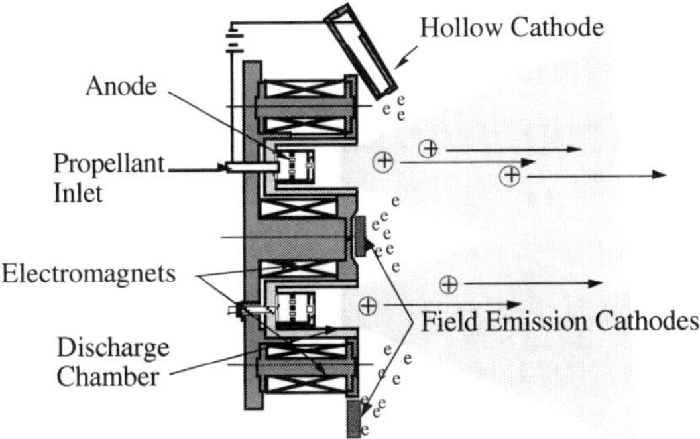

Fig. 1 Cross-sectional view of a Hall thruster showing hollow cathode and field emission cathode positions.

than 60 mm),[5,6] and colloid thrusters (<10 W), and field emission electric propulsion (FEEP)[7] systems, have been developed, microscale field emission ion sources have been fabricated,[8] and microscale colloid thrusters have been proposed. These electric thrusters require electron emission from cathodes for propellant ionization and ion beam neutralization at power levels below approximately 1 W. Larger-scale Hall and ion thrusters (~1 kW) use hollow cathodes; however, their heater and propellant requirements place lower limits on size and power scalability. Field emission (FE) cathodes have the potential of being more efficient than hollow cathodes because they do not require propellant or heaters. Since these cathodes are fabricated on silicon wafers using microfabrication techniques, they are easily scalable in size and power, enabling them to be used with both conventional and microscale electric propulsion systems. Because of their small size, they can be used internally or externally to Hall and ion thrusters. A cross-section of a conventional Hall thruster is shown in Fig. 1 with a hollow cathode and FE cathodes in two possible positions. In the internal position the cathode is mounted to the center electromagnet pole piece. In the external position the cathode is mounted to the side of the thruster.

A FE cathode is a promising alternative to thermionic cathodes. Thermionic cathodes require high operating temperatures for electrons to attain energy levels high enough to overcome the potential barrier between the emitter and vacuum. FE cathodes require electric fields in excess of 10^7 V/cm to lower the potential barrier for electrons to tunnel out of the emitter and into the vacuum.[9] Microtip field emission array (FEA) structures and planar negative electron affinity (NEA) thin-film FE sources are the most common types of FE cathodes. Silicon (Si) and molybdenum (Mo) microtip FEA cathodes and carbon or diamond-like carbon films are most commonly used, however, the combination of these technologies presents the most promising options.[10] The microtip cathodes consist of cones and gate electrodes with packing densities as high as 10^8 tips/cm^2. A single element of an array is shown in Fig. 2.

FEA cathodes have demonstrated impressive performance in ultra high vacuum (UHV) environments. Current densities of 2000 A/cm^2 have been measured from cathode areas approximately 1 mm^2 (Refs. 11, 12) and cathode lifetimes have exceeded 8000 h. Microtip field emission array cathodes have demonstrated emission currents of 100 mA, with less than 1 mW consumed through the gate

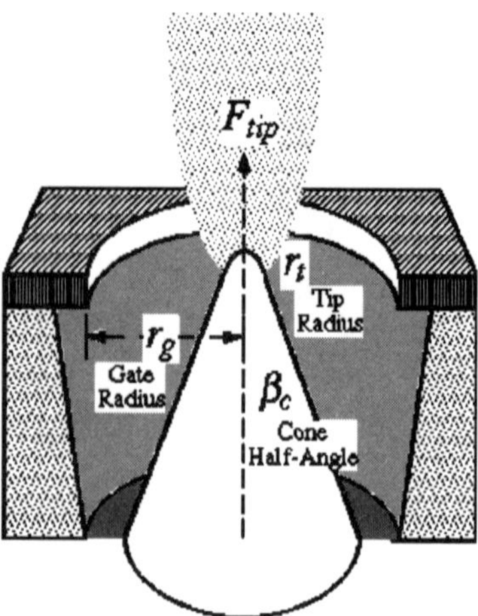

Fig. 2 Field emission microtip configuration (courtesy of Kevin Jensen, Naval Research Laboratory).

electrode.* For comparison, a hollow cathode would consume at least 2–7 W of keeper power in emitting this current.[13] While micropropulsion systems will need ~0.01 mA of current, larger-scale thrusters will require several cathodes to meet current requirements that exceed 1 A.

The two primary concerns with integrating FE cathode and electric propulsion (EP) technologies are cathode lifetime and space-charge–limited electron emission. While FE cathodes are typically operated in UHV environments (e.g., 10^{-9} Torr pressures or lower), the xenon (Xe) pressure in or near Hall and ion thrusters can reach $\sim 10^{-4}$ to 10^{-5} Torr.[14] However, this pressure depends on thruster operation and facility pressure. In the thruster environment, ions created near the electron emitting surface bombard and sputter the microtip emitting structure and significantly degrade performance and shorten cathode lifetime. In the thruster, ions from the discharge bombard the discharge cathode. Outside the thruster the ion beam interacts with the ambient neutrals to generate a charge-exchange (CEX) ion population that can further reduce cathode lifetime. A potential diagram for the thruster and cathode configuration is shown in Fig. 3. The effect of this environment on cathode lifetime depends on the operating voltage of the cathode V_g and the potential difference between the cathode gate electrode and local plasma ϕ_c. It has been shown that Mo or Si FE cathodes must operate at $V_g < 4$–37 V with ϕ_c at 20 V in this environment to avoid sputtering the electron emitting microtips. This issue is discussed in detail in Refs. 11 and 14.

The second concern with integrating these two technologies is space-charge–limited electron emission into a plasma. FE cathodes typically employ a triode configuration with a physical anode only a few millimeters away from the gate electrode to collect the emitted current. Hall and ion thrusters employ anodes and cathodes separated by a plasma with pressures as high as $\sim 10^{-4}$ Torr in the discharge chamber and 10^{-5} in the cathode region. In this environment, a plasma

*Private communication with C. Spindt, SRI International, 1999.

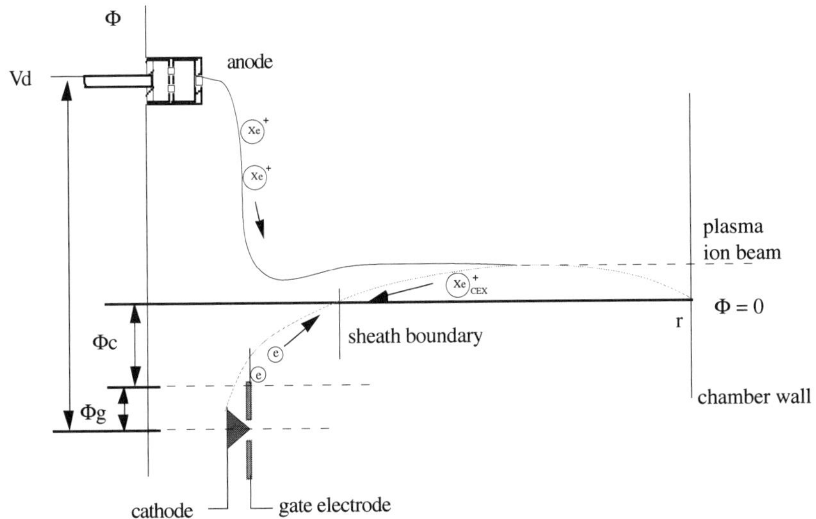

Fig. 3 Potential diagram for cathode, thruster, and plasma configuration.

anode will collect the emitted electron current at the cathode sheath boundary. The characteristics of the sheath determine the space-charge current limit. The thickness of this sheath and potential drop through it depend on the environment, cathode potentials, and characteristics of the emitted current. Hall thrusters typically operate in a configuration where the cathode and anode are floating. The cathode sheath voltage is typically ~20 V in a 1.35 kW Hall thruster system with a hollow cathode and 300 V anode–cathode voltage difference.[15] Hollow cathodes use propellant in addition to the primary supply to the anode to generate a plasma at the emitter surface to reduce negative space charge locally and increase electron transmission through the cathode sheath. A reduction in cathode sheath voltage improves the efficiency of the thruster because it increases the axial ion accelerating voltage. Low cathode potentials are critical when FE cathodes are used because of the lifetime limitations induced by ions sputtering the microtip structure and gate electrode after being accelerated from the plasma to the cathode. It is important that the gate electrode potential is lower than the plasma potential to retard the collection of thermalized plasma electrons while the cathode beam electrons escape to the plasma. It is anticipated that the field emitter gate electrode will be set at ~20 V below plasma potential to optimally facilitate this process. In an electric propulsion system plasma environment with virtual anodes, space-charge effects will limit the electron current that escapes through the sheath to the plasma. This concern must be properly addressed and current limits must be determined before FE cathodes can be designed and used for these applications.

Space-charge–limited electron emission problems have been studied extensively. The Childs–Langmuir relationship describes space-charge–limited currents in a vacuum diode. In the derivation, the current was limited by a zero electric field at the cathode surface because initial electron velocities were negligible. Fay et al.[16] determined emission current limitations in a parallel plate geometry when electrons are emitted with considerable velocities. In that case, positive electric fields at the cathode surface and virtual cathodes are stable solutions to the

Poisson equation, permitting larger currents to escape from the cathode to the anode. Virtual cathodes can form between the cathode and anode where a potential minimum occurs below cathode potential to limit the emitted current. Langmuir considered space-charge limitations for parallel plates separated by a plasma where ions contribute to negative space-charge neutralization to further increase space-charge–limited current. He claimed that electron and ion currents were limited in proportion to ion and electron masses when ion and electron initial velocities are negligible, but showed that these currents can increase with nonzero velocities at the sheath boundary.[17] Bohm modeled the sheath near the cathode and determined that for a cathode sheath to be stable, the ions must enter it with nonzero initial velocity.[18] To ensure that potential minima occur only at sheath boundaries for a nonemitting cathode, it was determined that $v_i > (T_e/m_i)^{1/2}$. Crawford and Cannara considered the case for space-charge–limited emission of a hot cathode into a plasma.[19] They determined a valid velocity range for ions entering the sheath, the velocity being limited by charge-exchange and momentum-exchange collisions in presheaths. Prewett and Allen[20] studied the same configuration and included a criterion derived by Andrews and Allen[21] to construct an analytical expression for the initial energy of ions entering a sheath. Using that expression in the sheath model, they achieved excellent correlation between theory and experiment. Goodfellow augmented this model by considering the initial thermal velocities of the electrons emitted at the cathode surface. Goodfellow used this sheath model with thermal models to predict cathode temperatures in a magnetoplasmadynamic thruster. He also received good correlation between experiment and theory.[22]

More detailed modeling can be done with three-dimensional particle simulation models. Recently, Wang and Lai developed a three-dimensional full particle simulation model to study the emission of a high-density ion beam into a space plasma.[23] They showed that when the emitted current exceeds a critical value, the potential profile becomes nonmonotonic with a positive potential peak near the emitter. At a sufficiently large beam current, the potential peak becomes a virtual anode that partly blocks the beam transmission. This behavior is similar to space-charge–limited current flow in a diode studied extensively by Fay et al.[16] However, beam divergence caused by space charge and interactions between the beam and the ambient plasma make the problem much more complex.

The objective of this study is to estimate the space-charge current limits on electron emission by FE cathodes in EP and electrodynamic tether (EDT) plasma environments. Once the current density limitations are understood, FE cathodes can be designed with the appropriate size and for the appropriate electron energies to emit the current required in the various plasma environments. Three environments were considered in this study for space-charge–limited emission from a field emission source of electrons into a plasma: a Hall thruster, a meso- or microscale ion thruster, and a tether in low Earth orbit (LEO). Because a large Hall thruster could use a FE cathode internally (i.e., on the center pole piece) or externally, as shown in Fig. 1, both conditions were considered in the analysis. A small or mesoscale Hall thruster requires an external cathode because of the thruster configuration anticipated. Most of the plasma parameters used to represent the internal and external environment of a Hall thruster were obtained by Domonkos et al.[16] in the plasma plume of a 1.35 kW Hall thruster. However, the electron number density in the external environment was obtained in another study from measurements of the ion current density.[24] These parameters were used because initial testing of these cathodes with Hall and ion thrusters could occur with such

Table 1 Characteristics of the plasma environments investigated

	n_{eo}, cm^{-3}	T_e, eV	λ_D, cm
Hall/ion thruster (external)[25]	8×10^8	5	0.06
Hall thruster (internal)[16]	8×10^{10}	1	0.003
Ion engine discharge chamber[26]	3×10^{11}	2–3	0.002
Tether at 250 km[27]	5×10^5	0.1	0.33

medium scale systems. An ion engine requires two cathodes: one cathode in the discharge chamber for propellant ionization and another cathode external to the thruster for ion beam neutralization. Plasma parameters used in the modeling of ion engine discharge chambers are estimates only for a mesoscale ion engine.* The external environment of an ion engine is similar to the external environment of a Hall thruster, and was modeled accordingly. There is also interest in using FE cathodes on tethers to emit electrons into a space plasma ~250–1000 km above the Earth. The parameters used for each of these environments are presented in Table 1.

A one-dimensional numerical and analytical model and a three-dimensional particle simulation model of field electron emission into a plasma are presented in this chapter. The three-dimensional particle simulation model uses a full particle-in-cell code similar to the model used by Wang and Lai in an ion beam study.[23] It takes into account beam divergence caused by space charge and virtual cathode formation. While this simulation model more accurately represents the problem being studied, it is more complicated and computationally expensive than the simpler one-dimensional model. Predictions on FE cathode electron emission limits for thruster and tether environments are presented and compared in this study for both models. The modeling results can be used to design FE cathodes for the current density required by the applications. Cathode configurations are suggested for thruster and tether applications.

II. One-Dimensional Cathode Sheath Model

A one-dimensional sheath model for planar and spherical sheath geometries is discussed in this section, and it was developed to predict the thickness of the sheath and emission current limitations for different cathode operating configurations. In the sheath model, the Andrews and Allen criterion was used to determine ion energy at the sheath boundary for a configuration with a strongly emitting cathode with initial electron energies greater than 20 eV, as is the case for field emission cathodes. One of the objectives of this study is to determine the optimal potential of the gate electrode with respect to the plasma potential that is required to transmit the desired electron current through the sheath when electrons are emitted with initial energies in excess of 20 eV into a plasma. The details of the presheath are not explored; however, it is assumed that the ions are accelerated through the presheath up to a high enough velocity to ensure monotonic potential profiles in

*The plasma number density was calculated assuming a xenon ion pressure of 10^{-5} Torr and temperature of 300 K.

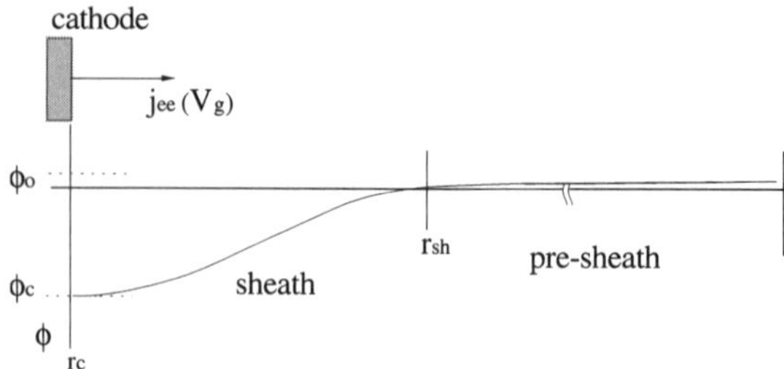

Fig. 4 Plasma–cathode potential configuration being modeled in this study.

the sheath. It is possible that a virtual cathode will form that will increase the upper limits on the space-charge–limited emission current and change the structure of the sheath. This model cannot predict virtual cathode formation. A reverse in the electric field polarity at the cathode surface will significantly increase the current collected by the cathode gate electrode, possibly to dangerously high current levels that could melt the gate electrode. This model is used to determine the space-charge current attainable in the limit of no electric field at the emitting surface. The potential configuration for all of the electrodes is shown in Fig. 3 whereas a possible potential profile in the sheath is shown in Fig. 4.

This one-dimensional sheath model was developed for planar and spherical cathode and sheath geometries. FE cathodes are planar, however, the structure of the sheath depends on the size of the cathode and the environment. With cathode dimensions greater than several electronic Debye lengths, the sheath structure is primarily planar. A cathode with dimensions smaller than a Debye length will have a sheath that is more spherical in geometry because the thickness of the sheath is typically several Debye lengths. Two different cathode and environment scenarios are illustrated in Fig. 5.

The one-dimensional sheath model described in this section was used to estimate cathode current limitations in plasma environments caused by space-charge effects. The effects of cathode operating voltages, sheath geometry, and cathode dimensions on the current were also investigated. A sheath model with a planar

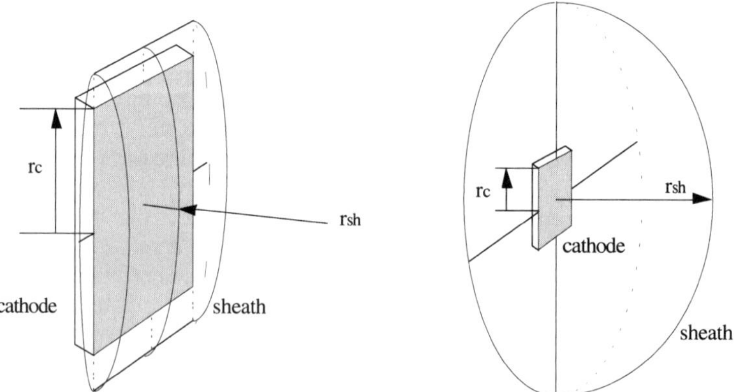

Fig. 5 Two different cathode sheath configurations demonstrating the necessity for a sheath model with different geometries for different environments.

cathode and sheath geometry predicts conservative limits on cathode current due to space-charge effects, whereas a spherical sheath model predicts upper limits on cathode current. All potentials represented with ϕ are defined with respect to the potential at the sheath boundary where quasi neutrality is assumed. The presheath accelerates the ions through ϕ_o to the sheath boundary. The potential difference between the plasma and cathode gate electrode is ϕ_c. This model is capable of predicting only monatonic potential profiles. Full recombination is assumed for the ions at the cathode surface. Potential magnitudes are used in the input and output of this model for simplicity; however, their true signs are inherent to the sheath equations. The planar and spherical sheath model derivations are included in Appendix A. The sheath model results are presented in the following section.

Two geometrical sheath configurations were investigated. The planar geometry in one dimension was the simplest geometry considered. The results of this model are most accurate when the sheath thickness is much smaller than the cathode dimensions. This model also most accurately represents the configuration of the cathode. As the cathode dimensions become smaller with respect to the sheath dimensions, the sheath becomes less planar and the beam eventually begins to expand radially in the sheath. When the cathode dimensions approach the electronic Debye length, the sheath approaches a spherical geometry. With a spherical sheath geometry, ion number densities vary as $1/r^2$, providing much better negative space-charge neutralization by the plasma ions than the planar geometry, thus permitting higher emitted electron current densities. The spherical sheath model was used to study the effect of cathode dimensions on current limitations, an investigation not possible with the planar model.

The planar model provided a conservative estimate of the space-charge limited cathode current density. The results of this model are shown in Table 2. The results show the effects of electrode voltages, and plasma electron temperature and number densities on space-charge current density limits. Current density objectives for tether and thruster environments are ~100 mA from a 1 cm^2 cathode. The results

Table 2 Space-charge–limited currents estimated by the planar sheath model for Hall and ion thruster and tether environments

n_{eo}, cm^{-3}	T_e, eV	V_g, V	ϕ_c, V	$j_{ee\,\max}$, mA/cm^2
		Hall (external)		
8×10^8	5	500	20	160
8×10^8	5	30	20	34
8×10^8	5	100	20	68
8×10^8	1	30	20	17
		Hall (internal)		
8×10^{10}	1	30	20	1,700
8×10^{10}	1	100	20	2,884
		Ion (internal)		
3×10^{11}	2	30	20	8,800
		Tether		
5×10^5	0.1	30	20	0.003
5×10^5	0.1	100	20	0.005
5×10^5	0.1	1000	20	0.016
5×10^5	0.1	30	100	0.002

of the planar model show that the lower limits on the emission current greatly exceed the requirements when the cathodes are used in the discharge chamber of an ion thruster or in the internal position of a Hall thruster. However, the planar sheath model results show that FE cathodes emitting into a tether environment will be severely limited by space-charge effects.

The effect of V_g and ϕ_c on the current limitations was also considered. The results of this investigation are also shown in Table 2. According to these modeling results, cathodes will require operating voltages that greatly exceed the sputter threshold of the Si or Mo cathodes if used externally to the Hall or ion thruster and for current densities greater than 100 mA/cm^2. Increasing the potential between the cathode and plasma decreased the cathode current limit, according to the results of the planar model. Increasing this potential only decreased the cathode current in the planar model. Increased cathode sheath potentials do not increase the electric field proportionately because increasing ϕ_c also increases the size of the sheath. Increasing ϕ_c also reduces the ion number density at the gate electrode to reduce negative space-charge neutralization by the ions. Increasing the electron energies at the gate electrode does increase the emission current limit from the cathode. FE cathode gate electrode voltage should not exceed 20 V below the plasma potential to maximize the current emitted from the cathode and minimize the damage caused by charge-exchange (CEX) ion bombardment. In the tether case where T_e is 0.1 eV, a lower ϕ_c should be sufficient (1–5 V). However, the optimal cathode voltage must be determined.

In the Hall thruster configuration where the cathode is external to the thruster, the sheath dimensions are comparable to the cathode dimensions, which are assumed to be approximately 1×1 cm. In this scenario, the sheath would be better represented by a geometry that considers some electron beam expansion in the sheath. In the tether environment, plasma densities are low enough that the cathode sheath dimensions could be much larger than the cathode dimensions, necessitating the use of a sheath model that is more spherical in geometry. A comparison of the results of the model for both sheath geometries is presented in Table 3. The planar model can be used to predict lower limits on the cathode current. Again, the planar and spherical geometries should bound the emission current range for these cases.

The primary environment investigated was that of a Hall thruster with $T_e = 5$ eV so that $\phi_c = 20$ V and $V_g = 30$ V. The sheath and cathode geometry and dimensions

Table 3 Space-charge–limited currents predicted by the planar and spherical sheath models in nondimensional units for the external Hall thruster environment

Geometry	ξ_c	ξ_{sh}	η_c	η_g	η_o	$J_{ee\,max}$
Planar	—	20.0	4	6	0.85	2.0
Spherical	40	48.6	4	20	1.34	31.0
Spherical	40	47.0	4	6	1.34	6.5
Spherical	20	27.0	4	6	1.22	7.4
Spherical	10	17.0	4	6	1.10	9.9
Spherical	4	10.4	4	6	0.95	17.6
Spherical	1	6.8	4	6	0.76	70.0
Spherical	0.4	5.4	4	6	0.65	168.0

Table 4 Field emission current limitations predicted by the sheath model in dimensional parameters for planar and spherical geometries for the Hall thruster configuration with an external cathode (V_g is at 30 V and ϕ_c is at 20 V)

Geometry	ξ_c	r_c, cm	j_{ee}, mA/cm^2	I_{ee}, mA
Planar	—	—	34	—
Spherical	20	1.2	126	570
Spherical	10	0.6	168	242
Spherical	4	0.24	299	69
Spherical	1	0.06	1190	17

strongly influence the results of the model, as data in Table 3 show. The planar case predicts larger sheath dimensions and lower current density limitations. The cases with spherical sheath geometry approach the results of the planar sheath case as the cathode dimensions increase.

Results in dimensional parameters from the sheath models for cathodes operating in the external cathode position of a Hall thruster are presented in Table 4 for different sheath geometries. While cathodes can emit much higher current densities into spherical sheaths, spherical sheath formation requires cathode dimensions that are smaller than the sheath thickness. With such small emitting cathode areas, total cathode current is sacrificed for current density.

Space-charge current limitations are also predicted using the sheath model to study a cathode emitting from a space tether into a LEO environment. Results with nondimensional parameters are shown in Table 3. Data in Table 5 show that the sheath dimensions are so large that the sheath geometry should be much more spherical than planar, even when ξ_c is 10. In the tether environment T_e is 0.1 eV,

Table 5 Field emission current limitations predicted by the sheath model in normalized parameters for a spherical sheath in a tether environment (T_e is 0.1 eV)

ξ_c	ξ_{sh}	η_c	η_g	η_o	$J_{ee\,max}$
10	64	100	100	0.53	40
10	100	200	200	0.52	91
10	104	200	300	0.52	120
9	103	200	300	0.52	145
5	100	200	300	0.52	430
1	97	200	300	0.52	8650
1	99	200	400	0.52	10400
1	101	200	600	0.52	13400
1	104	200	1000	0.52	18400
1	109	200	10000	0.53	63550
1	60.5	100	300	0.53	4925
1	37	50	300	0.55	2750
1	38	50	1000	0.56	5350
1	38.5	50	10000	0.56	17400

Table 6 Field emission current limitations predicted by the spherical sheath model for the tether environment (η_c at 200 and η_v at 300)

ξ_c	r_c, mm	V_c, V	V_g, V	j_{ee}, mA/cm^2	I_{ee}, mA
5	16.5	20	30	0.645	7.0
1	3.3	20	30	12.9	5.6
1	3.3	20	40	15.6	6.8
1	3.3	20	60	20.1	8.8
1	3.3	20	100	27.6	12.0
1	3.3	20	1000	95.3	41.5
1	3.3	5	30	4.1	1.8
1	3.3	5	100	8.0	3.5
1	3.3	5	1000	26.1	11.3

therefore η_g is 300 and η_c is 200 when V_g is 30 V and ϕ_c is 20 V. The current density emitted by the cathode greatly increases with decreasing cathode dimensions. Table 6 shows how the total current emitted by the cathode is affected by the size of the cathode. Smaller cathodes are capable of larger current densities, however, the emitting area is also considerably smaller. Therefore, the only advantage to shrinking the cathode size is the cathode size itself, overall current will not increase. Also shown in Tables 5 and 6 is the current gain attributable to increases in the potential of the gate electrode. Increasing the gate electrode from 30 to 100 V only increases the current by a factor of ~2. To satisfy the 2 A current requirements of the tether application, several hundred cathodes are required according to these modeling results.

The objective of the one-dimensional sheath model development was to quickly provide upper and lower limits on the current densities emitted from planar cathodes in different environments. To consider the true structure of the sheath and the effects of virtual cathode formation on space-charge limits, three-dimensional particle simulation models are required.

III. Three-Dimensional Particle Simulation Model

There are two inherent limitations in the one-dimensional model presented in the previous section. First, neither the one-dimensional planar nor the spherical solution actually represents the emission of a finite radius electron beam. In particular, beam divergence caused by space charge is not included in the formulation of the model. Second, the one-dimensional model only allows a monotonic potential profile. It does not allow for the nonmonotonic potential profile that appears when the current emitted exceeds the space-charge limits given by the Childs law. Because of the complexity of the problem, computer particle simulations are needed to provide a more accurate and complete solution to this problem.

Wang and Lai developed a three-dimensional full particle electrostatic PIC code to simulate the physics of ion beam emissions into the space plasma.[23] The particle simulation code developed by Wang and Lai was modified for this study. In this code, all charged particles (the emitted electrons, and the ambient electrons and ions) are treated as test particles. The trajectories of each test particle, the space charge, and the electric field are solved self-consistently from Poisson's equation and Newton's second law.

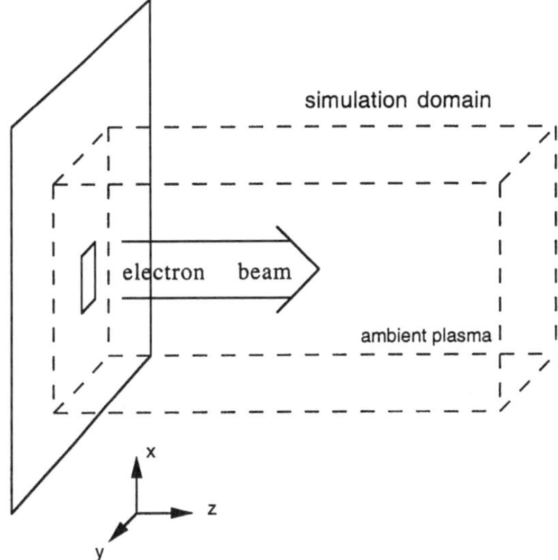

Fig. 6 Three-dimensional PIC simulation set-up.

The simulation set-up is shown in Fig. 6. The left-side boundary of the simulation domain was set to be the spacecraft surface and the emitter surface. All other domain boundaries are considered to be "open" boundaries. Test particles representing the ambient electrons and ions are loaded uniformly into the domain with a Maxwellian velocity distribution at the start of the simulation. Starting from $t = 0$, the beam electrons are injected into the simulation domain at every time step from the emitter in the z direction. The beam particles are injected to form a uniform beam at the emitter exit with charge density q_{bnbb} and beam velocity v_{bo}. At every time step, ambient electrons and ions are also injected into the domain from the open boundaries according to their thermal fluxes. Those test particles, which hit the spacecraft or emitter surfaces or flow out of the simulation domain, are deleted from the particle list. The simulation was performed until an equilibrium was reached.

For the results presented in this chapter, the simulations were typically run with $27 \times 27 \times 27$ cells. To minimize the effects of numerical noise, the number of initial ambient plasma particles was more than 50 per cell for the ions and electrons, and the number of emitted electrons was more than 100 per cell at thruster exit. The total number of test particles used varied from case to case but was typically more than one million. A series of test runs was also performed where the number of grid cells, the number of test particles, and the time step were varied to ensure that these simulation parameters do not affect the final solution.

The three-dimensional particle simulation model was applied to study field emission from a cathode in Hall thruster and tether environments. The simulation cases are listed with the results in Tables 7 and 8. The results from the three-dimensional model for a FE cathode operating in a thruster environment are shown in Table 7. The current emitted from the cathode is I_{ee}, whereas the current transmitted through the sheath is I_{trans}. The minimum potential in the cathode sheath with respect to the potential at the sheath boundary, the virtual cathode potential, is $\phi_{\min}$. In the Hall thruster case with the cathode in the external position, $n_{eo} = 8 \times 10^8$ cm^3,

Table 7 Current limitations and virtual cathode potentials predicted by the Wang simulation model for a cathode in a Hall thruster environment ($r_c = 4\lambda_D$, V_g at 30 V, ϕ_c at −20 V, T_e at 5 eV, and n_{eo} at 8×10^8/cm^3)

Case	j_{ee}, mA/cm^2	I_{ee}, mA	ϕ_{min}, V	I_{trans}, mA	I_{trans}/I_{ee}
1	34	7.5	−20	7.5	1
2	84	18.5	−34	18.5	1
3	125	27.5	−50	27.5	1
4	236	52	−75	34	0.67

$T_e = 5$ eV, $V_g = 30$ V, and $\phi_c = -20$ V. The field emitter area was 0.22 cm^2. For this condition and cathode size, the one-dimensional planar model yields a space-charge current density limit at 34 mA/cm^2 and the one-dimensional spherical model predicts a current density limit of 299 mA/cm^2. In the simulations, four different emission current values were considered: case 1, 34 mA/cm^2; case 2, 84 mA/cm^2; case 3, 125 mA/cm^2; and case 4, 236 mA/cm^2. The simulation results are shown in Figs. 7–9. Figure 7 shows the potential contours and beam electron vectors for these four current cases. Figure 8 compares the potential profile along the cathode axis for different electron beam current density levels. When $j_{ee} = 34$ mA/cm^2, the beam electron density is lower than the ambient density. Hence, the electron beam is readily neutralized by the ambient ions. As a result, there is very little beam divergence due to space charge and the results are similar to the results obtained in the one-dimensional planar model. In agreement with the planar model, the electric field at the cathode surface approaches zero at $J_b = 34$ mA/cm^2, the one-dimensional space-charge–limit current density. As the emitted current is further increased, the potential profiles become nonmonotonic and a negative potential peak develops in front of the cathode. While the one-dimensional planar model predicts an emission current limit of $J_b = 34$ mA/cm^2, it was shown that, for a finite-sized beam, the current emission limit is significantly higher than the one-dimensional planar model predictions because of beam divergence if virtual cathode formation is tolerated. The three-dimensional model results show that all of the current from the FE cathode will be transmitted out of the sheath, even when $j_{ee} = 125$ mA/cm^2 for the conditions shown in Table 7. In case 4, the emitted current density was further increased to $j_{ee} = 236$ mA/cm^2. In this case, the negative potential dip is high enough to block part of the current

Table 8 Current limitations and virtual cathode potentials predicted by the Wang simulation model for a cathode in a tether environment ($r_c = \lambda_D$, V_g at 30 V, ϕ_c at −20 V, T_e at 0.1 eV, and n_{eo} at 5×10^5/cm^3)

Case	j_{ee}, mA/cm^2	I_{ee}, mA	Φ_{min}, V	I_{trans}, mA	I_{trans}/I_{ee}
5	6.5	2.86	−42	2.86	1.0
6	9.1	3.88	−42	3.82	1.0
7	11.7	5.0	−47.6	4.62	0.92
8	13	5.7	−49	4.9	0.86
9	26	11.4	−60	5.8	0.51

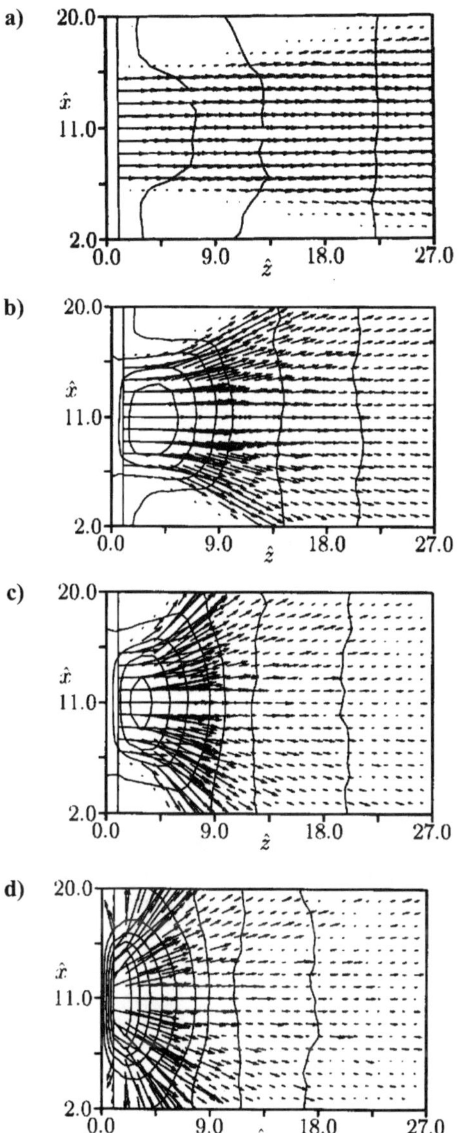

Fig. 7 Electron trajectories and potential contours for Hall thruster a) case 1: J_b = 34 mA/cm^2 with contours at −20, −15, −10, and −5 V; b) case 2: J_b = 84 mA/cm^2 with contours at −30, −25, −20, −15, −10, and −5 V; c) case 3: J_b = 125 mA/cm^2 with potential contours at −50, −40, −30, −20, −15, −10, and −5 V; and d) case 4: J_b = 236 mA/cm^2 with potential contours −70, −60, −50, −40, −30, −20, −10, and −5 V.

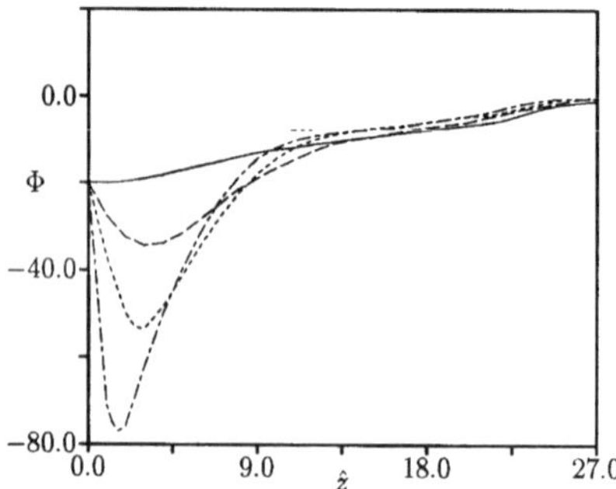

Fig. 8 Comparison of potential profiles along the field emitter emission center axis for Hall thruster cases 1 (solid), 2 (dashed), 3 (dotted), 4 (dot-dashed).

transmission. The potential hump has become a virtual cathode. Only about 67% of the current is transmitted through the virtual cathode. The reflection of the emitted electrons by the virtual cathode is obvious in the v_z vs phase plot shown in Fig. 9. The results of the three-dimensional model show that the maximum current density that can be emitted by the cathode is between 125 and 236 mA/cm^2. This current-density limit is within the range of limits predicted by the planar and spherical one-dimensional sheath models, which are 34 mA/cm^2 and 299 mA/cm^2, respectively.

The three-dimensional PIC model was also used to study current limits in tether applications. The plasma and cathode parameters used in the model for a FE cathode operating in a tether environment are $n_{eo} = 5 \times 10^5$, $T_e = 0.1$ eV, a cathode area of 0.44 cm^2, an initial electron energy V_g of 30 eV, and $\phi_c = -20$ V. In the tether application, the FE cathode will always be operated in an environment that has a much lower ambient plasma density than the beam density. For this condition, the one-dimensional planar model predicts a space-charge current limit of 3.0 μA/cm^2 and the one-dimensional spherical model predicts a space-charge limit of 12.8 mA/cm^2. In the three-dimensional particle simulation model, five different emission current densities were considered: case 5, 6.5 mA/cm^2; case 6, 9.1 mA/cm^2; case 7, 11.7 mA/cm^2; case 8, 13 mA/cm^2; and case 9, 26 mA/cm^2. The simulation results are shown in Figs. 10–12. Figure 10a–c shows the potential contours and beam electron vectors for cases 5, 8, and 9. Figure 11 shows

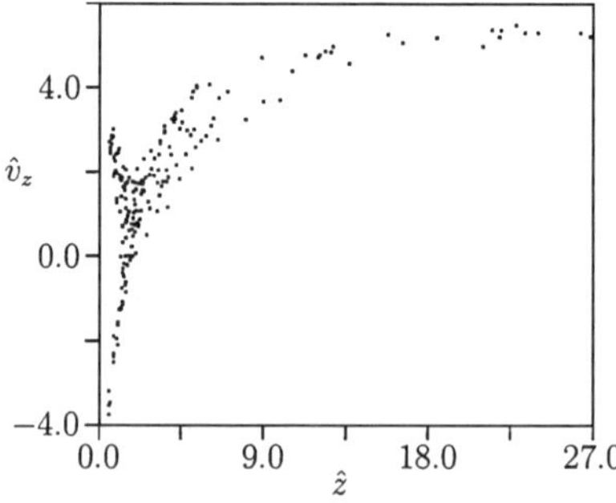

Fig. 9 V_z vs z-phase space plot for Hall thruster case 4.

a)

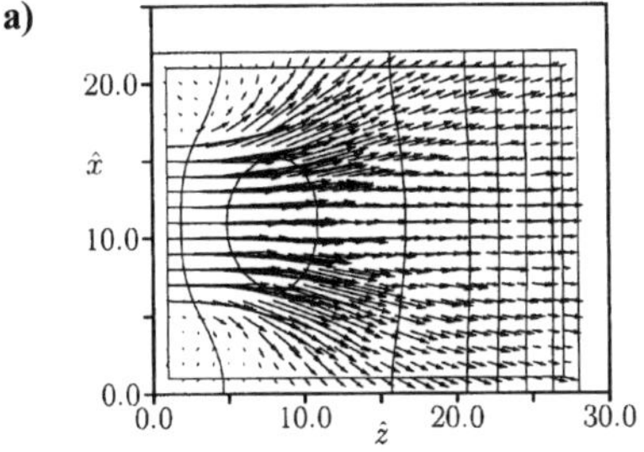

b)

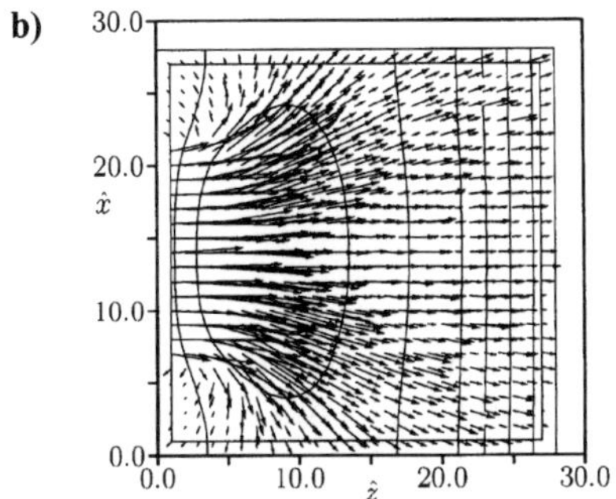

c)

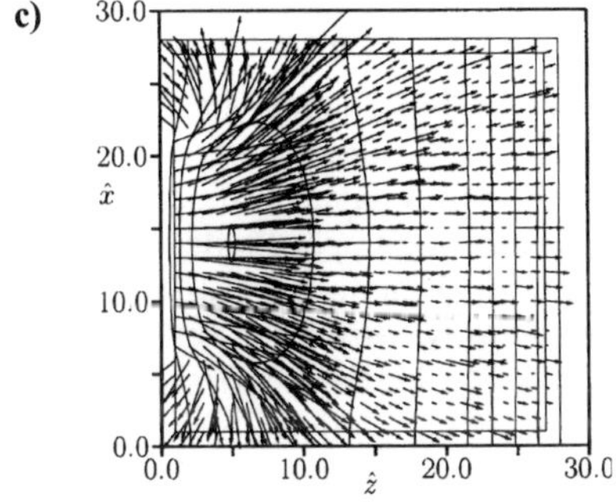

Fig. 10 Electron trajectories and potential contours for tether a) case 5: J_b = 6.5 mA/cm^2 and potential contours at −40, −30, −20, −15, −10, and −5 V; b) case 8: J_b = 13 mA/cm^2 and potential contours at −40, −30, −20, 15, −10, and −5 V; and c) case 9: J_b = 26 mA/cm^2 and potential contours at −50, −40, −30, −20, −15, −10, and −5 V.

a comparison of potential profiles along the cathode center axis showing virtual cathode structure and potential minimums. In each case, the beam density is much larger than the ambient density; hence a nonmonotonic potential profile develops. A virtual cathode appears in cases 7–9. In case 8, only 86% of the emitted current is transmitted through the sheath. In case 9, only about 50% of the emitted current is transmitted through the sheath. The reflection of the emitted electrons by the virtual cathode is obvious in the v_z vs phase plot in Fig. 12. The current density limit predicted by the three-dimensional model is between

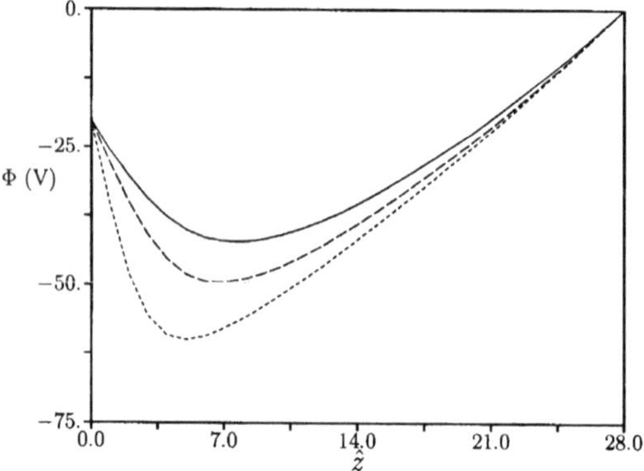

Fig. 11 Comparison of potential profiles along the field emitter axis for tether cases 5 (solid), 8 (dashed), and 9 (dotted).

9.1 mA/cm^2 and 11.7 mA/cm^2. Once again the three-dimensional model predicted current-density limits between those limits predicted by the one-dimensional planar, 0.003 mA/cm^2, and spherical, 12.9 mA/cm^2, models. In this case the solution obtained using the three-dimensional PIC model is closer to the solution obtained by the spherical one-dimensional sheath model than the planar model, as expected.

IV. Discussion

This chapter presents a one-dimensional sheath model that is capable of predicting upper and lower limits on the emission current density, which can be expected from a FE cathode in plasma environments and a three-dimensional PIC model that is capable of predicting upper limits on current densities much more accurately. The one-dimensional model provides immediate results whereas the three-dimensional model requires execution times greater than 1 h and is computationally more expensive. The one-dimensional model was used to study the effects of V_g and r_c on emission current limitations. Increasing V_g increases the j_{ee} limit. While decreasing r_c increases j_{ee}, I_{ee} from the cathode decreases. The optimal V_c

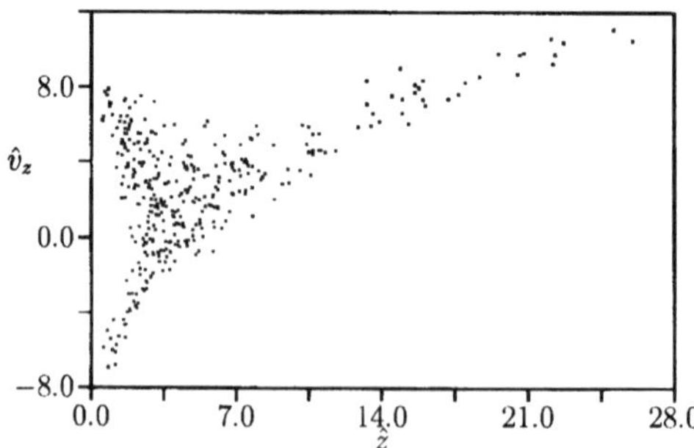

Fig. 12 V_z vs z-phase space plot for tether case 9.

Table 9 Current density limitations based on space-charge effects and recommended cathode configurations for thruster and tether applications

	V_g, V	J, mA/cm^2	I, A	Cathode area, cm^2
Hall/ion thruster (external)	30	>125	0.04–4.5	0.3–36
Hall thruster (internal)	30	>1000[a]	0.5–4.5	0.5–4.5
Ion thruster (internal)	30	>8000[a]	0.2–15	0.03–2
Tether	30	>9	2	222
Tether	100	~18	2	111
Tether	1000	~54	2	37

[a]In practice, the upper limit on current density will be further limited by cathode performance.

value depends upon the cathode operating configuration and environment. These models can be used to determine the maximum and minimum number of cathodes or emitting area required to supply the desired current or to estimate the exact number of cathodes required for any application.

The models were used to access the compatibility of FE cathode and electric propulsion and tether systems. Modeling results showed that the FE cathode current will not be space-charge–limited if operated in the discharge chamber of an ion engine or internal position of a Hall thruster. Current densities higher than 1.7 A/cm^2 can be emitted in these environments. The challenge in these scenarios is getting this much current from a FEA cathode with a gate electrode voltage, and electron energies at approximately 30 V. In the external position of a Hall or ion thruster, a cathode area of ~40 cm^2 is required if V_g is 30 V for 4.5 A. This current could be emitted from a segmented cathode ring around the thruster. Cathode area can be decreased by increasing V_g. However, the advantages of higher operating voltage to cathode dimensions must be weighed against cathode lifetime, as discussed in Refs. 11 and 14. The size of the cathode required is not prohibitively large for the Hall thruster application, which requires up to 4.5 A. The tether application could require a prohibitively large number of cathodes, or cathode area, if electrons are emitted at 30 eV. For the cathode and plasma scenario shown in Table 8, the maximum number of cathodes required to provide 2 A will be 700, with a total area of 300 cm^2. The number of cathodes required could be decreased by a factor of two if V_g, or electron energy, is increased from 30 V up to 100 V. Emitting electrons at energies ~1 keV should decrease the number of cathodes required to less than 100 with a total area less than 50 cm^2 to make these two technologies more compatible. Recommended cathode configurations are shown in Table 9.

If electron energies greater than 30 eV are required, they must be decoupled from the potential of the gate electrode to meet the lifetime requirements. This could be achieved with a cathode lens and ion repeller (CLAIR) structure that consists of three electrodes in addition to the gate electrode. The CLAIR design is discussed in detail elsewhere,[26] and is currently under development.

V. Conclusions

Two models were used to predict space-charge–limited electron emission from FE cathodes into a plasma. The one-dimensional sheath model, considering only radial variations in particle densities, was limited to monotonic sheath potential profiles whereas the three-dimensional numerical particle simulation model could consider virtual cathode formation in addition to three-dimensional beam expansion into a plasma.

The results of the sheath model show that the upper limit on the space-charge–limited current from a cathode positioned in the center of a Hall thruster or in the discharge chamber of an ion engine exceeds the current density requirements. When n_{eo} is below $8 \times 10^8/\text{cm}^3$, the current will be space-charge–limited with limits that depend upon the electron energy, cathode potential, and cathode dimensions. In the external position of a Hall or ion thruster, a cathode area of $\sim$40 cm^2 is required if V_g is 30 V for 4.5 A. Small thrusters operating at only 100 mA will only require a cathode with 1 cm^2 of emitting area. For the tether environment and cathode configuration shown in Table 8, the maximum number of cathodes required to provide 2 A will be 100 with a total area of approximately 40 cm^2 if the electrons are emitted with 1 keV energies. Electron energy must be decoupled from the gate electrode potential with an electron accelerating scheme like CLAIR to achieve these high electron energies.[10]

In the future, the one-dimensional model can be used to bound the cathode current-density limits almost immediately whereas the interactions of electron beams and spacecraft should be studied with the three-dimensional model to optimize the number of cathodes, operating voltages, and their placement on spacecraft for EP and EDT applications.

Appendix A: One-Dimensional Planar Sheath Model

In the planar sheath configuration, the one-dimensional form of Poisson's equation is

$$-\frac{\mathrm{d}^2\phi}{\mathrm{d}x^2} = \frac{e}{\varepsilon_o}(n_i - n_{ee} - n_e) \tag{A1}$$

The continuity and energy equations for the ions and electrons in the sheath are used to determine particle number densities. The ion number density at the sheath boundary is n_{io}. The ion number density, n_i, is

$$j_i = n_i e v_i = n_{io} e v_o(\phi_o) \tag{A2}$$

$$\tfrac{1}{2} m_i v_i^2 = e(\phi + \phi_o) \tag{A3}$$

$$n_i = n_{io}\left(1 + \frac{\phi}{\phi_o}\right)^{-\frac{1}{2}} \tag{A4}$$

Plasma electron densities are defined by the Boltzmann relationship for a Maxwellian distribution of electrons:

$$n_e = n_{eo} \exp\left(\frac{-e\phi}{kT_e}\right) \tag{A5}$$

It is assumed that the electrons are Maxwellian with temperature T_e, and the number density at the sheath boundary is n_{eo}. The beam electron number densities are defined as

$$j_{ee} = n_{ee} e v_{ee} \tag{A6}$$

$$\tfrac{1}{2} m_e v_{ee}^2 = e(V_g + \phi_c - \phi)^{-\frac{1}{2}} \tag{A7}$$

$$n_{ee} = \frac{j_{ee}}{e\sqrt{2e/m_e}} (V_g + \phi_c - \phi)^{-\frac{1}{2}} \tag{A8}$$

The condition of quasi neutrality at the sheath boundary is used to define the ion number density at the sheath boundary in terms of electron number densities. At $\phi = 0$, $n_{io} = n_{eo} + n_{ee}$, and

$$n_{io} = n_{eo} \left(1 + \frac{j_{ee}}{n_{eo} e\sqrt{2e/m_e}} (V_g + \phi_c)^{-\frac{1}{2}} \right) \tag{A9}$$

Poisson's equation then can be written as

$$-\frac{\mathrm{d}^2(-\phi)}{\mathrm{d}x^2} = \frac{\rho}{\varepsilon_o}$$

$$= \frac{e}{\varepsilon_o} \left(n_{io} \left(1 + \frac{\phi}{\phi_o} \right)^{-\frac{1}{2}} - \frac{j_{ee}}{e(2e/m_e)^{\frac{1}{2}}} (V_g + \phi_c - \phi)^{-\frac{1}{2}} - n_{eo} \exp\left(\frac{-e\phi}{kT_e} \right) \right) \tag{A10}$$

The parameters used to normalize the Poisson equation are

$$\eta = \frac{e\phi}{kT_e} \tag{A11}$$

$$\xi = \frac{x}{\lambda_D} \tag{A12}$$

$$\lambda_D = \left(\frac{\varepsilon_o k T_e}{n_{eo} e^2} \right)^{\frac{1}{2}} \tag{A13}$$

$$\bar{\rho} = \frac{\rho}{e n_{eo}} \tag{A14}$$

$$j_e = n_{eo} e \sqrt{\frac{2kT_e}{m_e}} \tag{A15}$$

and

$$J_{ee} = \frac{j_{ee}}{j_e} \tag{A16}$$

where λ_D is the electronic Debye length. The normalized form of Poisson's equation in this model is

$$\frac{\mathrm{d}^2\eta}{\mathrm{d}\xi^2} = \left[1 + J_{ee}(\eta_g + \eta_c)^{-\frac{1}{2}}\right] \left(1 + \frac{\eta}{\eta_o} \right)^{-\frac{1}{2}} - J_{ee}(\eta_g + \eta_c - \eta)^{-\frac{1}{2}} - \exp(-\eta) \tag{A17}$$

Integrating Poisson's equation once and applying the first of the following two boundary conditions,

$$\frac{d\eta}{d\xi}(\eta = 0) = 0 \tag{A18}$$

and

$$\eta(\xi = 0) = \eta_c \tag{A19}$$

the electric field in this region is determined to be

$$\left(\frac{d\eta}{d\xi}\right)^2 = 4\eta_o\left[1 + J_{ee}(\eta_g + \eta_c)^{-\frac{1}{2}}\right]\left[\left(1 + \frac{\eta}{\eta_o}\right)^{\frac{1}{2}} - 1\right]$$

$$+4J_{ee}\left[(\eta_g + \eta_c - \eta)^{\frac{1}{2}} - (\eta_g + \eta_c)^{\frac{1}{2}}\right] + 2[\exp(-\eta) - 1] \quad (2) \tag{A20}$$

The ion energy at the sheath boundary can be approximated using the criterion derived by Andrews and Allen[22] that

$$\frac{d}{d\eta}\left(\frac{d^2\eta}{d\xi^2}\right)\bigg|_{\eta=0} = 0 \tag{A21}$$

so that

$$\eta_o = \frac{1 + J_{ee}(\eta_g + \eta_c)^{-\frac{1}{2}}}{2 - J_{ee}(\eta_g + \eta_c)^{-\frac{3}{2}}} \tag{A22}$$

Notice that in the limit of zero cathode emission, η_o –> 1/2. This result is consistent with the model developed by Bohm.[19] The upper limit on the emission current is being defined when the electric field at the cathode surface becomes zero,

$$0 = 4\eta_o\left[1 + J_{ee\,\max}(\eta_g + \eta_c)^{-\frac{1}{2}}\right]\left[(1 + \eta_c/\eta_o)^{\frac{1}{2}} - 1\right]$$

$$+4J_{ee\,\max}\left[(\eta_g)^{\frac{1}{2}} - (\eta_g + \eta_c)^{\frac{1}{2}}\right] + 2(\exp(-\eta_c) - 1) \tag{A23}$$

The maximum emission current possible is

$$J_{ee\,\max} = \frac{\eta_o\left[(1 + \eta_c/\eta_o)^{\frac{1}{2}} - 1\right] + \frac{1}{2}[\exp(-\eta_c) - 1]}{\eta_o(\eta_g + \eta_c)^{-\frac{1}{2}}\left[1 - (1 + \eta_c/\eta_o)^{\frac{1}{2}}\right] + (\eta_g + \eta_c)^{\frac{1}{2}} - (\eta_g)^{\frac{1}{2}}} \tag{A24}$$

Equations (A22) and (A24) then can be solved simultaneously to estimate $J_{ee\,\max}$ for any combination of η_c and η_g. $J_{ee\,\max}$ is the emission current density that corresponds to a zero electric field at the cathode surface. This value represents the lower limit on the emission current density capability of FE cathodes emitting electrons into a plasma. If electron beam expansion occurs in the sheath, virtual cathodes form in the sheath, or plasma instabilities develop, larger emission currents can be possible. Emission current limitations predicted by this model agree with results of the Goodfellow sheath model, which provided the basis for this model. In the case considered, η_c was 10, η_g was 0.3, and both models predicted that $J_{ee\,\max}$ was 0.7. Goodfellow used a similar sheath model combined with thermal models to predict cathode temperatures in a magnetoplasmadynamic thruster system and achieved good correlation between experiment and theory. Results of this model were also consistent with current limitations predicted by Prewett and Allen for cases where initial electron energies are negligible.

Appendix B: One-Dimensional Spherical Sheath Model

The sheath model was also developed for a spherical cathode and sheath geometry. The spherical cathode is emitting electrons radially outward into plasma where the electron number density is n_{eo}. The charged particle number density changes with potential variations in the sheath and with a $1/r^2$ relationship inside of the sheath because of the geometry.

Plasma electron number density is given by the Boltzmann equation (A5), assuming a Maxwellian distribution of energies. The number density of electrons emitted from the cathode is derived using the continuity and energy equations,

$$\eta_{ee} = \frac{j_{ee}}{e(2e/m_e)}(\phi_g + \phi_c - \phi)^{\frac{1}{2}}\left(\frac{r_c}{r}\right)^2 \tag{A25}$$

The continuity and energy equations are used to determine the ion number density,

$$n_i = n_{io}\left(1 + \frac{\phi}{\phi_o}\right)^{-\frac{1}{2}}\left(\frac{r_{\text{sh}}}{r}\right)^2 \tag{A26}$$

The initial ion energy at the sheath boundary is $e\phi_o$, and the ion number density at the sheath boundary is $n_{io} = n_{eo} + n_{ee}$, which is equal to

$$n_{io} = n_{eo}\left(1 + \frac{j_{ee}}{en_{eo}(2e/m_e)^{\frac{1}{2}}}(\phi_g + \phi_c)^{-\frac{1}{2}}\left(\frac{r_c}{r_{\text{sh}}}\right)^2\right) \tag{A27}$$

where quasi neutrality is assumed. The ion number density in the sheath is

$$n_i = n_{eo}\left(1 + \frac{j_{ee}}{en_{eo}(2e/m_e)^{\frac{1}{2}}}(\phi_g + \phi_c)^{-\frac{1}{2}}\left(\frac{r_c}{r_{\text{sh}}}\right)^2\right)\left(1 + \frac{\phi}{\phi_o}\right)^{-\frac{1}{2}}\left(\frac{r_{\text{sh}}}{r}\right)^2 \tag{A28}$$

Poisson's equation can then be written for this cathode configuration and environment as

$$-\nabla^2\phi = -\left(\frac{\mathrm{d}^2(-\phi)}{\mathrm{d}r^2} + \frac{2}{r}\frac{\mathrm{d}(-\phi)}{\mathrm{d}r}\right) = \frac{\rho}{\varepsilon_o}\frac{e}{\varepsilon_o}\left[n_{io}\left(1 + \frac{\phi}{\phi_o}\right)^{-\frac{1}{2}}\left(\frac{r_{\text{sh}}}{r}\right)^2\right.$$
$$\left. - \frac{j_{ee}}{e(2e/m_e)^{\frac{1}{2}}}(\phi_g + \phi_c - \phi)^{-\frac{1}{2}}\left(\frac{r_c}{r}\right)^2 - n_{eo}\exp\left(-\frac{e\phi}{kT_e}\right)\right] \tag{A29}$$

Normalization of Poisson's equation describing the sheath is performed with Eqs. (A11–A16),

$$\frac{\mathrm{d}^2\eta}{\mathrm{d}\xi^2} + \frac{2}{\xi}\frac{\mathrm{d}\eta}{\mathrm{d}\xi} = \bar{\rho} = \left[1 + J_{ee}(\eta_g + \eta_c)^{-\frac{1}{2}}\left(\frac{\xi_c^2}{\xi_{\text{sh}}^2}\right)\right]\left(1 + \frac{\eta}{\eta_o}\right)^{-\frac{1}{2}}\left(\frac{\xi_{\text{sh}}^2}{\xi^2}\right)$$
$$- J_{ee}(\eta_g + \eta_c - \eta)^{-\frac{1}{2}}\left(\frac{\xi_c^2}{\xi^2}\right) - \exp(-\eta) \tag{A30}$$

The Andrews and Allen criterion is evaluated at the sheath boundary to derive the relationship for the nondimensional voltage drop in the presheath,

$$\eta_o = \frac{1 + J_{ee}(\eta_g + \eta_c)^{-\frac{1}{2}}\left(\frac{\xi_c}{\xi_{sh}}\right)^2}{2 - J_{ee}(\eta_g + \eta_c)^{-\frac{3}{2}}\left(\frac{\xi_c}{\xi_{sh}}\right)^2} \tag{A31}$$

Equation (A30) is evaluated numerically using a Runge–Kutta fourth-order method for second-order differential equations. The initial conditions at the sheath boundary are

$$\eta(\xi = \xi_{sh}) = 0 \tag{A32}$$

and

$$\frac{d\eta}{d\xi}(\xi = \xi_{sh}) = 0 \tag{A33}$$

The cathode sheath problem is not very well defined, knowing only the potential at the cathode and sheath boundary and the electric field at the sheath boundary. The position of the sheath boundary is not known a priori. The much simpler nature of the Poisson equation for the planar sheath geometry does not require this input parameter. Therefore, the structure of the sheath can be determined for any J_{ee}, η_c, and η_g combinations. In spherical coordinates Poisson's equation is more complicated, requiring more input parameters. Initial guesses are made for the thickness of the sheath, ξ_{sh}. It is assumed that the maximum current is emitted when the electric field at the cathode surface is zero and the potential at the cathode surface is equivalent to the defined cathode potential. The sheath thickness is adjusted until these conditions are met at the cathode surface to estimate $J_{ee\,\max}$. A potential profile for the sheath can be obtained only for limiting cases. Potential profiles and $J_{ee\,\max}$ are obtained in this model for the limiting case of a zero electric field at the cathode surface.

Acknowledgments

The authors would like to gratefully acknowledge the Ballistic Missile Defense Organization, the Air Force Office of Scientific Research (Mitat Birkan), and the Jet Propulsion Laboratory, California Institute of Technology, under contract by NASA, for their support of this research.

References

[1]Mueller, J., "A Review and Applicability Assessment of MEMS-Based Microvalve Technologies for Microspacecraft Propulsion," AIAA Paper 99-2725, June 1999.

[2]Mueller, J., "Thruster Options for Microspacecraft: A Review and Evaluation of Existing Hardware and Emerging Technologies," AIAA Paper 97-3058, July 1997.

[3]Belikov, M. B., Gorshkov, O. A., Rizakhanov, R. N, Shagayda, A. A., and Khartov, S. A., "Hall-Type Low- and Mean Power Thrusters Output Parameters," AIAA Paper 99-2571, June 1999.

[4]Khayms, V., and Martinez-Sanchez, M., "Fifty-Watt Hall Thruster for Microsatellites," Micropropulsion for Small Spacecraft, Progress in Astronautics and Aeronautics, Vol. 187, edited by M. Micci and A. Ketsdever, AIAA, Reston, VA, 2000, Chap. 9.

[5]Sohl, G., Fosnight, V. V., and Goldner, S. J., "Cesium Electron Bombardment Ion Microthrustors," AIAA Paper 67-81, Jan. 1967.

[6]Gorshkov, O., Muravlev, V. A., Grigoryan, V. G., and Minakov, V. I., "Research in Low-Power Ion Thrusters with Slit-Type Grid Systems," AIAA Paper 99-2855, June 1999.

[7]Marcuccio, S., Giannelli, S., and Andrenucci, M., "Attitude and Orbit Control of Small Satellites and Constellations with FEEP Thrusters," International Electric Propulsion Conf., (IEPC) Paper 97-188, Aug. 1997.

[8]Brodie, I., "Vacuum Microelectronic Devices," *Proceedings of the IEEE*, Vol. 82, No. 7, July 1994, pp. 1006–1034.

[9]Gomer, R., "Field Emission, Field Ionization, and Field Desorption," *Surface Science*, Vol. 300, No. 1-3, 1994, pp. 129–152.

[10]Marrese, C. M., "A Review of Field Emission Cathode Technologies for Electric Propulsion Systems and Instruments," IEEE Aerospace Conf. 2000, Paper 382, March 2000.

[11]Murphy, R. A., Harris, C. T., Matthews, R. H., Graves, C. A., Hollis, M. A., Kodis, M. A., Shaw, J., Garven, M., Ngo, M. T., and Jensen, K. L., IEEE International Conf. on Plasma Science (ICOPS), 1997.

[12]Spindt, C. A., and Brodie, I., Technical Digest of the 1996 IEEE International Electron Devices Meeting (IEDM), 12.1.1 (1996); also Spindt, C. A., Holland, C. E., Schwoebel, P. R., and Brodie, I., IEEE International Conf. on Plasma Science, 1997.

[13]Patterson, M. J., Grisnik, S. P., and Soulas, G. C., "Scaling of Ion Thrusters to Low Power," International Electric Propulsion Conf. (IEPC), Paper 97-098, 1997.

[14]Marrese, C. M., Polk, J. E., Jensen, K. L., Gallimore, A. D., Spindt, C., Fink, R. L., Tolt, Z. L., and Palmer, W. D., "Performance of Field Emission Cathodes in Xenon Electric Propulsion System Environments," *Micropropulsion for Small Spacecraft*, Progress in Astronautics and Aeronautics, Vol. 187, edited by M. Micci and A. Ketsdever, AIAA, Reston, VA, 2000, Chap. 11 (this volume).

[15]Domonkos, M. T., Gallimore, A. D., Marrese, C. M., and Haas, J. M., "Very-Near-Field Plume Investigation of the Anode Layer Thruster," *Journal of Propulsion and Power*, Vol. 16, No. 1, 2000, pp. 91–98.

[16]Fay, C. E., Samuel, A. L., and Shockley, W., "On the Theory of Space Charge Between Parallel Plane Electrodes," *Bell System Technical Journal.*

[17]Langmuir, I., "The Interaction of Electron and Positive Ion Space Charges in Cathode Sheaths," *Physics Review*, Vol. 33, June 1929.

[18]Bohm, D., *Characteristics of Electrical Discharges in Magnetic Fields*, edited by A. Guthrie and R. K. Wakerling, McGraw-Hill, New York, 1949.

[19]Crawford, F. W., and Cannera, A. B., "Structure of the Double Sheath in a Hot Cathode Plasma," *Journal of Applied Physics* Vol. 36, No. 10, 1965.

[20]Prewett, P. D., and Allen, J. E., "The Double Sheath Associated with a Hot Cathode," *Proceedings of the Royal Society of London, Series A.*, Vol. 348, 1976.

[21]Andrews, J. G., and Allen, J. E., "Theory of a Double Sheath Between Two Plasmas," *Proceedings of the Royal Society of London, Series A.*, Vol. 320, 1971, pp. 459–472.

[22]Goodfellow, K. D., "A Theoretical and Experimental Investigation of Cathode Processes in Electric Thrusters," Ph.D. Dissertation, Univ. of Southern California, Los Angeles, CA, 1996.

[23]Wang, J. J., and Lai, S. T., "Virtual Anode in Ion Beam Emission in Space: Numerical Simulations," *Journal of Spacecraft and Rockets*, Vol. 34, No. 6, 1997, pp. 829–836.

[24]Marrese, C. M., "Compatibility of Field Emission Cathode and Electric Propulsion Technologies," Ph.D. Dissertation, Univ. of Michigan, Ann Arbor, MI, 1999.

[25]Chen, F. F., *Plasma Physics and Controlled Fusion*, Plenum Press, New York, 1984.

[26]Marrese, C. M., Gallimore, A. D., Mackie, W. A., and Evans, D., "A Cathode to Operate in an Oxygen-Rich Environment," Space Technology and Applications International Forum (STAIF) Paper 224, Jan. 1997.

Chapter 19

Review and Applicability Assessment of MEMS-Based Microvalve Technologies for Microspacecraft Propulsion

Juergen Mueller*
Jet Propulsion Laboratory, California Institute of Technology
Pasadena, California

I. Introduction

MICROSPACECRAFT concepts are experiencing growing attention within the aerospace community. Several reasons may be named as the motivations behind this trend, such as reduced mission cost due to the use of smaller and cheaper launch vehicles needed for such microspacecraft, as well as the possibility to explore new and unique mission scenarios enabled by the use of microspacecraft. For example, microspacecraft constellations charting entire regions of space may be envisioned. The measurement of particle and field distributions around a planetary object or even within the heliopause at the edge of our solar system may be performed more efficiently with such a constellation, providing a larger return of data than can be collected along the trajectory of a single larger craft.

Using such a "fleet" of microspacecraft, each spacecraft equipped with its own set of experiments, will also increase mission reliability since the loss of one or even a few microspacecraft will not jeopardize the entire mission. Mission scenarios may be envisioned where small microprobes are released from a larger spacecraft to perform particularly risky parts of the mission. For example, probes may be released into Saturn's ring system to allow for a close-up survey of the system, while a larger craft serves as a communications node to Earth, staying safely behind.†

A few tens of kilograms or less is currently being envisioned as a typical mass target for microspacecraft. Spacecraft with a mass of 10 kg may be no larger than a "shoebox" or "basketball." Even smaller microspacecraft, ranging in mass around

Copyright © 2000 by the American Institute of Aeronautics and Astronautics, Inc. The U.S. Government has a royalty-free license to exercise all rights under the copyright claimed herein for Governmental purposes. All other rights are reserved by the copyright owner.

*Advanced Propulsion Technology Group.

†West, J., Personal communication, Jet Propulsion Lab., California Inst. of Technology, Pasadena, CA, Fall 1995.

Table 1 Definition and classifications of microspacecraft

Designation	S/C mass, kg	S/C power, W	S/C dimension, m	Comments
Microspacecraft (AF/European definition)	10–100	10–100	0.3–2	Micropropulsion concepts beneficial due to weight/size savings, possibly enabling based on performance requirements (e.g., very small impulse bits for ultrafine spacecraft pointing). Low end of mass range; see below.
Class I Microspacecraft (<10 kg Nanosat)	5–20	5–20	0.2–0.4	Use miniature "conventional" components, possibly MEMS/ microfabricated. Conventional integration (e.g., feed lines) still possible; higher level of integration between components/subsystem desirable.
Class II Microspacecraft	1–5	1–5	0.1–0.2	MEMS/microfabricated components; high level of integration between components and subsystems required (subsystems on a chip?).
Class III Microspacecraft (Picosat)	<1	<1	<0.1	All MEMS/microfabricated. Very high level of integration between subsystems and within subsystems required.

1 kg, are being studied that in turn may be no larger than a "softball."[1] Table 1 shows an attempt to classify microspacecraft and distinguish various degrees of miniaturization and integration required to realize them.[2] A photograph of a 7-kg ground demo functional model of one such microspacecraft design is shown in Fig. 1.[3] It shows the MTD (Microspacecraft Technology Development) II model that was assembled and ground-tested at the Jet Propulsion Laboratory (JPL). While the MTD II craft was not designed for spaceflight, it allows for testing of microspacecraft technologies and their integration in a hardware environment on the ground.

To enable the construction of such microspacecraft, each subsystem will have to be reduced in size and adapted in function to meet the new and unique requirements of such a craft. For example, components of a micropropulsion subsystem for such a microspacecraft will have to be reduced in size to fit within the spacecraft envelope, requiring extensive miniaturization. Furthermore, thrust levels and impulse bits will have to be reduced. Thrust levels for attitude control of a microspacecraft may be of the order of a few millinewtons or less and impulse bits as little as 10^{-6} Ns may be required. Thrust levels and impulse bits that low require the control of very low propellant flow rates. Microvalves will be required to control those flows.

Different microvalve concepts are currently under investigation. Conventionally machined, miniature solenoid valves are one valve option being studied at present. Several valve manufacturers in the United States, such as Moog, Inc., Marotta Scientific Controls Inc., and Kaiser–Marquardt, Inc., have developed, or are in the

Fig. 1 MTD II ground demo spacecraft model.[3]

process of developing, this type of valve and have achieved impressive degrees of miniaturization to date.[4]

Paralleling efforts in other spacecraft subsystem areas, entirely microfabricated propulsion components, machined from silicon using microelectromechanical systems (MEMS) fabrication techniques, have been studied recently.[2] Potentially significant additional mass and volume savings could be achieved if these microfabricated thruster components could be tightly integrated by chip-to-chip bonding with other MEMS-based components, such as valves, filters, regulators, and sensors, as well as the control electronics required to drive these devices. Apart from the potential of offering mass and volume reductions over conventionally integrated propulsion systems, such a highly integrated MEMS-based propulsion system would also have minimal external interfaces, easing and reducing the cost of integration of the propulsion subsystem into the microspacecraft bus.[2] The latter point is of particular interest in the case of microspacecraft designs due to their small size.

Propulsion systems featuring such a high degree of miniaturization and integration will likely require suitably microfabricated MEMS microvalve technology. Repeatedly, previously developed commercial MEMS valves are being cited as examples of valves that may be applicable for use in such systems. In this chapter, following a brief review of valve design requirements as currently assumed for microspacecraft propulsion systems, presently available MEMS valves are reviewed and evaluated in view of microspacecraft applications. As a result of this review, several technology needs are identified, pointing to the requirement for substantial additional development efforts if this valve technology is to be considered a candidate for future micropropulsion designs.

II. Microspacecraft Valve Requirements

An attempt is made in this section to present a set of representative requirements for valves suitable for use on future microspacecraft. Besides obvious restrictions with respect to size and weight, power consumption, voltage requirements, and actuation time (defined here as the time to open the valve fully), as well as leakage, valve seating pressures, and filtration requirements need to be considered.

It should be noted that, given the preliminary nature of microspacecraft designs, no clear valve design guidelines have been established yet and many of the requirements listed will certainly be subject to further review as microspacecraft designs progress and become more concrete. Also, the set of requirements presented here is not complete. Mission-specific requirements such as vibrational and thermal requirements are ill defined at this point. In addition, without the knowledge of a concrete overall propulsion system layout, the requirements listed here have to be somewhat generic.

However, despite these limitations, the list of requirements provided below may be considered adequate in the context of the scope of this study. As shown below, the specifications of many of the MEMS valve types reviewed here are falling far out of the range of requirements listed in this chapter, so that the level of detail at which these requirements are presented here is thought to be sufficient.

A. Size and Weight

With current, conventional machining techniques it is possible to machine solenoid valves having a cylindrical envelope about 1 cm in diameter and 1 cm in height or slightly less. MEMS valve technology, even when individually packaged, should stay within this envelope. Current miniature solenoid valve masses are as low as about 10 g or slightly less. MEMS valves, fully packaged, should stay at least within this mass margin or should weigh significantly less. For MEMS valves, the package may easily weigh more than the silicon valve mechanism. Thus, weight and volume savings for MEMS valve technology will most likely occur when several valves will be required to be assembled into a system and direct silicon-to-silicon bonding can be exploited in the integration of those components.

B. Power Consumption

Available power levels on microspacecraft will be severely limited.[2] The power consumption of a microvalve should probably not exceed a few watts. Power consumption may also need to be constrained to prevent valve thermal management problems, such as excessive self-heating of the valve and propellant. If possible, latching valve mechanisms should be explored that require power only during the actual opening or closing process.

C. Voltage

Typically, current spacecraft have bus voltages of 28 V. For microspacecraft, bus voltages are expected to be much lower. The MTD II design shown in Fig. 1 has a maximum bus voltage of ± 15 V. Voltages of no more than 5 V are expected in future microspacecraft designs.[2] Microvalves to be used on such craft should be able to operate with these voltages. The possibility exists to provide MEMS-based transformer technology for valve actuation mechanisms that require higher voltages. However, in view of the size and weight constraints alluded to above, a tight integration of this power conditioning circuitry with the valve concept would be required in that case.

D. Minimum Valve Cycle Time

Minimum valve cycle times (defined here as the minimum time required to open and reclose the valve) are an important valve performance parameter for propulsion applications in view of minimum impulse bit requirements. Impulse bit (I_{bit}) is defined as

$$I_{\text{bit}} = \int_{t_{\text{on}}}^{t_{\text{off}}} F(t)\,\mathrm{d}t \tag{1}$$

where t_{on} and t_{off} are the times at which the valve opens and closes, respectively, and $F(t)$ is the thrust force of the rocket engine, typically time dependent over a valve cycle. For microspacecraft applications, the impulse bit has to be minimized since otherwise the rate of turn of the spacecraft becomes too high, too many thruster firings will be required to maintain a certain dead band (pointing accuracy), thus wasting fuel, and the attitude of the spacecraft may be difficult to control. Impulse bit requirements as low as 10^{-6} Ns have been estimated in the past.[2] Again, these estimates are to be considered preliminary at this point.

If the thruster can be scaled down to provide a thrust level of 1 mN, valve cycle times of 1 ms must follow given Eq. (1), assuming a constant thrust level over the valve cycle for the sake of simplicity of the argument at this point. Valve cycle times in the 1- to 10-ms range are achievable with fast-acting miniature solenoid valves today. Using MEMS technologies, nozzle throat diameters can likely be scaled down to a point where only a fraction of a millinewton can be provided, so that valve cycle times in the range of 1–10 ms appear acceptable.

E. Pressure Requirements

Valve pressure requirements will be determined by the propellant tank (feed) pressure, the basic type of feed system (blow down or regulated), and the location of the valve in the feed system. The highest valve pressure, at the propellant tank, may range up to 300 psi in liquid propellant systems. Gaseous propellant systems (as well as pressurization systems for the expulsion of liquid propellants) may require storage pressures ranging up into the 10,000-psi range to mitigate the size of the tank.

F. Leakage

Every valve has a certain degree of internal leakage through the valve seat. For space-qualified valves on conventional spacecraft, leak rates of about 10^{-3}–10^{-4} scc/s GHe (gaseous helium) have been found to be adequate. In general, leak rate concerns are much more severe for gaseous than for liquid propellants. Leak rate requirements will also be more severe for microspacecraft than for conventional-sized craft. This is because the overall propellant supply onboard a microspacecraft will be limited. For a given mission profile (defined by the delta-v of the mission), the required propellant mass scales with the spacecraft mass,

$$M_p \propto M_{\text{S/C}} \tag{2}$$

following the rocket equation. Furthermore, the propellant fraction xM_p lost due to leakage scales with the leak rate LR(t) and mission duration Δt,

$$xM_p \propto \mathrm{LR}(t)\Delta t \tag{3}$$

Here, the leak rate LR(t) may be a function of time due to the fact that propellant tank pressures may change, affecting leak rates, or due to a larger number of contaminates that may locate themselves on the valve seat as the mission wears on and more propellant flows across the seat, depositing these contaminates and, potentially, increasing leak rates. Given Eq. (2), Eq. (3) implies that the required leakage rate scales with the spacecraft mass. If a leak rate of 10^{-3} scc/s GHe is acceptable for a conventional 500-kg-class spacecraft, leak rates will have to be reduced to about 10^{-6} to 10^{-5} scc/s for microspacecraft ranging in the 1- to 10-kg class to result in the same mass fraction of propellant lost due to leakage. If, on the other hand, the same leakage rate was to be maintained, a correspondingly larger fraction of the propellant would be lost due to leakage, and consequently a larger amount of propellant would have to be loaded to offset this loss.

G. Liquid Propellant Compatibility

Achieving leak rates as low as specified with the limited actuation forces available for MEMS valves will be an extraordinary challenge. Therefore, the estimations performed here may imply the use of liquid propellants for long-duration microspacecraft missions for which leak rates are substantially lower than for gaseous propellants (besides other advantages such as higher storage densities, resulting in a reduced tank volume and mass). Therefore, MEMS microvalves will likely have to be compatible with liquid propellants. Some of these propellants (e.g., hydrazine) may not be compatible with typical silicon-based MEMS valve designs. In such cases, impermeable inert film coatings would need to be explored, or propellant-wetted portions of the valve would have to be constructed from other, compatible materials, potentially resulting in "hybrid" (silicon plus other materials) MEMS designs.

H. Valve Seating Forces

Internal valve leakage through valve seats can be reduced by increasing forces exerted by the valve mechanism onto the valve seat. In the case of soft seats, contaminants that may settle on the valve seat may be pushed deep into the seat material where they no longer can provide a leakage path, whereas in the case of hard seats stronger sealing forces may crush contaminates, thus reducing leakage. For conventional soft-seat valves, valve seating pressures of several hundred to several thousand pounds per square inch for high-pressure valve applications are typical. However, most presently available MEMS valves feature harder seats. For hard-seat applications, seating pressures well in excess of 100,000 psi are desirable.

Given these requirements, and the limited actuation forces available for MEMS valves, using MEMS valves in space propulsion applications is sometimes regarded as a futile attempt. Note, however, that seating pressures, rather than total actuator forces, are crucial in this application. Since MEMS does offer the opportunity to machine extremely narrow valves seats, seating pressures may be increased through a reduction of valve seating area alone. In addition, narrower seats will reduce the probability of contamination since less area of the valve will be contamination sensitive.

I. Filtration

It would seem prudent to integrate a filter at microvalving inlet ports, as is common practice in today's thruster valves to prevent contamination of the valve. No clear design rules for filter rating determination exist. In general, the contaminant particle size has to be significantly smaller than the valve stroke and seat width, requiring adequate filtration upstream of the valve with a filter rating correspondingly lower than the seat width and valve stroke.

III. MEMS Microvalve Survey

In this section, different MEMS valve technologies currently available or under considerable development are reviewed, including MEMS valve mechanisms based on thermopneumatic, bimorph, memory alloy, electrostatic, piezoelectric, and electromagnetic actuation. Several pneumatic valve concepts also exist; however, they are of considerably lesser interest for space applications since a separate gas supply would be needed to operate these valves, leading to system complexities and added weight. MEMS check valves have been fabricated and may see use in specific applications. Normally open valves are not considered in this review since they will require power to be held closed, leading to high power consumptions over the course of a mission and reliability concerns (loss of power will cause the valve to open). Following this review, these valve technologies will be evaluated in terms of the requirements listed above.

A. Thermopneumatic Valves

Thermopneumatically actuated valves were first designed and built by Angell and Zdeblick at Stanford University in the late 1970s and early 1980s.[5–7] Later, Zdeblick founded Redwood Microsystems Corp. and produced this valve type commercially.[7] The principle of operation is illustrated in Fig. 2. A liquid is trapped inside a cavity that is being formed by a recess in a silicon wafer and a Pyrex cover wafer anodically bonded to the silicon. The whole assembly is bonded to a glass substrate via a fulcrum joint fabricated into the silicon wafer. An electric heater, deposited onto the Pyrex wafer rather than the silicon wafer for better thermal insulation, heats the fluid to its boiling point. Virtually any fluid can be used and operating parameters of the valve will change with the choice of fluid. Redwood uses a class of so-called 3M Fluorinert liquids with boiling points ranging between 56 and 253°C.[7] The increasing vapor pressure inside the cavity causes a thin silicon membrane to bow outward. This "ballooning" effect causes the poppet of the valve to raise off the seat, opening the valve. Besides the normally closed valve design shown in Fig. 2, a normally open valve has also been fabricated.[5,6]

In theory this valve type can be used with any liquid or gas that is compatible with the valve materials used. However, heat transfer from the actuator cavity into a liquid propellant could have an effect on the operation of the valve, draining energy to heat and actuate the valve, or cause forced-convective cooling, potentially causing deactuation of the valve. These effects would be much more pronounced with higher-conductivity liquid propellants than gaseous propellants. Valve performances are listed in Table 2. Valve operation at pressures up to 3000 psi has been reported,[5] however, for a normally open valve. Actuation forces of up to 20 N have apparently been demonstrated in the case of such a normally open valve version.[5] Valve operation at 100 psi appears to be more typical.[5,7] Valve strokes of up to

Table 2 Typical Redwood valve performance characteristics

Parameter	Representative performance data
Pressure, psi	100 (3000 psi for NO valve)
Power, W	2
Weight, g	4.5
Size, cm^3	$0.63 \times 0.66 \times 0.2$
Response time, ms	400
Stroke, μm	150
Flow rate, sccm	Up to 15,000
Reference nos.	5, 7

150 μm can be reached,[5] and power levels to open the valve range between 0.5 and 2 W.[5,7] The large valve strokes allow for a considerable flow rate capability. Valves have been built that are able to handle up to 15 slpm N_2 at 100 psi,[5] however, 2 slpm (2000 sccm) is a more typical flow rate.[4] The response time of the valve is slow, about 400 ms at power levels of 2 W according to Ref. 7. Higher power levels will allow for faster valve actuation, however, as a result closing speeds are even slower due to the longer times required to cool the valve. When packaged, the valve weighs about 4.5 g and fits into a volume of $0.63 \times 0.66 \times 0.2\ cm^3$.[5,7]

Issues with this valve technology include its limited operating temperature range. A typical Redwood valve actuates around 50–60°C,[7] limiting its operating temperatures to values less than that. As mentioned, actuator fluids with higher boiling

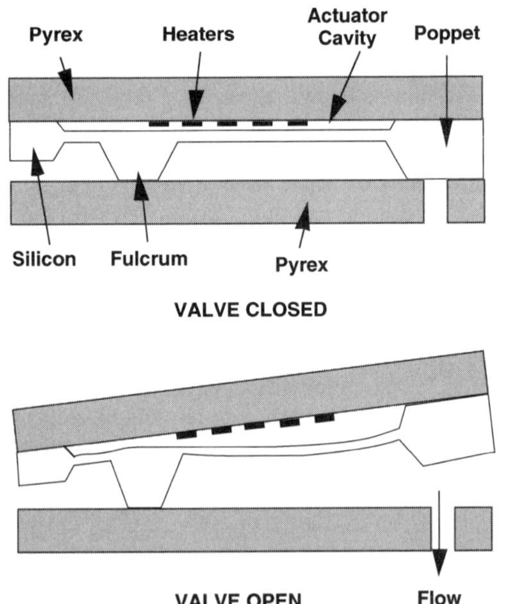

Fig. 2 Thermopneumatic valve concept. Adapted from Ref. 7.

points could be used that would extend the operating temperature range of the valve, however, likely at the expense of higher power values to actuate the valve.

Another limitation of this valve is the use of silicon-to-glass valve seats. Any contaminant that may locate itself on the seat may cause the valve to stay open and cause leakage. Unlike soft-seat materials, which may embed the contaminant particle, or harder seating materials, which would allow knife-edge seals to be fabricated to crush contaminants, the flat poppet in this valve, combined with the fact that no thermopneumatic forces are exerted in the closed state, appears to be relatively vulnerable to contamination. Note also that the cantilevered poppet movement does not provide a self-aligning seat design.

Attempts are under way at the California Institute of Technology[8] to address seat issues through the use of soft-seat silicon rubber. In similar devices explored at Caltech the actuator cavity is sealed with a silicone rubber membrane that has been molded in place.[8] Only a normally open valve has been manufactured so far. As the working fluid in the actuator cavity expands upon heating, the silicon rubber membrane expands until it touches the seat, sealing the valve. The valve was operated against 20-psi pressure at power levels as low as 0.28 W. Valve strokes were 100 μm,[8] although the membrane is capable of significant larger deflections, "ballooning" up to 1-mm diameters.[8] Unfortunately, silicone rubber is permeable to the working fluids used in the experiments (ispropanol and PF5060, an industrial version of Fluorinert). Future work thus focuses on proper sealing of the rubber material through additional coatings. Also, as mentioned, normally open valve versions are of little interest to the space community.

B. Bimorph Valves

Several types of bimorph valves have been explored to date and have been made available commercially in the past. Hewlett–Packard[7,9] and IC Sensors[7,10,11] in the United States and Robert Bosch GmbH[12] in Germany have conducted work with this valve type. As with other thermally actuated valve concepts in this section, the bimorph valves are suitable only for gaseous operation. The Hewlett–Packard concept is shown in Fig. 3 to illustrate the concept. This valve concept was developed by Barth et al.[9] It features a nickel–silicon bimorph membrane actuator. The nickel and silicon membrane thicknesses are both between 25 and 50 μm. As can be seen by inspecting Fig. 3 more closely, the membrane is bent slightly outward

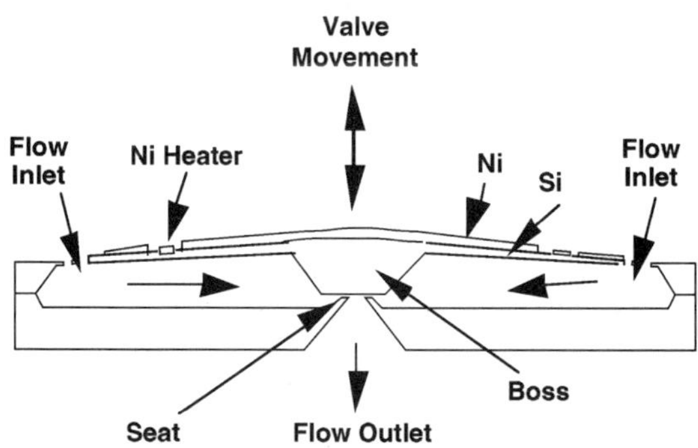

Fig. 3 Bimorph valve, Hewlett–Packard concept. Adapted from Ref. 9.

in the closed position through the use of a central boss that is slightly higher than the sealing ring along the perimeter of the valve body, thus providing spring forces aiding in sealing the valve. The valve appears to have been fabricated through a series of anisotropic etches, combined with silicon fusion bonding steps and metal deposition sequences.

Actuation of the valve follows by passing an electric current through a heater, also made from nickel and deposited onto the membrane. As the membrane heats up, aided by the excellent thermal conduction through the silicon material, the nickel layer, due to its higher coefficient of thermal expansion, extends relative to the silicon membrane and bends the membrane outward, thus opening the valve. Flow inlet occurs along the perimeter of the valve, through gaps in the membrane. The membrane design is elaborate, featuring torsion bars reducing the force requirements to flex it.

The Hewlett–Packard valve has been operated at pressures between 5–200 psi and flow rates between 0.1 and 1000 sccm.[9] Valve response times are about 100 ms,[9] with total valve cycle times being longer, due to the time required to cool the valve. Power requirements are of the order of 1 W (Ref. 9) and voltage requirements are up to 15 V to open[9] the valve at 100 psi. No leak rate information is given in the literature. Valve strokes are of the order of 50–100 μm.[9] The valve seat is small, consisting of a 20-μm-wide rim surrounding a 200-μm^2 orifice.[9] The reasons for this design, however, are thermal in nature: by reducing the contact area between the seat and the valve poppet (boss), heat losses into the remainder of the valve structure are minimized, reducing the power requirements to actuate the valve.

Another bimorph valve type was developed by Jerman at IC Sensors Corp.[7,10,11] The valve relies on the same principle as the Hewlett–Packard valve shown in Fig. 3. However, in the IC Sensors valve, the valve inlet is located off to the side of the membrane at one location along its perimeter. Flow exits the valve through the lower wafer, as in the case of the Hewlett–Packard valve. The membrane thickness is typically about 10 μm. A thick aluminum layer (5 μm) deposited onto the top silicon wafer forms a bimorph structure with the underlying silicon membrane. In the IC Sensors valve design, diffused doped silicon resistors inside the silicon membrane act as heating elements for the bimorph membrane structure. As in the case of the Hewlett–Packard valve, the aluminum layer, due to its higher coefficient of thermal expansion (CTE), expands to a higher degree upon heating than the underlying silicon layer, thus bowing the membrane upward, raising the boss, and opening the valve. Besides a normally closed valve configuration, a normally open valve has also been fabricated.[11]

The IC Sensors valve has been operated at pressures up to 50 psig.[10] Reported leak rates are somewhat ambiguous. At 30-psi inlet pressure, leak rates of 3×10^{-4} scc/s have been reported,[10] while at 5-psi inlet pressures the reported leak rate was 5×10^{-4} scc/s,[10] i.e., higher than for the higher pressure value. Maximum flow rates of up to 150 sccm were reported.[10] Valve response times range between 100 and 300 ms to open the valve fully,[10] depending on the power level. Additional time is required to close the valve by cooling, leading to total valve cycle times of about 250–450 ms.[10] Power requirements for this valve are given as 0.5 W.[7] The total package weight of the IC Sensors valve is 5.8 g.[7]

Both bimorph valves, as the previously introduced thermopneumatically valves, suffer from the risk of unintended valve opening if valve temperatures rise too

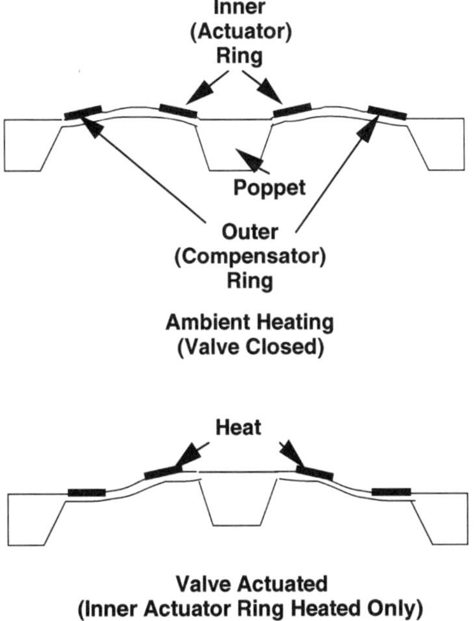

Fig. 4 Bimorph valve concept by Bosch with thermal compensation. Adapted from Ref. 12.

high, causing the bimorph actuation mechanism to go into effect. A third bimorph valve type was developed at the automotive supplier Robert Bosch Company[12] in Germany. Here, two separate aluminum heater rings are deposited onto the silicon membrane as shown in Fig. 4. If the valve heats up due to ambient heat influx, both heater rings go into tension, causing the membrane to buckle without lifting the poppet off the seat. Actuation of the valve is achieved by just heating the inner ring, causing the same bimorph actuation as in the case of the previously mentioned valves.

The Bosch valve has demonstrated flow rates of 5 sccm at 10- to 100-kPa pressure.[12] Power requirements for the valve are of the order of 1 W at the required flow rate.[12] Total valve cycle times are 100 ms, with the actuation time to open the valve fully being about 50 ms.[12] The valve stroke is small, only 8 μm. The valve chip size is $1 \times 0.6 \times 0.13$ cm^3. All three bimorph valve performances are listed in Table 3.

C. Shape-Memory Alloy Valves

Shape-memory alloy valves have been developed by the Microflow and TiNi alloy companies.[7,13–15] Microflow no longer exists and the Microflow valve design, with some changes, is now being marketed by the TiNi Company.[7] A schematic of the valve design is shown in Fig. 5. The valve consists of three silicon wafers. The first silicon wafer features the valve seat and the valve outlet. The second wafer features the shape-memory alloy actuator and silicon poppet. The third

Table 3 Typical performance characteristics of bimorph valves

Parameter	Representative performance data		
	Hewlett–Packard	IC Sensors	Robert Bosch GmbH
Pressure, psi	5–200	1–50	0.15–15
Power, W	1	0.5	1
Weight, g	5.8	5.8	—
Size, cm^3	$2.3 \times 1.7 \times 0.6$	$2.7 \times 2.3 \times 1.1$	$1 \times 0.6 \times 0.13$
Response time, ms	100	100–300	50
Stroke, μm	50–100	25	8
Flow rate, sccm	0.1–1000	Up to 150	5
Reference No(s).	9	7, 11	12

wafer contains a silicon spring that pushes the shape-memory actuator and silicon poppet onto the valve seat back into its closed position.

Shape-memory alloy can be deformed plastically at low temperatures, as accomplished by the spring action in the valve shown in Fig. 5. Upon heating above its so-called transition temperature, the alloy "remembers" its original, or parent, state and returns to this state. The parent state is established through a previous high-temperature anneal of the material during fabrication.[14] According to Ref. 14, the shape-memory actuator is formed by sputter deposition from a nickel–titanium

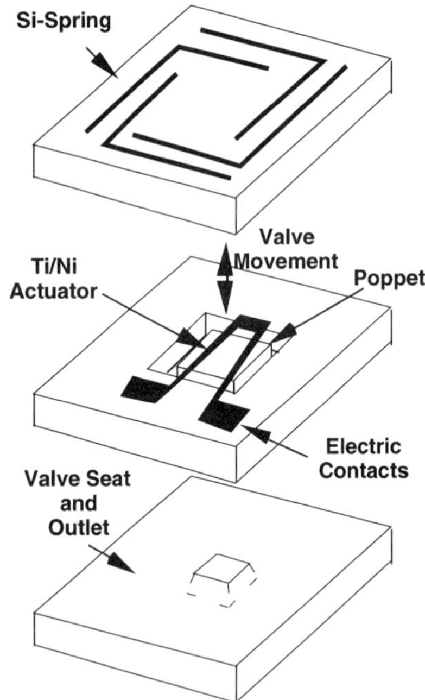

Fig. 5 Shape-memory alloy valve concept (TiNi Company). Adapted from Ref. 13.

Table 4 Typical performance characteristics of shape-memory alloy valves

Parameter	Representative performance data
Pressure, psi	100–400
Power	0.3–2
Response time, ms	1 to open 20 to close
Flow rate, sccm	Up to 6000
Leak rate, sccm	0.01
Reference Nos.	7, 13, 15

target. The resulting film then behaves like the bulk shape-memory alloy material. Heating of the Ti/Ni actuator is accomplished by passing a resistive current through it. Actuator film thicknesses range between 4 and 10 μm according to Ref. 14. The silicon spring returns the valve to its closed position after the current through the Ti/Ni membrane has been switched off.

Shape-memory alloy valves have been operated between 100 and 400 psi of inlet pressure,[7,13] have achieved maximum flow rates of 6000 sccm, and have response times of about 1 ms to open and 20 ms to close.[15] Overall valve cycle times are long again due to the time required for cooling the valve, causing it to close. Power requirements are quoted as 0.3–2 W.[7,13] Leak rates of 0.01 sccm have been measured.[15] Valve performances are listed in Table 4.

D. Electrostatic Valves

Several types of electrostatic valves have been studied. One of the more promising designs was developed by the Massachusetts Institute of Technology (MIT) in collaboration with the aforementioned Robert Bosch Company in Germany.[16–19] The Bosch Company is a major European automotive supplier and required a microvalve design suitable for use in hydraulic systems with a pressure handling capability of up to 15 MPa, or about 2000 psi.[19] The conceptual valve design is shown schematically in Fig. 6, adapted from Refs. 16–18. The moving part of the valve is equipped with a flange featuring the valve seat. This flange is about equal in size (3–3.5 mm in diameter) to the membrane to which the valve stem is attached. Pressure forces acting on the membrane, exerted by the fluid to be controlled, almost cancel each other with pressure forces acting on the flange. (The fact that the cavity underneath the membrane may be evacuated, and higher pressures may exist at the valve outlet, need to be taken into account when trying to balance the sum of these pressure forces.) Thus, only small actuation forces are required to move the valve to its open position, even for relatively large fluid pressures.

As a result of this design approach, the cavity underneath the membrane may be evacuated. This is an important design feature in view of the electrostatic actuation of the valve. In the case of electrostatic actuation, a voltage difference is applied between the silicon membrane and the silicon substrate. Insulation between these

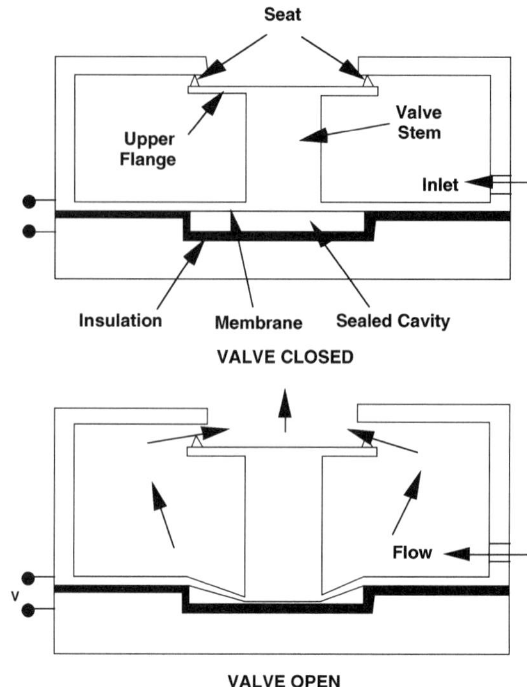

Fig. 6 Electrostatic valve concept by MIT/Bosch. Adapted from Refs. 16–18.

two layers is accomplished by a thin silicon dioxide layer. Since the cavity is evacuated, electrostatic actuation of the valve is possible even when operating the valve with electrically conducting liquids. The shallow cavity limits the valve stroke to 5 μm, which is the depth of the cavity.

Besides using an electrostatic actuation mechanism, the valve may also be pneumatically actuated by alternately pressurizing or evacuating the cavity. While this actuation mechanism may be suitable for the automotive industry, space-operated valves typically do not rely on pneumatic actuation due to system complexities and associated weight penalties.

This valve type has been successfully operated at pressures up to 60 psi, too low for space applications and requiring actuation voltages higher than 200 V. Since actuation is accomplished electrostatically, power consumption of this valve should be very low in the open valve position, determined only by small leakage currents through the insulating oxide layer and the applied voltage (higher power levels, although only over comparably short actuation times, will be required to open the valve). However, no power values were found in the literature. Leak rates were estimated to be lower than 6×10^{-3} scc/s at 35 psi obtained with a pneumatically actuated valve version.[18] Accurate determination of leak rates was not possible, however, due to measuring instrument limitations. Note that while the pressure-balancing effect allows the valve to be operated at higher pressures than may have been possible otherwise, the same effect also limits the sealing forces of the valve since the pressure forces of the liquid are no longer being exploited in sealing the valve.

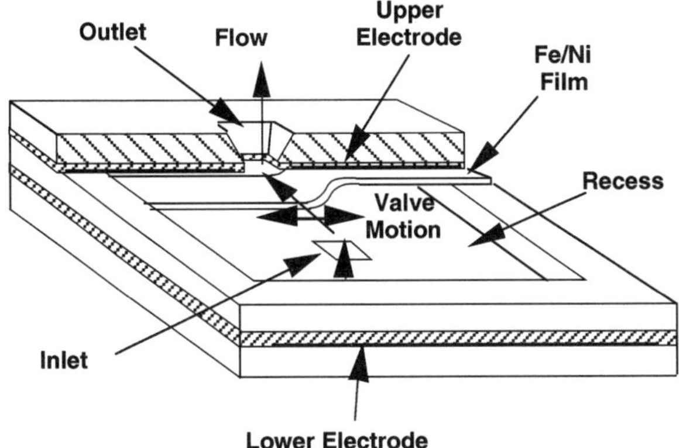

Fig. 7 Electrostatic valve concept by Hitachi. Adapted from Ref. 20.

Other electrostatic valve designs are given in the literature.[20,21] Figure 7 shows an electrostatic valve design by Hitachi, Ltd.,[20] developed for use in molecular beam epitaxy equipment. Here, a thin, oversized Fe–Ni film is placed between two silicon wafers coated with an insulating silicon oxide layer. The silicon wafers contain two embedded electrodes. Since the Fe–Ni film is oversized, it forms an S-shaped structure when placed between the two wafers. Applying an electrostatic potential to one of the two embedded electrodes while keeping the film grounded at all times moves the film toward the electrode to which the potential has been applied. As a result, the S-shaped turn moves across the gap, opening or closing the valve (compare with Fig. 7). Applied voltages are of the order of 100 V. The valve to flow about 10 sccm at pressures of up to 1 atm, suitable for the intended application but far out of the realm of pressures required for space propulsion applications.

Several normally open electrostatic valve designs also exist. One such design was developed by Ohnstein et al.[21] at Honeywell and is shown in Fig. 8. Here, a cantilevered beam is deflected by applying an electrostatic force between an electrode formed by a conductive layer embedded in the beam and another electrode embedded in the silicon material surrounding the seat. These metal electrodes are formed between silicon nitride passivation layers (shaded regions). Since the beam is formed through the removal of a sacrificial layer located between it and the valve

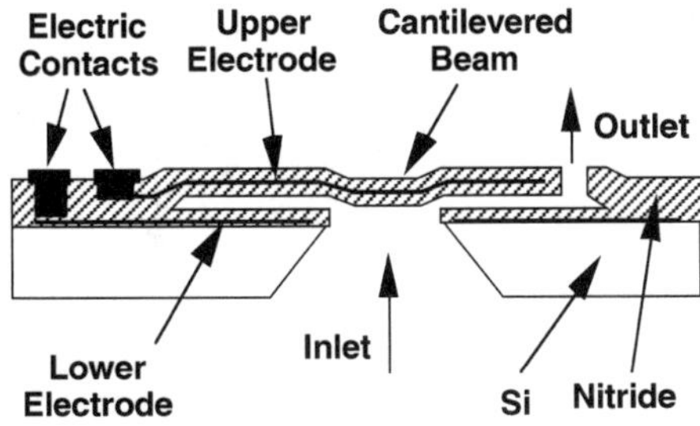

Fig. 8 Electrostatic valve concept by Honeywell. Adapted from Ref. 21.

seat, the valve is normally open and requires a constant force to bend the beam downward to close the valve.

The valve could be closed with 30 V applied against pressures of about 2 psi and held closed against pressures of about 14 psi. At a voltage of 30 V, valve leakage was 6×10^{-2} sccm. The pressure handling capability is very poor in view of microspacecraft applications, and a normally open valve concept, even considering the low power consumption (only a small leakage current flows between the electrodes), is not very useful for space applications since this valve would fail open in the case of a power failure.

Another normally open electrostatic valve is mentioned here because of useful system design aspects that were developed to ease its integration. Kluge et al.,[22] at the Fraunhofer Institute for Solid State Technology in Germany, provided their normally open electrostatically actuated valve, relying on a similar principle of membrane deflection as in the case of the Ohnstein design, with an appropriate transformer circuitry. Even though the valve requires 200 V to actuate, the transformer circuitry requires only 5 V and provides the actuation voltage to the valve. The power consumption of the valve is 0.5 mW, with the transformer circuitry requiring another 72 mW. The valve operates at pressures up to 10 atm and is able to conduct flow rates of up to 700 sccm at these pressures.[22] The chip is $0.6 \times 0.6 \times 0.1$ cm^3 in size. The valves are assembled on a wafer level using a low-temperature silicon fusion bonding technique.[22]

All electrostatic valves are still in the very early stages of their development. In can be noted, however, that even though the valves are fast (an advantage for space applications), electrostatic valves are unable to operate at very high pressure levels due to the limited forces that can be provided by electrostatic means and thus appear not to be very useful for space propulsion applications.

E. Piezoelectric Valves

A piezoelectric microvalve design by Esahi et al.[23] is shown in Fig. 9. The valve consists of a Pyrex wafer featuring the valve outlet and the valve seat, a silicon wafer, featuring a so-called movable valve mesa and integrated knife-edge sealing ring, and the piezoelectric actuator mechanism. The piezoelectric actuator is placed between the valve mesa air-fixed silicon plate, bonded to the valve body

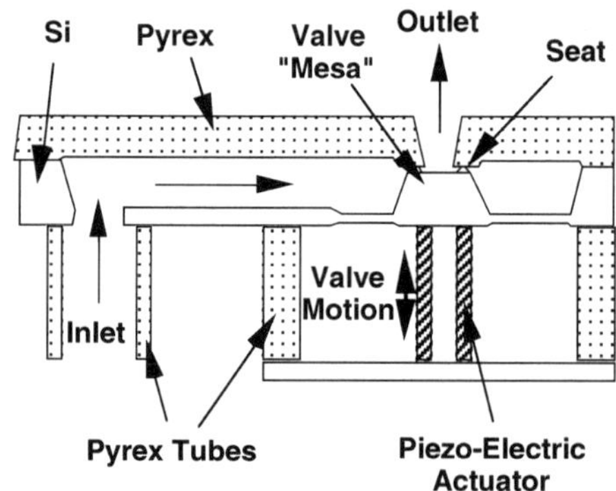

Fig. 9 Piezoelectric valve concept by Esashi et al.[23] Adapted from Ref. 23.

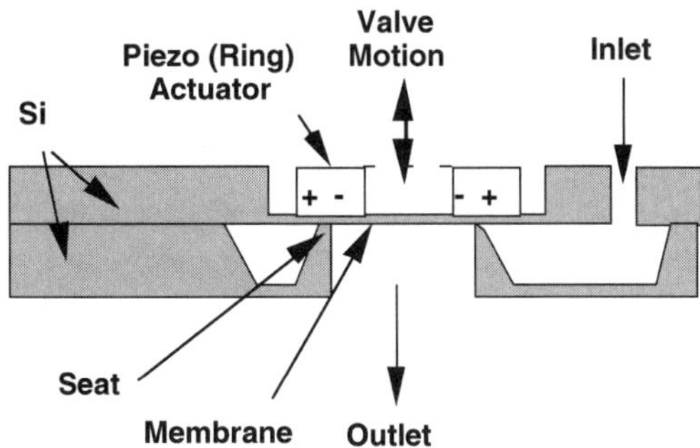

Fig. 10 Piezoelectric valve concept by Stenmark et al.[24–26] Adapted from Ref. 24.

via a 9-mm-long glass tube, using epoxy. The valve is normally closed, and upon applying voltage to the piezoelectric actuator it contracts, thus pulling the movable valve mesa off the valve seat, opening the valve.

Piezoelectric actuation mechanisms are characterized by high voltage requirements comparable to those found for electrostatic valves. The type of piezoelectric valve described above requires between 50 and 100 V to open the valve, depending on the pressure and flow rate. Flow rates through the valve were varied between 0.1 and 90 sccm by varying the voltage between 0 and 100 V at a pressure of 0.75 atm. At 0 V the valve leaks at flow rates of about 0.1 sccm or less, depending on pressure.

Other piezoelectric valves are currently being explored in Europe and in the United States. At ACR Electronic Company, in collaboration with Uppsala University, both in Sweden, piezoelectric microvalves are currently being studied under funding by the European Space Agency (ESA) for use as valves in a micromachined cold-gas thruster quad.[24–26] This valve is still under development and no performance data are available yet. One option being explored in the Swedish study is the use of stacked piezoelectric actuators. Due to this stacking approach, the deflections of each piezoelectric element in the stack are additive, allowing large deflections to be obtained with much lower applied voltages of only 25 V. An added benefit of this arrangement is that piezoelectric actuators provide more force at smaller deflections.[11] Since the deflection of each element is small, relatively high actuation forces may be obtained.

Figure 10 shows a schematic of this valve based on information provided in Refs. 24–26. As can be seen on comparing Figs. 9 and 10, the ACR/Uppsala piezoelectric valve[24–26] appears to be much more compact and robust than the valve by Esashi et al.[23] The Swedish valve also relies on a silicon-membrane deflection effected by a ring-shaped piezoactuator, causing a poppet connected to the silicon membrane to lift off the seat, thus opening the valve. A similar valve concept is also under development at JPL.[27] This activity is in its earliest development stages. One focus area in that study is the development of unique valve seat designs.[27]

F. Electromagnetic Valves

Several types of electromagnetically actuated valves have been studied.[28–31] However, due to current limitations of MEMS machining techniques in providing

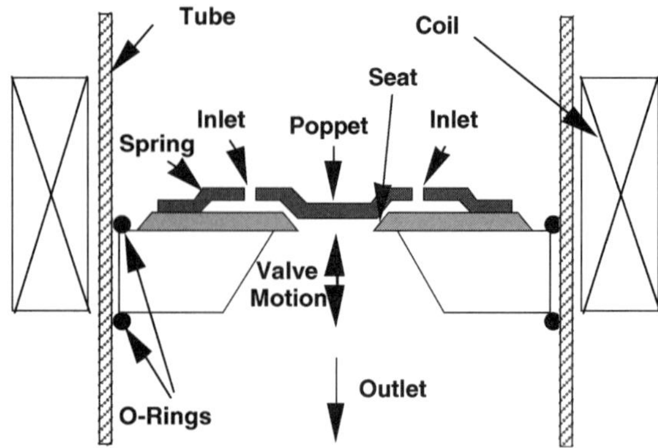

Fig. 11 Electromagnetic valve concept by Yanagisawa et al.[28] Adapted from Ref. 28.

coils with a sufficient number of turns, these valve types typically use external coils or permanent magnets,[28–31] resulting in MEMS-hybrid valve versions. One such type is shown in Fig. 11.[28] This valve type features a valve cap (poppet), integrated with a spiral-shaped spring, fabricated using argon ion beam sputter-deposited thin-film magnetic NiFe material. The spring is connected to the valve body, featuring the seat and the valve outlet. The valve is inserted into a tube and an electromagnetic coil is placed over the outside of the tube. The magnetic field of the coil interacts with the magnetic poppet and moves it up or down, depending on the field and coil current direction. Flow enters the valve through the spring.

The valve is fabricated using a sacrificial layer technique, leaving a gap between the poppet and the valve seat upon removal of this layer, resulting in a normally open valve state. However, valve poppet and spring configurations can be fabricated where the Ni/Fe film experiences a compressive stress, achieved by properly adjusting the argon ion beam energy in the sputter deposition process of the film. In this case, the poppet is pressed onto the valve seat in its nonactuated state (normally closed). Actuating the coil will lift the poppet off the seat, providing proper magnetic field direction.

As can be seen on inspecting Fig. 11, this valve requires a rather specific packaging arrangement, limiting its use in tightly integrated propulsion packages as discussed in Ref. 2. Similar valve concepts were developed by Pourahmadi et al.[29] and Smith et al.[30] No performance data were found in the literature for either of these valve types.

Another electromagnetic valve type, more amenable to integration but still featuring a MEMS-hybrid design approach, was developed by Bosch et al.[31] of Daimler–Chrysler Aerospace, formerly known as Deutsche Aerospace. This valve type is one of very few MEMS valves specifically developed for space applications (others are the Swedish and JPL piezoelectric valves). The Daimler valve concept was targeted for use in an ion engine feed system. A schematic view of the valve is shown in Fig. 12.

The valve consists of a bonded wafer pair. The top wafer features the valve inlet and a recess about 10 μm deep. The recess wall is coated with a conducting electrode and an insulating layer (not specified). The lower wafer features a membrane suspended by four cantilevers. Gold-deposited current paths run along the

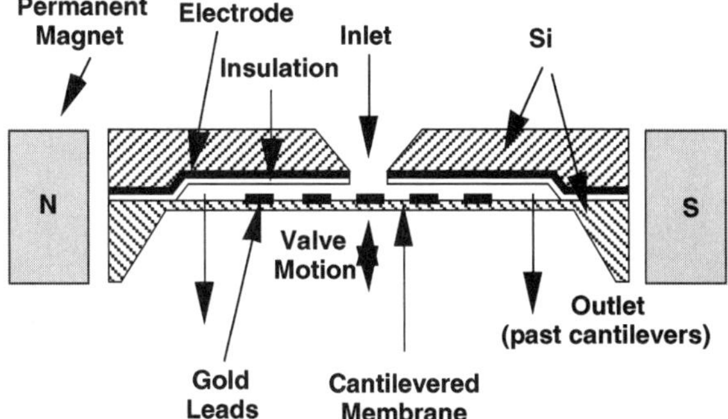

Fig. 12 Electromagnetic valve concept by Daimler–Chrysler/Germany. Adapted from Ref. 31.

top surface of the membrane, carrying a current pointing into or out of the plane of the paper in Fig. 12. This current interacts with a magnetic field generated by two strong, external permanent magnets that are placed next to the chip but not integrated with it. The resulting Lorentz force will cause the membrane to move either up or down depending on the current direction, thus closing or opening the valve. An electrostatic potential may be applied between the electrode embedded in the cavity wall and the conducting current paths on top of the membrane, thus holding the valve in a closed position.

Weak electrostatic forces used to hold the valve in its closed position, in combination with the inlet flow impinging directly onto the membrane, lead to a poor pressure handling capability. The valve can be operated against pressures of 160 mbar (about 2 psi) only and held closed with 30 V applied across the electrodes up to a pressure of merely 300 mbar (about 4 psi). Power requirements for this valve are low, however, ranging around 50 mW, and voltage requirements have been limited to 30 V. Valve strokes are of the order of 10–15 μm and valve response times of less than 1 ms have been estimated.

Thus, although specifically designed with space applications in mind, this valve concept does not appear to meet this goal, with the exception of very low-pressure applications. While the electromagnetic actuation mechanism is an interesting approach, relying on electrostatic forces to keep the valve closed severely compromises the valve design with respect to its pressure handling capability, in addition to providing a potentially severe failure mode, as loss of power would cause the valve to fail open, causing propellant leakage.

G. Check Valves

Check valves are normally closed in the absence of a pressure differential and open only when a pressure differential exists in the free flow direction. Check valves fit specific applications in propulsion systems, and are typically used in bipropellant systems upstream of the propellant tanks, preventing propellant vapors from migrating upstream into the pressurization system for these tanks, where fuel and oxidizer may mix and possibly lead to explosions. Check valves cannot replace command-controlled valves such as the valves discussed in previous sections. Given that bipropellant systems, due to their complexity, high part count, and

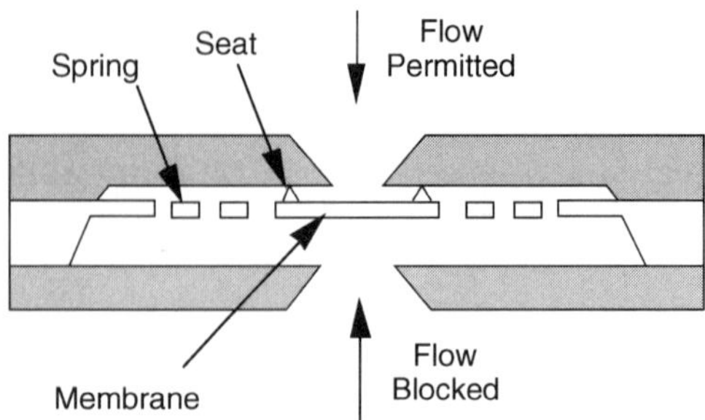

Fig. 13 Check valve concept.

associated weight and volume requirements, may not be ideal candidates for microspacecraft systems, the development of MEMS-based versions of check valves may not be an urgent requirement for microspacecraft.

Briefly, MEMS-based check valves may follow very simple designs.[32,33] One concept is illustrated in Fig. 13. This valve features a spring-loaded, suspended membrane. If flow enters the valve through the top wafer, the membrane is pushed downward, away from the inlet, and the valve will allow flow to pass through it, past the membrane suspensions. If the flow direction is reversed, however, the membrane will be pushed against a valve seat (in this case constituted simply by the flat silicon substrate surface of the top wafer in Fig. 13) and the valve will seal.

The concept shown is merely representative of others. In Ref. 32, for example, the concept is based on a cantilevered beam acting in much the same way as the suspended membrane in Fig. 13. The problem with a cantilevered beam approach is that the beam may not press evenly against the valve seat (not self-aligning) and gaps may form between the beam and the seat, allowing flow to pass even in the closed valve position. However, no leakage data were found for the cantilevered beam check valve concept.[32]

H. Isolation Valves

Isolation valves, such as the commonly used pyrovalves in conventional feed systems, are one-time opening valves (normally closed type) or one-time closing valves (normally open type). Thus, they cannot replace the function of a valve allowing for repeated actuation, but serve critical functions in a propulsion system nonetheless. Isolation valves serve to seal the propulsion system during launch, for example, where valves designed for repeated actuation may shutter, leading to leakage, or seal a propulsion system during long, inactive interplanetary cruises, providing zero leak rates.

A MEMS-based version of such an isolation valve is currently being developed at JPL.[34,35] This valve is silicon-based and fits on a chip $1 \times 1 \times 0.05$ cm^3 in size. A photograph of an early valve prototype is shown in Fig. 14 and the valve concept is shown in Fig. 15. In this valve concept, flow is prevented from exiting the valve prior to actuation by a doped silicon barrier blocking the flow. This barrier is an integral part of the valve structure, machined by etching it into place, and does not feature any seals that may be compromised through contamination

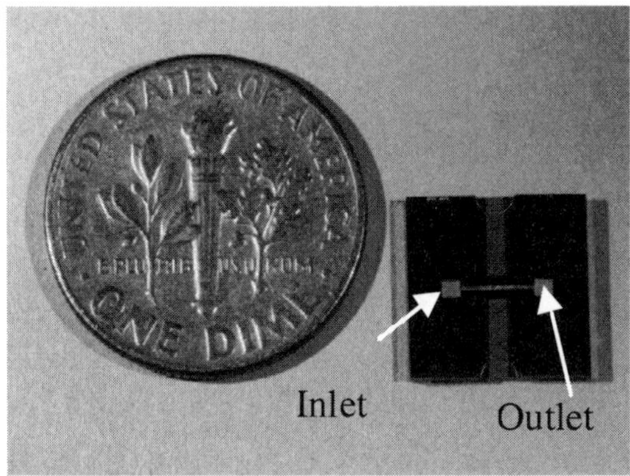

Fig. 14 Micro-isolation valve test chip.

or vibrations experienced by the valve. To actuate the valve, an electric current is passed through the narrow barrier (10–50 μm thick). As a result of the heat dissipation of the current passing through the barrier, causing it to melt and fracture, and the upstream propellant pressure, the barrier is blown away, opening the valve.

The micro-isolation valve is still in its earliest development phases and current emphasis is on proving the feasibility of the valve. Sufficient pressure handling capability, demonstration of valve actuation, and trapping of barrier debris within the valve body, avoiding the contamination of downstream flow components, are considered major milestones in proving the feasibility of this concept. Of these, the first two milestones have recently been accomplished. Valves have been fabricated featuring burst pressures of up to 3000 psig[34] and valves were recently successfully fired, opening within less than 0.5 ms.[35] Valve debris was detected on the downstream side of the opened barrier, as expected, however, this debris appears to stick to channel wall surfaces, a fact that may be exploited in the next crucial step of valve development, seeking to demonstrate debris trapping within

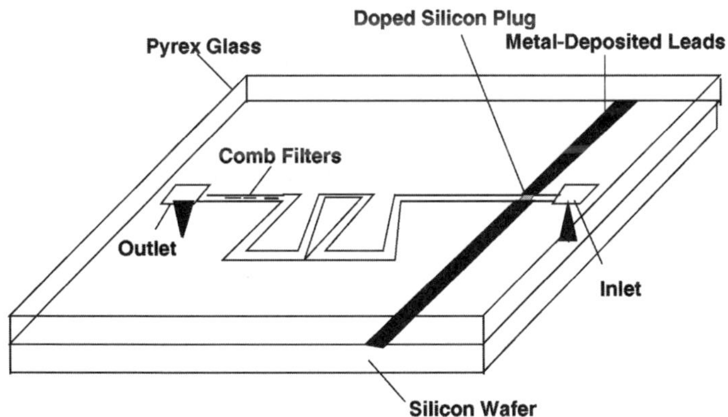

Fig. 15 Schematic of the micro-isolation valve concept.

the isolation valve body. A more detailed description of the results obtained may be found in a companion paper.[35]

I. Pneumatic Valves

Work has been performed on a series of pneumatically actuated MEMS-based valves.[36–39] One of these concepts was introduced earlier, representing a version of the otherwise electrostatically actuated MIT/Bosch Company valve. However, as noted earlier, pneumatically actuated valves are of very limited interest to space applications, as their actuating mechanism requires additional flow management, leading to system complexities and weight penalties. Therefore, pneumatically actuated valves were not considered in this study.

IV. Evaluation of State-of-the-Art MEMS Valves and Future Technology Needs

A. Evaluation of State-of-the-Art Technology

In Table 5 the results of an evaluation of the MEMS valve technologies reviewed in this study is given in view of microspacecraft propulsion applications. The evaluation is guided along the valve requirements for microspacecraft listed in Section II. Given the aforementioned preliminary character of these requirements, the results of this evaluation are kept rather qualitative.

Prior to discussing the specifics of this evaluation, a few notes are in order, putting the results obtained in this evaluation in perspective. As a quick glance at Table 5 reveals, there appears to be no MEMS valve technology existent today meeting all the evaluation criteria listed in Section II. This is not to be understood as a criticism of the valve technology surveyed. Each one of these valves represents a remarkable progress in miniaturization over technology available only a few years ago. The technology produced is testimony to the creativity and originality of their respective innovators. However, most of these valves were designed with terrestrial applications in mind, ranging from tasks within the medical field, semiconductor processing equipment, or automotive applications, among others. Space propulsion applications are unique and differ substantially in their requirements from those aforementioned applications, posing severe constraints on leakage, required actuation times, and robustness. It is therefore not surprising that state-of-the-art MEMS valve technology is lacking with regard to many of these design criteria.

Specifically, while most MEMS valves fairly easily match the mass, volume, and power requirements defined in Section II, several (thermopneumatic, bimorph, and shape memory-alloy) do poorly with respect to valve actuation times, which would lead to long thruster on-times and wide impulse bits. These types of valves also suffer from the risk of uncommanded valve opening if the valve gets too hot due to ambient heating, initiating the actuation mechanism. Only one bimorph design[12] addresses the latter concern.

Several valves have excellent response times (electrostatic, piezoelectric, and electromagnetic valves). However, both electrostatic and piezoelectric valves have very high voltage requirements at their current stage of development, requiring dedicated power conditioning circuitry. Both electrostatic and piezoelectric valves offer only small valve strokes at the present time, which may pose limitations

Table 5 Evaluation of MEMS valve technology for microspacecraft applications[a]

	Thermopneumatic	Bimorph	Shape-memory alloy	Electrostatic	Piezoelectric	Electromagnetic
Size and weight	Excellent	Excellent	Excellent	Excellent	Excellent	Excellent
Power	Good	Good	Good	Excellent	Excellent	Excellent
Voltage	Acceptable	Good	Unknown	Poor	Poor	Acceptable
Cycle time	Poor	Poor	Poor	Excellent	Excellent	Excellent
Pressure	Marginal	Marginal	Marginal	Poor	Unknown	Unknown
Leakage	Poor	Poor	Poor	Poor	Unknown	Unknown
Seating pressures	Acceptable	Acceptable	Acceptable	Poor	Good	Good

[a]Ratings: excellent, good, acceptable, marginal, and poor.

to the use of these valves with very viscous fluids. Electrostatic valves also do very poorly with respect to providing adequate seating forces, resulting in severe concerns with respect to leakage and pressure handling capability.

Leakage concerns, however, are not limited to electrostatic valves alone. All valves considered here do relatively poorly with respect to leakage and rate at best marginally with respect to pressure handling capability, pointing to the need for improved valve seat designs and improved actuation mechanisms.

B. Future Technology Needs

Given the current limitations of MEMS valve designs, new valve developments are clearly needed if this valve technology is to be applied in microspacecraft propulsion systems. While size, weight, and power requirements currently pose no challenge when resorting to microfabricated valves, the following areas of concern stand out when seeking to improve current MEMS valve designs.

1. *Valve Cycle Times*

To achieve fast valve actuation and cycle times, only piezoelectric or electromagnetic valve approaches appear to be appropriate at this stage. Thermally actuated valves are too slow and may overcome their limitations in this regard only to some extent during the opening cycle of the valve if high power levels are applied, shortening the opening cycle. To achieve short actuation times during opening, however, the actuator will have to be well insulated to cause rapid temperature increases. This in turn will lengthen the time required to cool and close the valve. Electrostatic valves, although fast, are unable to provide the required sealing forces. Thus, it appears that future MEMS valve research activities, aimed at providing valves for micropropulsion applications, should target either piezoelectric or electromagnetic actuator mechanisms.

2. *Seating Pressures*

Seating forces or, more precisely, seating pressures of MEMS valves need to be increased to reduce valve leakage and enable higher-pressure operations. While piezoelectric actuators are known to deliver relatively high forces, these are typically delivered only over very short valve strokes. In the case of electromagnetic actuators the need to increase valve actuation forces immediately translates into a high number of turns for the coil. Fabricating such coils using MEMS techniques may pose major challenges. Note, however, that MEMS techniques also offer advantages that may be exploited in achieving high seating pressures. Using MEMS, very narrow valve seats can easily be fabricated, potentially resulting in high seating pressures even if seating forces are limited. Seat design, discussed next, will thus play an important role in future MEMS valve research.

3. *Seat Design*

As discussed seat design may aid in obtaining high seating pressures by resorting to very narrow, "knife-edge" seals. Narrower valve seats will also decrease the likelihood of seat contamination. Knife-edge seating techniques, however, will require the use of very hard materials and self-aligning seats to achieve good valve closure. Such "hard-seat" techniques may provide enough seating pressure to crush contaminates, thus sealing the valve. Another approach may be to resort to "soft-seat" designs. In this approach, the contaminates are not being crushed, as in the previously described hard-seat design, but instead are being embedded in the seat material. One of the soft-seat technologies currently under investigation is the use of silicone rubber materials.[8,36,39] Silicone rubber has shown excellent adhesion to silicon and silicon nitride. One of the problems encountered with silicone rubber material is its permeability with respect to various liquids. Studies are under way to develop composite membranes using silicone rubber and appropriate sealing films.[8]

4. *Integration Aspects*

Future microvalves may have to be tightly integrated with other components in micropropulsion feed systems.[2] This is either due to the desire to achieve very compact propulsion modules, requiring the integration of various flow components, such as thrusters and filters with valves, or due to the need for actuation voltages that may exceed microspacecraft bus voltages, requiring power conditioning chips to be integrated with the valve chips. Piezo-electric valves, for example, typically require voltages in the range of 100–200 V. The integration of power conditioning circuitry with the valve body on a chip level may still allow for an extremely compact microvalve structure to be realized, even for these elevated operating conditions.

5. *Material Compatibility*

Microvalves, due to the use of microfabrication approaches in their construction, may face unique material compatibility issues between the materials of construction and the propellants used. This will be the case particularly for silicon-based microvalves and propellants such as hydrazine, for example. Silicon may be used for reasons of microfabrication heritage and ease of integration between flow components and power conditioning electronics, while hydrazine has traditionally been used for many attitude propulsion applications and may be used for primary propulsion applications on microspacecraft as well.[2] In cases such as these, detailed material compatibility studies will be required, and special erosion-resistant coating techniques (e.g., silicon dioxide), as well as the use of new MEMS materials, may have to be explored.

6. *Additional Valve Types*

A typical propulsion system usually features valves of several functionalities, such as bistable thruster valves, which can be repeatably opened and closed and fail shut in a power-off situation, isolation valves, fill and drain valves, and latch valves, not requiring any power in the open or close position. The valve types currently available all appear to fall in the category of bistable thruster valves, with the exception of one isolation valve reviewed. Development of other valve types, in particular also low power-consuming latching valves, will eventually be needed for microspacecraft propulsion feed systems.

V. Conclusions

State-of-the-art MEMS valve technologies were reviewed in view of microspacecraft propulsion applications. The MEMS valves were evaluated against a set of requirements defined in this study. None of the valve types considered met all the requirements. This is because virtually all MEMS valve technologies existent today were developed for terrestrial applications, in the medical field or semiconductor and automotive industries, for example. Space propulsion requirements are unique and pose severe challenges with respect to valve leakage, actuation times, seating forces, and pressure handling capability. MEMS valves considered for space propulsion will also have to meet severe mass, volume, and power constraints, which, however, are already being met by many commercially available microvalves today.

Piezoelectric and electromagnetic valve actuation mechanisms appear to be the most promising approaches to date due to the achievable fast valve actuation times. However, presently available valve technology based on these actuation mechanisms is severely lacking with regard to leak rate and pressure handling requirements and will require significant additional development efforts to be suitable for use in microspacecraft propulsion systems. These efforts may include actuator design improvements, such as stacked piezoelectric elements or microfabricated high-turn electromagnetic coils, as well as improvements in valve seat design, using soft-seat materials or knife-edge hard-seat designs, chip-level integration of valves with driver and power conditioning electronics, and appropriate coatings of valve internal components exposed to propellants to avoid potential valve material erosion concerns.

The design challenges facing the development of a MEMS valve suitable for applications in space propulsion feed systems may seem daunting. However, the use of MEMS-based design approaches may also provide benefits over more conventional, non-MEMS fabrication techniques that may be exploited in an effort to overcome those challenges. The aforementioned "knife-edge" hard seats, for example, may be manufactured to much smaller dimensions than is possible with conventional fabrication techniques, thus reducing the probability of seat contamination and reducing actuator force requirements. The tight, on-the-chip integration between valve components and driver/power conditioning circuitry would not be possible using nonmicrofabrication methods. Finally, chip-to-chip bonding between flow components, such as MEMS-based thrusters, filters, sensors, and valves, and the required driver and power conditioning circuitry would allow propulsion modules to be realized that may be substantially smaller than are obtainable with any other fabrication method available today. A microfabricated valve would constitute a keystone in such a design approach, which appears to make the further development of MEMS-based valves well worth the associated technical risks.

Acknowledgments

The research described in this chapter was carried out by the Jet Propulsion Laboratory, California Institute of Technology, under a contract with NASA.

References

[1]West, J., "Microelectromechanical Systems (MEMS)/Nanotechnology Studies," Jet Propulsion Lab. Internal Document, JPL-D-13302, Pasadena, CA, Jan. 12, 1996.

[2]Mueller, J., "Thruster Options for Microspacecraft: A Review and Evaluation of State-of-the-Art and Emerging Technologies," *Micropropulsion for Small Spacecraft*, Progress in Astronautics and Aeronautics, Vol. 187, edited by M. Micci and A. Ketsdever, AIAA, Reston, VA, 2000, Chap. 3 (this volume).

[3]Jones, R., "JPL Microspacecraft Technology Development (MTD) Program," Jet Propulsion Lab. Internal Document, Pasadena, CA, May 31, 1996.

[4]Strand, L., Toews, H., Schwartz, K., and Milewski, R., "Extended Duty Cycle Testing of Spacecraft Propulsion Miniaturized Components," AIAA Paper 95-2810, San Diego, CA, July 1995.

[5]Zdeblick, M. J., Anderson, R., Jankowski, J., Kline-Schoder, B., Christel, L., Miles, R., and Weber, W., "Thermopneumatically Actuated Microvalves and Integrated Electro-Fluidic Circuits," Technical Digest, Solid-State Sensor and Actuator Workshop, Hilton Head, SC, 1994, pp. 251–255.

[6]Zdeblick, M. J., and Angell, J. B., "A Microminiature Electro-to-Fluidic Valve," *Transducers '87*, 1987, pp. 827–829.

[7]Barth, P. W., " Silicon Microvalves for Gas Flow Control," *Transducers '95*, Stockholm, Sweden, June 25–29, 1995, p. 276.

[8]Yang, X., Grosjean, C., Tai, Y. C., and Ho, C. M., "A MEMS Thermopneumatic Silicone Membrane Valve," MEMS '97, Nagoya, Japan, 1997.

[9]Barth, P. W., Beatty, C., Field, L., Baker, J., and Gordon, G., "A Robust Normally-Closed Silicon Microvalve," Technical Digest, Solid-State Sensor and Actuator Workshop, Hilton Head, SC, 1994, p. 248.

[10]Jerman, H., *IEEE Proceedings*, 91CH2817-5/91, 1991, Inst. of Electrical and Electronics Engineers, pp. 1045–1048.

[11]Jerman, H., "Electrically-Activated Micromachined Diaphragm Valves," *Technical Digest, Solid State Sensor and Actuator Workshop*, Hilton Head, SC, June 4–7, 1990, p. 67.

[12]Franz, J., Baumann, H., and Trah, H., "A Silicon Microvalve with Integrated Flow Sensor," Transducers '95, Stockholm, Sweden, June 1995.

[13]Ray, C., Sloan, C., Johnson, D., Busch, J., and Petty, B., "A Silicon-Based Shape Memory Alloy Microvalve," *Materials Research Society Symposium Proceedings*, Vol. 276, 1992, pp. 161–166.

[14]Busch, J., and Johnson, D., *IEEE Proceedings*, CH2832-4/90, Inst. of Electrical and Electronics Engineers, 1990, pp. 40–41.

[15]Johnson, D., and Bokaie, M., "Valves for Instrumentation and Propulsion Systems in Microspacecraft," *Proceedings, 9th Advanced Space Propulsion Workshop*, JPL D-15671, Jet Propulsion Laboratory, Pasadena, CA, March 1998.

[16]Huff, M., Mettner, M., Lober, T., and Schmidt, M., "A Pressure-Balanced Electrostatically-Actuated Microvalve," *Technical Digest, Solid State Sensor and Actuator Workshop*, Hilton Head, SC, June 1990, p. 123.

[17]Huff, M., Gilbert, J., and Schmidt, M., "Flow Characteristics of a Pressure-Balanced Microvalve," *Digest of Technical Papers*, Transducers '93, Yokohama, Japan, June 1993.

[18]Huff, M., and Schmidt, M., "Fabrication, Packaging, and Testing of a Wafer-Bonded Microvalve," *Technical Digest, Solid State Sensor and Actuator Workshop*, Hilton Head, SC, June 1992, p. 194.

[19]Mettner, M., Huff, M., Lober, T., and Schmidt, M., "How to Design a Microvalve for High-Pressure Application."

[20]Shikida, M., Sato, K., Tanaka, S., Kawamura, Y., and Fujisaki, Y., "Electrostatically-Actuated Gas Valve with Large Conductance," *Digest of Technical Papers*, Transducers '93, Yokohama, Japan, June 1993.

[21]Ohnstein, T., Fukiura, T., Ridley, J., and Bonne, U., "Micromachined Silicon Microvalve," IEEE Proceedings, CH2832-4/90, Inst. of Electrical and Electronics Engineers, 1990, pp. 95–98.

[22]Kluge, S., Klink, G., and Woias, P., "A Fast-Switching, Low-Power Pneumatic Microvalve with Electrostatic Actuation Made by Silicon Micromaching," *American Laboratory*, March 1998, pp. 17–18.

[23]Esahi, M., Shoji, S., and Nakano, A., "Normally-Closed Microvalve and Micropump Fabricated on a Silicon Wafer," *Sensors and Actuators*, Vol. 20, 1989, pp. 163–169.

[24]Stenmark, L., and Lang, M., "Micro Propulsion Thrusters and Technologies," Proceedings, *Second European Spacecraft Propulsion Conference*, ESTEC, Noordwijk, The Netherlands, ESA SP-398, May 1997, pp. 399–405.

[25]Stenmark, L., Lang, M., Köhler, J., and Simu, U., "Micro Machined Propulsion Components," *Proceedings, Second Round Table on Micro/Nano Technologies for Space*, ESTEC, Noordwijk, The Netherlands, ESA WPP-132, Oct. 1997, pp. 69–76.

[26]Stenmark, L., "Micro Machined Cold Gas Thrusters," *Proceedings, Workshop on Low Cost Spacecraft Propulsion Technologies for Small Satellites*, ESA-ESTEC, Noordwijk, The Netherlands, March 1998.

[27]Chakraborty, I., Tang, W., Bame, D., and Tang, T., "MEMS Micro-Valve for Space Applications," *Technical Digest, Transducers '99*, Sendai, Japan, June 1999, pp. 1820–1823.

[28]Yanagisawa, K., Kuwano, H., and Tago, A., "An Electromagnetically Driven Microvalve," *Digest of Technical Papers, Transducers '93*, Yokohama, Japan, June 1993.

[29]Pourahmadi, F., Christel, L, Petersen, K., Mallon, J., and Bryzek, J., "Variable Flow Microvalve Structure Fabricated with Silicon Fusion Bonding," *Technical Digest, Solid-State Sensor and Actuator Workshop*, Hilton Head, SC, June 1990.

[30]Smith, R., Bower, R., and Collins, S., "The Design and Fabrication of a Magnetically Actuated Micromachined Flow Valve," *Sensors and Actuators A*, Vol. 24, 1990, pp. 47–53.

[31]Bosch, D., Heimhofer, B., Mück, G., Seidel, H., Thumser, U., and Welser, W., "A Silicon Microvalve with Combined Electromagnetic/Electrostatic Actuation," *Sensors and Actuators A*, Vol. 37–38, 1993, pp. 684–692.

[32]Tiren, J., Tenerz, L., and Hök, B., "A Batch-Fabricated Non-Reverse Valve with Cantilever Beam Manufactured by Micromachining of Silicon," *Sensors and Actuators A*, Vol. 18, 1989, pp. 389–396.

[33]Lin, Y. C., Hesketh, P., Boyd, J., Lunte, S., and Wilson, G., "Characteristics of a Polyimide Microvalve," *Technical Digest, Solid State Sensor and Actuator Workshop*, Hilton Head, SC, June 1996, pp. 113–116.

[34]Mueller, J., Vargo, S., Forgrave, J., Bame, D., Chakraborty, I., and Tang, W., "Micro-Isolation Valve Concept: Initial Results of a Feasibility Study," *Micropropulsion for Small Spacecraft*, Progress in Astronautics and Aeronautics, Vol. 187, edited by M. Micci and A. Ketsdever, AIAA, Reston, VA, 2000, Chap. 17 (this volume).

[35]Mueller, J., Vargo, S., Bame, D., and Tang, W., "Proof-of-Concept Demonstration of a Micro-Isolation Valve," AIAA Paper 99-2726, 35th Joint Propulsion Conf., Los Angeles, CA, June 1999.

[36]Bousse, L., Dijkstra, E., and Guenat, O., "High-Density Arrays of Valves and Interconnects for Liquid Switching," *Technical Digest, Solid State Sensor and Actuator Workshop*, Hilton Head, SC, June 1996, p. 272.

[37]Sim, D., Kurabayashi, T., and Esashi, M., "Bakable Silicon Pneumatic Microvalve," Transducers '95, Stockholm, Sweden, June 1995.

[38]Lisec, T., Kreutzer, M., and Wagner, B., "A Bistable Pneumatic Microswitch for Driving Fluidic Components," Transducers '95, Stockholm, Sweden, June 1995.

[39]Vieider, C., Öhman, O., and Elderstig, H., "A Pneumatically Actuated Micro Valve with a Silicon Rubber Membrane for Integration with Fluid-Handling Systems," Transducers '95, Stockholm, Sweden, June 1995.

Author Index

PROGRESS IN ASTRONAUTICS AND AERONAUTICS SERIES VOLUMES

***1. Solid Propellant Rocket Research (1960)**
Martin Summerfield
Princeton University

***2. Liquid Rockets and Propellants (1960)**
Loren E. Bollinger
Ohio State University
Martin Goldsmith
The Rand Corp.
Alexis W. Lemmon Jr.
Battelle Memorial Institute

***3. Energy Conversion for Space Power (1961)**
Nathan W. Snyder
Institute for Defense Analyses

***4. Space Power Systems (1961)**
Nathan W. Snyder
Institute for Defense Analyses

***5. Electrostatic Propulsion (1961)**
David B. Langmuir
Space Technology Laboratories, Inc.
Ernst Stuhlinger
NASA George C. Marshall Space Flight Center
J. M. Sellen Jr.
Space Technology Laboratories, Inc.

***6. Detonation and Two-Phase Flow (1962)**
S. S. Penner
California Institute of Technology
F. A. Williams
Harvard University

***7. Hypersonic Flow Research (1962)**
Frederick R. Riddell
AVCO Corp.

***8. Guidance and Control (1962)**
Robert E. Roberson
Consultant
James S. Farrior
Lockheed Missiles and Space Co.

***9. Electric Propulsion Development (1963)**
Ernst Stuhlinger
NASA George C. Marshall Space Flight Center

***10. Technology of Lunar Exploration (1963)**
Clifford I. Cumming
Harold R. Lawrence
Jet Propulsion Laboratory

***11. Power Systems for Space Flight (1963)**
Morris A. Zipkin
Russell N. Edwards
General Electric Co.

***12. Ionization in High-Temperature Gases (1963)**
Kurt E. Shuler, Editor
National Bureau of Standards
John B. Fenn,
Associate Editor
Princeton University

***13. Guidance and Control–II (1964)**
Robert C. Langford
General Precision Inc.
Charles J. Mundo
Institute of Naval Studies

***14. Celestial Mechanics and Astrodynamics (1964)**
Victor G. Szebehely
Yale University Observatory

***15. Heterogeneous Combustion (1964)**
Hans G. Wolfhard
Institute for Defense Analyses
Irvin Glassman
Princeton University
Leon Green Jr.
Air Force Systems Command

***16. Space Power Systems Engineering (1966)**
George C. Szego
Institute for Defense Analyses
J. Edward Taylor
TRW Inc.

***17. Methods in Astrodynamics and Celestial Mechanics (1966)**
Raynor L. Duncombe
U.S. Naval Observatory
Victor G. Szebehely
Yale University Observatory

***18. Thermophysics and Temperature Control of Spacecraft and Entry Vehicles (1966)**
Gerhard B. Heller
NASA George C. Marshall Space Flight Center

***19. Communication Satellite Systems Technology (1966)**
Richard B. Marsten
Radio Corporation of America

---
*Out of print.

***20. Thermophysics of Spacecraft and Planetary Bodies: Radiation Properties of Solids and the Electromagnetic Radiation Environment in Space (1967)**
Gerhard B. Heller
NASA George C. Marshall Space Flight Center

***21. Thermal Design Principles of Spacecraft and Entry Bodies (1969)**
Jerry T. Bevans
TRW Systems

***22. Stratospheric Circulation (1969)**
Willis L. Webb
Atmospheric Sciences Laboratory, White Sands, and University of Texas at El Paso

***23. Thermophysics: Applications to Thermal Design of Spacecraft (1970)**
Jerry T. Bevans
TRW Systems

***24. Heat Transfer and Spacecraft Thermal Control (1971)**
John W. Lucas
Jet Propulsion Laboratory

25. Communication Satellites for the 70's: Technology (1971)
Nathaniel E. Feldman
The Rand Corp.
Charles M. Kelly
The Aerospace Corp.

26. Communication Satellites for the 70's: Systems (1971)
Nathaniel E. Feldman
The Rand Corp.
Charles M. Kelly
The Aerospace Corp.

27. Thermospheric Circulation (1972)
Willis L. Webb
Atmospheric Sciences Laboratory, White Sands, and University of Texas at El Paso

28. Thermal Characteristics of the Moon (1972)
John W. Lucas
Jet Propulsion Laboratory

***29. Fundamentals of Spacecraft Thermal Design (1972)**
John W. Lucas
Jet Propulsion Laboratory

***30. Solar Activity Observations and Predictions (1972)**
Patrick S. McIntosh
Murray Dryer
Environmental Research Laboratories, National Oceanic and Atmospheric Administration

***31. Thermal Control and Radiation (1973)**
Chang-Lin Tien
University of California at Berkeley

***32. Communications Satellite Systems (1974)**
P. L. Bargellini
COMSAT Laboratories

***33. Communications Satellite Technology (1974)**
P. L. Bargellini
COMSAT Laboratories

***34. Instrumentation for Airbreathing Propulsion (1974)**
Allen E. Fuhs
Naval Postgraduate School
Marshall Kingery
Arnold Engineering Development Center

***35. Thermophysics and Spacecraft Thermal Control (1974)**
Robert G. Hering
University of Iowa

36. Thermal Pollution Analysis (1975)
Joseph A. Schetz
Virginia Polytechnic Institute
ISBN 0-915928-00-0

***37. Aeroacoustics: Jet and Combustion Noise; Duct Acoustics (1975)**
Henry T. Nagamatsu, Editor
General Electric Research and Development Center
Jack V. O'Keefe, Associate Editor
The Boeing Co.
Ira R. Schwartz, Associate Editor
NASA Ames Research Center
ISBN 0-915928-01-9

***38. Aeroacoustics: Fan, STOL, and Boundary Layer Noise; Sonic Boom; Aeroacoustics Instrumentation (1975)**
Henry T. Nagamatsu, Editor
General Electric Research and Development Center
Jack V. O'Keefe, Associate Editor
The Boeing Co.
Ira R. Schwartz, Associate Editor
NASA Ames Research Center
ISBN 0-915928-02-7

***39. Heat Transfer with Thermal Control Applications (1975)**
M. Michael Yovanovich
University of Waterloo
ISBN 0-915928-03-5

---
*Out of print.

***40. Aerodynamics of Base Combustion (1976)**
S. N. B. Murthy, Editor
J. R. Osborn,
Associate Editor
Purdue University
A. W. Barrows
J. R. Ward,
Associate Editors
Ballistics Research Laboratories
ISBN 0-915928-04-3

***41. Communications Satellite Developments: Systems (1976)**
Gilbert E. LaVean
Defense Communications Agency
William G. Schmidt
CML Satellite Corp.
ISBN 0-915928-05-1

***42. Communications Satellite Developments: Technology (1976)**
William G. Schmidt
CML Satellite Corp.
Gilbert E. LaVean
Defense Communications Agency
ISBN 0-915928-06-X

***43. Aeroacoustics: Jet Noise, Combustion and Core Engine Noise (1976)**
Ira R. Schwartz, Editor
NASA Ames Research Center
Henry T. Nagamatsu,
Associate Editor
General Electric Research and Development Center
Warren C. Strahle,
Associate Editor
Georgia Institute of Technology
ISBN 0-915928-07-8

***44. Aeroacoustics: Fan Noise and Control; Duct Acoustics; Rotor Noise (1976)**
Ira R. Schwartz, Editor
NASA Ames Research Center
Henry T. Nagamatsu,
Associate Editor
General Electric Research and Development Center
Warren C. Strahle,
Associate Editor
Georgia Institute of Technology
ISBN 0-915928-08-6

***45. Aeroacoustics: STOL Noise; Airframe and Airfoil Noise (1976)**
Ira R. Schwartz, Editor
NASA Ames Research Center
Henry T. Nagamatsu,
Associate Editor
General Electric Research and Development Center
Warren C. Strahle,
Associate Editor
Georgia Institute of Technology
ISBN 0-915928-09-4

***46. Aeroacoustics: Acoustic Wave Propagation; Aircraft Noise Prediction; Aeroacoustic Instrumentation (1976)**
Ira R. Schwartz, Editor
NASA Ames Research Center
Henry T. Nagamatsu,
Associate Editor
General Electric Research and Development Center
Warren C. Strahle,
Associate Editor
Georgia Institute of Technology
ISBN 0-915928-10-8

***47. Spacecraft Charging by Magnetospheric Plasmas (1976)**
Alan Rosen
TRW Inc.
ISBN 0-915928-11-6

***48. Scientific Investigations on the Skylab Satellite (1976)**
Marion I. Kent
Ernst Stuhlinger
NASA George C. Marshall Space Flight Center
Shi-Tsan Wu
University of Alabama
ISBN 0-915928-12-4

***49. Radiative Transfer and Thermal Control (1976)**
Allie M. Smith
ARO Inc.
ISBN 0-915928-13-2

***50. Exploration of the Outer Solar System (1976)**
Eugene W. Greenstadt
TRW Inc.
Murray Dryer
National Oceanic and Atmospheric Administration
Devrie S. Intriligator
University of Southern California
ISBN 0-915928-14-0

***51. Rarefied Gas Dynamics, Parts I and II (two volumes) (1977)**
J. Leith Potter
ARO Inc.
ISBN 0-915928-15-9

***52. Materials Sciences in Space with Application to Space Processing (1977)**
Leo Steg
General Electric Co.
ISBN 0-915928-16-7

*Out of print.

***53. Experimental Diagnostics in Gas Phase Combustion Systems (1977)**
Ben T. Zinn, Editor
Georgia Institute of Technology
Craig T. Bowman, Associate Editor
Stanford University
Daniel L. Hartley, Associate Editor
Sandia Laboratories
Edward W. Price, Associate Editor
Georgia Institute of Technology
James G. Skifstad, Associate Editor
Purdue University
ISBN 0-915928-18-3

***54. Satellite Communication: Future Systems (1977)**
David Jarett
TRW Inc.
ISBN 0-915928-18-3

***55. Satellite Communications: Advanced Technologies (1977)**
David Jarett
TRW Inc.
ISBN 0-915928-19-1

***56. Thermophysics of Spacecraft and Outer Planet Entry Probes (1977)**
Allie M. Smith
ARO Inc.
ISBN 0-915928-20-5

***57. Space-Based Manufacturing from Nonterrestrial Materials (1977)**
Gerald K. O'Neill, Editor
Brian O'Leary, Assistant Editor
Princeton University
ISBN 0-915928-21-3

***58. Turbulent Combustion (1978)**
Lawrence A. Kennedy
State University of New York at Buffalo
ISBN 0-915928-22-1

***59. Aerodynamic Heating and Thermal Protection Systems (1978)**
Leroy S. Fletcher
University of Virginia
ISBN 0-915928-23-X

***60. Heat Transfer and Thermal Control Systems (1978)**
Leroy S. Fletcher
University of Virginia
ISBN 0-915928-24-8

***61. Radiation Energy Conversion in Space (1978)**
Kenneth W. Billman
NASA Ames Research Center
ISBN 0-915928-26-4

***62. Alternative Hydrocarbon Fuels: Combustion and Chemical Kinetics (1978)**
Craig T. Bowman
Stanford University
Jorgen Birkeland
Department of Energy
ISBN 0-915928-25-6

***63. Experimental Diagnostics in Combustion of Solids (1978)**
Thomas L. Boggs
Naval Weapons Center
Ben T. Zinn
Georgia Institute of Technology
ISBN 0-915928-28-0

***64. Outer Planet Entry Heating and Thermal Protection (1979)**
Raymond Viskanta
Purdue University
ISBN 0-915928-29-9

***65. Thermophysics and Thermal Control (1979)**
Raymond Viskanta
Purdue University
ISBN 0-915928-30-2

***66. Interior Ballistics of Guns (1979)**
Herman Krier
University of Illinois at Urbana–Champaign
Martin Summerfield
New York University
ISBN 0-915928-32-9

***67. Remote Sensing of Earth from Space: Role of "Smart Sensors" (1979)**
Roger A. Breckenridge
NASA Langley Research Center
ISBN 0-915928-33-7

***68. Injection and Mixing in Turbulent Flow (1980)**
Joseph A. Schetz
Virginia Polytechnic Institute and State University
ISBN 0-915928-35-3

---

*Out of print.

***69. Entry Heating and Thermal Protection (1980)**
Walter B. Olstad
NASA Headquarters
ISBN 0-915928-38-8

***70. Heat Transfer, Thermal Control, and Heat Pipes (1980)**
Walter B. Olstad
NASA Headquarters
ISBN 0-915928-39-6

***71. Space Systems and Their Interactions with Earth's Space Environment (1980)**
Henry B. Garrett
Charles P. Pike
Hanscom Air Force Base
ISBN 0-915928-41-8

***72. Viscous Flow Drag Reduction (1980)**
Gary R. Hough
Vought Advanced Technology Center
ISBN 0-915928-44-2

***73. Combustion Experiments in a Zero-Gravity Laboratory (1981)**
Thomas H. Cochran
NASA Lewis Research Center
ISBN 0-915928-48-5

***74. Rarefied Gas Dynamics, Parts I and II (two volumes) (1981)**
Sam S. Fisher
University of Virginia
ISBN 0-915928-51-5

***75. Gasdynamics of Detonations and Explosions (1981)**
J. R. Bowen
University of Wisconsin at Madison
N. Manson
Universite de Poitiers
A. K. Oppenheim
University of California at Berkeley
R. I. Soloukhin
Institute of Heat and Mass Transfer, BSSR Academy of Sciences
ISBN 0-915928-46-9

***76. Combustion in Reactive Systems (1981)**
J. R. Bowen
University of Wisconsin at Madison
N. Manson
Universite de Poitiers
A. K. Oppenheim
University of California at Berkeley
R. I. Soloukhin
Institute of Heat and Mass Transfer, BSSR Academy of Sciences
ISBN 0-915928-47-7

***77. Aerothermodynamics and Planetary Entry (1981)**
A. L. Crosbie
University of Missouri-Rolla
ISBN 0-915928-52-3

***78. Heat Transfer and Thermal Control (1981)**
A. L. Crosbie
University of Missouri-Rolla
ISBN 0-915928-53-1

***79. Electric Propulsion and Its Applications to Space Missions (1981)**
Robert C. Finke
NASA Lewis Research Center
ISBN 0-915928-55-8

***80. Aero-Optical Phenomena (1982)**
Keith G. Gilbert
Leonard J. Otten
Air Force Weapons Laboratory
ISBN 0-915928-60-4

***81. Transonic Aerodynamics (1982)**
David Nixon
Nielsen Engineering & Research, Inc.
ISBN 0-915928-65-5

***82. Thermophysics of Atmospheric Entry (1982)**
T. E. Horton
University of Mississippi
ISBN 0-915928-66-3

***83. Spacecraft Radiative Transfer and Temperature Control (1982)**
T. E. Horton
University of Mississippi
ISBN 0-915928-67-1

***84. Liquid-Metal Flows and Magneto-hydrodynamics (1983)**
H. Branover
Ben-Gurion University of the Negev
P. S. Lykoudis
Purdue University
A. Yakhot
Ben-Gurion University of the Negev
ISBN 0-915928-70-1

---
*Out of print.

***85. Entry Vehicle Heating and Thermal Protection Systems: Space Shuttle, Solar Starprobe, Jupiter Galileo Probe (1983)**
Paul E. Bauer
McDonnell Douglas Astronautics Co.
Howard E. Collicott
The Boeing Co.
ISBN 0-915928-74-4

***86. Spacecraft Thermal Control, Design, and Operation (1983)**
Howard E. Collicott
The Boeing Co.
Paul E. Bauer
McDonnell Douglas Astronautics Co.
ISBN 0-915928-75-2

***87. Shock Waves, Explosions, and Detonations (1983)**
J. R. Bowen
University of Washington
N. Manson
Universite de Poitiers
A. K. Oppenheim
University of California at Berkeley
R. I. Soloukhin
Institute of Heat and Mass Transfer, BSSR Academy of Sciences
ISBN 0-915928-76-0

***88. Flames, Lasers, and Reactive Systems (1983)**
J. R. Bowen
University of Washington
N. Manson
Universite de Poitiers
A. K. Oppenheim
University of California at Berkeley
R. I. Soloukhin
Institute of Heat and Mass Transfer, BSSR Academy of Sciences
ISBN 0-915928-77-9

***89. Orbit-Raising and Maneuvering Propulsion: Research Status and Needs (1984)**
Leonard H. Caveny
Air Force Office of Scientific Research
ISBN 0-915928-82-5

***90. Fundamentals of Solid-Propellant Combustion (1984)**
Kenneth K. Kuo
Pennsylvania State University
Martin Summerfield
Princeton Combustion Research Laboratories, Inc.
ISBN 0-915928-84-1

91. Spacecraft Contamination: Sources and Prevention (1984)
J. A. Roux
University of Mississippi
T. D. McCay
NASA Marshall Space Flight Center
ISBN 0-915928-85-X

92. Combustion Diagnostics by Nonintrusive Methods (1984)
T. D. McCay
NASA Marshall Space Flight Center
J. A. Roux
University of Mississippi
ISBN 0-915928-86-8

93. The INTELSAT Global Satellite System (1984)
Joel Alper
COMSAT Corp.
Joseph Pelton
INTELSAT
ISBN 0-915928-90-6

94. Dynamics of Shock Waves, Explosions, and Detonations (1984)
J. R. Bowen
University of Washington
N. Manson
Universite de Poitiers
A. K. Oppenheim
University of California at Berkeley
R. I. Soloukhin
Institute of Heat and Mass Transfer, BSSR Academy of Sciences
ISBN 0-915928-91-4

95. Dynamics of Flames and Reactive Systems (1984)
J. R. Bowen
University of Washington
N. Manson
Universite de Poitiers
A. K. Oppenheim
University of California at Berkeley
R. I. Soloukhin
Institute of Heat and Mass Transfer, BSSR Academy of Sciences
ISBN 0-915928-92-2

96. Thermal Design of Aeroassisted Orbital Transfer Vehicles (1985)
H. F. Nelson
University of Missouri-Rolla
ISBN 0-915928-94-9

97. Monitoring Earth's Ocean, Land, and Atmosphere from Space—Sensors, Systems, and Applications (1985)
Abraham Schnapf
Aerospace Systems Engineering
ISBN 0-915928-98-1

---
*Out of print.

98. Thrust and Drag: Its Prediction and Verification (1985)
Eugene E. Covert
Massachusetts Institute of Technology
C. R. James
Vought Corp.
William F. Kimzey
Sverdrup Technology AEDC Group
George K. Richey
U.S. Air Force
Eugene C. Rooney
U.S. Navy Department of Defense
ISBN 0-930403-00-2

99. Space Stations and Space Platforms—Concepts, Design, Infrastructure, and Uses (1985)
Ivan Bekey
Daniel Herman
NASA Headquarters
ISBN 0-930403-01-0

100. Single- and Multi-Phase Flows in an Electromagnetic Field: Energy, Metallurgical, and Solar Applications (1985)
Herman Branover
Ben-Gurion University of the Negev
Paul S. Lykoudis
Purdue University
Michael Mond
Ben-Gurion University of the Negev
ISBN 0-930403-04-5

101. MHD Energy Conversion: Physiotechnical Problems (1986)
V. A. Kirillin
A. E. Sheyndlin
Soviet Academy of Sciences
ISBN 0-930403-05-3

102. Numerical Methods for Engine-Airframe Integration (1986)
S. N. B. Murthy
Purdue University
Gerald C. Paynter
Boeing Airplane Co.
ISBN 0-930403-09-6

103. Thermophysical Aspects of Re-Entry Flows (1986)
James N. Moss
NASA Langley Research Center
Carl D. Scott
NASA Johnson Space Center
ISBN 0-930430-10-X

***104. Tactical Missile Aerodynamics (1986)**
M. J. Hemsch
PRC Kentron, Inc.
J. N. Nielson
NASA Ames Research Center
ISBN 0-930403-13-4

105. Dynamics of Reactive Systems Part I: Flames and Configurations; Part II: Modeling and Heterogeneous Combustion (1986)
J. R. Bowen
University of Washington
J.-C. Leyer
Universite de Poitiers
R. I. Soloukhin
Institute of Heat and Mass Transfer, BSSR Academy of Sciences
ISBN 0-930403-14-2

106. Dynamics of Explosions (1986)
J. R. Bowen
University of Washington
J.-C. Leyer
Universite de Poitiers
R. I. Soloukhin
Institute of Heat and Mass Transfer, BSSR Academy of Sciences
ISBN 0-930403-15-0

***107. Spacecraft Dielectric Material Properties and Spacecraft Charging (1986)**
A. R. Frederickson
U.S. Air Force Rome Air Development Center
D. B. Cotts
SRI International
J. A. Wall
U.S. Air Force Rome Air Development Center
F. L. Bouquet
Jet Propulsion Laboratory, California Institute of Technology
ISBN 0-930403-17-7

***108. Opportunities for Academic Research in a Low-Gravity Environment (1986)**
George A. Hazelrigg
National Science Foundation
Joseph M. Reynolds
Louisiana State University
ISBN 0-930403-18-5

109. Gun Propulsion Technology (1988)
Ludwig Stiefel
U.S. Army Armament Research, Development and Engineering Center
ISBN 0-930403-20-7

*Out of print.

110. Commercial Opportunities in Space (1988)
F. Shahrokhi
K. E. Harwell
University of Tennessee Space Institute
C. C. Chao
National Cheng Kung University
ISBN 0-930403-39-8

111. Liquid-Metal Flows: Magnetohydrodynamics and Application (1988)
Herman Branover
Michael Mond
Yeshajahu Unger
Ben-Gurion University of the Negev
ISBN 0-930403-43-6

112. Current Trends in Turbulence Research (1988)
Herman Branover
Micheal Mond
Yeshajahu Unger
Ben-Gurion University of the Negev
ISBN 0-930403-44-4

113. Dynamics of Reactive Systems Part I: Flames; Part II: Heterogeneous Combustion and Applications (1988)
A. L. Kuhl
R&D Associates
J. R. Bowen
University of Washington
J.-C. Leyer
Universite de Poitiers
A. Borisov
USSR Academy of Sciences
ISBN 0-930403-46-0

114. Dynamics of Explosions (1988)
A. L. Kuhl
R & D Associates
J. R. Bowen
University of Washington
J.-C. Leyer
Universite de Poitiers
A. Borisov
USSR Academy of Sciences
ISBN 0-930403-47-9

115. Machine Intelligence and Autonomy for Aerospace (1988)
E. Heer
Heer Associates, Inc.
H. Lum
NASA Ames Research Center
ISBN 0-930403-48-7

116. Rarefied Gas Dynamics: Space Related Studies (1989)
E. P. Muntz
University of Southern California
D. P. Weaver
U.S. Air Force Astronautics Laboratory (AFSC)
D. H. Campbell
University of Dayton Research Institute
ISBN 0-930403-53-3

117. Rarefied Gas Dynamics: Physical Phenomena (1989)
E. P. Muntz
University of Southern California
D. P. Weaver
U.S. Air Force Astronautics Laboratory (AFSC)
D. H. Campbell
University of Dayton Research Institute
ISBN 0-930403-54-1

118. Rarefied Gas Dynamics: Theoretical and Computational Techniques (1989)
E. P. Muntz
University of Southern California
D. P. Weaver
U.S. Air Force Astronautics Laboratory (AFSC)
D. H. Campbell
University of Dayton Research Institute
ISBN 0-930403-55-X

119. Test and Evaluation of the Tactical Missile (1989)
Emil J. Eichblatt Jr.
Pacific Missile Test Center
ISBN 0-930403-56-8

120. Unsteady Transonic Aerodynamics (1989)
David Nixon
Nielsen Engineering & Research, Inc.
ISBN 0-930403-52-5

121. Orbital Debris from Upper-Stage Breakup (1989)
Joseph P. Loftus Jr.
NASA Johnson Space Center
ISBN 0-930403-58-4

122. Thermal-Hydraulics for Space Power, Propulsion and Thermal Management System Design (1990)
William J. Krotiuk
General Electric Co.
ISBN 0-930403-64-9

---
*Out of print.

123. Viscous Drag Reduction in Boundary Layers (1990)
Dennis M. Bushnell
Jerry N. Hefner
NASA Langley Research Center
ISBN 0-930403-66-5

***124. Tactical and Strategic Missile Guidance (1990)**
Paul Zarchan
Charles Stark Draper Laboratory, Inc.
ISBN 0-930403-68-1

125. Applied Computational Aerodynamics (1990)
P. A. Henne
Douglas Aircraft Company
ISBN 0-930403-69-X

126. Space Commercialization: Launch Vehicles and Programs (1990)
F. Shahrokhi
University of Tennessee Space Institute
J. S. Greenberg
Princeton Synergetics Inc.
T. Al-Saud
Ministry of Defense and Aviation Kingdom of Saudi Arabia
ISBN 0-930403-75-4

127. Space Commercialization: Platforms and Processing (1990)
F. Shahrokhi
University of Tennessee Space Institute
G. Hazelrigg
National Science Foundation
R. Bayuzick
Vanderbilt University
ISBN 0-930403-76-2

128. Space Commercialization: Satellite Technology (1990)
F. Shahrokhi
University of Tennessee Space Institute
N. Jasentuliyana
United Nations
N. Tarabzouni
King Abulaziz City for Science and Technology
ISBN 0-930403-77-0

***129. Mechanics and Control of Large Flexible Structures (1990)**
John L. Junkins
Texas A&M University
ISBN 0-930403-73-8

130. Low-Gravity Fluid Dynamics and Transport Phenomena (1990)
Jean N. Koster
Robert L. Sani
University of Colorado at Boulder
ISBN 0-930403-74-6

131. Dynamics of Deflagrations and Reactive Systems: Flames (1991)
A. L. Kuhl
Lawrence Livermore National Laboratory
J.-C. Leyer
Universite de Poitiers
A. A. Borisov
USSR Academy of Sciences
W. A. Sirignano
University of California
ISBN 0-930403-95-9

132. Dynamics of Deflagrations and Reactive Systems: Heterogeneous Combustion (1991)
A. L. Kuhl
Lawrence Livermore National Laboratory
J.-C. Leyer
Universite de Poitiers
A. A. Borisov
USSR Academy of Sciences
W. A. Sirignano
University of California
ISBN 0-930403-96-7

133. Dynamics of Detonations and Explosions: Detonations (1991)
A. L. Kuhl
Lawrence Livermore National Laboratory
J.-C. Leyer
Universite de Poitiers
A. A. Borisov
USSR Academy of Sciences
W. A. Sirignano
University of California
ISBN 0-930403-97-5

134. Dynamics of Detonations and Explosions: Explosion Phenomena (1991)
A. L. Kuhl
Lawrence Livermore National Laboratory
J.-C. Leyer
Universite de Poitiers
A. A. Borisov
USSR Academy of Sciences
W. A. Sirignano
University of California
ISBN 0-930403-98-3

---

*Out of print.

135. Numerical Approaches to Combustion Modeling (1991)
Elaine S. Oran
Jay P. Boris
Naval Research Laboratory
ISBN 1-56347-004-7

136. Aerospace Software Engineering (1991)
Christine Anderson
U.S. Air Force Wright Laboratory
Merlin Dorfman
Lockheed Missiles & Space Company, Inc.
ISBN 1-56347-005-0

137. High-Speed Flight Propulsion Systems (1991)
S. N. B. Murthy
Purdue University
E. T. Curran
Wright Laboratory
ISBN 1-56347-011-X

138. Propagation of Intensive Laser Radiation in Clouds (1992)
O. A. Volkovitsky
Yu. S. Sedenov
L. P. Semenov
Institute of Experimental Meteorology
ISBN 1-56347-020-9

139. Gun Muzzle Blast and Flash (1992)
Günter Klingenberg
Fraunhofer-Institut für Kurzzeitdynamik, Ernst-Mach-Institut
Joseph M. Heimerl
U.S. Army Ballistic Research Laboratory
ISBN 1-56347-012-8

140. Thermal Structures and Materials for High-Speed Flight (1992)
Earl. A. Thornton
University of Virginia
ISBN 1-56347-017-9

141. Tactical Missile Aerodynamics: General Topics (1992)
Michael J. Hemsch
Lockheed Engineering & Sciences Company
ISBN 1-56347-015-2

142. Tactical Missile Aerodynamics: Prediction Methodology (1992)
Michael R. Mendenhall
Nielsen Engineering & Research, Inc.
ISBN 1-56347-016-0

143. Nonsteady Burning and Combustion Stability of Solid Propellants (1992)
Luigi De Luca
Politecnico di Milano
Edward W. Price
Georgia Institute of Technology
Martin Summerfield
Princeton Combustion Research Laboratories, Inc.
ISBN 1-56347-014-4

144. Space Economics (1992)
Joel S. Greenberg
Princeton Synergetics, Inc.
Henry R. Hertzfeld
HRH Associates
ISBN 1-56347-042-X

145. Mars: Past, Present, and Future (1992)
E. Brian Pritchard
NASA Langley Research Center
ISBN 1-56347-043-8

146. Computational Nonlinear Mechanics in Aerospace Engineering (1992)
Satya N. Atluri
Georgia Institute of Technology
ISBN 1-56347-044-6

147. Modern Engineering for Design of Liquid-Propellant Rocket Engines (1992)
Dieter K. Huzel
David H. Huang
Rocketdyne Division of Rockwell International
ISBN 1-56347-013-6

148. Metallurgical Technologies, Energy Conversion, and Magnetohydrodynamic Flows (1993)
Herman Branover
Yeshajahu Unger
Ben-Gurion University of the Negev
ISBN 1-56347-019-5

149. Advances in Turbulence Studies (1993)
Herman Branover
Yeshajahu Unger
Ben-Gurion University of the Negev
ISBN 1-56347-018-7

150. Structural Optimization: Status and Promise (1993)
Manohar P. Kamat
Georgia Institute of Technology
ISBN 1-56347-056-X

---
*Out of print.

151. Dynamics of Gaseous Combustion (1993)
A. L. Kuhl
Lawrence Livermore National Laboratory
J.-C. Leyer
Universite de Poitiers
A. A. Borisov
USSR Academy of Sciences
W. A. Sirignano
University of California
ISBN 1-56347-060-8

152. Dynamics of Heterogeneous Gaseous Combustion and Reacting Systems (1993)
A. L. Kuhl
Lawrence Livermore National Laboratory
J.-C. Leyer
Universite de Poitiers
A. A. Borisov
USSR Academy of Sciences
W. A. Sirignano
University of California
ISBN 1-56347-058-6

153. Dynamic Aspects of Detonations (1993)
A. L. Kuhl
Lawrence Livermore National Laboratory
J.-C. Leyer
Universite de Poitiers
A. A. Borisov
USSR Academy of Sciences
W. A. Sirignano
University of California
ISBN 1-56347-057-8

154. Dynamic Aspects of Explosion Phenomena (1993)
A. L. Kuhl
Lawrence Livermore National Laboratory
J.-C. Leyer
Universite de Poitiers
A. A. Borisov
USSR Academy of Sciences
W. A. Sirignano
University of California
ISBN 1-56347-059-4

155. Tactical Missile Warheads (1993)
Joseph Carleone
Aerojet General Corporation
ISBN 1-56347-067-5

156. Toward a Science of Command, Control, and Communications (1993)
Carl R. Jones
Naval Postgraduate School
ISBN 1-56347-068-3

***157. Tactical and Strategic Missile Guidance Second Edition (1994)**
Paul Zarchan
Charles Stark Draper Laboratory, Inc.
ISBN 1-56347-077-2

158. Rarefied Gas Dynamics: Experimental Techniques and Physical Systems (1994)
Bernie D. Shizgal
University of British Columbia
David P. Weaver
Phillips Laboratory
ISBN 1-56347-079-9

159. Rarefied Gas Dynamics: Theory and Simulations (1994)
Bernie D. Shizgal
University of British Columbia
David P. Weaver
Phillips Laboratory
ISBN 1-56347-080-2

160. Rarefied Gas Dynamics: Space Sciences and Engineering (1994)
Bernie D. Shizgal
University of British Columbia
David P. Weaver
Phillips Laboratory
ISBN 1-56347-081-0

161. Teleoperation and Robotics in Space (1994)
Steven B. Skaar
University of Notre Dame
Carl F. Ruoff
Jet Propulsion Laboratory, California Institute of Technology
ISBN 1-56347-095-0

162. Progress in Turbulence Research (1994)
Herman Branover
Yeshajahu Unger
Ben-Gurion University of the Negev
ISBN 1-56347-099-3

163. Global Positioning System: Theory and Applications, Volume I (1996)
Bradford W. Parkinson
Stanford University
James J. Spilker Jr.
Stanford Telecom
Penina Axelrad,
Associate Editor
University of Colorado
Per Enge,
Associate Editor
Stanford University
ISBN 1-56347-107-8

164. Global Positioning System: Theory and Applications, Volume II (1996)
Bradford W. Parkinson
Stanford University
James J. Spilker Jr.
Stanford Telecom
Penina Axelrad,
Associate Editor
University of Colorado
Per Enge,
Associate Editor
Stanford University
ISBN 1-56347-106-X

---
*Out of print.

165. Developments in High-Speed Vehicle Propulsion Systems (1996)
S. N. B. Murthy
Purdue University
E. T. Curran
Wright Laboratory
ISBN 1-56347-176-0

166. Recent Advances in Spray Combustion: Spray Atomization and Drop Burning Phenomena, Volume I (1996)
Kenneth K. Kuo
Pennsylvania State University
ISBN 1-56347-175-2

167. Fusion Energy in Space Propulsion (1995)
Terry Kammash
University of Michigan
ISBN 1-56347-184-1

168. Aerospace Thermal Structures and Materials for a New Era (1995)
Earl A. Thornton
University of Virginia
ISBN 1-56347-182-5

169. Liquid Rocket Engine Combustion Instability (1995)
Vigor Yang
William E. Anderson
Pennsylvania State University
ISBN 1-56347-183-3

170. Tactical Missile Propulsion (1996)
G. E. Jensen
United Technologies Corporation
David W. Netzer
Naval Postgraduate School
ISBN 1-56347-118-3

171. Recent Advances in Spray Combustion: Spray Combustion Measurements and Model Simulation, Volume II (1996)
Kenneth K. Kuo
Pennsylvania State University
ISBN 1-56347-181-7

172. Future Aeronautical and Space Systems (1997)
Ahmed K. Noor
NASA Langley Research Center
Samuel L. Venneri
NASA Headquarters
ISBN 1-56347-188-4

173. Advances in Combustion Science: In Honor of Ya. B. Zel'dovich (1997)
William A. Sirignano
University of California
Alexander G. Merzhanov
Russian Academy of Sciences
Luigi De Luca
Politecnico di Milano
ISBN 1-56347-178-7

174. Fundamentals of High Accuracy Inertial Navigation (1997)
Averil B. Chatfield
ISBN 1-56347-243-0

175. Liquid Propellant Gun Technology (1997)
Günter Klingenberg
Fraunhofer-Institut für Kurzzeitdynamik, Ernst-Mach-Institut
John D. Knapton
Walter F. Morrison
Gloria P. Wren
U.S. Army Research Laboratory
ISBN 1-56347-196-5

176. Tactical and Strategic Missile Guidance Third Edition (1998)
Paul Zarchan
Charles Stark Draper Laboratory, Inc.
ISBN 1-56347-279-1

177. Orbital and Celestial Mechanics (1998)
John P. Vinti
Gim J. Der, Editor
TRW
Nino L. Bonavito, Editor
NASA Goddard Space Flight Center
ISBN 1-56347-256-2

178. Some Engineering Applications in Random Vibrations and Random Structures (1998)
Giora Maymon
RAFAEL
ISBN 1-56347-258-9

179. Conventional Warhead Systems Physics and Engineering Design (1998)
Richard M. Lloyd
Raytheon Systems Company
ISBN 1-56347-255-4

180. Advances in Missile Guidance Theory (1998)
Joseph Z. Ben-Asher
Isaac Yaesh
Israel Military Industries—Advanced Systems Division
ISBN 1-56347-275-9

181. Satellite Thermal Control for Systems Engineers (1998)
Robert D. Karam
ISBN 1-56347-276-7

---

*Out of print.

182. Progress in Fluid Flow Research: Turbulence and Applied MHD (1998)
Yeshajahu Unger
Herman Branover
Ben-Gurion University of the Negev
ISBN 1-56347-284-8

183. Aviation Weather Surveillance Systems (1999)
Pravas R. Mahapatra
Indian Institute of Science
ISBN 1-56347-340-2

184. Flight Control Systems (2000)
Rodger W. Pratt, Editor
Loughborough University
ISBN 1-56347-404-2

185. Solid Propellant Chemistry, Combustion, and Motor Interior Ballistics (2000)
Vigor Yang
Pennsylvania State University
Thomas B. Brill
University of Delaware
Wu-Zhen Ren
China Ordnance Society
ISBN 1-56347-442-5

186. Approximate Methods for Weapons Aerodynamics (2000)
Frank G. Moore
ISBN 1-56347-399-2

187. Micropropulsion for Small Spacecraft (2000)
Michael M. Micci
Pennsylvania State University
Andrew D. Ketsdever
Air Force Research Laboratory, Edwards Air Force Base
ISBN 1-56347-448-4

---

*Out of print.